普通高等教育“十二五”规划教材

Electric Machinery

电 机 学

（英汉双语）

刘慧娟　范　瑜　主编

机 械 工 业 出 版 社

本书共分两大部分，第一部分为电机学的英文部分，第二部分为相应的中文部分，两部分内容基本一致。每部分包括6章：第0章为绪论；后面5章包括直流电机、变压器、交流电机的共同理论、感应电机和同步电机。在直流电机与感应电机两章中添加了电动机的机械特性、电动机的起动、制动和调速等“电力拖动”的内容。为加强学生对相关知识点的理解，各章中配有相应的例题；为便于学生学习，每一章后附有本章内容小结和相应数量的习题。

本书可作为普通高等院校电气工程及其自动化专业和其他电气类、自动化类等专业双语教学的教材，亦可供相关专业技术人员参考。

图书在版编目（CIP）数据

电机学=Electric machinery：汉英对照/刘慧娟，范瑜主编．—北京：机械工业出版社，2014.4（2025.7重印）
普通高等教育“十二五”规划教材
ISBN 978-7-111-45221-8

Ⅰ.①电… Ⅱ.①刘… ②范… Ⅲ.①电机学—双语教学—高等学校—教材—汉、英 Ⅳ.①TM3

中国版本图书馆CIP数据核字（2013）第307205号

机械工业出版社（北京市百万庄大街22号 邮政编码100037）
策划编辑：王雅新 责任编辑：王雅新 王 琪
版式设计：常天培 责任校对：杜雨霏
封面设计：张 静 责任印制：张 博
北京建宏印刷有限公司印刷
2025年7月第1版第7次印刷
184mm×260mm·22.75印张·576千字
标准书号：ISBN 978-7-111-45221-8
定价：59.80元

电话服务
客服电话：010-88361066
010-88379833
010-68326294

网络服务
机 工 官 网：www.cmpbook.com
机 工 官 博：weibo.com/cmp1952
金 书 网：www.golden-book.com
机工教育服务网：www.cmpedu.com

前　言

随着世界经济一体化的不断发展，我国与世界经济、科技、文化的合作和交流日渐频繁，国际化的经济与科技环境对我国专业技术人员的国际化水平提出了更高的要求，特别是专业英语的交流应用能力。

近年来，国家教育部陆续颁发了一系列高等教育教学改革文件，要求推动双语教学课程建设，切实提高大学生的专业英语水平和使用英语从事科研的能力，并将双语教学列为高等院校教育质量评估中的一项重要指标。可见，培养既掌握专业知识又掌握专业外语的高素质复合型人才是目前中国高等教育发展的重要目标之一。

本书的主编在与美国、中国香港多所大学多年的合作研究与交流中，深切体会到专业英语的直接交流会极大地帮助我们发挥自身的科研潜能，增强科研竞争能力。同时，在北京交通大学的多年教学中，也发现国内本科生和研究生的专业英语的交流应用能力亟待提高。因此，坚持专业基础课程的双语教学，必将对提高我国专业技术人员的英语交流应用能力起到积极的推动作用。

本书是编者在总结多年电机学课程双语教学工作经验的基础上，结合电气工程及其自动化专业和电气类其他专业的电机学教学大纲，对多本英文原版电机学教材的教学内容和表述方法进行梳理、调整和更新，以适应21世纪我国教学改革的需要而编写的。

本书是高等院校电气工程及其自动化专业和其他电气类、自动化类等专业的主干课“电机学”双语教学的教材，内容涉及本学科一些最基本的理论和分析方法，是“电机学”和“电力拖动基础”两门课程主要内容的有机结合。

本书共分两大部分，第一部分为电机学的英文部分，第二部分为相应的中文部分，两部分内容基本一致。

本教材的主要特点是：

1）以直流电机、变压器、感应电机和同步电机4种典型通用电机为研究对象，着重讲解其工作原理和稳态运行特性，以及各类电机的共同性问题，突出本专业本科教学应掌握的基本概念、基本理论和基本分析方法，提高学生解决工程实际问题的能力。

2）本书在讲述电机理论的基础上，加入了交、直流电动机的电力拖动内容，为学习后续课程和解决今后遇到的工程问题打下了相应的基础。

3）将“磁路”等电机分析的基础知识放在绪论中，供读者学习参考。

4）本书的每一章之后都有本章小结和相应的习题，以帮助读者加强对本章内容的理解和掌握。

5）各章的内容具有相对独立性，可根据实际需要和学时决定取舍，各章的次序在具体讲授时也可以改变。

本书的编写方针是结合国情、博采众长、削枝强干、推陈出新，目标是编写一本取材精、科学性强、概念清、便于教学的简明电机学双语教学教材。

本书的英文部分由美国俄亥俄州立大学 Longya Xu 教授和香港理工大学的傅为农博士主审，中文部分由哈尔滨理工大学汤蕴璆教授主审。全书由北京交通大学范瑜教授、刘慧娟副教授、刘瑞芳副教授、张威老师、郭芳老师和桂俊峰老师共同编写。刘慧娟、范瑜担任主编，全书内容由刘慧娟统稿和策划。具体分工为：范瑜编写中英文绪论；刘慧娟编写中英文的第 1 章；桂俊峰编写中英文的第 2 章，张威编写中英文的第 3 章；刘瑞芳编写中英文的第 4 章；郭芳编写中英文的第 5 章。Longya Xu 教授、傅为农博士与汤蕴璆教授对全书做了非常仔细的审阅，并提出了许多宝贵的修改意见和建议，在此表示衷心的感谢！

本书的编写得到了北京交通大学电气学院领导的关怀和支持，得到了各位同仁的热心帮助，在此一并表示谢意。

由于编者水平有限，编写时间仓促，书中难免有不妥和错漏之处，恳请读者提出宝贵意见，以便在再版中修正。

编　者

CONTENTS

中文部分

Chapter 0 Introduction

0.1 The role of electrical machines in our lives

Electrical energy is the most widely used secondary energy and it is suitable for mass production, centralized management, long-distance transmission, flexible allocation and automatic control. Electrical machines are mechanical energy conversion machineries associated with production, transformation, distribution, utilization and the control of electrical energy; it plays an important role in various areas of the national economy.

1. Electrical machines are the main equipment in the production, transmission and distribution of electrical energy

In power plants, the rotation of rotors in generators is driven by steam turbines, hydraulic turbines, diesel or other power machineries. Rotors are used to convert mechanical energy, which has been converted from primary energies (fuel combustion heat, the potential energy of the water, the nuclear fission of atomic energy, wind energy, solar energy and tidal energy, etc) to electrical energy. Since the generator plays a critical role in the production of electricity, it is the most important electrical equipment in the power systems. In general, power plants are far away from load centers; therefore, high-voltage transmission is adopted in order to transmit power to remote places economically. Voltage of generators ranges from 10.5kV to 20kV and the typical transmission voltages are 110kV, 220kV, 330kV, 500 kV or higher. So step-up transformers are needed to increase the generator voltages before transmission. In China, the common values of high transmission voltage are 110kV, 220kV and 500kV. It can even go up to 1000kV. When electricity power reaches the load center, for the safety reason, various step-down transformers are needed to reduce the transmission voltages. The total capacity of transformers required by the power system usually is 7 to 8 times of the total capacity of the power generation equipment. Therefore, generators and transformers are the main electrical equipment in power plants and substations in the electric power industry.

2. Electrical machines are the motive equipment of the production machines and equipment

Various types of electrical machines are widely used in industry, agriculture, transportation and everyday life to drive production machines and equipment. In the machinery industry, for example, workhorses are driven and controlled by one or more electrical motors of different capacities. Grinder motors have a rotational speed of several tens of thousands per minute or even higher. Some machine tools need multi-speed motors. In the metallurgy industry, blast furnaces, converters and open-hearth-furnaces need more than one motor to drive. A large rolling mill may need a 5000 kW DC motor or even higher powered motors for proper functioning. Almost all machines in the world need a variety of AC and DC motors to be prime movers today. Machines can be used for different purposes such as power irrigation, sideline products processing in agriculture, blowers, crane, transportation and transfer in enterprises, mining and more transport equipments, urban rail transport, Maglev, high speed train traction, electric vehicles, papermaking machinery, medical equipment, as well as household appliances.

3. Electrical machines are an important element in the automatic control systems

With the development of science and technology, the level of automation in industry, agriculture and national defense facilities has continued to rise. Various control motors are used as detectors, amplifiers and solver components. Such motors are usually with small power, wide variety of output and high precision. Artillery and radar automatic positioning, satellite launch and flight control, the ship rudder automatic control, automatic control of the machining, the operation control of the computer, automatic recording instruments monitoring all depend on various control motors. Therefore precision control motors of different capacities are important elements in the whole automatic control systems.

An electrical machine is an electromechanical energy conversion device. When it is used to convert mechanical energy to electrical energy, it is called a generator. When it is used to convert electrical energy to mechanical energy, it is called a motor. Since all electrical machines can convert one type of energy to another, they can be either generators or motors.

In an electrical machine, there are magnetic circuits and electrical circuits that produce induction electromotive force and torque according to the electromagnetic induction law and the law of electromagnetic force to achieve electromechanical energy conversion.

0.2 Types of electrical machines

Electrical machines can be classified into four types according to their functions.

1) Generator: a device that can convert mechanical energy to electrical energy.

2) Motor: a device that can convert electrical energy to mechanical energy.

3) Transformer, phase shifter, frequency machine, variable flow machine: devices that can convert electrical energy to other forms of electrical energy, namely voltage, phase, frequency, and current of electrical power.

4) Control motor: part of the automatic control systems. It is used to generate and transmit signals. It can also be used as a servo element.

Depending on the type of applied current, rotating electrical machines can be divided into two types: DC machines and AC machines. AC machines can be further divided into three types: synchronous machines, induction machines (also called asynchronous machines) and AC commutator machines.

0.3 Materials used in internal structure of electrical machines

Materials used in electrical machines can be classified into the following four categories.

1) Electrically conductive materials: They are used in the electrical circuit system of the elec-

trical machines. In order to reduce line losses, conductive materials of small resistivity are used. The most commonly used materials are copper and aluminum.

2) Magnetically permeable (ferromagnetic) materials: They are used in the magnetic circuit system of the electrical machines. In order to produce a strong magnetic field and reduce the iron losses, the magnetically permeable materials should have a high magnetic permeability and low iron loss coefficient. Silicon steel is commonly used in AC magnetic circuit while steel plate and cast steel are used in DC magnetic circuit.

3) Insulating materials: They are used as the isolation bodies between conductors and iron cores. High dielectric strength and high heat-resistant strength are essential qualities of the materials. Insulating materials of electrical machines are divided into five insulation grades: A, E, B, F, and H. Their maximum working temperatures are 105℃, 120℃, 130℃, 155℃, 180℃ respectively.

4) Structural materials: They are used to support and connect all members of the machine. Materials with good mechanical strength, and light weight and easy to process are used, which are commonly cast steel, cast iron, steel, aluminum and engineering plastics.

0.4 Magnetic circuits

0.4.1 Magnetic fields

Magnetic fields are the fundamental mechanism by which energy is converted from one form to another in all electrical machines.

There are four basic principles to describe how magnetic fields are used in electrical machines:

1) A current-carrying wire can produce a magnetic field in the area around it.

2) A time-changing magnetic field can induce an *Induction Electromotive Force* (emf) in a coil of wire when it passes through the coil. (Transformers are based on this principle.)

3) A current-carrying wire experiences an induced force in the presence of magnetic field. (Motors are based on this principle.)

4) A moving wire experiences an *Induced Electromotive Force* (emf) in the presence of magnetic field. (Generators are based on this principle.)

Some specific quantities are used to describe magnetic fields, such as flux density $\boldsymbol{B}$ [Tesla, T], magnetic field intensity $\boldsymbol{H}$ [Ampere per meter, A/m], magnetic flux $\boldsymbol{\Phi}$ [Weber, Wb], magnetomotive force (mmf) $\boldsymbol{F}$ [Ampere turn, A], magnetic reluctance R_m [Ampere-turns per Weber, A/Wb, H^{-1}], magnetic conductance (permeance) Λ_m [Henry, H], and magnetic flux linkage Ψ [Weber-turn].

1. Specific quantities used in magnetic fields

(1) Flux density $\boldsymbol{B}$ [Tesla, T]

A current-carrying wire can produce a magnetic field in the area around it, and the flux density is used to describe the strength and the direction of the magnetic fields.

(2) Magnetic field intensity $\boldsymbol{H}$ [Ampere per meter, A/m]

Magnetic field intensity $\boldsymbol{H}$ is another quantity used to describe the strength and the direction of the magnetic fields. The relationship between $\boldsymbol{B}$ and $\boldsymbol{H}$ is expressed as

$$\boldsymbol{H} = \frac{\boldsymbol{B}}{\mu} \tag{0-1}$$

where μ is the magnetic permeability of the material. The permeability of vacuum is the magnetic constant μ_0, and $\mu_0 = 4\pi \times 10^{-7}$H/m. This enables us to write Eq. 0-1 in the approximate form: $\boldsymbol{H} = 800\ 000\boldsymbol{B}$. And the $\boldsymbol{B}$-$\boldsymbol{H}$ curve of vacuum is a straight line. A vacuum never saturates, no matter how great the flux density may be. Nonmagnetic materials such as copper, paper, rubber and air have $\boldsymbol{B}$-$\boldsymbol{H}$ curves almost identical to that of vacuum.

The flux density in a magnetic material is given by $\boldsymbol{B} = \mu_0\mu_r\boldsymbol{H}$. Where μ_r is the *relative permeability* of a material, and it is the ratio of the permeability in the material to μ_0. The *relative permeability* μ_r of cast steel is about 1000 and in silicon iron it is about 6000 ~ 7000.

$$\mu_r = \frac{\mu}{\mu_0} \tag{0-2}$$

(3) Flux Φ [Weber, Wb]

The total flux Φ flowing in a cross section A is expressed as

$$\Phi = \int_A \boldsymbol{B} \cdot \mathrm{d}\boldsymbol{A} \tag{0-3}$$

Assuming that the flux density $\boldsymbol{B}$ is constant throughout hence constant $\boldsymbol{A}$, and $\boldsymbol{B}$ is perpendicular to section $\boldsymbol{A}$, then

$$\Phi = \boldsymbol{BA} \quad \text{or} \quad \boldsymbol{B} = \frac{\Phi}{\boldsymbol{A}} \tag{0-4}$$

(4) Magnetomotive force (mmf) F [Ampere turn, A]

A current-carryingcoil can produce a magnetic field, if the turns of the coil is N, the current is I, the total magnetomotive force (mmf) F in the coil can be expressed as:

$$F = NI \tag{0-5}$$

(5) Magnetic reluctance R_m [Ampere-turns per Weber, A/Wb, H^{-1}]

Reluctanceis is similar to resistance in the circuit. If the magnetic permeability of the material in the magnetic circuit is μ, the cross-section of the magnetic circuit is A, the length of the magnetic circuit is l, then the magnetic circuit reluctance R_m can be calculated by

$$R_m = \frac{l}{\mu A} \tag{0-6}$$

(6) Magnetic conductance (permeance) Λ_m [Henry, H]

The reciprocal of reluctance R_m is called magnetic permeance Λ_m, and

$$\Lambda_m = \frac{1}{R_m} = \frac{\mu A}{l} \tag{0-7}$$

(7) Magnetic flux linkage Ψ [Weber-turn]

The magnetic flux linkage Ψ is defined as the number of coil turns N multiplied by the flux Φ, and

$$\Psi = N\Phi \tag{0-8}$$

2. Production of magnetic field

The basic law governing the production of magnetic field induced by current is Ampere's Law:

$$\oint \boldsymbol{H}d\boldsymbol{l} = I_{net} \tag{0-9}$$

where $\boldsymbol{H}$ is the magnetic field intensity vector produced by the current I_{net} and $d\boldsymbol{l}$ is a differential element of length along the path of integration. $\boldsymbol{I}$ is measured in amperes and $\boldsymbol{H}$ is measured in Ampere-turns per meter.

Fig. 0-1 shows a rectangular ferromagnetic core with a N-turn current carrying winding wrapped around one leg of the core. The mean path length of the core is l_c [m] and the cross-sectional area of the core is A [m^2]. The current in the winding is I [A].

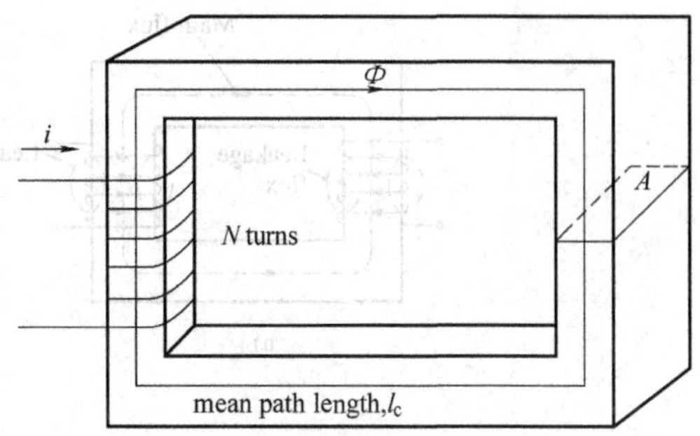

Fig. 0-1 A simple magnetic core

According to Ampere's law, the magnitude of the induced magnetic field is proportional to the total amount of current flowing through the winding with N turns around the ferromagnetic material as shown. Since the core is made of ferromagnetic materials, it is assumed that the majority of the magnetic field will be confined to the core.

When the current-carrying coil with current i cuts the path of integration N times, the current passing within the path of integration I_{net} is Ni. Hence Ampere's Law becomes

$$Hl_c = Ni \tag{0-10}$$

where H is the magnitude of the magnetic field intensity vector $\boldsymbol{H}$. Therefore, the magnitude of the magnetic field intensity in the core due to the applied current is

$$H = \frac{Ni}{l_c} \tag{0-11}$$

The magnetic field intensity $\boldsymbol{H}$ (Ampere turns per meter) is known as the "effort" required to induce a magnetic field. The strength of the magnetic field flux produced in the core also depends on the material of the core. According to Eq. 0-1, the flux density in the core is

$$B = \mu H = \frac{\mu Ni}{l_c} = \frac{\mu_0 \mu_r Ni}{l_c} \tag{0-12}$$

Therefore, the flux in the ferromagnetic core is

$$\Phi = BA = \frac{\mu_0 \mu_r NiA}{l_c} \tag{0-13}$$

0.4.2 Magnetic circuits

The path which the magnetic flux is passing is called magnetic circuit. Fig. 0-2 shows the magnetic circuits of transformer and 4-pole DC machine respectively.

In Fig. 0-2, the main (mutual) flux is passing in the core and coupling with two windings, and its path is called main magnetic circuit; the leakage flux is only coupling with one winding, and its path is called leakage magnetic circuit. The coils which carry current (DC or AC) to produce a magnetic flux are called excitation winding, and the current is called excitation current: If the excitation current is direct current, the magnetic circuit is called DC magnetic circuit, such as DC machines' magnetic circuit and if the excitation current is AC, the magnetic circuit is called AC magnetic circuit, such as transformer's magnetic circuit and induction machine's magnetic circuit.

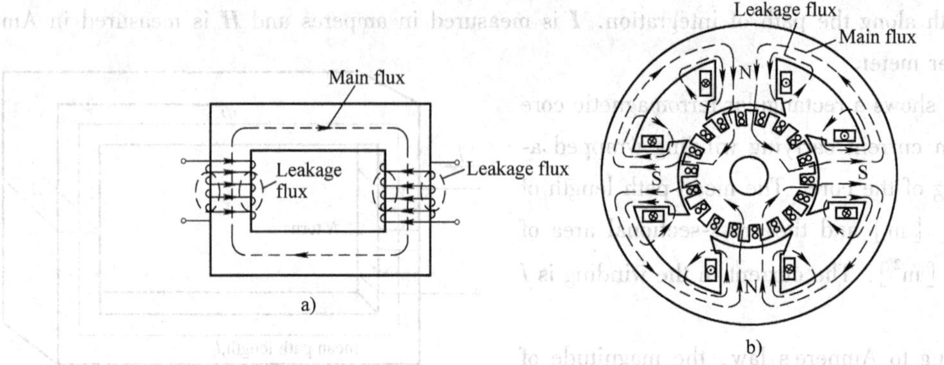

Fig. 0-2 Magnetic circuits of transformer and 4-pole DC machine

From Eq. 0-13, we can see that current in a coil which wrapped around a core produces a magnetic flux in the core. This is in some sense analogous to voltage in electrical circuit producing current flow. It is possible to define a "magnetic circuit" whose behavior is governed by equations analogous to those for an electric circuit. The magnetic circuit model of magnetic behavior is often applied in the design of electrical machines and transformers to simplify the design process.

A simplified electric circuit and a simplified magnetic circuit are shown in Fig. 0-3a and b respectively. In Fig. 0-3a, the voltage source (electromotive force) U drives the current I to flow in the circuit and it passes through the resistance R. The relationship between these quantities is given by Ohm's law, i. e. $U = IR$.

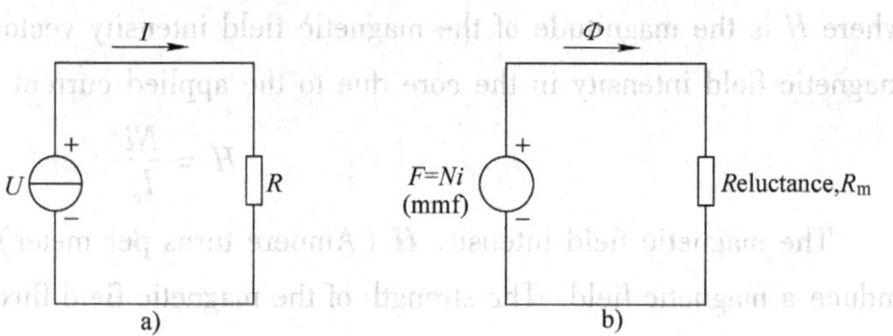

Fig. 0-3 Magnetic circuit

a) A simplified electric circuit

b) The magnetic circuit analogous to a transformer core

Referring to the magnetic circuit analogy, F is denoted as magnetomotive force (mmf) which is similar to electromotive force (emf) in an electrical circuit. The magnetomotive force F of the magnetic circuit is the prime mover which pushes magnetic flux to go around a ferromagnetic core at the value of Ni (refer to ampere's law). The relationship between magnetomotive force F and flux Φ is

$$F = \Phi R_m \tag{0-14}$$

where R_m is the reluctance of the magnetic circuit measured in Ampere-turns per Weber.

The polarity of mmf determines the direction of flux. The 'right hand curl' rule is used to determine the direction of flux; If the fingers of your right hand curl in the direction of the current flowing in a coil of wire, the thumb direction will point at the positive direction of mmf.

According to Eq. 0-7, the permeance Λ_m of a magnetic circuit is the reciprocal of its reluctance, and Eq. 0-14 can be rewritten as

$$\Phi = F\Lambda_{m} \tag{0-15}$$

According to Eq. 0-6, the reluctance of the core shown in Fig. 0-1 can be calculated. The element R_m in the magnetic circuit analogy is similar to the concept of electrical resistance. It basically represents the material resistance in the magnetic flux flow. Reluctance in this analogy obeys the Series and Parallel Rules.

The equivalent reluctance of reluctances in series is

$$R_{meq} = R_{m1} + R_{m2} + R_{m3} + \cdots \tag{0-16}$$

Similarly, reluctances in parallel can be combined according to the following equation

$$\frac{1}{R_{meq}} = \frac{1}{R_{m1}} + \frac{1}{R_{m2}} + \frac{1}{R_{m3}} + \cdots \tag{0-17}$$

Magnetic circuit approach helps to simplify the calculations related to the magnetic field in ferromagnetic materials; however, the resulting solution is rather inaccurate due to the assumptions behind the approach (within 5% of the real answer). Possible reasons of inaccuracy are:

1) The magnetic circuit is formed with assumption that all flux are confined within the core, but in reality a small fraction of the flux escapes from the core into the low-permeability surrounding air. The small fraction of flux is called leakage flux.

2) The reluctance calculation is made under the assumption that both the path length and cross sectional area of the core are in certain mean values. This is acceptable if the core is just a block of ferromagnetic material with no corners. However, the assumption will affect the calculation of practical ferromagnetic core which has corners due to its design.

3) In ferromagnetic materials, the permeability varies with the amount of flux formed in the material. The material permeability is not constant and it is non-linear in nature.

4) If there are air gaps in a ferromagnetic core, the effective cross-sectional area of the air gap will be larger than that of the iron core on both sides. The extra effective area is caused by the "fringing effect" of the magnetic field at the air gap (as shown in Fig. 0-4).

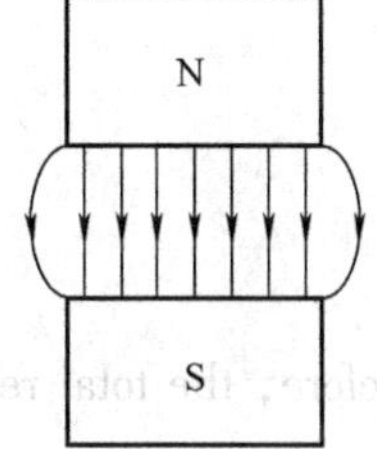

Fig. 0-4 The fringing effect of a magnetic field at an air gap

Example 0-1 A ferromagnetic core is shown in Fig. 0-5. Three sides of this core are of uniform width (15cm), while the fourth side is somewhat thinner (10cm). The depth of the core (into the page) is 10cm, and the other dimensions are shown in the figure. There is a 200-turn coil wrapped around the left side of the core. Assuming the relative permeability μ_r is 2500, how much flux will be produced by a 1A input current?

Solution:

Three sides of the core have the same cross-sectional area (*CSA*), while the 4th side has a different area. Thus the core can be divided into 2 regions: ① the single thinner side, ② the other 3 sides taken together. The magnetic circuit corresponding to this core is shown in Fig. 0-5b.

The mean length of region ① is 45cm, and its cross-sectional area is $10\text{cm} \times 10\text{cm} = 100\text{cm}^2$. Therefore, the reluctance in the first region is

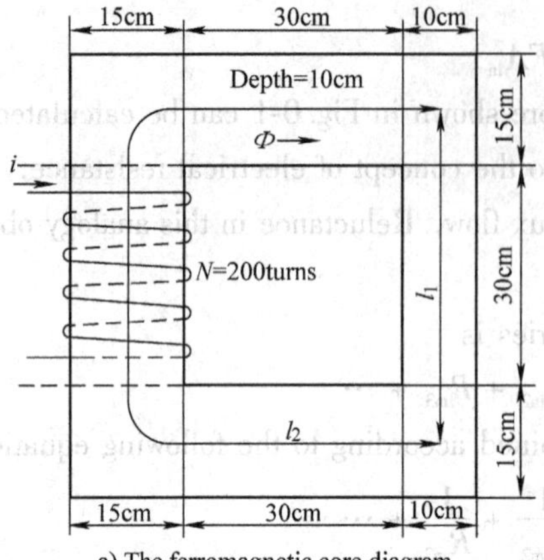

a) The ferromagnetic core diagram

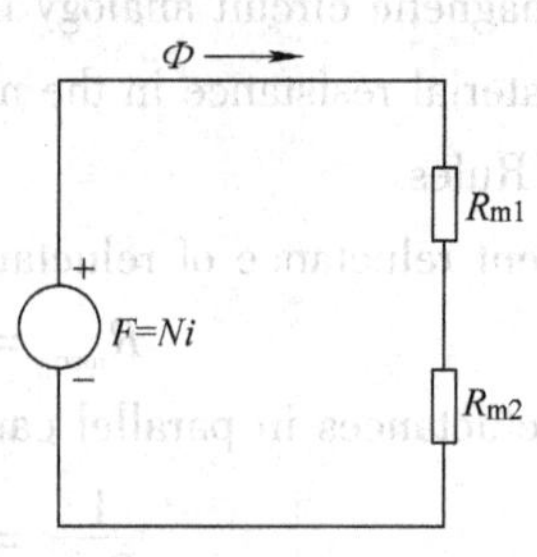

b) The magnetic circuit of the ferromagnetic core

Fig. 0-5 The diagram of Example 0-1

$$R_{m1}=\frac{l_1}{\mu A_1}=\frac{l_1}{\mu_r\mu_0 A_1}$$

$$=\frac{0.45\text{m}}{2500\times 4\pi\times 10^{-7}\times 0.01\text{m}^2}$$

$$=14\,300\text{H}^{-1} \tag{0-18}$$

The mean length of region ② is 130cm, and its cross-sectional area is $15\text{cm}\times 10\text{cm}=150\text{cm}^2$. Therefore, the reluctance in the second region is

$$R_{m2}=\frac{l_2}{\mu A_2}=\frac{l_2}{\mu_r\mu_0 A_2}$$

$$=\frac{1.3\text{m}}{2500\times 4\pi\times 10^{-7}\times 0.015\text{m}^2}$$

$$=27\,600\text{H}^{-1} \tag{0-19}$$

Therefore, the total reluctance in the core is

$$R_{meq}=R_{m1}+R_{m2}$$

$$=14\,300\text{H}^{-1}+27\,600\text{H}^{-1}$$

$$=41\,900\text{H}^{-1} \tag{0-20}$$

The total magnetomotive force is

$$F=Ni=200\text{turns}\times 1.0\text{A}=200\text{A}\cdot\text{t} \tag{0-21}$$

Finally, the total flux in the core is given by

$$\Phi=\frac{F}{R_{meq}}=\frac{200\text{A}\cdot\text{t}}{41\,900\text{H}^{-1}}$$

$$=0.004\,8\text{Wb} \tag{0-22}$$

Example 0-2 Fig. 0-6 shows a ferromagnetic core whose mean path length is 40cm. There is a small gap of 0.05cm in the structure of the otherwise whole core. The cross-sectional area of the core is 12cm^2, the relative permeability of the core is 4000, and the coil of wire on the core has 400 turns. Assume that fringing in the air gap increases the effective *CSA* of the gap by 5%. Find:

1) The **total reluctance** of the flux path (iron plus air gap)

2) The **current** required to produce a flux density of 0. 5T in the air gap.

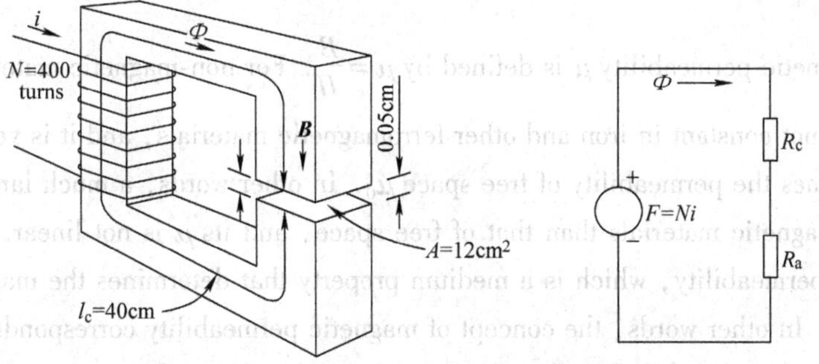

a) The ferromagnetic core diagram   b) The magnetic circuit of the ferromagnetic core

Fig. 0-6 The diagram of Example 0-2

Solution:

The magnetic circuit corresponding to this core is shown in Fig. 0-6b

1) the reluctance of the core is

$$R_c = \frac{l_c}{\mu A_c} = \frac{l_c}{\mu_r \mu_0 A_c}$$

$$= \frac{0.4\text{m}}{4000 \times 4\pi \times 10^{-7} \times 0.0012\text{m}^2}$$

$$= 66\ 300\text{H}^{-1} \tag{0-23}$$

The effective area of the air gap is $1.05 \times 12\text{cm}^2 = 12.6\text{cm}^2$, so the reluctance of the air gap is

$$R_a = \frac{l_a}{\mu_0 A_a}$$

$$= \frac{0.0005\text{m}}{4\pi \times 10^{-7} \times 0.00126\text{m}^2}$$

$$= 316\ 000\text{H}^{-1} \tag{0-24}$$

Therefore, the total reluctance of the flux path is

$$R_{eq} = R_c + R_a$$

$$= 66\ 300\text{H}^{-1} + 316\ 000\text{H}^{-1}$$

$$= 382\ 300\text{H}^{-1} \tag{0-25}$$

Note that the air gap contributes most of the reluctance even though it is 800 times shorter than the core.

2) The **current** required to produce a flux density of 0. 5T in the air gap is

$$i = \frac{BAR_{eq}}{N}$$

$$= \frac{0.5\text{T} \times 0.001\ 26\text{m}^2 \times 383\ 200\text{H}^{-1}}{400\text{turns}}$$

$$= 0.602\text{A} \tag{0-26}$$

Notice that, since the air-gap flux was required, the effective air-gap was used in the above equation.

0.4.3 Magnetic behaviors of ferromagnetic materials

In Eq. 0-1, magnetic permeability μ is defined by $\mu = \frac{B}{H}$. For non-magnetic material, μ is constant, however, it is not constant in iron and other ferromagnetic materials, and it is very high which can be up to 6000 times the permeability of free space μ_0. In other words, a much larger value of B is produced in ferromagnetic materials than that of free space, and its μ is not linear.

Thus, it is the permeability, which is a medium property that determines the magnetic characteristics of a medium. In other words, the concept of magnetic permeability corresponds to the ability of a material to permit the magnetic flux flow through it.

Fig. 0-7 shows typical magnetization curve (*B-H* curve) and μ curve of a ferromagnetic material. The curve represents the effect on B when there is an increase of DC current flowing through a coil wrapped around the ferromagnetic core.

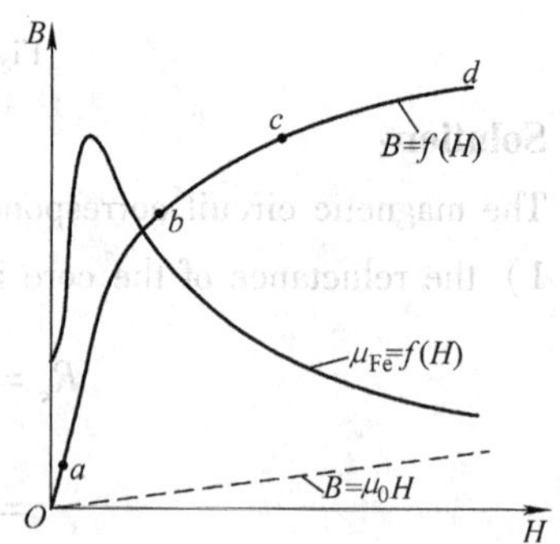

Fig. 0-7 A DC magnetization curve and μ curve of a ferromagnetic material

Whena graph of the flux density B produced in the core versus the field intensity H producing it is plotted, the resulting plot is called a *saturation curve* or a *magnetization curve*. At the beginning, a small increase in H leads to a large increase in the resulting B. After a certain point, further increase in H has relatively small effect on B. In the last part of the graph, B is independent of H, i. e. further increase will not lead to any change of B. The region in which the curve flattens out is called saturation region, and the core is said to be saturated. The region where B changes rapidly is called the unsaturated region. The transition region is called the 'knee' of the curve.

The μ curve of a ferromagnetic material (as shown in Fig. 0-7) shows that as magnetizing intensity H increased, the permeability μ is also increased in the unsaturated region. It then drops gradually as the core becomes heavily saturated.

The main advantage of using ferromagnetic materials to make cores in electric machines and transformers is that more flux can be produced for a given mmf than that of air (free space). Generators and motors rely on magnetic flux to produce voltage and torque, therefore the greater the amount of magnetic flux, the higher the voltage and torque. Normally, they operate near the knee of the magnetization curve, and the flux in their cores is not linearly related to the mmf producing it. Non-linearity gives the peculiar behaviors of machines.

0.4.4 Energy losses in a ferromagnetic core

1. Hysteresis losses

Transformers and most electric motors operate with alternating current. In those devices, the flux in the iron changes continuously in both value and direction.

Now let's move on the discussion to the application of AC current applied to the coil wrapped around a core. As shown in Fig. 0-8, assume that the flux in the core is initially zero, as the current increases for the first time, the flux in the core traces out path Oa, which gives the saturation curve shown in Fig. 0-7. The flux density reaches a value B_m at a magnetic field intensity H_m.

If the current is reduced to zero gradually, the flux density B does not change according to the original curve. Instead, it moves along the curve ab above Oa. As the reverse current increases, the flux traces out path bcd. Later, when the reverse current decrease to zero, and the forward current increases again, the flux traces out path $defa$. Noted that the amount of flux in the core not only depends on the amount of current applied to the coils which wrapped around an iron core, but also the previous history of the flux in the iron core. This phenomenon is called *hysteresis*. The closed curve $abcdefa$ traced out in Fig. 0-8 is called *hysteresis loop*.

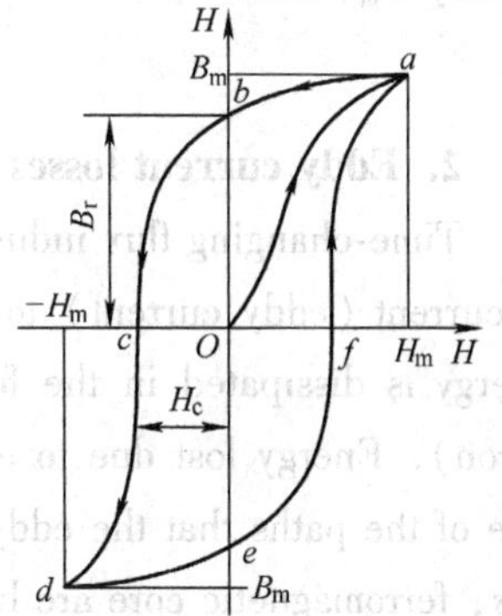

Fig. 0-8 A typical Hysteresis loop when AC current is applied

Noted that, when H is reduced to zero, the corresponding substantial flux density is called residual flux density or *residual induction* (B_r). This characteristic explained how permanent magnets are produced. To force the flux to zero, an amount of mmf must be applied in the opposite direction. The magnetic field intensity required to reduce the flux to zero is called *coercive force* (H_c).

Material having small B_r and H_c (narrow hysteresis loop) is called *soft magnetic* (SM) material and its typical *hysteresis loop* is shown in Fig. 0-9a. If a type of material has large B_r and H_c (wide hysteresis loop), it is called *hard magnetic* (HM) material; and the typical *hysteresis loop* is shown in Fig. 0-9b.

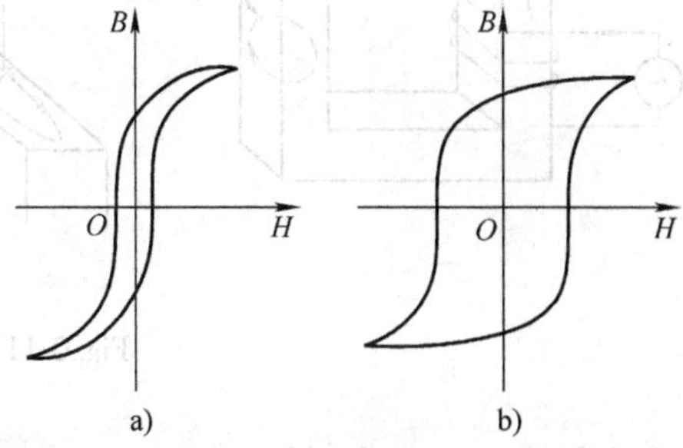

Fig. 0-9 A typical Hysteresis loop of soft magnetic materials and hard magnetic materials

To describe a hysteresis loop (Fig. 0-8), the flux moves successively from $+B_m$, $+B_r$, 0, $-B_m$, $-B_r$, 0 and $+B_m$ corresponding respectively to points a, b, c, d, e, f and a of Fig. 0-8. The magnetic material absorbs energy during each cycle and and at the same time, the energy is dissipated as heat. This phenomenon is called *hysteresis loss*. Hysteresis loss in an iron core is the energy required to accomplish the reorientation of domains during each cycle of the alternating current applied to the core. It is shown that the area enclosed in the hysteresis loop formed by applying AC current to the core is directly proportional to the energy lost in a given AC cycle. Thus the smaller the applied magnetomotive force to the core, the smaller the area of the hysteresis loop, and so the resulting hysteresis loss is also smaller. Fig. 0-10 illustrates this point.

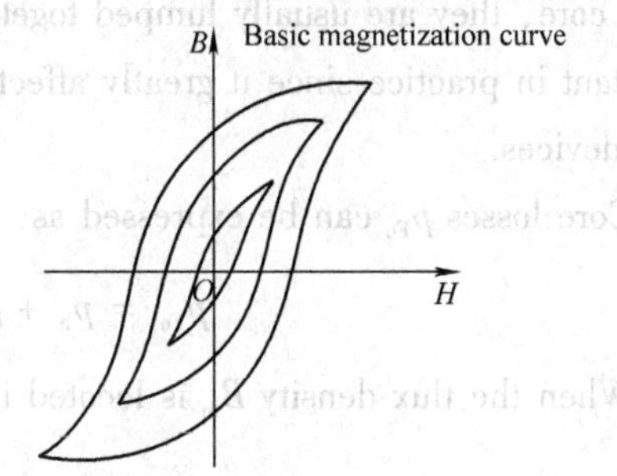

Fig. 0-10 The effect of the size of magnetomotive force excursions on the magnitude of the hysteresis loss

Hysteresis loss p_h is proportional to the frequency f, core volume V and the magnitude of flux density B_m, i. e.

$$p_h = C_h f B_m^n V \tag{0-27}$$

2. Eddy current losses

Time-changing flux induces voltage within a ferromagnetic core. Induced voltage causes swirls of current (eddy current) to flow within the core (as shown in Fig. 0-11). At the same time, energy is dissipated in the form of heat because eddy current is flowing in a resistive material (iron). Energy lost due to eddy current is called *eddy current loss*, and it is proportional to the size of the paths that the eddy currents follow through within the core. To reduce the *eddy current loss*, ferromagnetic core are broken up into small strips or laminations, and built up the core out of these strips. An insulating oxide or resin is used between the strips to minimize the area of current paths.

Eddy current losses p_e can be calculated by:

$$p_e = C_e \Delta^2 f^2 B_m^2 V \tag{0-28}$$

where C_e is the eddy current loss coefficient; Δ is the thickness of silicon.

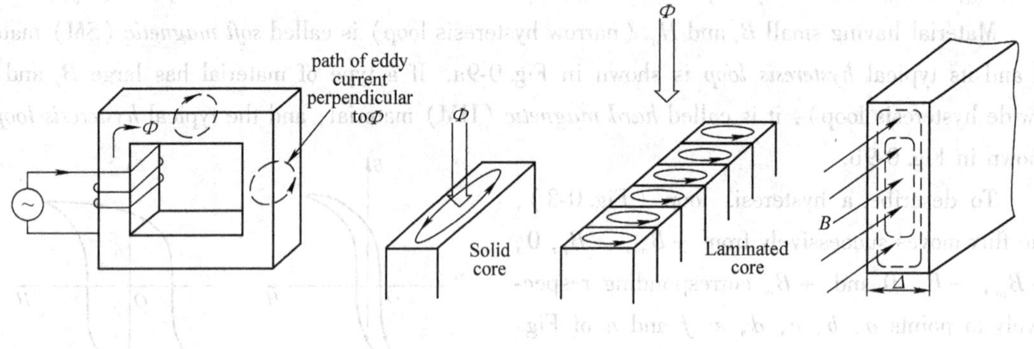

Fig. 0-11 Eddy current loss

Both hysteresis and eddy current losses cause heating in the core material, and both losses must be considered in the design of any machine or transformer. Since both losses occur within the metal of the core, they are usually lumped together and called *core losses* p_{Fe}. Core losses p_{Fe} is extremely important in practice since it greatly affects operating temperatures, efficiencies and ratings of magnetic devices.

Core losses p_{Fe} can be expressed as

$$p_{Fe} = p_h + p_e = C_h f B_m^n V + C_e \Delta^2 f^2 B_m^2 V \tag{0-29}$$

When the flux density B_m is located in the range of $1T < B_m < 1.8T$, Eq. 0-29 can be rewritten

$$p_{Fe} \approx C_{Fe} f^{1.3} B_m^2 G \tag{0-30}$$

where C_{Fe} is the core loss coefficient; G is the weight of core.

0.5 Related laws

0.5.1 Faraday's law of electromagnetic induction

It is now time to examine the various ways in which an existing magnetic field affects its surrounding. The first major effect to be considered is called *Faraday's law*. The operation of transformers and generators is on the basis of it. Faraday's law states that if a flux passes through a coil of N turns, a voltage, which is directly proportional to the *rate of change* in the flux with respect to time will be induced in the coil. It can be expressed in equation:

$$e = -N\frac{\mathrm{d}\Phi}{\mathrm{d}t} \tag{0-31}$$

where e is the voltage induced in the coil; N is the number of turns of the coil; Φ is the flux passing through the coil.

The minus sign in the equation is given by *Lenz's law*. Lenz's law states that the direction of the voltage buildup in the coil in a way that if the coil ends are short circuited, it will produce current that would cause a flux opposing to the change of original flux. Since the induced voltage opposes the change that causes it, a minus sign is added in Eq. 0-31.

If the flux presented in each turn of the coil is not the same, Eq. 0-31 can be rewritten by

$$e = -\sum_{i=1}^{N} e_{\mathrm{i}} = -\sum_{i=1}^{N}\frac{\mathrm{d}\Phi_{\mathrm{i}}}{\mathrm{d}t} = -\frac{\mathrm{d}}{\mathrm{d}t}\left(\sum_{i=1}^{N}\Phi_{\mathrm{i}}\right) = -\frac{\mathrm{d}\psi}{\mathrm{d}t} \tag{0-32}$$

where ψ is the flux linkage of the coil, its unit is Weber-turn, and

$$\psi = \sum_{i=1}^{N}\Phi_{\mathrm{i}} \tag{0-33}$$

When a conductor with proper orientation moves through a magnetic field (as shown in Fig. 0-12), a voltage will be induced in it. When movement direction of the conductor and direction of the field are perpendicular to each other, the voltage induced in the wire is in direction perpendicular to the movement and field direction. Magnitude of the voltage can be calculated by

$$e = Blv \tag{0-34}$$

where B is the flux density of the magnetic field [T]; l is the active length of the conductor in the magnetic field [m]; v is the relative speed of the conductor [m/s].

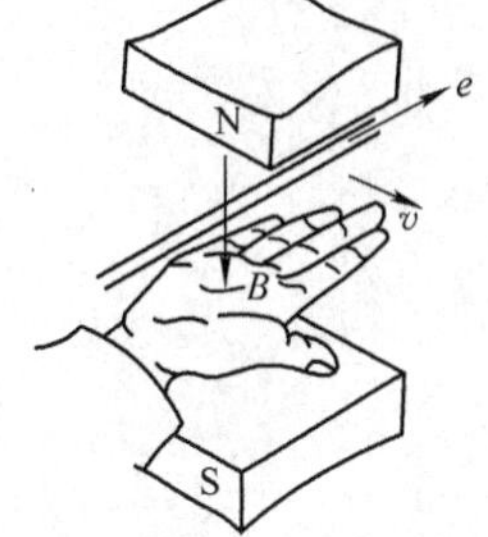

Fig. 0-12 Right-hand rule

The direction of induced voltage can be determined by *right hand rule* (Fig. 0-12). If the direction of the flux density B goes through the palm of right hand vertically and the thumb of the right hand points at the direction of the conductor moving, the direction of the voltage induced in the conductor is shown by where the middle finger of the right hand is pointing at.

0.5.2 Production of induction force on a wire

The second effect of a magnetic field on its surrounding is that it induces a force on a current-carrying wire within the field. The basic concept involved is illustrated in Fig. 0-13. The figure shows a uniform magnetic field of flux density B with a conductor in it. The conductor is l meters long and it contains a current of i amperes. The force induced on the conductor is given by

$$f = Bli \tag{0-35}$$

where B is the flux density of the magnetic field [T]; l is the active length of the conductor in the magnetic field [m]; i is the current of the conductor [A].

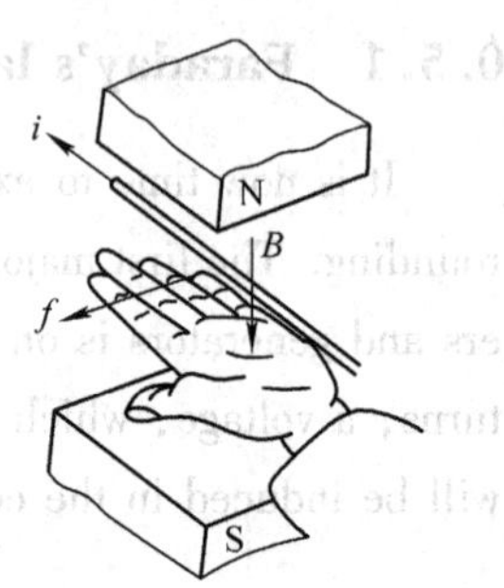

Fig. 0-13 Left-hand rule

The direction of the force is given by *left-hand rule* (Fig. 0-13). If the direction of the flux density B goes through the palm of left hand vertically and the index finger of the left hand points at the direction of the current in the conductor, the direction of the resultant force on the conductor is the same as the direction that the thumb is pointing at.

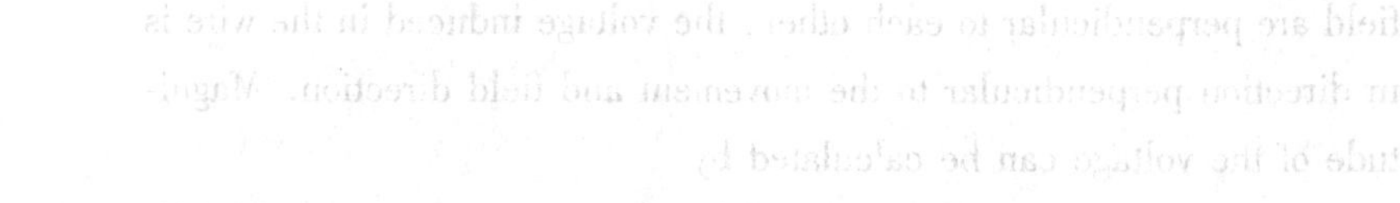

Chapter 1 Direct-Current Machines

1.1 Introduction to DC Machines

DC machines are electromechanical energy—conversion devices. They can be operated either as generators or motors (because mechanical energy and electrical energy conversion is reversible with a DC machine). When a DC machine is operated as a generator, mechanical energy will be converted to DC electrical energy. When it is used as a motor, electrical energy will be converted to mechanical energy. In principle, DC machines are similar to AC machines, since both DC and AC machines have AC voltages and currents within their windings. However, only DC machines have DC outputs because of their unique commutation mechanism that converts the internal AC voltages to DC voltages at their terminals. DC machines are also known as *commutating machines* that are derived from the name of the mechanism—commutator.

DC machines are characterized by their versatility. They can be designed to have a wide variety of volt-ampere or speed-torque characteristics for both dynamic and steady-state operation by means of various combinations of shunt-, series-, and separately-excited field windings. Because the speed of them can be easily controlled, DC machines have been frequently used in applications requiring a wide range of motor speeds or precise control of motor output. However, the popularity of DC machine systems has been largely reduced in recent years. This is because solid-state AC motor drives have been developed rapidly to replace DC machines in many applications previously associated almost exclusively with DC machines. Nevertheless, the versatility and simplicity of the drive systems of DC machines ensure their position still in a wide variety of applications.

In this chapter, we will firstly introduce the construction, the electric circuit and the magnetic fields of DC machines. Then, we will investigate the operational principles and characteristics of various types of DC machines, beginning with a description of how the DC voltage is obtained. The influence of armature windings *magnetomotive force* (mmf) on machine behavior will be explored together with its effect on commutation and its external characteristics. Then the procedures of finding machine performance through the use of the governing equations, equivalent circuit, power-flow diagram and magnetization curve will be outlined and illustrated. And the motor speed-torque curves and various methods for speed control will be analyzed. Finally, The topic of commutation maintains its preeminence in the study of DC machines because they are almost useless without good commutation.

1.1.1 Construction of DC machines

DC machines are composed of stationary components (or stator) and rotating components (or rotor, armature). The stator consists of field poles (or main poles), commutating poles (or interpoles), brushers and stator frames. The frame provides physical and mechanical support to both the stator and rotor. The rotor consists of an armature core, armature windings, a commutator, a shaft and a cooling fan. There is an air-gap between the stator and rotor (armature). Fig. 1-1 shows various parts of a DC machine. The cross section of the DC machine is shown in Fig. 1-2.

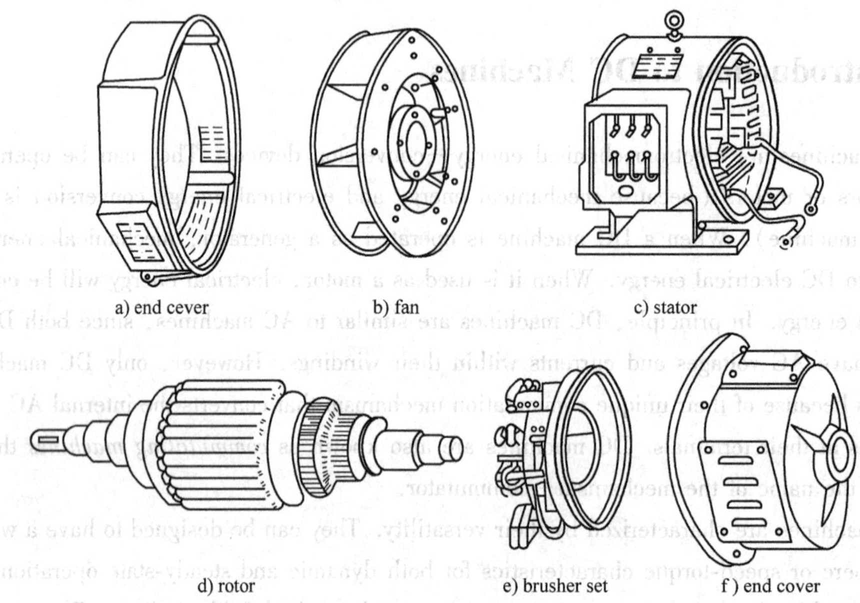

Fig. 1-1 Different parts of a DC machine

The number of poles of a DC machine ranges from 2 to 24, depending on the physical size of the machine. The bigger the DC machine is, the more poles it will have. By using a multi-pole design, we can reduce the dimensions and cost of large machines, and also improve their performances. The field windings of a multi-pole machine are connected in a way that all poles have opposite magnetic polarities to their adjacent poles (Fig. 1-2). The field windings are insulated from the pole pieces to prevent short-circuit.

The flux in the field poles is basically originated from a set of salient poles which is composed of stationary electromagnet mounted on the frame. The field windings, which are mounted on the poles, carry the DC exciting current. In some DC machines, the flux is created by permanent magnets. The main pole is shown in Fig. 1-3.

The frame is usually made of solid cast steel, whereas the pole pieces are made of stacked iron laminations. The commutating poles (or interpoles) (Fig. 1-4) in DC machines are located between the main poles (Fig. 1-2).

The brushes of DC machines are made of carbon, graphite, metal graphite, or a mixture of carbon and graphite (Fig. 1-5). DC machines possess an equal number of brush sets to the number of poles. The brush sets, as suggested by the name, are composed of one or more brushes, the design of brush sets depends upon the current capacity of the machine (Fig. 1-5). The brush sets are spaced around the commutator with equal intervals.

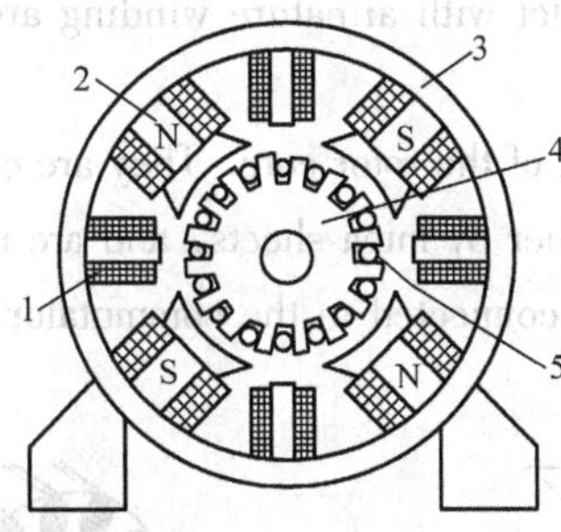

Fig. 1-2 The cross section of a DC machine
1—commutating pole 2—main pole 3—yoke
4—armature 5—armature winding

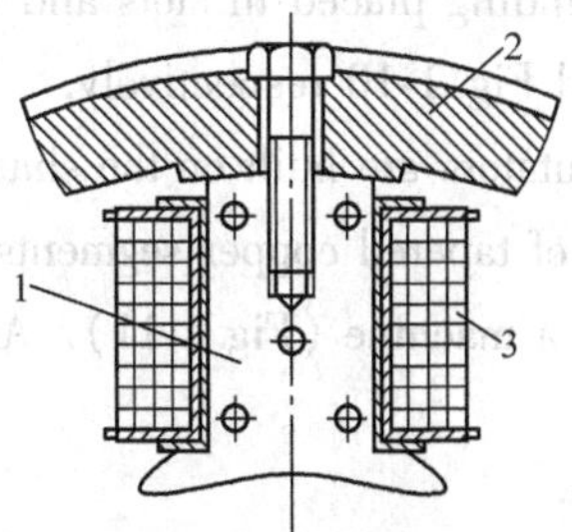

Fig. 1-3 The field pole of a DC machine
1—field core 2—yoke
3— field coil

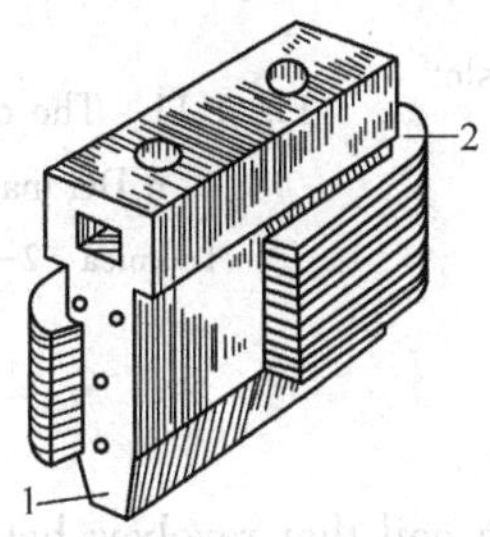

Fig. 1-4 The commutating pole of a DC machine
1—coil 2—commutating core

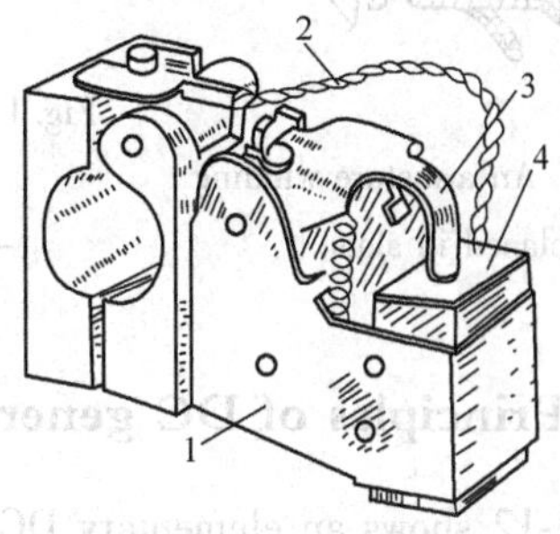

Fig. 1-5 The brush of a DC machine
1—brush holder 2—copper leader 3—spring 4—brusher

The air gap is the short space between the armature and the pole pieces.

Armature is the inner and rotating part of DC machines. It consists of commutators, iron cores, and armature windings (Fig. 1-6). Fig. 1-7 shows a complete DC armature with a cooling fan. Slots are skewed in order to reduce noise. The armature is keyed to a shaft and rotates within the field poles. The iron core is composed of slotted iron laminations that are stacked to form a solid cylindrical core. Laminations are coated individually with an insulating film so that they are not in contact electrically with each other. The slots are lined up to provide the space needed to insert an armature conductor. A lamination of an armature is shown in Fig. 1-8.

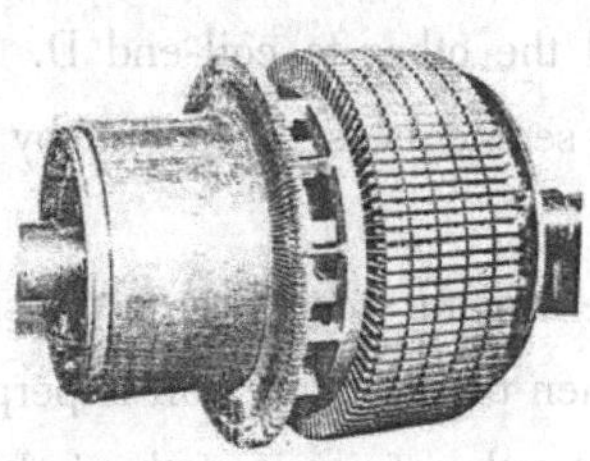

Fig. 1-6 An armature core and a commutator

Fig. 1-7 An armature with skewed slots and a ventilating fan

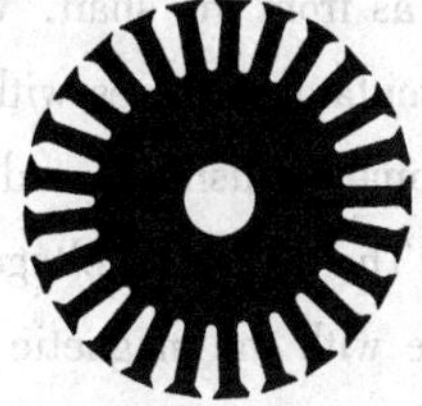

Fig. 1-8 An armature lamination with slots

Armature conductors are placed in slots of the armature laminations. They are insulated from the iron core by several layers of paper or mica and are firmly held in place by fiber slot sticks. An armature winding placed in slots and the cross section of a slot with armature winding are shown in Fig. 1-9 and Fig. 1-10 respectively.

Commutators are built on the shaft of the rotor at one end of the rotor core. They are composed of a assembly of tapered copper segments insulated from each other by mica sheets, and are mounted on the shaft of a machine (Fig. 1-11). Armature conductors are connected to the commutator segments.

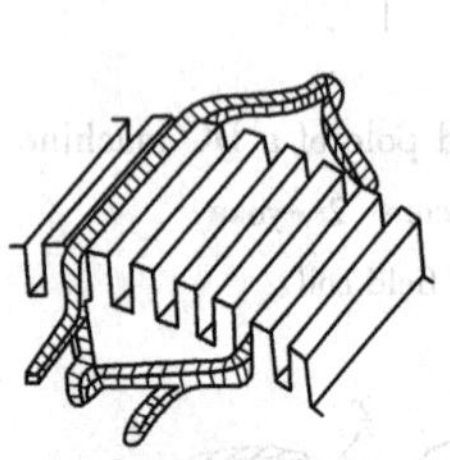

Fig. 1-9 An armature winding placed in slots

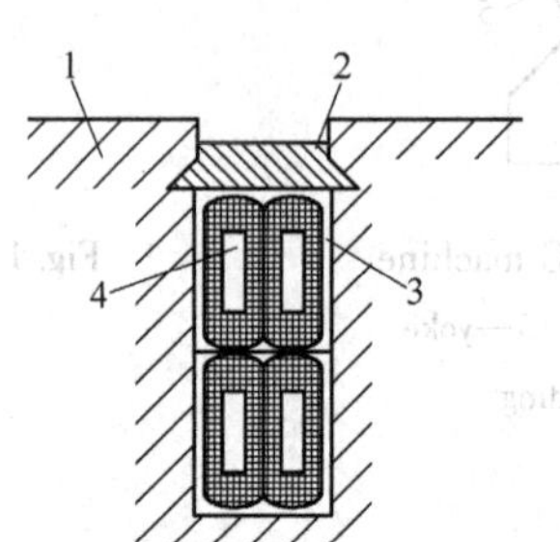

Fig. 1-10 The cross section of a slot with armature windings

1—iron teeth 2—fiber slot stick 3—insulation 4—conductor

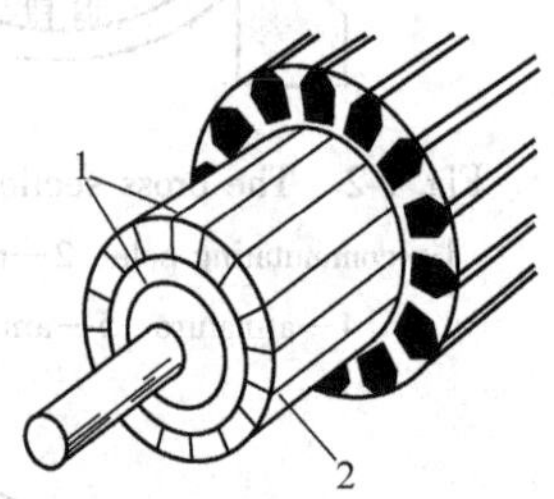

Fig. 1-11 The commutator of a DC machine

1—mica 2—segment

1.1.2 Principles of DC generators

Fig. 1-12 shows an elementary DC generator composed of a coil that revolves between a pair of N, S poles of a permanent magnet at 60r/min. The rotation is due to an external driving torque from a device such as a motor (not shown). The coil is connected to two slip rings mounted on the shaft. The slip rings are connected to an external load by means of two stationary brushes x and y. Brushes x, y can switch from one slip ring to another every time when the polarity of the field pole to meet is about to change. Brush x will always be positive and brush y will be negative. This configuration can be achieved by using a commutator (Fig. 1-12). A commutator in its simplest form is composed of a slip ring that is cut in half, with each segment insulated from the other as well as from the shaft. One segment is connected to coil-end A and the other to coil-end D. The commutator revolves with the armature coil and the voltage between segments is picked up by two stationary brushes x and y.

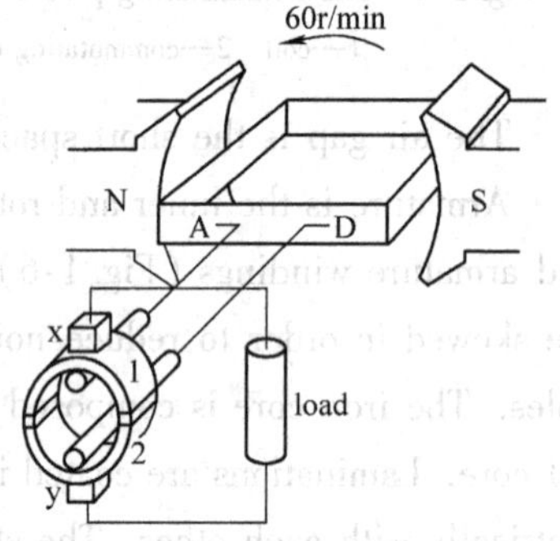

Fig. 1-12 An elementary DC generator

The induced voltage is therefore maximum (20V, say) when the armature coil lies on the same plane with the magnetic field as shown in Fig. 1-12. No flux is cut when the armature coil is perpendicular to the field, consequently the voltage is zero momentarily. Another feature of the induced voltage in the armature coil is that its polarity changes every time when the coil makes a half turn. The coil in our example revolves at a uniform speed, therefore the voltage can be represented as a function of the angle of rotation. The wave shape depends upon the shape of the N, S poles. We assume that the sinusoidal wave shown below can be generated by the designed poles.

The voltage between brushes x and y pulsates but the voltage between the two brushes never changes the polarity (Fig. 1-13). The alternating voltage in the coil is therefore rectified by the commutator.

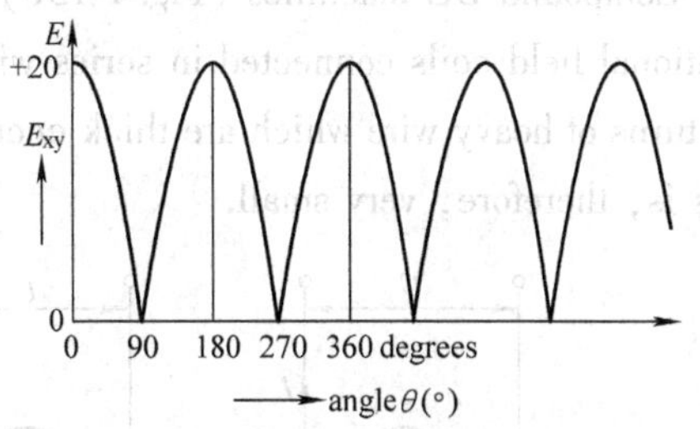

Fig. 1-13 The DC voltage between brushes x and y

Due to the constant polarity of the voltage between brushes, current in the external load always flows in the same direction. The machine shown in Fig. 1-12 is called a direct-current generator or dynamo.

1.1.3 Principles of DC motors

Direct-current motors are built in the same way as that of generators, consequently, a DC machine can be operated either as a motor or a generator. Fig. 1-14 shows an elementary DC motor composed of a coil and a pair of N, S poles of permanent magnet. The coil initially at rest is connected to a DC voltage U by means of a switch (Fig. 1-14). A magnetic field is created by the permanent magnets.

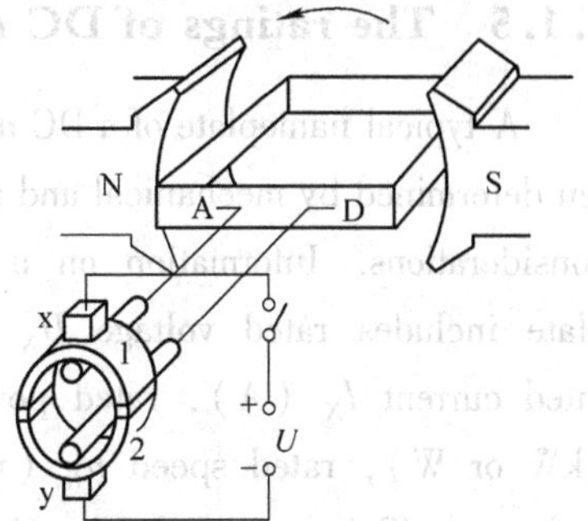

Fig. 1-14 Schematic diagram of an elementary DC motor

As soon as the switch is closed, current will flow into the coil. The coil will immediately be subjected to a force because it is immersed in the magnetic field created by the permanent magnets. The force produces a powerful torque, causing the coil to rotate.

From Fig. 1-13 and Fig. 1-14, we can find that the DC generator and DC motor have the same construction, when the external condition (DC voltage or driving torque) changes, the operational mode of the machine will change corresponding and this is called electric machine's reversible operation principle.

1.1.4 The excitation of DC machines

There are various types of DC machines with different field excitations. Instead of using permanent magnets to create magnetic field, we can use a pair of electromagnets (field poles) to create magnetic field. When a DC field current in such a DC machine is supplied by an independent source as shown in Fig. 1-15a (such as a battery or another generator also called an exciter), the DC machine is said to be excited separately.

A shunt-excited DC machine is a machine whose field winding is connected in parallel with armature terminals, so that the machine can be self-excited (Fig. 1-15b). The principal advantage of this connection is that the DC machine can eliminate the need for an external excitation source.

A series-excited DC machine is a machine whose field winding is connected in series with armature terminals as shown in Fig. 1-15c. Series-excited DC machines can also be self-excited.

Compound DC machines (Fig. 1-15d) are similar to shunt DC machines except that they have additional field coils connected in series with armatures. These series field coils are composed of a few turns of heavy wire which are thick enough to carry armature current. The resistance of the series coils is, therefore, very small.

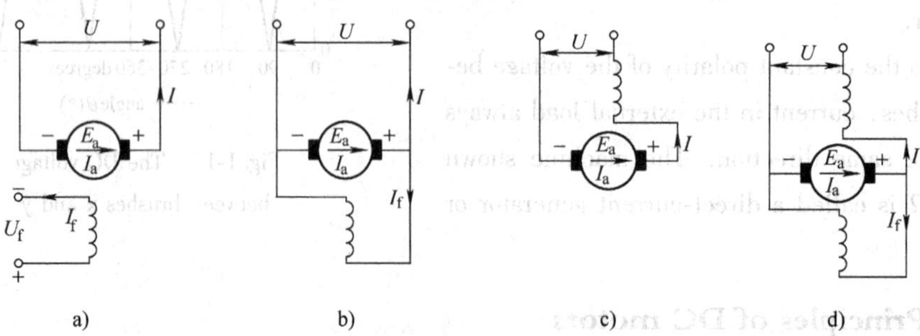

Fig. 1-15 The excitation types of DC machines

1.1.5 The ratings of DC machines

A typical nameplate of a DC machine is shown in Fig. 1-16. Ratings of electric machines are often determined by mechanical and thermal considerations. Information on a nameplate includes rated voltage U_N (V), rated current I_N (A), rated power P_N (kW or W), rated speed n_N (r/min) and rated efficiency η_N. Information such as the excitation type, excited voltage U_f (V) and excited current I_f (A) are also included.

DC Motor			
Type	Z_3—95	Products number	7001
Construction type		Excitation type	Separately excited
Power	30kW	Excitation voltage	220V
Voltage	220V	Insulation class	B
Current	160.5A	Operation mode	Continuous
Speed	750r/min	Weight	685kg
Standard number	JB 1104—68	Date	

Fig. 1-16 A typical nameplate of a DC machine

For electric machines, the rated power is the same as the output power. For DC generators, P_N is the electrical power generated from the armature windings, and

$$P_N = U_N I_N \quad \text{(DC generator)} \tag{1-1}$$

For DC motors, P_N is the mechanical power generated from the shaft, and

$$P_N = U_N I_N \eta_N \quad \text{(DC motor)} \tag{1-2}$$

Example 1-1 Calculate the rated current and rated input power of a DC generator with rated power $P_N = 10\text{kW}$, rated voltage $U_N = 230\text{V}$, rated speed $n_N = 2850\text{r/min}$, rated efficiency $\eta_N = 0.85$.

Solution:

According to Eq. 1-1, the rated current of the DC generator is

$$I_N = P_N / U_N = 10 \times 1000/230\text{A} = 43.48\text{A}$$

The rated input power is

$$P_1 = P_N/\eta_N = 10 \times 1000/0.85\text{W} = 11\,765\text{W}$$

Example 1-2 Calculate the rated current and rated input power of a DC motor with rated power $P_N = 17\text{kW}$, rated voltage $U_N = 220\text{V}$, rated speed $n_N = 1500\text{r/min}$, rated efficiency $\eta_N = 0.83$.

Solution:

According to Eq. 1-2, the rated input power of the DC motor is

$$P_1 = P_N/\eta_N = 17 \times 1000/0.83\text{W} = 20\ 482\text{W}$$

Its rated current is

$$I_N = P_1/U_N = 20\ 482/220\text{A} = 93.1\text{A}$$

1.2 Armature windings

Armature windings are classified according to the sequence of their connections to the commutator segments. There are two basic sequences of armature winding connections—*lap windings and wave windings*. The simplest types of lap and wave winding are simplex lap and wave winding respectively.

Some terms are employed for better description of the construction of DC armature windings. They are winding pitch (y), pole number ($2p$), pole pitch (τ), slot number (Q), winding element number (S), commutator bar number (K), commutator pitch (y_c), the first pitch (or coil pitch) (y_1) and the second pitch (y_2).

Elements of a lap winding and a wave winding with two turns are shown in Fig. 1-17a, b respectively. The *winding element* is defined as one or more turns which its beginning and end are connected to different commutator bars. The winding element consists of at least 1 turn. It may, however, consist of 2, 3, or more turns in series. A turn has 2 conductors in series, and the two conductors are placed in different slots as shown in Fig. 1-9. The conductor may be solid. It may also consist of several insulated strands in parallel, which are either round or rectangular in shape.

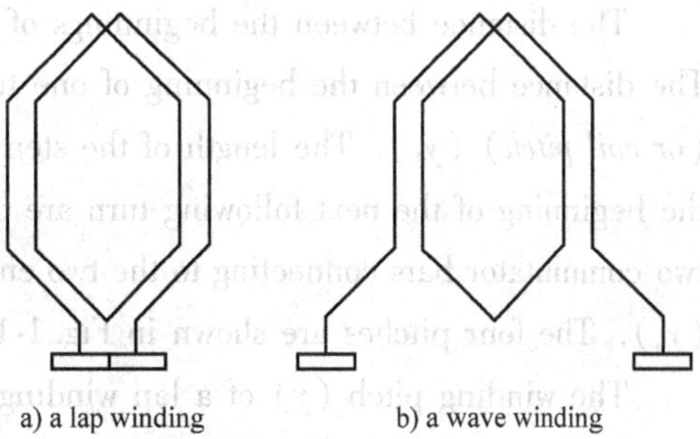

Fig. 1-17 Element of a lap and a wave winding

There is a configuration that two coil sides are placed one above the other in each slot of a winding (Fig. 1-9). Such windings are called 2-layer windings to differentiate it from the single-layer windings. 2-layer windings have equal number of coils and slots. Single-layer windings have half the number of slots. The number of conductors per coil side varies from case to case.

Fig. 1-18a shows a lap winding. Turn *a* consists of conductors a'-a'', turn *b* consists of conductors b'-b'' and so forth. The end a'' of turn *a* is connected with the beginning b' of turn *b* which lies under the same pole as the beginning of turn *a*; the end of turn *b* (b'') is connected with the beginning of turn *c* (c') and so forth. The end of the winding is then connected with the beginning of the first turn and the winding is closed.

As mentioned, it is important that the end of one turn and the beginning of the following turn must lie under poles with opposite polarity, otherwise the emf of two consecutive turns will not add.

The wave winding shown in Fig. 1-18b is a closed winding, it is similar to a lap winding. The first wave is connected to 1, 10, 19 commutator bars; the second wave is connected to 19, 9, 18

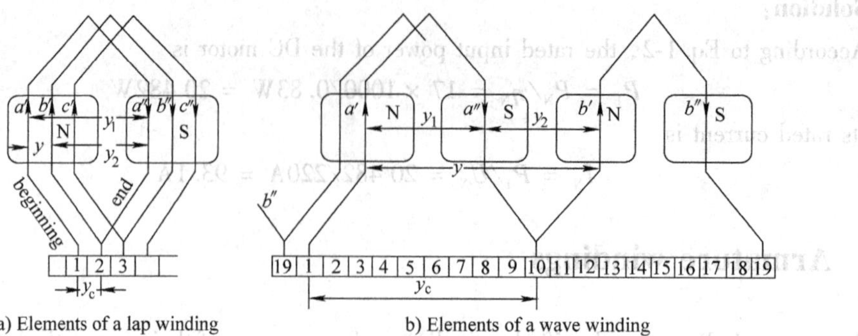

a) Elements of a lap winding

b) Elements of a wave winding

Fig. 1-18 Winding pitch of a lap and a wave winding

commutator bars and so forth. The last wave is connected to 12, 2, 11 commutator bars. The last turn is connected to commutator bars 11 and 1 and close the winding.

It is clear that the beginning and the end of each winding element must be connected to a commutator bar to give a complete winding circuit. Therefore the number of winding elements must be the same as the number of *commutator bars* (K) for both lap windings and wave windings, putting it as an equation, it is $K = S$.

Winding pitch y, the first pitch y_1, the second pitch y_2 and the commutator pitch y_c

The distance between the beginnings of two consecutive turns is called the *winding pitch* (y). The distance between the beginning of one turn to the end of the same turn is called *the first pitch* (*or coil pitch*) (y_1). The length of the step necessary to follow through from the end of one turn to the beginning of the next following turn are called *the second pitch* (y_2). The distance between the two commutator bars connecting to the two ends of an element winding is called the commutator pitch (y_c). The four pitches are shown in Fig. 1-18.

The winding pitch (y) of a lap winding is equal to the difference between the first pitch (y_1) and the second pitch (y_2). The winding pitch (y) of a wave winding is equal to the sum of the first pitch (y_1) and the second pitch (y_2).

$$y = y_1 - y_2 \quad \text{(lap winding)} \tag{1-3}$$

$$y = y_1 + y_2 \quad \text{(wave winding)} \tag{1-4}$$

$$y = y_c \quad \text{(lap and wave winding)} \tag{1-5}$$

The *first pitch* (*coil pitch*, y_1) of lap and wave windings must be equal or approximately equal to the *pole pitch* τ ($\tau = Q/2p$ in slot pitches).

$$y_1 = \tau \pm \varepsilon \quad \text{(lap and wave winding)} \tag{1-6}$$

In Eq. 1-6, when the coil pitch y_1 is equal to the pole pitch τ ($\varepsilon = 0$), the winding is defined as full-pitch winding; if y_1 is greater than τ ($\varepsilon > 0$), the winding is defined as a long-pitch winding; and if y_1 is less than τ ($\varepsilon < 0$), the winding is defined as short-pitch winding, and is also called fractional pitch winding.

The four pitches, y, y_1, y_2 and y_c can be expressed in several ways. They can be expressed in any linear dimension measures at the surface of armatures, in slot pitches, in degrees or in commutator bars (commutator pitches). In this text, these pitches will be measured in slots (or commutator bars).

For simplex lap windings in which coils contain one or more turns of wire with the two ends of each coil connected to the adjacent commutator bars (Fig. 1-18a), if the end of a coil is connected to the bar after the starting bar, the winding is a progressive lap winding and $y=1$; if the end of the coil is connected to the bar before the starting bar, the winding is a retrogressive lap winding and $y=-1$.

$$y = y_c = \pm 1 \quad \text{(simplex lap winding)} \tag{1-7}$$

In wave windings, the end of the winding element goes to a slot under the next pole of the same polarity and is connected there to commutator bar $(1+y)$, which is located about 2 pole pitches distant (measured on the commutator) from the starting commutator bar 1. The commutator bar reaches the nearby position of the starting bar after progressing through a number of winding elements which are equal to the number of pole pairs p. The winding pitch y of wave windings must somewhat deviate from the original pole pitch (measured in terms of commutator bars) 2 times, or else the starting bar 1 will be reached after progressing through p elements. The latter will result in a closed circuit at once, thus a series connection of all coils of armature paths can not be achieved. Therefore, in wave winding, $y \neq K/p$, but

$$y = y_c = (K \pm 1)/p \quad \text{(simplex wave winding)} \tag{1-8}$$

The minus or plus sign in Eq. 1-8 depends upon the direction that the winding elements brought to by the progress through p winding elements. The '+' indicates that the winding goes to the commutator bar on the left of the starting bar and '−' indicates that the winding goes to the bar on the right of the starting bar, respectively. If the winding is brought to the left of the starting bar, it is called *retrogressive*; if to the right, *progressive*.

1.2.1 Simplex lap windings

Let us use the following example to explain the composition of simplex lap windings.

Example 1-1 For a 4-pole machine with full-pitch progressive simplex lap winding having the same number of winding elements, slots and commutator bars, i. e., $Q=S=K=16$,

1. Pitches calculation

For the full-pitch progressive simplex lap winding, $y=y_c=1$.

According to Eq. 1-6: $y_1=\tau=Q/2p=16/4=4$,

According to Eq. 1-3: $y_2=y_1-y=4-1=3$

2. Winding elements connection

According to the result above, the winding elements connection should be presented as the figures shown below (Fig. 1-19). You may have discovered that the lap winding is a closed winding.

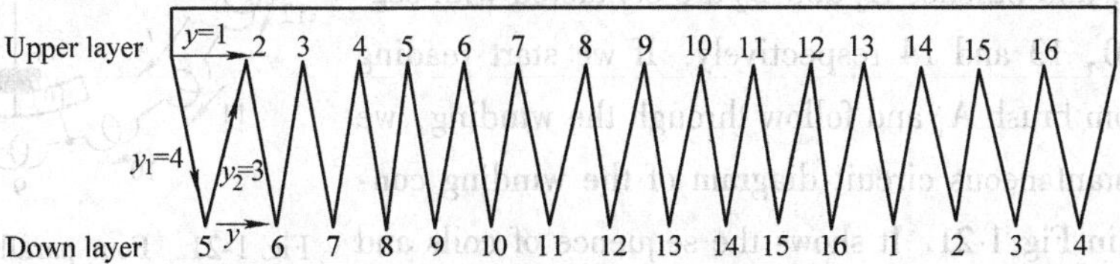

Fig. 1-19 Diagram of the winding elements connection

3. Developed winding diagram

Developed winding connection with 4-poles is shown in Fig. 1-20. In Fig. 1-20, the upper and lower layers are shown side by side. The upper layer is represented by solid line and the lower layer is represented by dotted line. According to Fig. 1-19, the upper layer of the element 1 is connected to commutator bar 1; the lower layer of element 1 is connected to the upper layer of element 2 and commutator bar 2, and so forth. The lower layer of the last element 16 is then connected with the upper layer of element 1 and commutator bar 1, and the winding is closed.

4. Placing magnetic poles and brushes

The magnetic poles are placed evenly on the stator with N and S pole alternately (Fig. 1-20), and the pole width is usually equal to 0.75τ. The axis between two adjacent N and S poles is called *neutral axis or quadrature-axis*.

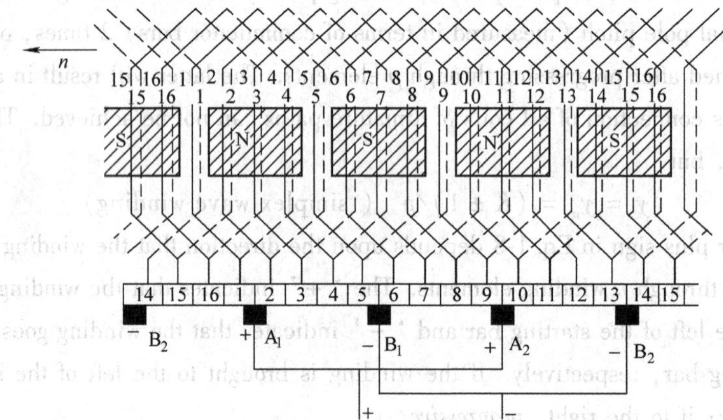

Fig. 1-20 Developed simplex lap winding diagram

The brushes are positioned around the commutator in a way that they are connected with the conductors lying at the middle points of the two magnetic poles (N and S). As shown in Fig. 1-20, the two conductors of coil 1, 5, 9, 13 are lying at the middle points of N and S poles and coil 1, 5, 9, 13 are shorted by brush A_1, B_1, A_2 and B_2 respectively. The emf induced in the conductors of coil 1, 5, 9, 13 is so small that the circulating current flowing through the brushes is negligible. In general, lap windings have equal number of brushes and poles.

5. Instantaneous circuit diagram

In the moment of shown in Fig. 1-20, brush A_1 is contacted with segments 1 and 2, bursh B_1 is contacted with segments 5 and 6, and burshes A_2 and B_2 are contacted with segments 9 and 10, 13 and 14 respectively. If we start reading the diagram from brush A_1 and follow through the winding, we can get the instantaneous circuit diagram of the winding connection shown in Fig. 1-21. It shows the sequence of coils and the direction of the induced emf.

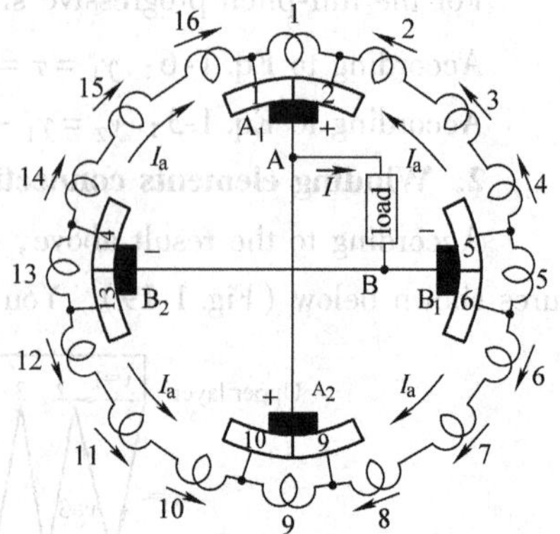

Fig. 1-21 Four parallel circuits of the winding shown in Fig. 1-20

6. Parallel branches of windings

If we start reading the diagram (Fig. 1-20) from brush A_1 and follow through the winding, we can have Fig. 1-21. It shows the sequence of coils and the direction of the induced emf. From Fig. 1-21, we can find that conductors lying between poles (coil 1, 5, 9, 13) do not contribute to the induced emf and the number of parallel branches is equal to the number of poles.

An interesting discovery of simplex lap windings is that the number of parallel current path ($2a$) is the same throughout the machine and so the number of poles on the machine.

$$2a = 2p \qquad \text{(simplex lap winding)} \tag{1-9}$$

1.2.2 Simplex wave windings

Let us use the following example to explain the composition of simplex wave windings.

Example 1-2 For a 4-pole machine with short-pitch *retrogressive* simplex wave winding, noted that the number of winding elements, slots and commutator bars is the same with each other, i. e., $Q = S = K = 15$.

1. Pitches calculation

For the short-pitch *retrogressive* simplex wave winding, according to Eq. 1-8,

$$y = y_c = (K - 1)/p = (15 - 1)/2 = 7.$$

According to Eq. 1-6, $y_1 = \tau - \varepsilon = Q/2p - \varepsilon = 15/4 - \varepsilon = 3$.

According to Eq. 1-4, $y_2 = y - y_1 = 7 - 3 = 4$.

2. Winding elements connection

According to the result obtained above, the winding elements connection should be presented as the figure shown below (Fig. 1-22). You may have discovered that the wave winding is a closed winding.

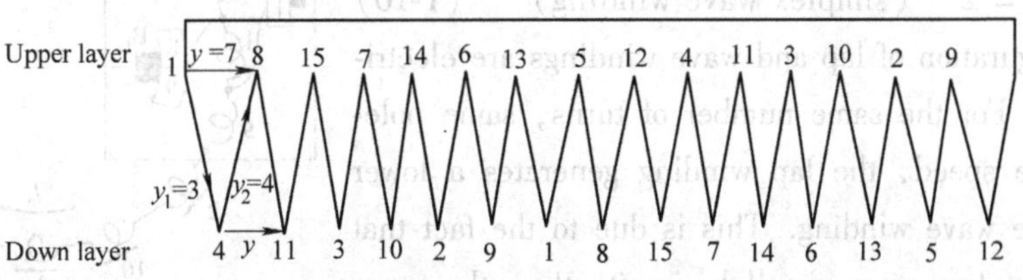

Fig. 1-22 Diagram of the winding elements connection

3. Developed winding diagram

Developed winding connection with 4-pole is shown in Fig. 1-23. In Fig. 1-23, the upper and lower layers are shown side by side. The upper layer is represented by solid line and the lower layer is represented by dotted line. According to Fig. 1-22, the first wave is connected to commutator bar 1, 8, 15; the second wave is connected to commutator bar 15, 7, 14; the third wave is connected to commutator bar 14, 6, 13; the fourth, fifth and sixth wave are connected to commutator bars (13, 5, 12); (12, 4, 11); (11, 3, 10) respectively. The last wave is connected to commutator bar 10, 2, 9, and the last turn is connected to commutator bar 9 and 1, and the winding is closed.

The brushes are attached on the commutator in a way that they are connected with the conductors lying between two magnetic poles (N and S). As shown in Fig. 1-23b, brush A_1, B_1, A_2 and B_2 are placed on commutator bars (4 and 5), (8 and 9), 12, (15 and 1) respectively. In general, wave windings have equal number of brushes and poles.

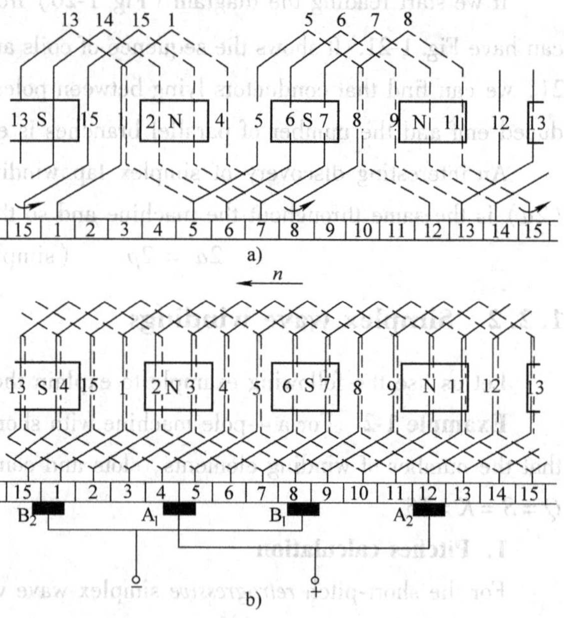

Fig. 1-23 Developed simplex wave winding diagram

4. Parallel branches of windings

If we see the winding diagram Fig. 1-23 from the position brush B_2 and go through the winding, Fig. 1-24 can be deduced from the route of winding which is shown in Fig. 1-23. It shows the sequence of coils and the direction of the induced emf. From Fig. 1-24, we can find that conductors lying between poles (coil 15, 1, 4, 5, 8, 9, 12) do not contribute to the induced emf and the number of parallel branches, which is two, does not equal to the number of poles.

In simplex wave windings, there are only two branches ($2a$) regardless of the number of poles. This feature of wave windings explains why it is designated as a series winding, and the lap winding is designated as a parallel winding.

$$2a = 2 \quad \text{(simplex wave winding)} \qquad (1\text{-}10)$$

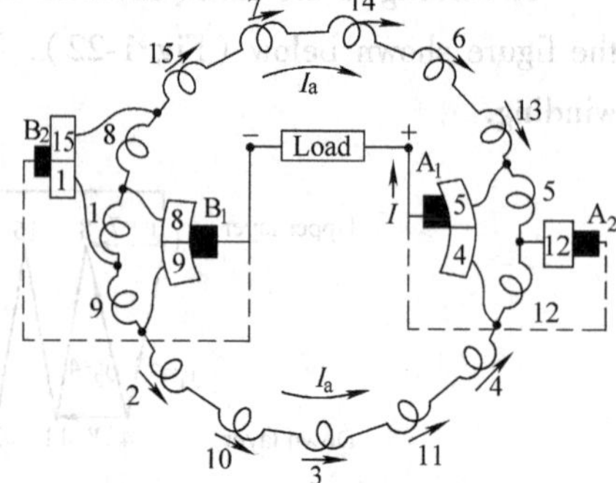

Fig. 1-24 The two parallel circuits of the winding shown in Fig. 1-23

The configuration of lap and wave windings are electrically different. For the same number of turns, same pole-flux, and same speed, the lap winding generates a lower voltage than the wave winding. This is due to the fact that the lap winding has more parallel circuits than the wave winding, i. e., fewer coils are connected in series in each path of the lap winding than in the wave winding, thus a lower voltage is generated.

1.3 Magnetic fields of DC machines

A DC machine has 2 basic windings: a field winding which is placed on the poles and an armature winding. When the armature is not loaded, the induced magnetic flux is only determined by the field winding and the operation is called no-load. When both windings carry current (i. e. the armature is loaded), the induced magnetic flux is determined by the resultant magnetomotive force (mmf) of both the field and armature windings.

1.3.1 Magnetic field in no-load condition

Figs. 1-25 and 1-26 show the main flux path of a 2-pole machine and a 6-pole machine respectively. The magnetic flux of each pole has its path through the stator yoke, pole body, air gap between pole body and armature, armature teeth, armature core, then back through the armature teeth, air gap, and pole body to the stator yoke. The equivalent closed magnetic flux line passes through the air gap, teeth and pole body twice. The pole flux is divided into two parts in the stator yoke as well as in the armature core.

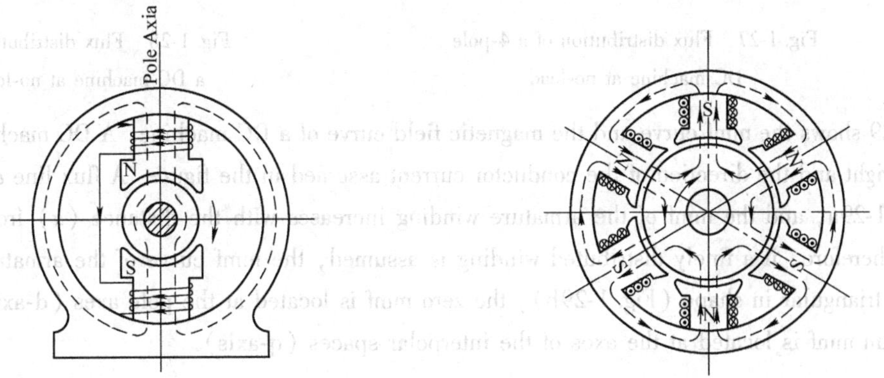

Fig. 1-25 Main flux paths of a 2-pole DC machine Fig. 1-26 Main flux paths of a 6-pole DC machine

Fig. 1-27 shows half of a 4-pole machine at no-load. It shows that most flux lines cross the air gap and enter the armature through the teeth. The flux lines interlinks with armature winding and contribute to the armature emf, thus they represent a main flux (mutual flux) Φ_0. Some of the lines go from pole tip to pole tip and from pole body to pole body without passing through the air gap and armature. These lines do not interlink with the armature winding and do not contribute to the armature emf, thus they represent leakage flux Φ_σ. The pole-leakage flux can make up 2% to 8% of the useful flux Φ_0.

It is mentioned that the number of ampere-turns necessary to drive the main flux through the air gap is the largest among the 5 components (armature core, armature teeth, air gap, pole body and stator yoke) because the reluctance of the air gap is the largest among the 5 parts of the magnetic circuit.

From Fig. 1-27, we can see that the length of the air gap between the pole and the armature is not uniform. The air gap length is larger under the pole shoe than those under other parts of the pole. So the flux density in the air gap between the pole shoe and the armature is smaller than that of the air gap between the pole face and armature. Fig. 1-28 shows the flux density curve of a DC machine at no-load, it is a flat-top shaped curve.

1.3.2 Magnetic field in loaded condition

1. mmf and flux density of armature windings

To illustrate the armature winding mmf, it is assumed that only armature windings carry current.

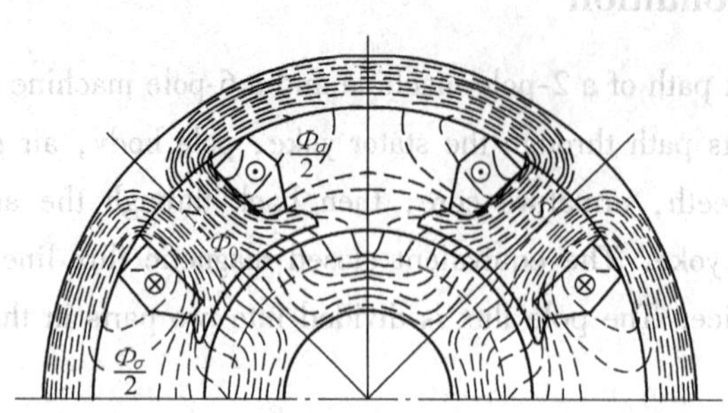

Fig. 1-27 Flux distribution of a 4-pole DC machine at no-load

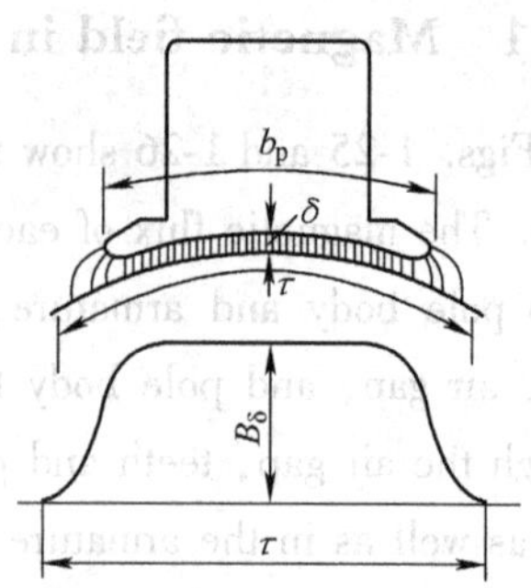

Fig. 1-28 Flux distribution of a DC machine at no-load

Fig. 1-29 shows the mmf curve and the magnetic field curve of a DC machine. A DC machine moving to the right and the direction of the conductor current assumed in the figure. A flux line *aa* is shown in Fig. 1-29a, and the mmf of the armature winding increases with the distance (x) from the pole axis. Therefore, if a finely distributed winding is assumed, the mmf curve of the armature winding will be triangular in shape (Fig. 1-29b), the zero mmf is located at the pole axes (d-axis) and the maximum mmf is located at the axes of the interpolar spaces (q-axis).

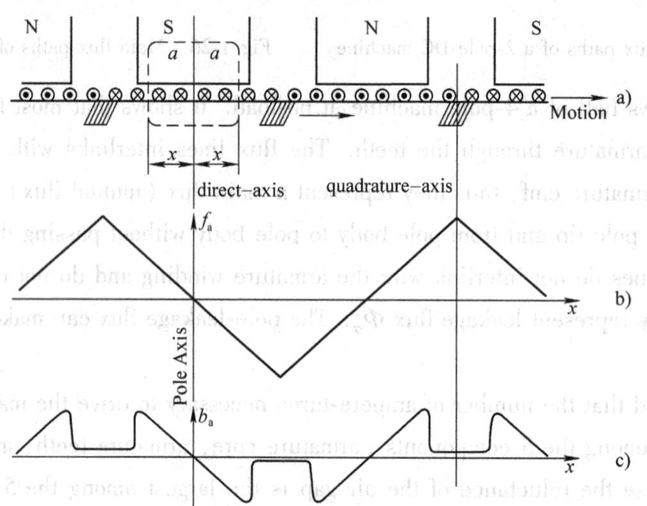

Fig. 1-29 Mmf and field curve of a DC machine with a finely distributed winding

Field curves, i. e., the flux density distribution B_{ax} as a function of distance x, are determined by the mmf F_{ax} curve and the length of the air-gap δ_x (or the magnetic reluctance) for the different tubes of force, i. e. $B_{ax} = \mu_0 F_{ax}/\delta_x$. Since the magnetic reluctance is very high in the interpolar space, the flux density in the q-axis is very small. The approximated shape of the field curve which corresponds to the mmf curve in Fig. 1-29b is shown in Fig. 1-29c.

In Fig. 1-29, it is assumed that the brushes are in the neutral axis (or quadrature axis). If the brushes are shifted away from the neutral axis, the shape of the mmf curve (Fig. 1-29b) will not change, but its zero-crossing and maximum points will not coincide with the pole axis (or direct axis) and axis of the interpolar spaces. The shape of the field curve, therefore, will be different from the one shown in Fig. 1-29c.

In accordance with Faraday's law for induced emf, all conductors lying under north pole will induce emf in one direction and all those under south poles will induce emf in the opposite direction. If the brushes lie in the neutral axis, the same rule as that of the voltage also applies to current, thus, all conductors above (or below) the neutral axis will induce current flowing in the same direction (Fig. 1-30a). If the brushes shift away from the neutral axis (Fig. 1-30b), Faraday's law for emf will no longer hold for current. This is due to the fact that brushes position determines the parallel circuits. Since the current direction is the same for all conductors within the parallel circuits, all conductors above (or below) the line determined the brushes must have the currents of the same direction.

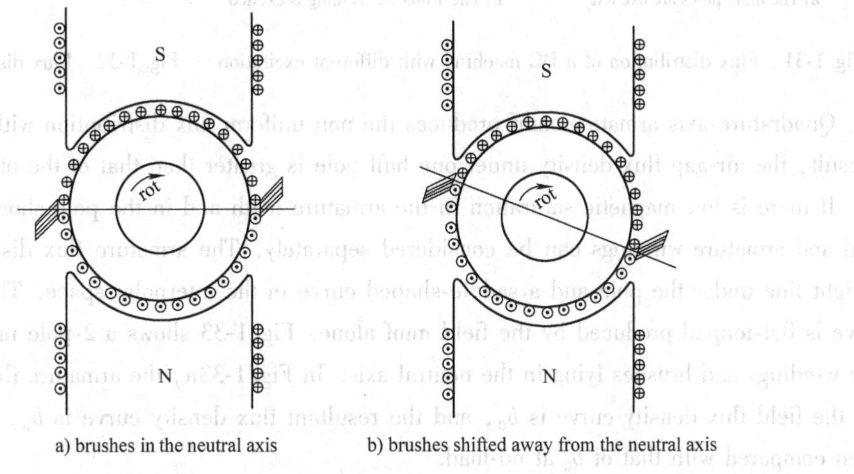

Fig. 1-30 Current distribution of a 2-pole DC machine

2. Both field windings and armature windings carry current and Armature reaction

When the armatures are loaded, the magnetic flux distribution is determined jointly by the mmfs of field and armature windings. The distribution is very different from the flux distribution at no-load. The effect of the armature mmf upon the flux distribution is called *armature reaction*.

(1) Quadrature-axis armature reaction

Fig. 1-31a shows the flux distribution of a 2-pole generator when only the field winding carries current. Fig. 1-31b shows the flux distribution when only the armature carries current and the brushes are in the neutral axis. The direction of the armature flux is orthogonal with respect to the main flux. Furthermore, the armature mmf strengthens half of the pole and weakens the other half. When the iron is saturated, the weakening effect of the armature mmf is greater than that of the strengthening effect, resulting in a reduction of the total main flux within a pole.

Fig. 1-32 shows that the quadrature-axis armature flux weakens the leading pole half and strengthens the trailing pole half in a generator. In a motor, the current and induced emf flow in opposite directions (Fig. 1-33b). The armature mmf does not only weaken the pole flux and the air-gap flux. It also distorts them and causes a displacement of the neutral zone.

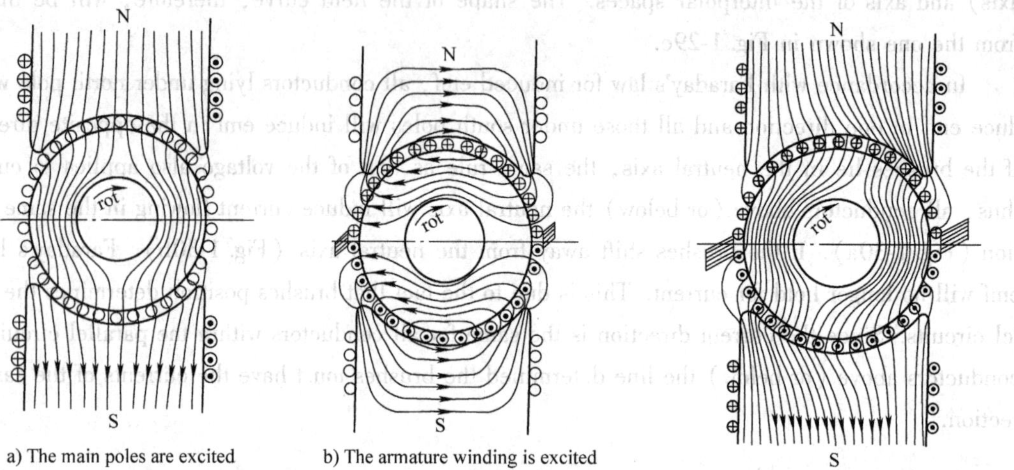

Fig. 1-31 Flux distribution of a DC machine with different excitation

Fig. 1-32 Flux distribution under load

Quadrature-axis armature mmf produces the non-uniform flux distribution within a pole, and as a result, the air-gap flux density under one half pole is greater than that of the other half pole.

If there is low magnetic saturation in the armature teeth and in the pole shoe, the mmfs of the field and armature windings can be considered separately. The armature flux distribution is then a straight line under the pole and a saddle-shaped curve in the interpolar space. The flux distribution curve is flat-topped produced by the field mmf alone. Fig. 1-33 shows a 2-pole machine with armature windings and brushes lying in the neutral axis. In Fig. 1-33a, the armature flux density curve is b_a, the field flux density curve is b_0, and the resultant flux density curve is b_δ, distorted in shape when compared with that of b_0 at no-load.

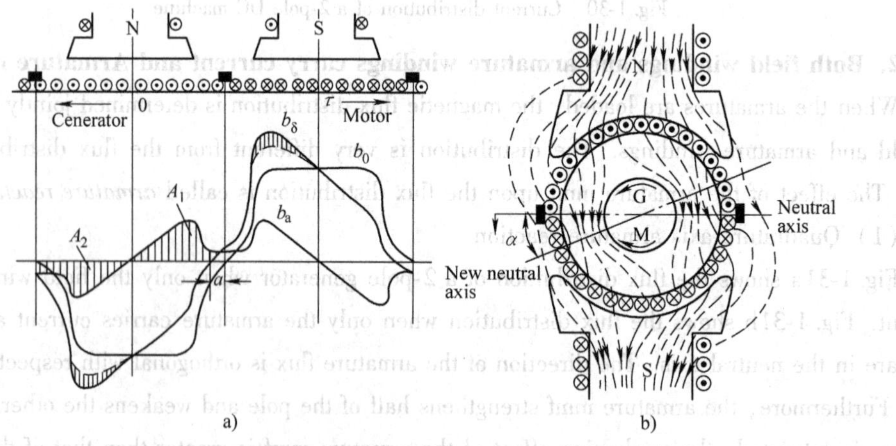

Fig. 1-33 Field curve at load and brushes in the neutral axis

Since the saturation of iron is low, the armature flux and the axis of abscissa are symmetrical. Also, the area under curve b_0 is equal to the area under curve b_δ. That means the magnitude of the air-gap flux, which is proportional to the area of the flux density curve, remains the same under load as that of no-load.

If the armature teeth and pole shoes are saturated, the area lying above curve b_0 between curve b_δ and curve b_0 will be smaller than that of lying below curve b_0 (Fig. 1-33a). As a result of armature reaction, the resultant air-gap flux under load will be less than that of no-load and the neutral axis will shift α degree to the left/right (Fig. 1-33b).

(2) Demagnetizing effect of armatures

When brushes deviate from the neutral axis, part of the armature conductors will produce crossed magnetic field (quadrature-axis armature reaction), and the remaining part will produce an mmf which its effect will act along the pole axis (direct-axis armature reaction). This direct-axis armature reaction will either weaken or strengthen the mmf of the field winding, and consequently alter the flux Φ. Fig. 1-34 shows the armature reaction when brushes diverge from the neutral axis.

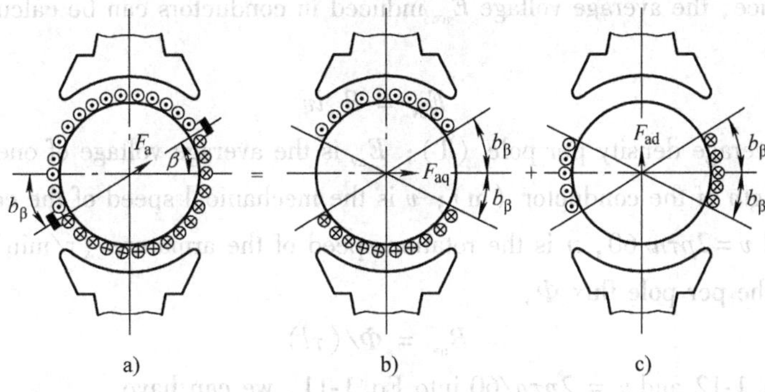

Fig. 1-34 Quadrature-axis and direct-axis mmf at load and brushes from the neutral axis

In general, flux can be weakened under two conditions: ① brushes of generators are placed in the direction of rotation. ② brushes of motors are displaced in the opposite direction of rotation. Contrarily, flux can be strengthened by ① brushes of generators are displaced in the opposite direction of rotation and ② brushes of motors are displaced in the direction of rotation. Fig. 1-35 shows the flux density curves when brushes deviate from the neutral axis. Curve 1 shows the total armature mmf, curve 2 shows the quadrature-axis armature mmf and curve 3 shows the direct-axis armature mmf.

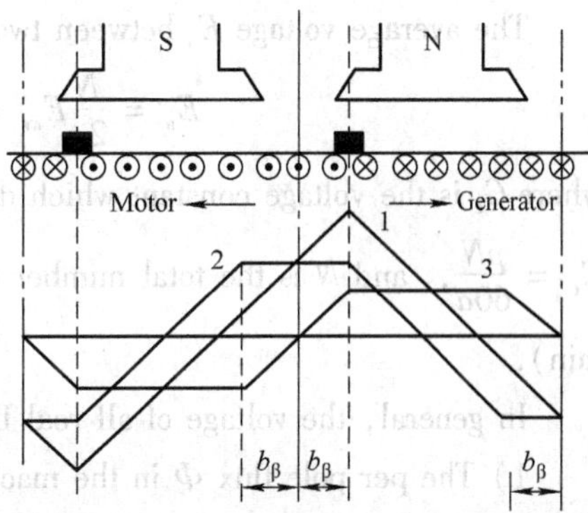

Fig. 1-35 Field curve at load and brushes from the neutral axis

1.4 Induced voltage and electromagnetic torque

1.4.1 Induced voltage

Although the ultimate purpose of DC machines is to generate DC voltage, it is evident that the speed voltage produced in armature-winding is an alternating voltage when the armature winding coils rotate within the DC flux distribution produced by the stationary field winding. Therefore, the armature-winding alternating voltage is to be rectified if DC voltage is desired. The armature AC voltage rectification can be carried out by commutators mechanically.

Voltage generated from DC machines is the same as the DC voltage between two brushes, i. e., the total voltage of a parallel circuit.

The air gap flux density under one pole varies with distance from the pole. However, the same curve can be produced under each pole whatever the distance (Fig. 1-36). The induced voltage of each parallel circuit remains constant because the number of coils between brushes is always the same irrespective of the armature position.

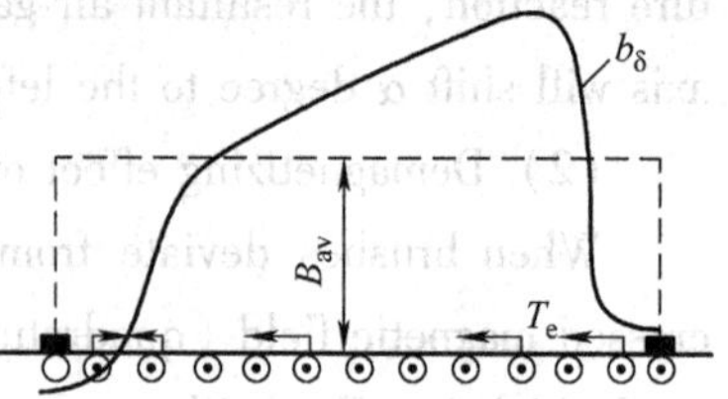

Fig. 1-36 The air gap flux density curve under one pole

From Fig. 1-36, we can see that when the armature rotates for one-pole distance, the average voltage E_{av} induced in conductors can be calculated by the following equation:

$$E_{av} = B_{av} l v \tag{1-11}$$

where B_{av} is the average density per pole (T); E_{av} is the average voltage of one conductor (V); l is the effective length of the conductor (m); v is the mechanical speed of the conductor cutting the field (m/s), and $v = 2p\tau n/60$, n is the rotation speed of the armature (r/min).

Considering the per pole flux Φ,

$$B_{av} = \Phi/(\tau l) \tag{1-12}$$

Substitute Eq. 1-12 and $v = 2p\tau n/60$ into Eq. 1-11, we can have

$$E_{av} = 2p\Phi n/60 \tag{1-13}$$

The average voltage E_a between two brushes can be calculated using Eq. 1-14.

$$E_a = \frac{N}{2a}E_{av} = \frac{N}{2a}2p\Phi\frac{n}{60} = \frac{pN}{60a}\Phi n = C_e\Phi n \tag{1-14}$$

where C_e is the voltage constant which depends upon the construction parameters of DC machines, $C_e = \frac{pN}{60a}$, and N is the total number of conductors; n is the rotation speed of the armature (r/min).

In general, the voltage of all real DC machines depends on the same factors:

1) The per pole flux Φ in the machine.

2) The speed (n) of the machine's rotor.

3) The constant C_e with respect to the construction of DC machines.

If the saturation of iron is low, the per pole flux Φ can be expressed $K_f I_f$, where K_f is a constant and I_f is the excitation current. Eq. 1-14 can be rewritten as follows:

$$E_a = C_e\Phi n = K_f C_e I_f n = G_{af} I_f \Omega \tag{1-15}$$

where $G_{af} = \frac{60}{2\pi}C_e K_f = C_T K_f$; Ω is the mechanical speed of rotor, rad/s, and $\Omega = 2\pi n/60$.

1.4.2 Electromagnetic torque

Electromagnetic torque produced by DC machines can be calculated from the basic law of force of conductors in the magnetic field.

In Fig. 1-36, when the armature rotates one-pole distance, the average force f_{av} produced by conductors can be calculated by the following equation:

$$f_{av} = B_{av} l I_c \tag{1-16}$$

where B_{av} is the average density per pole (T); l is the effective length of the conductor (m); I_c is the current per conductor in ampere; f_{av} is the average force in newton.

For DC machines, the current per conductor (i. e., the total armature current I_a divided by the number of parallel paths $2a$) has to be introduced as I_c.

$$I_c = I_a/(2a) \tag{1-17}$$

The torque (in newton-meter) produced by a single conductor of smooth armature can be calculated as follows. Noted that the armature diameter D_a should be expressed in meter.

$$T_{av} = \frac{D_a}{2} f_{av} = \frac{D_a}{2} B_{av} l I_c \tag{1-18}$$

where T_{av} is the average torque of one conductor (N · m); D_a is the armature diameter (m), and $D_a = 2p\tau/\pi$.

Conductors lying under a pole of a given polarity carry currents flowing in the same direction (Fig. 1-30a) and their torques add. Conductors lying under poles of different polarities also add their torques. Thus the total torque can be obtained by adding up the torques of all N conductors around the armature. The calculation is shown below:

$$T_e = N B_{av} l D_a I_a / (2a \times 2) \tag{1-19}$$

Substitute $B_{av} = \Phi/\tau l$ and $D_a = 2p\tau/\pi$ into Eq. 1-19, we can have

$$T_e = \frac{pN}{2\pi a} \Phi I_a = C_T \Phi I_a \tag{1-20}$$

where C_T is the torque constant with respect to the construction parameters of the machine, $C_T = \frac{pN}{2\pi a}$. Noted that the flux Φ per pole is expressed in weber (Wb); the armature current I_a is expressed in in ampere (A), and the torque T_e is expressed in newton-meter (N · m).

In general, the electromagnetic torque of a machine depends on three factors:

1) The per pole flux Φ in the machine.

2) The armature (or rotor) current I_a.

3) The constant C_T with respect to the construction of the DC machine.

If the saturation of iron is low, the per pole flux Φ can be expressed as $K_f I_f$, where K_f is a constant and I_f is the excitation current. Eq. 1-20 can be rewritten as follows:

$$T_e = C_T \Phi I_a = K_f I_f C_T I_a = G_{af} I_f I_a \tag{1-21}$$

where $G_{af} = \frac{60}{2\pi} C_e K_f = C_T K_f$.

Comparing Eq. 1-14 and Eq. 1-20, you may have discovered that the relationship between the two constants $C_e = \frac{pN}{60a}$ and $C_T = \frac{pN}{2\pi a}$ is expressed as:

$$C_T = \frac{30}{\pi} C_e = 9.55 C_e \tag{1-22}$$

Example 1-3 Calculate the induced voltage of a DC motor having a simplex lap winding with the pole pair $p=3$, the total conductor number $N=398$, the per pole flux $\Phi=2.1\times10^{-2}$Wb when speed $n_1=1500$r/min and $n_2=500$r/min.

Solution:

According to the simplex lap winding, the parallel current paths a equal the pole pair p, i. e. $a=p=3$.

According to Eq. 1-14, when speed $n_1=1500$r/min, the induced voltage is

$$E_a=\frac{pN}{60a}\Phi n_1=\frac{3\times398}{60\times3}\times2.1\times10^{-2}\times1500\text{V}=208.95\text{V}$$

When speed $n_2=500$r/min, the induced voltage is

$$E_a=\frac{pN}{60a}\Phi n_1=\frac{3\times398}{60\times3}\times2.1\times10^{-2}\times500\text{V}=69.65\text{V}$$

Example 1-4 Calculate the induced torque of a DC motor having a simplex lap winding with the pole pair $p=3$, the total conductor number $N=398$, the per pole flux $\Phi=2.1\times10^{-2}$Wb when armature current $I_1=10$A and $I_2=15$A.

Solution:

According to the simplex lap winding, the parallel current paths a equal the pole pair p, i. e. $a=p=3$.

According to Eq. 1-20, when armature current $I_1=10$A, the induced torque is

$$T_e=\frac{pN}{2\pi a}\Phi I_a=\frac{3\times398}{2\pi\times3}\times2.1\times10^{-2}\times10\text{N}\cdot\text{m}=13.3\text{N}\cdot\text{m}$$

When armature current $I_2=15$A, the induced torque is

$$T_e=\frac{pN}{2\pi a}\Phi I_a=\frac{3\times398}{2\pi\times3}\times2.1\times10^{-2}\times15\text{N}\cdot\text{m}=20\text{N}\cdot\text{m}$$

Example 1-5 Calculate the rated current and the induced voltage of a DC generator having a simplex lap winding with the pole pair $p=3$, the total conductor number $N=780$, the per pole flux $\Phi=1.3\times10^{-2}$Wb. Its rated power $P_N=17$kW, rated voltage $U_N=230$V and rated speed $n_N=1500$r/min.

Solution:

According to Eq. 1-3, rated current $I_N=P_N/U_N=17\times1000/230\text{A}=73.91\text{A}$.

the armature winding is simplex lap winding, the parallel current paths a equal the pole pair p, i. e. $a=p=3$.

According to Eq. 1-14, when speed $n_N=1500$r/min, the induced voltage is

$$E_a=\frac{pN}{60a}\Phi n_1=\frac{3\times780}{60\times3}\times1.3\times10^{-2}\times1500\text{V}=253.5\text{V}$$

1.5 Equivalent circuits of DC machines

1.5.1 Equivalent circuits of DC generators

1. Separately excited DC generators

Field current of separately excited DC generators is supplied by a separate external DC voltage source. Its equivalent circuit is shown in Fig. 1-37. In the circuit, voltage U represents the voltage measured at the terminals of the generator and current I represents the current flowing in the lines connected to the terminals. The internal generated voltage is E_a and the armature current is I_a. It is clear that the armature current is equal to the line current in a separately excited generator.

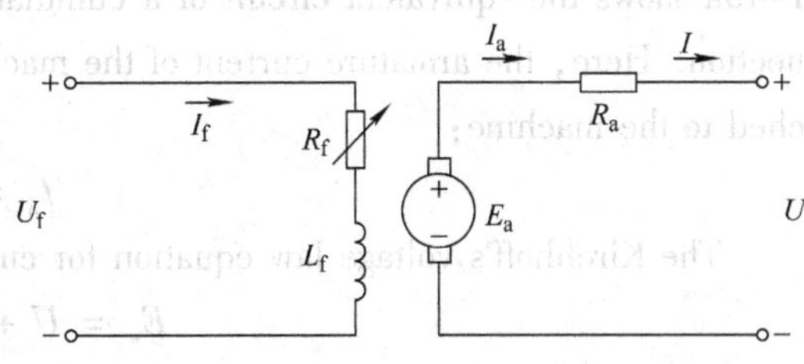

Fig. 1-37 The equivalent circuit of a separately excited DC generator

$$I_a = I \tag{1-23}$$

The Kirchhoff's voltage law equation for armature circuits of separately excited DC machine is

$$\left.\begin{aligned} E_a &= U + I_a R_a \\ U_f &= I_f R_f \end{aligned}\right\} \tag{1-24}$$

2. Shunt DC generators

Shunt DC generators supply their own field current by connecting fields directly across the terminals of machines. The equivalent circuit of a shunt DC generator is shown in Fig. 1-38. In the circuit, the armature current of the machine supplies both the field circuit and the load attached to the machine:

$$I_a = I + I_f \tag{1-25}$$

The Kirchhoff's voltage law equation for armature circuits of shunt DC generators is

$$\left.\begin{aligned} E_a &= U + I_a R_a \\ U &= I_f R_f \end{aligned}\right\} \tag{1-26}$$

3. Series DC generators

The fields of series DC generators are connected with armatures in series. The equivalent circuit of a series DC generator is shown in Fig. 1-39. Here, the armature current, field current and line current all have the same value.

$$I_a = I = I_f \tag{1-27}$$

The Kirchhoff's voltage law equation for series DC generator is

$$E_a = U + I_a(R_a + R_f) \tag{1-28}$$

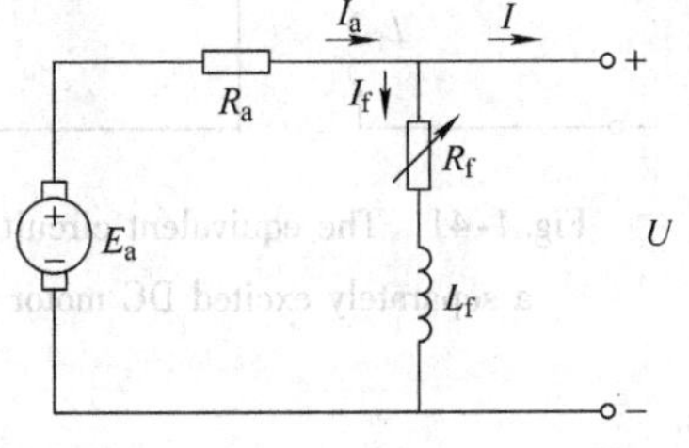

Fig. 1-38 The equivalent circuit of a shunt DC generator

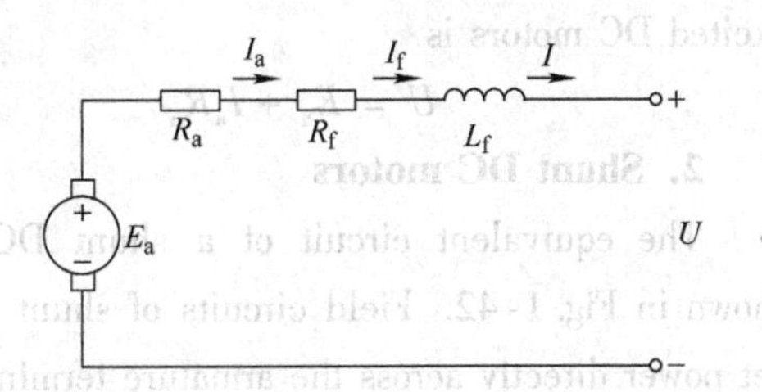

Fig. 1-39 The equivalent circuit of a series DC generator

4. Cumulatively compound DC generators

Cumulatively compound DC generators have both series fields and shunt fields. The two fields are connected so that the magnetomotive forces (mmf) from the two fields can be added up. Fig. 1-40a shows the equivalent circuit of a cumulatively compound DC generator in "long-shunt" connection. Here, the armature current of the machine supplies both the field circuit and the load attached to the machine:

$$I_a = I + I_f \tag{1-29}$$

The Kirchhoff's voltage law equation for cumulatively compound DC generator is

$$\left.\begin{aligned} E_a &= U + I_a(R_a + R_{sf}) \\ I_f &= U/R_f \end{aligned}\right\} \tag{1-30}$$

There is another way to hook up a cumulatively compound DC generator. It is the "short-shunt" connection. Fig. 1-40b shows the equivalent circuit of a cumulatively compound DC generator in "short-shunt" connection.

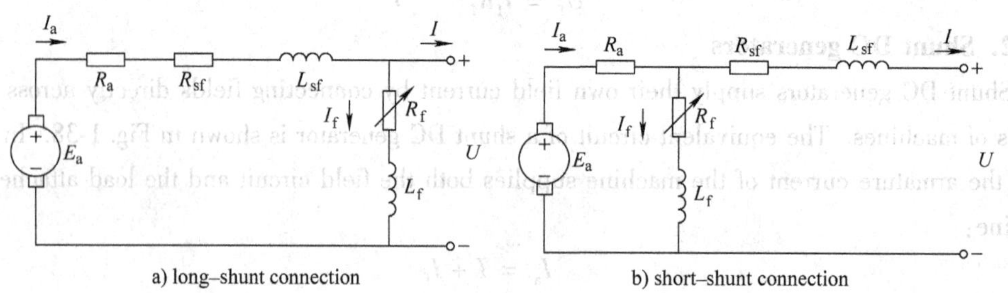

Fig. 1-40 The equivalent circuit of a cumulatively compound DC generator

1.5.2 Equivalent circuits of DC motors

1. Separately excited DC motors

The equivalent circuit of a separately excited DC motor is shown in Fig. 1-41. Field circuits of separately excited DC motors are supplied by separated constant-voltage power supplies.

$$\left\{\begin{aligned} I_a &= I \\ I_f &= U_f/R_f \end{aligned}\right. \tag{1-31}$$

The Kirchhoff's voltage law equation for separately excited DC motors is

$$U = E_a + I_aR_a \tag{1-32}$$

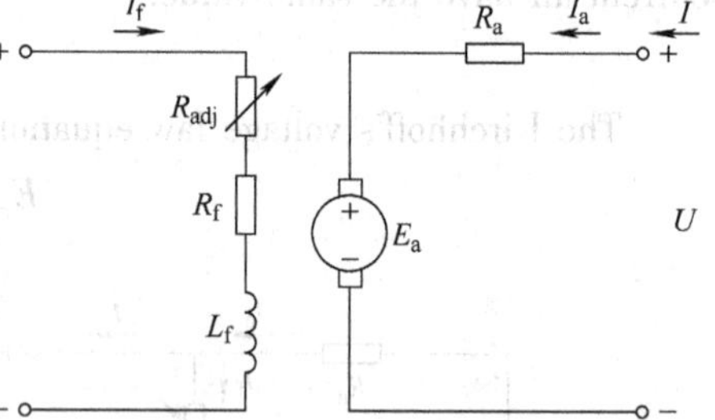

Fig. 1-41 The equivalent circuit of a separately excited DC motor

2. Shunt DC motors

The equivalent circuit of a shunt DC motor is shown in Fig. 1-42. Field circuits of shunt DC motors get power directly across the armature terminals of motors.

$$\left\{\begin{aligned} I_a &= I - I_f \\ I_f &= U/R_f \end{aligned}\right. \tag{1-33}$$

The Kirchhoff's voltage law equation for shunt DC motors is

$$U = E_a + I_aR_a \tag{1-34}$$

3. Series DC motors

The equivalent circuit of a series DC motor is shown in Fig. 1-43. In a series DC motor, the armature current, field current and line current are all the same. The Kirchhoff's voltage law equation for series DC motors is

$$U = E_a + I_a(R_a + R_f) \tag{1-35}$$

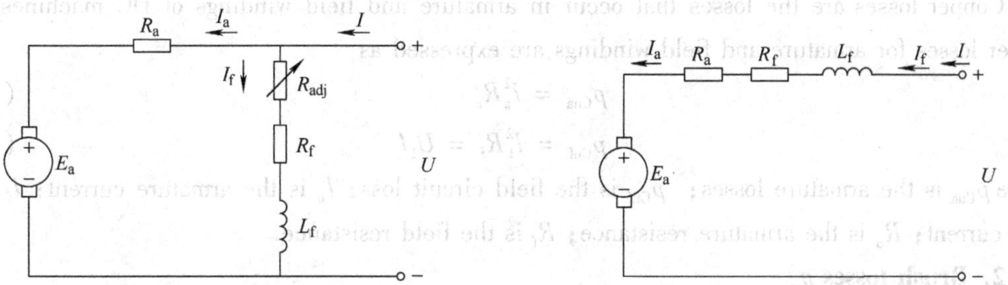

Fig. 1-42 The equivalent circuit of a shunt DC motor Fig. 1-43 The equivalent circuit of a series DC motor

4. Cumulatively compound DC motors

Compound DC motors carry both series fields and shunt fields. In a cumulative compound motor, the mmf of the two fields can be added up. Noted that shunt field is always stronger than series field. Fig. 1-44 shows the equivalent circuit of a compound motor.

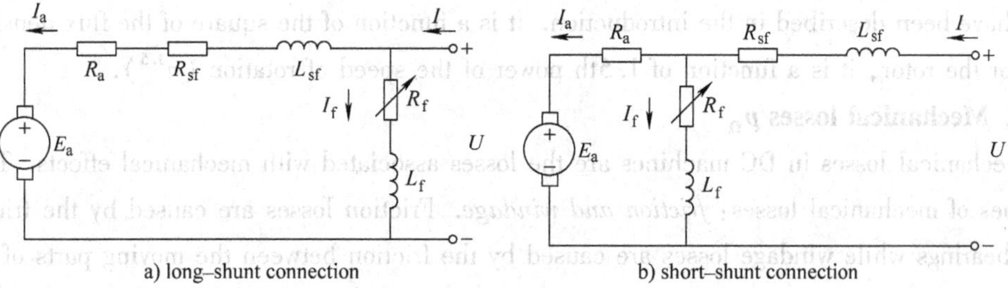

Fig. 1-44 The equivalent circuit of a cumulatively compound DC motor

1.6 Power flow and losses in DC machines

DC generators convert mechanical power to electrical power while DC motors do the opposite. In both cases, only part of the power input can be successfully converted into useful output because there is always some loss associated with the conversion process.

The efficiency of a DC machine is defined by the following equation

$$\eta = \frac{P_{out}}{P_{in}} \times 100\% \tag{1-36}$$

where P_{out} is the output power and P_{in} is the input power of the DC machine. The difference between input and output power is the losses ($\sum p_{loss}$). Therefore,

$$\eta = \frac{P_{out}}{P_{in}} \times 100\% = \frac{P_{in} - \sum p_{loss}}{P_{in}} \times 100\% \tag{1-37}$$

1.6.1 Losses in DC machines

Losses that occur in DC machines can be divided into five basic categories: Electrical or copper losses (I^2R losses), Brush losses, Core losses, Mechanical losses and Stray load losses.

1. Electrical or copper losses (I^2R losses)

Copper losses are the losses that occur in armature and field windings of DC machines. The copper losses for armature and field windings are expressed as

$$p_{Cua} = I_a^2 R_a \tag{1-38}$$

$$p_{Cuf} = I_f^2 R_f = U_f I_f \tag{1-39}$$

where p_{Cua} is the armature losses; p_{Cuf} is the field circuit loss; I_a is the armature current; I_f is the field current; R_a is the armature resistance; R_f is the field resistance.

2. Brush losses p_c

Brush losses are the power losses across the contact potential at the brushes of DC machines, and are expressed as

$$p_c = 2\Delta U_s I_a \tag{1-40}$$

where I_a is the armature current; $2\Delta U_s$ is the brush voltage drop of a pair of brushes.

3. Core losses p_{Fe}

Core losses are the hysteresis losses and eddy current losses occurring in metal of motors. These losses have been described in the introduction. It is a function of the square of the flux density (B^2) and, for the rotor, it is a function of 1.5th power of the speed of rotation ($n^{1.5}$).

4. Mechanical losses p_Ω

Mechanical losses in DC machines are the losses associated with mechanical effects. There are two types of mechanical losses: *friction and windage*. Friction losses are caused by the friction between bearings while windage losses are caused by the friction between the moving parts of DC machines and the air inside the motor's casing. These losses vary as the cube of the speed of rotation of the machine. In DC machines, the sum of core losses p_{Fe} and mechanical losses p_Ω is called no-load losses $p_0 = p_{Fe} + p_\Omega$.

5. Stray load losses p_s

Stray load losses are the losses that cannot be placed in one of the previous categories. No matter how carefully people categorize losses, some can still escape from the above categorization. For those losses that fail to be categorized, they are grouped as stray losses. For most machines, stray losses are conventionally taken to be 1 percent of full load.

1.6.2 The power-flow of DC generators

One of the most convenient techniques to illustrate power losses is the power-flow diagram. A power-flow diagram of a DC generator is shown in Fig. 1-45. In the figure, mechanical power P_1 is applied to a DC generator, and stray losses p_s, mechanical losses p_Ω and core losses p_{Fe} are subtracted. After they have been subtracted, the remaining power P_e is ideally converted from mechanical to electrical form. The mechanical power P_e that is converted is given by.

$$P_e = T_e\Omega \tag{1-41}$$

where Ω is the mechanical angular velocity in radians per second (r/s). $\Omega = \frac{2\pi n}{60}$.

The resulting electric power produced is given by

$$P_e = E_a I_a \tag{1-42}$$

P_e is called the converted power or the electromagnetic power. However, this is not the power that appears at the machine's terminals. Before the input reaches the terminals, copper losses p_{Cua}, p_{Cuf} and brush losses p_c must be subtracted.

$$P_1 = T_1\Omega = p_\Omega + p_{Fe} + p_s + P_e \tag{1-43}$$

$$P_e = p_{Cua} + p_{Cuf} + p_c + P_2 \tag{1-44}$$

$$P_2 = UI \tag{1-45}$$

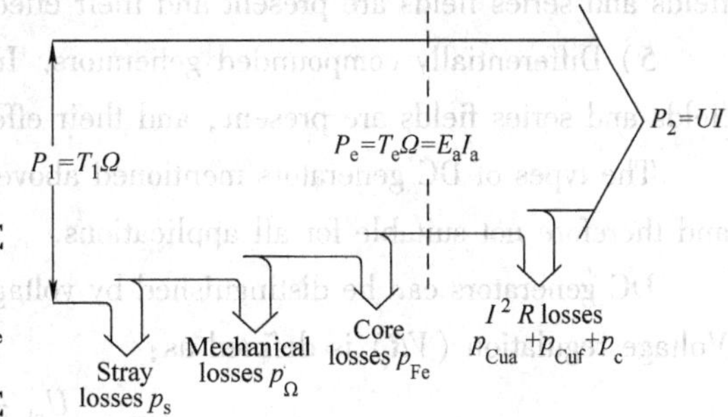

Fig. 1-45 Power-flow diagram of a DC generator

In Fig. 1-45, T_1 is the input torque of the DC generator supplied by a prime mover, $T_1 = \frac{P_1}{\Omega}$, T_e is the electromagnetic torque produced in the DC generator, $T_e = \frac{P_e}{\Omega}$, and T_0 is the no-load torque, $T_0 = \frac{p_0}{\Omega}$.

$$T_1 = T_e + T_0 \tag{1-46}$$

1.6.3 The power-flow of DC motors

In the case of DC motors, the power-flow diagram is simply reversed with that of DC generators. The power-flow diagram for a motor is shown in Fig. 1-46.

In Fig. 1-46, T_2 is the output torque of the DC motor, $T_2 = \frac{P_2}{\Omega}$, T_e is the electromagnetic torque produced in the DC motor, $T_e = \frac{P_e}{\Omega}$, and T_0 is the no-load torque, $T_0 = \frac{p_0}{\Omega}$.

The torque equation of a DC generator is given by

$$T_e = T_2 + T_0 \tag{1-47}$$

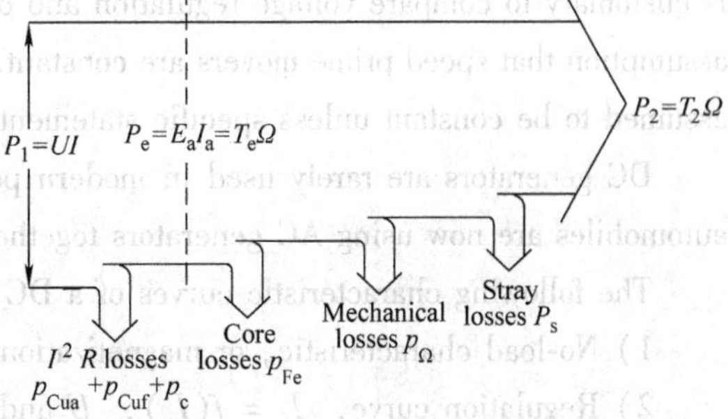

Fig. 1-46 Power-flow diagram of a DC motor

1.7 Operational characteristics of DC generators

DC generators are DC machines being used as generators. As mentioned before, there is no difference between generators and motors in terms of their internal construction. The only difference between them is the direction of power flow. There are five major types of DC generators which are classified according to the ways of production of field flux.

1) Separately excited generators: In separately excited generators, field flux is derived from separate power sources independent of the generator itself.

2) Shunt generators: In shunt generators, field flux is created by connecting field circuits directly across the terminals of generators.

3) Series generators: In series generators, field flux is produced by connecting field circuits and armatures of generators in series.

4) Cumulatively compounded generators: In cumulatively compounded generators, both shunt fields and series fields are present and their effects can be added up.

5) Differentially compounded generators. In differentially compounded generators, both shunt fields and series fields are present, and their effects are subtractive.

The types of DC generators mentioned above differ in terminal (voltage-current) characteristics and therefore not suitable for all applications.

DC generators can be distinguished by voltage, power rating, efficiency and voltage regulation. Voltage regulation (VR) is defined as:

$$VR = \frac{U_{nl} - U_{fl}}{U_{fl}} \times 100\% \tag{1-48}$$

where U_{nl} is the no-load terminal voltage of generators; U_{fl} is the full-load terminal voltage of generators.

VR helps to estimate the shape of generator's voltage-current characteristic. A positive voltage regulation means a drooping characteristic, and a negative voltage regulation means a rising characteristic.

All generators are driven by mechanical power, which is usually called the prime mover of generators. Prime movers for DC generators can be steam turbines, diesel engines or even electric motors. Since prime movers vary in speed and the speed will affect the output voltage of generators, it is customary to compare voltage regulation and output characteristic of different generators under the assumption that speed prime movers are constant. Throughout this chapter, the speed of generators is assumed to be constant unless specific statements are made.

DC generators are rarely used in modern power systems. Even the DC power systems used in automobiles are now using AC generators together with rectifiers to produce DC power.

The following characteristic curves of a DC generator will be discussed in the next section.

1) No-load characteristic, or magnetization curve, $U_0 = f(I_f)$; n = constant.

2) Regulation curve, $I_f = f(I_a)$, U and n = constant.

3) Terminal curve, or external curve, $U = f(I_a)$, n = constant.

1.7.1 Operational characteristics of separately excited DC generators

Separately excited DC generators are generators whose field current is supplied by separate external DC voltage sources. The equivalent circuit of separately excited DC generators is shown in Fig. 1-37. In the circuit, voltage U represents the voltage measured at the terminals of the generator and current I represents the current flowing in the lines connected to the terminals. The internal generated voltage is E_a, and the armature current is I_a. It is clear that the armature current is equal to the line current I in a separately excited generator.

1. Magnetization curve of separately excited DC generators

(1) Field flux vs exciting current

When a separately excited DC generator runs at no-load (open armature circuit), a change in the exciting current will cause a corresponding change in the induced voltage.

Let us increase the exciting current I_f gradually. It is observed that mmf of the field increases with the increasing current. It leads to an increase in the flux Φ per pole. If we plot Φ as a function of I_f, we can draw the saturation curve shown in Fig. 1-47. This curve can be obtained whether or not the generator is turning.

When the exciting current is very small, the flux is small and the iron in the machine is unsaturated. A very small mmf is needed to establish the flux through the iron, with the result that the mmf developed by the field coils is almost entirely available to drive the flux through the air gap. Because the permeability of air is constant, the flux increase is directly proportion to the exciting current, as shown by the linear portion **0*a*** in the saturation curve (Fig. 1-47).

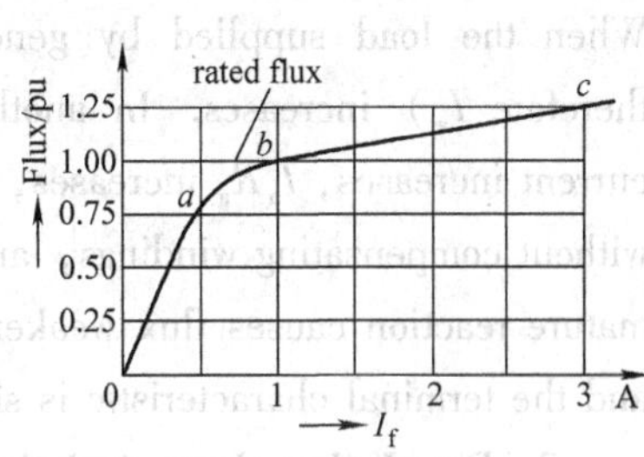

Fig. 1-47 Flux per pole versus exciting current I_f

However, as the exciting current increases gradually, the iron in the field and the armature begin to saturate. Thus, a large increase in mmf is needed to produce a small flux, as shown by portion ***bc*** of the curve (Fig. 1-47). The machine is now said to be saturated. Saturation of the iron begins to be important when we reach the so-called "knee" ***ab*** of the saturation curve.

(2) Induced voltage vs exciting current

Internal generated voltage E_a of DC machines can be calculated by Equation $E_a = C_e \Phi n$. From the equation, E_a is directly proportional to flux in machines and speed of rotation of machines.

How is internal generated voltage E_a related to field current in the machine? If we drive generators at constant speed, $U_0 = E_0$ is directly proportional to flux Φ. Consequently, we can draw a curve whose shape is identical to the saturation curve (Fig. 1-47) by plotting U_0 as a function of I_f. The resulting curve is shown in Fig. 1-48. It is called magnetization curve (or no-load saturation curve) of generators.

Rated voltage U_N of DC generators is usually a little above the knee of the curve, and induced voltage varies with exciting current. By reversing the current, the flux reverse and so gives the polarity of the induced voltage as shown in Fig. 1-48.

(3) Induced voltage vs speed

Induced voltage is directly proportional to speed under a given exciting current. This relationship is shown in Eq. 1-14.

If we reverse the direction of rotation, polarity of induced voltage also reverses. However, if we reverse both theflowing direction of exciting current and the direction of rotation, polarity of induced voltage remains the same.

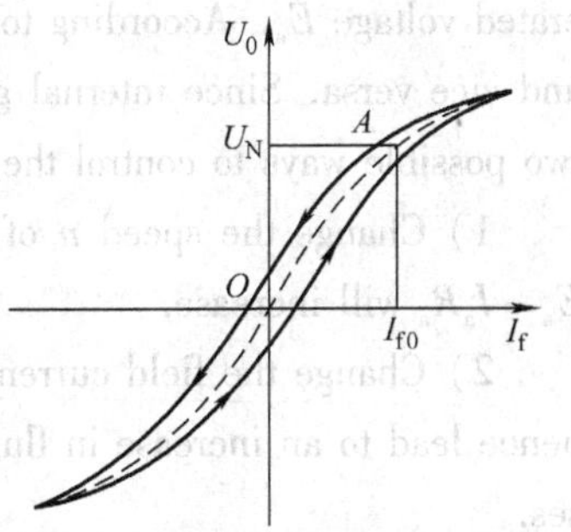

Fig. 1-48 Saturation curve of a DC generator

2. Terminal characteristic of separately excited DC generators

Terminal characteristic of a machine is a plot showing the relationship of different output quantities. For generators, output quantities are terminal voltage U and line current I. The terminal characteristic of a separately excited generator $U = f(I_a)$ is thus a plot of U versus I_a under a constant speed n. By Kirchhoff's voltage law, the terminal voltage is

$$U = E_a - I_a R_a \tag{1-49}$$

Since induced voltage E_a is independent of I_a, terminal characteristic of separately excited generators is a straight line. When the load supplied by generators increases, I (and therefore I_a) increases. In another words, when armature current increases, $I_a R_a$ increases, thus will lead to a decrease in terminal voltage. In generators without compensating windings, an increase in I_a causes an increase in armature reaction, and armature reaction causes flux weakening. Therefore, this flux weakening causes a decrease in E_a, and the terminal characteristic is shown in Fig. 1-49.

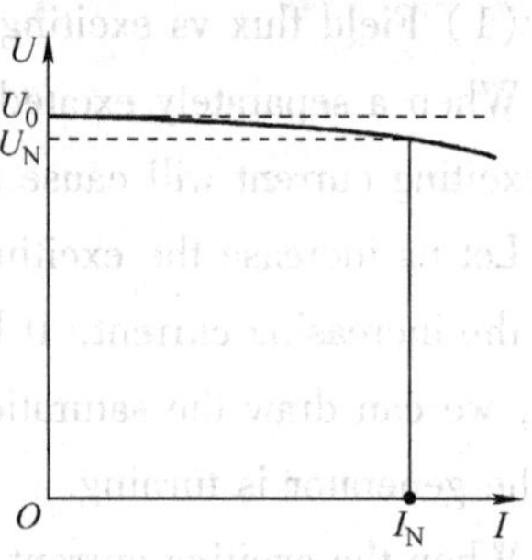

Fig. 1-49 The terminal characteristic of a separately excited generator

3. Regulation characteristic of separately excited DC generators

Quantities for generators include terminal voltage, line current and field current. Regulation characteristic of separately excited generators $I_f = f(I_a)$ is thus a plot of I_f versus I_a under a constant speed n and a constant terminal voltage U.

When the load supplied by generators increases, I (and therefore I_a) increases. In another word, when armature current increases, $I_a R_a$ increases. In order to keep a constant terminal voltage U, the field current must increase to cause an increase in flux. As shown in Fig. 1-50, the regulation characteristic of a separately excited generator increases with an increasing line current.

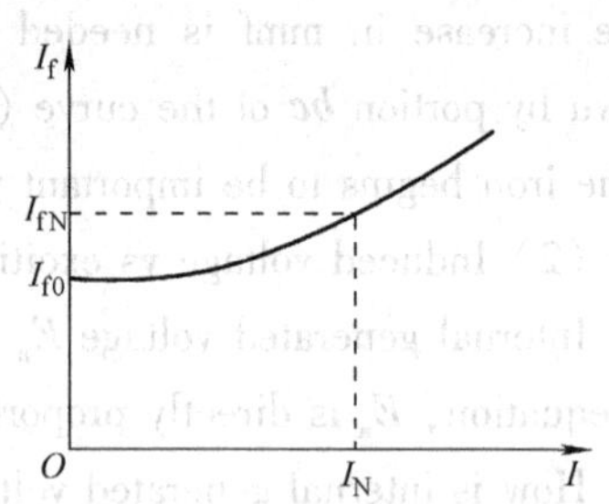

Fig. 1-50 The regulation characteristic of a separately excited generator

4. Voltage control for separately excited DC generators

Terminal voltage of separately excited DC generators can be controlled by adjusting internal generated voltage E_a. According to kirchhoff's voltage law $E_a = U + I_a R_a$, if E_a increases, U increases, and vice versa. Since internal generated voltage E_a is given by the equation $E_a = C_e \Phi n$, there are two possible ways to control the voltage of separately excited DC generators:

1) Change the speed n of rotation. When n increases, $E_a = C_e \Phi n$ will increase, so as $U = E_a - I_a R_a$ will increase.

2) Change the field current I_f. When R_f decreases, field current I_f increases ($I_f = U_f / R_f$) and hence lead to an increase in flux Φ. It leads to an increase in $E_a = C_e \Phi n$, so $U = E_a - I_a R_a$ increases.

In many applications, the speed range of prime movers is quite limited, so the terminal voltage is most commonly controlled by changing field current.

1.7.2 Operational characteristics of shunt DC generators

Shunt DC generators supply their own field current by connecting field directly across the terminals. The equivalent circuit of a shunt DC generator is shown in Fig. 1-38. In the circuit, the armature current of the machine is the power supply of both the field circuit and the load attached to the machine.

This type of generator has a distinct advantage over the separately excited DC generator since no external power supply is needed to create field circuit. However, it raises another question: if the generator supplies its own field current, how does it get the initial field flux to initiate the circuit at the moment when it is on?

1. The voltage buildup in shunt DC generators

Assuming that the generator in Fig. 1-38 has no load connected to it and the prime mover starts to turn the shaft of the generator, how is initial voltage produced at the terminals?

The voltage buildup in DC generators depends on the presence of residual flux in the poles of generators. When a generator starts to turn, an internal voltage will be generated which is given by $E_a = C_e \Phi_{res} n$.

The internal voltage first appears at the terminals of the generator (it may only be a volt or two). It causes a current to flow into the field coil ($I_f = U_f/R_f$). The field current produces a magnetomotive force (mmf) in the poles and increases the flux in them. It leads to an increase in E_a and terminal voltage U. When U increases, I_f increases. This leads to the increase of flux Φ. Eventually, E_a and other quantities will also increase.

The voltage buildup behavior is shown in Fig. 1-51. Notice that it is the effect of magnetic saturation which eventually limits the terminal voltage of generators. In Fig 1-51, curve 1 is the magnetization curve of generators $U_0 = f(I_f)$, curve 2 is the volt-ampere characteristic of field circuit (or field resistance line) $U_f = I_f R_f$.

What if a shunt generator is started but no voltage is built up? What might have gone wrong? There are several possibilities that can cause the failure of the initial voltage buildup. They are:

1) *There is no residual magnetic flux* in the generator to start the process. If the residual flux $\Phi_{res} = 0$, then $E_a = 0$, and no voltage can ever build up. If this problem occurs, you should disconnect the field from the armature circuit and connect it directly to an external DC source such as a battery. The current produced by the external DC source flows into the circuit and leave a residual flux in the poles, thus the generator can produce initial voltage using the residual flux. This procedure is known as "flashing the field".

2) Either *the direction of rotation of the generator* or *the connections of the field has been reversed*. In both cases, the residual flux produces an internal generated voltage E_a. The voltage E_a produces a field current, which will then produce a flux opposing to the residual flux. Eventually, the flux decreases below Φ_{res} and no voltage can ever build up.

If this problem occurs, you can fix it by reversing the direction of rotation, the field connections, or by flashing the field with the opposite magnetic polarity.

3) *The field resistance is greater than the critical resistance*. To understand this problem, you

can refer to Fig. 1-52. Normally, shunt generators will build up to the point where the magnetization curve intersects the field resistance line. If the strength of field resistance increase, and the field resistance line is the same as curve 3 in Fig. 1-51, the field resistance line is nearly parallel to the magnetization curve. In that case, the voltage of generators can fluctuate widely without bringing significant change to the value of R_f or I_a. This value of the resistance is called the *critical resistance*. If R_f is larger than the critical resistance (curve 4), the steady-state operating voltage remains at the residual level and never be able to build up. The solution to this problem is the reduction of R_f.

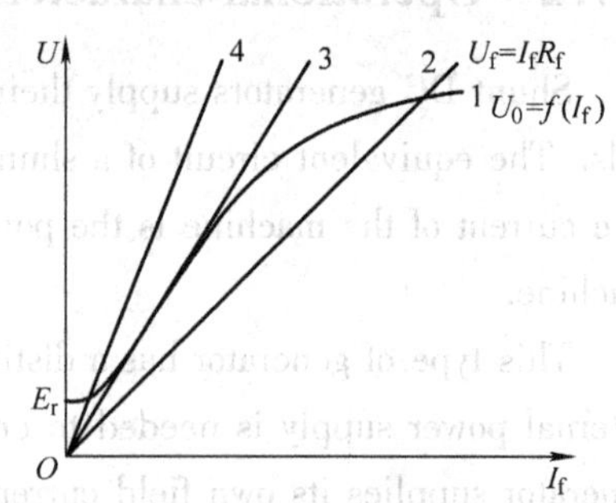

Fig. 1-51 The Voltage built up when starting a shunt DC generator

Since the voltage of the magnetization curve is a function of shaft speed, the critical resistance also depends upon the speed. In general, the lower the shaft speed, the lower the critical resistance.

2. Terminal characteristics of shunt DC generators

Terminal characteristics of shunt DC generators differ from that of separately excited DC generators since the amount of field current depends on the terminal voltage. To understand the terminal characteristics of shunt DC generators, we can start by observing current and terminal voltage of generators before and after loaded. As the load on the generator increases, I increases and so as $I_a = I_f + I$ increases. An increase in I will increase the armature resistance voltage drop (I_aR_a), causing $U = E_a - I_aR_a$ to decrease. This is exactly the same behavior observed in separately excited generators. However, for shunt DC generators, when U decreases, field current I_f decreases too. This causes flux Φ to decrease. The weakening flux leads to a decrease in E_a, which will then decrease the terminal voltage of generators $U = E_a - I_aR_a$. The resulting terminal characteristic is shown in Fig. 1-52. Notice that the slope of voltage drop-off is steeper than just that of the I_aR_a drop in a separately excited generator. In other words, the voltage regulation of this generator is worse than the voltage regulation of the same piece of equipment connected separately excited.

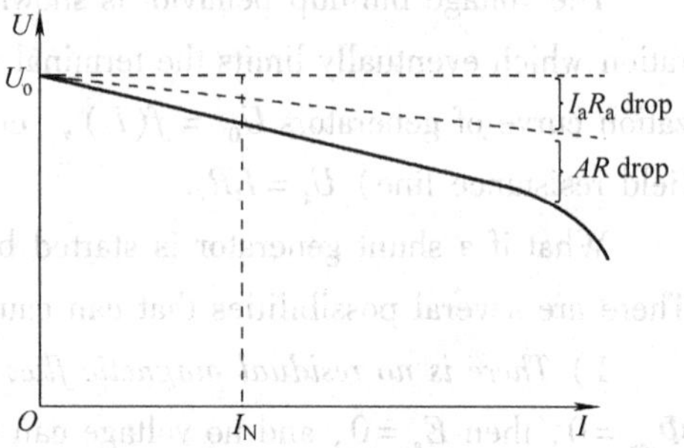

Fig. 1-52 The terminal characteristic of a shunt DC generator

3. Regulation characteristics of shunt DC generators

Regulation curve of shunt DC generators is the same as that of separately excited DC generators because the voltage drop produced by field current is very small. The voltage drop is so small that it can be neglected when compared with the voltage drop caused by the load current.

4. Voltage control of shunt DC generators

Just like separately excited generators, there are two ways to control the voltage of shunt DC generators.

1) Change shaft speed n of generators.

2) Change field resistor of generators, thus change the field current.

Changing field resistor is the most popular method used to control terminal voltage in real shunt generators. Field current $I_f = U/R_f$ increases with decreasing field resistor R_f. When I_f increases, flux Φ increases and it causes internal generated voltage E_a to increase. Eventually, the terminal voltage of generators will increase.

Example 1-6 A shunt DC generator with rated power $P_N = 9\text{kW}$, rated voltage $U_N = 115\text{V}$, rated speed $n_N = 1450\text{r/min}$, armature resistance $R_a = 0.15\Omega$, is turning at rated operation state. The total resistance of field circuit R_{fN} is 33Ω, the core loss is 410W, the mechanical loss is 101W, the stray loss makes up 0.5 percent of the rated power. Calculate the following:

1) The induced torque of the generator.

2) The efficiency of the generator turning at rated operation state.

Solution:

1) the rated current of the machine is

$I_N = P_N/U_N = 9000/115\text{A} = 78.26\text{A}$

The rated field current is

$I_f = U_N/R_{fN} = 115/33\text{A} = 3.48\text{A}$

The armature current is

$I_a = I_N + I_f = 78.26 + 3.48\text{A} = 81.74\text{A}$

The electromagnetic power is

$P_e = P_N + p_{Cuf} + p_{Cua} = 9000 + 115 \times 3.48 + 81.74^2 \times 0.15\text{W} = 10\,402.4\text{W}$

The induced torque is

$T_e = P_e/\Omega = 10\,402.4 \times 60/\ (2 \times 3.14 \times 1450)\ \text{N} \cdot \text{m} = 68.54\text{N} \cdot \text{m}$

2) The efficiency of the machine is

$\eta_N = P_N/(P_e + p_{Fe} + p_\Omega + p_{ad}) = 9000/(10\,402.4 + 410 + 101 + 0.005 \times 9000) = 82.13\%$

1.7.3 Operational characteristics of compound DC generators

Cumulatively compound DC generators contain both shunt and series fields. A cumulatively compound DC generator is connected in a way that the two magnetomotive forces from two separate fields are additive. Fig. 1-40a shows an equivalent circuit of a cumulatively compound DC generator in "long-shunt" connection. Fig. 1-40b shows an equivalent circuit of a cumulatively compound DC generator in "short-shunt" connection.

1. Terminal characteristics of cumulatively compound DC generators

To understand the terminal characteristics of a cumulatively compound DC generator, it is necessary to understand the competing effects that occur within the machine.

Load current I increases with increasing load. Since $I_a = I_f + I$ increases, armature current I_a increases when load current I increases. At this point, two processes have already occurred in the generator:

1) When I_a increases, $I_a(R_a + R_{sf})$ voltage drop also increases. It causes a decrease in terminal voltage $U = E_a - I_a(R_a + R_{sf})$.

2) When I_a increases, the series field magnetomotive force (mmf) also increases. This increases the total magnetomotive force, thus increases the flux in the generator. The increasing flux in the generator casues E_a to increase, Eventually, $U = E_a - I_a(R_a + R_{sf})$ will increase.

These two effects go against each other. One tends to increase U while the other one tends to decrease U. Which is the predominant effect? It all depends on the number of series turns placed on the poles of the machine. This can be illustrated by investigating several cases:

(1) *Fewer series turns*

If there are only a few series turns, the resistive voltage drop effect is dominant. The voltage falls off just as in a shunt generator, but not quite as steeply (Fig. 1-53). This type of construction, where the full-load terminal voltage is less than the no-load terminal voltage, is called *undercompounded*.

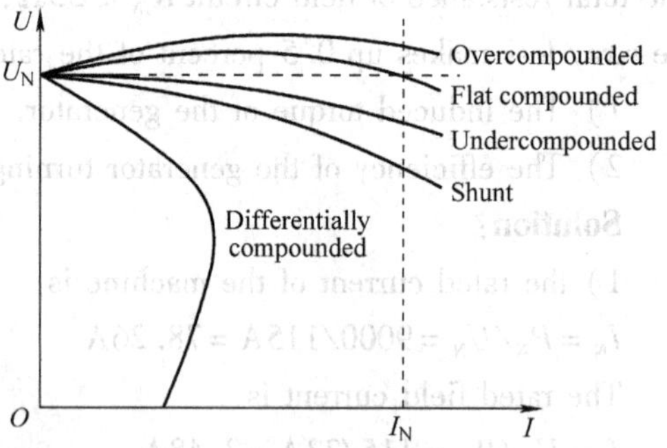

Fig. 1-53 The terminal characteristics of a cumulatively compounded DC generator

(2) *More series turns*

If there are a few more series turns of wire on the poles, at first the flux-strengthening effect will outweigh the resistive drop and the terminal voltage will increase with the increasing load. However, as the load continues to increase, magnetic saturation sets in and the resistive drop becomes stronger than the flux strengthening effect. In this case, the terminal voltage first increases and then falls as the load continue to increase. If U at no load is equal to U at full load, the generator is called *flat-compounded*.

(3) *A lot of series turns are added*

If a lot of series turns are added to the generator, the flux-strengthening effect predominates for a longer time before the resistive drop takes over. The result is that the full-load terminal voltage is actually higher than the no-load terminal voltage. If U at a full load exceeds U at no load, the generator is called *over-compounded*.

The above possibilities are illustrated in Fig. 1-53.

2. Voltage control for cumulatively compound DC generators

Similar to shunt DC generators, there are two ways to control the voltage of cumulatively compound DC generators.

(1) Change the speed n of the generator

An increase in n causes $E_a = C_e n\Phi$ to increase. Eventually the terminal voltage $U = E_a - I_a(R_a + R_{sf})$ will increase.

(2) Change the field current

If the field resistor R_f decreases, the field current $I_f = U/R_f$ increases. It leads to an increase in the machine's flux Φ, causing the internal generated voltage E_a to increase. The increase in E_a will eventually cause the terminal voltage of the generator to increase.

1.8 Operational characteristics of DC motors

DC motors are often compared by their speed regulations. The *speed regulation* (*SR*) of a motor is defined by

$$SR = \frac{n_{nl} - n_{fl}}{n_{fl}} \times 100\% \tag{1-50}$$

where n_{nl} is the no-load speed of the motor; n_{fl} is the full-load speed of the motor.

It is a rough measure of the shape of a motor's torque-speed characteristic, a positive speed regulation means that a motor's speed drops with an increasing load, whereas a negative speed regulation means a motor's speed increases with an increasing load. The magnitude of the speed regulation indicates how steep the slope of the torque-speed curve is approximately.

DC motors are driven by a DC power supply. Unless otherwise specified, the input voltage to a DC motor is assumed to be constant. Because of this assumption, it simplifies the analysis of motors and the comparison between different types of motors.

There are four major types of DC motors in general:

1) The separately excited DC motor.

2) The shunt DC motor.

3) The series DC motor.

4) The compounded DC motor.

1.8.1 The separately excited DC motor and shunt DC motor

The equivalent circuit of a separately excited DC motor is shown in Fig. 1-41. A separately excited DC motor is a motor which field circuit is supplied from a separate constant-voltage power supply. The equivalent circuit of a shunt DC motor is shown in Fig. 1-42. A shunt DC motor is a motor which field circuit receives its power directly from the armature terminals of the motor. When the supply of the voltage towards a motor is assumed to be constant, there is no difference in terms of the characteristic of these two machines.

In a DC motor, there are several quantities, such as terminal voltage U, field current I_f, armature current I_a, speed n, produced torque T_e and efficiency η. The following characteristic curves of a DC motor will be considered.

1) Speed characteristic curve $n = f(I_a)$ Speed n is defined as a function of armature current I_a; I_f = constant, U = constant and there is no external resistance in the armature circuit.

2) Torque characteristic curve $T_e = f(I_a)$ Torque T_e is defined as a function of armature current I_a; I_f = constant, U = constant and there is no external resistance in the armature circuit.

3) Efficiency characteristic curve $\eta = f(I_a)$ Efficiency η is defined as a function of armature current I_a; I_f = constant, U = constant and there is no external resistance in the armature circuit.

4) Terminal characteristic curve $n = f(T_e)$ Speed n is defined as a function of torque T_e; I_f = constant and U = constant.

1. The speed characteristic curve $n = f(I_a)$

The speed characteristic curve $n = f(I_a)$ of a shunt DC motor is a plot of the speed n as a function of armature current I_a; where the field current I_f and the terminal voltage U are constant and there is no external resistance in the armature circuit.

The voltage equation for a shunt motor is $U = E_a + I_aR_a$, the induced voltage $E_a = C_e n\Phi$, thus

$$U = C_e n\Phi + I_aR_a \tag{1-51}$$

Finally, solving for the motor's speed yields

$$n = \frac{E_a}{C_e\Phi} = \frac{U}{C_e\Phi} - \frac{R_a}{C_e\Phi}I_a \tag{1-52}$$

If the armature reaction is neglected, the equation stated above can be expressed as a straight line equation with a negative slope.

$$n = n_0 - \beta I_a \tag{1-53}$$

where $n_0 = \dfrac{U}{C_e\Phi}$ is the ideal no-load speed, i. e. the speed is at $I_a = 0$; therefore $\beta = \dfrac{R_a}{C_e\Phi}$ is the slope of the straight line.

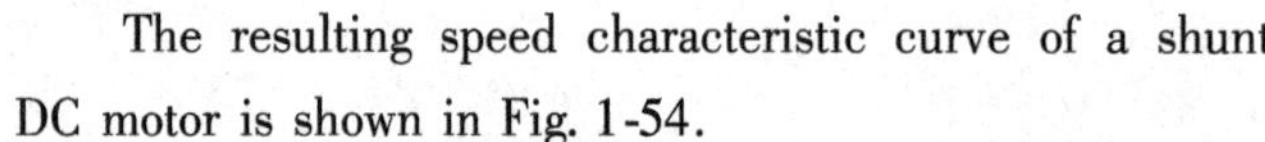

The resulting speed characteristic curve of a shunt DC motor is shown in Fig. 1-54.

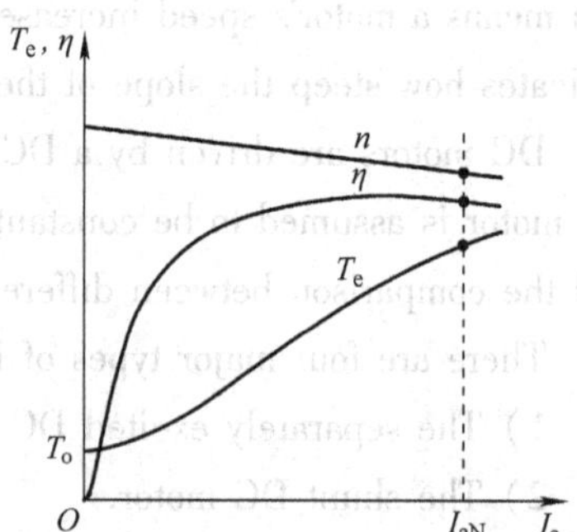

Fig. 1-54 The operation characteristic curves of a shunt DC motor

2. The electromagnetic torque characteristic curve $T_e = f(I_a)$

The torque characteristic curve $T_e = f(I_a)$ of a shunt DC motor is a plot of the torque T_e, which refers as a function of armature current I_a; the field current I_f and the terminal voltage U are constant and there is no external resistance in the armature circuit.

According to Eq. 1-20 the torque T_e of the motor is proportional to its armature current I_a; $T_e = C_T\Phi I_a$. For a constant field current the torque does not increase linearly with the armature current; however, if the latter increases, the armature reaction will increase and the resultant flux Φ decrease. When the motor is running at no-load condition, its armature current $I_a = I_{a0}$, and its torque $T_e = T_0$. Fig. 1-54 shows the developed torque curve with a constant field current. The developed torque curve deviates from a straight line because of the non-linearity of the magnetization curve.

3. The efficiency characteristic curve $\eta = f(I_a)$

The efficiency characteristic curve $\eta = f(I_a)$ of a shunt DC motor is a plot of the efficiency η as a function of armature current I_a; the field current I_f and the terminal voltage U are constant where there is no external resistance in the armature circuit.

According to Eq. 1-36, the efficiency η of the motor is proportional to its output power P_2, and inversely proportional to its input power P_1.

$$\eta = \frac{P_2}{P_1} \times 100\% = \frac{P_1 - \sum p_{\text{loss}}}{P_1} \times 100\% \tag{1-54}$$

The resulting efficiency characteristic curve of a shunt DC motor is shown in Fig. 1-54.

When the shaft load is light, the efficiency of the motor is low; when the load increases, the efficiency of the motor will increase with an increasing power output. Finally, when the load is heavy, the efficiency will decrease with an increasing load, because the copper losses increase quickly with an increasing armature current.

4. The terminal characteristic curve $n = f(T_e)$

The terminal characteristic of a shunt DC motor is a plot of its torque T_e versus speed n; where the field current I_f and the terminal voltage U are constant.

Assuming the load on the shaft of a shunt motor is increased, thereby the load torque T_2 will exceed the induced torque T_e in the machine, and the motor will slow down. When the motor slows down, its internal generated voltage drops ($E_a = C_e n\Phi$), so the armature current in the motor $I_a = (U - E_a)/R_a$ increases. As the armature current rises, the induced torque in the motor increases ($T_e = C_T \Phi I_a$). Finally the induced torque will equal to the load torque at a low mechanical speed of rotation n.

The output characteristic of a shunt DC motor can be derived from the induced voltage and torque equations of the motor plus Kirchhoff's voltage law.

The voltage equation for a shunt motor is $U = E_a + I_a R_a$, and the induced voltage $E_a = C_e n\Phi$, so

$$U = C_e n\Phi + I_a R_a \tag{1-55}$$

Since $T_e = C_T \Phi I_a$, armature current I_a can be expressed as

$$I_a = \frac{T_e}{C_T \Phi} \tag{1-56}$$

Combining Eq. 1-55 and Eq. 1-56 produces

$$U = C_e n\Phi + \frac{T_e}{C_T \Phi} R_a \tag{1-57}$$

Finally, solving for the motor's speed yields

$$n = \frac{U}{C_e \Phi} - \frac{R_a}{C_e C_T \Phi^2} T_e \tag{1-58}$$

If the armature reaction is neglected, the equation stated above can be expressed as a straight line equation with a negative slope.

$$n = n_0 - \beta' T_e \tag{1-59}$$

where $n_0 = \dfrac{U}{C_e \Phi}$ is the ideal no-load speed, i. e. the speed at $T_e = 0$; $\beta' = \dfrac{R_a}{C_e C_T \Phi^2}$ is the slope of the straight line.

The real no-load speed can be calculatedas below

$$n_0' = n_0 - \frac{R_a}{C_e C_T \Phi^2} T_0 \tag{1-60}$$

where T_0 is the no-load torque, $T_0 = \dfrac{p_0}{\Omega}$.

The resulting torque-speed characteristic of a shunt DC motor is shown in Fig. 1-55a.

According to Eq. 1-58, when the speed drop Δn can be expressed as

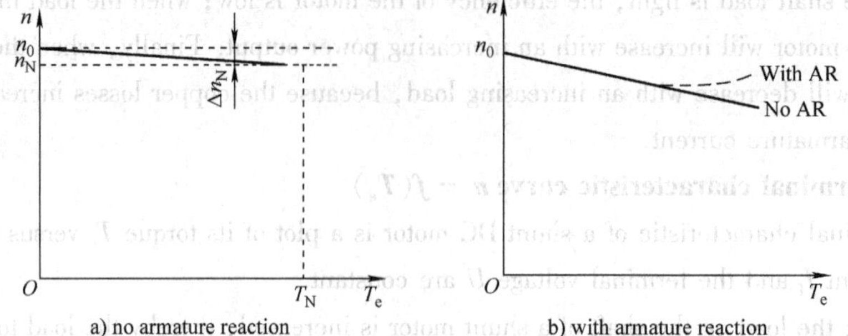

a) no armature reaction b) with armature reaction

Fig. 1-55 The terminal characteristic of a shunt DC motor

$$\Delta n = \frac{R_a}{C_e C_T \Phi^2} T_e = \beta' T_e \tag{1-61}$$

According to Eq. 1-59, the rated speed regulation $\Delta n_N\%$ is defined as

$$\Delta n_N = \frac{n_0 - n_N}{n_N} \times 100\% \tag{1-62}$$

where n_0 is the ideal no-load speed; n_N is the rated speed of the motor.

It is important to realize that, in order to keep the relationship between the speed of the motor and torque linear, the other terms in this expression must keep constant even when the load changes. The terminal voltage is supplied by the DC power source, which is assumed to be constant. If it is not constant, the voltage variations will affect the shape of the torque-speed curve.

Other effect from the motor can also affect the shape of the speed curve, this effect is called armature reaction. If a motor has an armature reaction, when its load increases, the flux-weakening effects will reduce its flux. As Eq. 1-58 shows, the effect of a reduction in flux is to increase the motor's speed at any given load over the speed. The motor would run freely without armature reaction. The torque-speed characteristic of a shunt DC motor with armature reaction is shown in Fig. 1-55b. If a motor has compensating windings, obviously there will be no flux-weakening problems in the machine, and the flux in machine will be constant.

5. The man-made terminal characteristic curve of a shunt DC motor

In Eq. 1-58, if either the field current I_f or the terminal voltage U is not constant, or there is an external resistance R_S in the armature circuit, then there will have three types of man-made terminal characteristics.

(1) External resistance R_S is added in the armature circuit

According to Eq. 1-58, the terminal characteristic for $U = U_N$, $\Phi = \Phi_N$, $R = R_a + R_S$ can be expressed as

$$n = \frac{U_N}{C_e \Phi_N} - \frac{R_a + R_S}{C_e C_T \Phi_N^2} T_e \tag{1-63}$$

The resulting man-made torque-speed characteristic of a shunt DC motor is shown in Fig. 1-56.

(2) The terminal voltage U is not keep at the rated value

According to Eq. 1-58, the terminal characteristic for $R = R_a$, $\Phi = \Phi_N$ can be expressed as

$$n = \frac{U}{C_e \Phi_N} - \frac{R_a}{C_e C_T \Phi_N^2} T_e \tag{1-64}$$

The resulting man-made torque-speed characteristic of a shunt DC motor in this case is shown in Fig. 1-57.

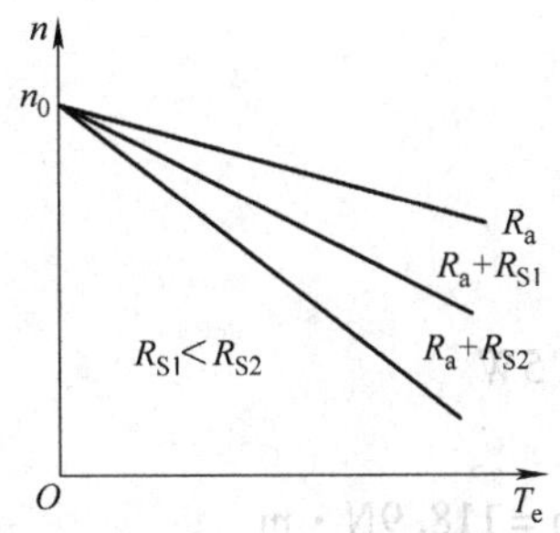

Fig. 1-56 The terminal characteristic curve for $R = R_a + R_S$

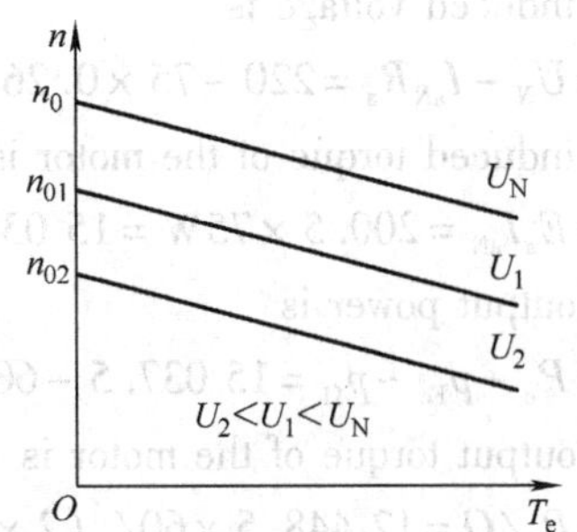

Fig. 1-57 The terminal characteristic curve for $U \neq U_N$

(3) The field current I_f is not keep at the rated value

According to Eq. 1-58, the terminal characteristic for $U = U_N$, $R = R_a$ can be expressed as

$$n = \frac{U_N}{C_e\Phi} - \frac{R_a}{C_e C_T \Phi^2} T_e \qquad (1\text{-}65)$$

The resulting man-made torque-speed characteristic of a shunt DC motor in this case is shown in Fig. 1-58.

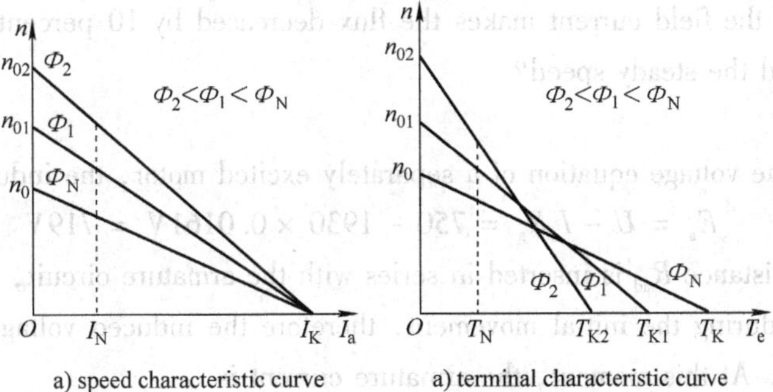

Fig. 1-58 The terminal characteristic curve for $I_f \neq I_{fN}$

Example 1-7 When a shunt DC motor with a rated voltage $U_N = 220\text{V}$, rated armature current $I_{aN} = 75\text{A}$, rated speed $n_N = 1000\text{r/min}$, and the total resistance of the armature circuit $R_a = 0.26\Omega$, as the motor turns to a rated operation state, the total resistance of the field circuit is $R_{fN} = 91\Omega$, the core loss is 600W and the mechanical loss is 1989W. Find:

1) the output torque of the motor turning at rated operation state?

2) the rated efficiency of the motor?

Solution:

1) the field current is

$I_f = U_N / R_{fN} = 220/91\text{A} = 2.42\text{A}$

The line current of the motor is

$I_N = I_{aN} + I_f = 75 + 2.42\text{A} = 77.42\text{A}$

The input power of the motor is

$P_1 = U_N I_N = 200.5 \times 77.42\text{W} = 17\,032.4\text{W}$

The induced voltage is

$E_a = U_N - I_{aN} R_a = 220 - 75 \times 0.26\text{V} = 200.5\text{V}$

The induced torque of the motor is

$P_e = E_a I_{aN} = 200.5 \times 75\text{W} = 15\,037.5\text{W}$

The output power is

$P_2 = P_e - p_{Fe} - p_\Omega = 15\,037.5 - 600 - 1989\text{W} = 12\,448.5\text{W}$

The output torque of the motor is

$T_2 = P_2/\Omega = 12\,448.5 \times 60/\ (2 \times 3.14 \times 1000)\ \text{N} \cdot \text{m} = 118.9\text{N} \cdot \text{m}$

2) the rated efficiency of the motor is

$\eta_N = P_2/P_1 = 12\,448.5/17\,032.4 = 73.1\%$

Example 1-8 A separately excited DC motor with a rated power is $P_N = 1325\text{kW}$, rated voltage is $U_N = 750\text{V}$, rated armature current is $I_a = 1930\text{A}$, rated speed $n_N = 200\text{r/min}$ and armature resistance $R_a = 0.0161\Omega$. when the motor is turning at 200r/min and the loading is on full load. Work out the following calculation:

1) Inserting a resistance $R_{ad} = 0.0746\Omega$ in series with the armature circuit, calculate the armature current at the inserting moment and also calculate the armature current and the speed at the steady operation respectively?

2) Decreasing the field current makes the flux decreased by 10 percent; calculate the steady armature current and the steady speed?

Solution:

According to the voltage equation of a separately excited motor, the induced voltage E_a is

$$E_a = U - I_a R_a = 750 - 1930 \times 0.0161\text{V} = 719\text{V}$$

1) When a resistance R_{ad} is inserted in series with the armature circuit, the speed of the motor cannot be changed during the initial movement, therefore the induced voltage cannot be changed, namely, $E_a = 719\text{V}$. At this moment, the armature current is

$$I_a = (U - E_a)/(R_a + R_{ad}) = (750 - 719)/(0.0161 + 0.0743)\text{A} = 343\text{A}$$

Since the load of the motor is full load, the steady induced torque of the motor will resume to the same value. The armature current will go back to $I_a = 1930\text{A}$, and the induced voltage E'_a is at a steady operation state is

$$E'_a = U - I_a(R_a + R_{ad}) = 750 - 1930 \times (0.0161 + 0.0743)\text{V} = 575\text{V}$$

The speed n' at the steady operation state is

$$n' = nE'_a/E_a = 200 \times 575/719\text{r/min} = 160\text{r/min}$$

2) When the flux is decreased by 10 percent, namely $\Phi/\Phi'' = 1/0.9$, Since the load of the motor is full load, so the steady induced torque of the motor will resume to the same value, and the armature current I''_a, and the induced voltage E''_a are

$$I''_a = I_a \Phi/\Phi'' = 1930 \times 1/0.9\text{A} = 2145\text{A}$$

$$E''_a = U - I''_a R_a = 750 - 2145 \times 0.0161\text{V} = 715\text{V}$$

Then the steady speed of the motor is

$$n'' = nE''_a\Phi/\ E_a\Phi'' = 200 \times 715/(719 \times 0.9)\text{r/min} = 221\text{r/min}$$

Example 1-9 A separately excited DC motor with rated voltage $U_N = 220\text{V}$; rated speed $n_N = 1500\text{r/min}$; rated line current $I_N = 41.1\text{A}$, and the armature resistance $R_a = 0.4\Omega$, is running at a full load condition. Work out the following calculation:

1) Inserting a resistance $R_{ad} = 1.65\Omega$ series with the armature circuit, calculate the steady speed of the motor?

2) Decrease the voltage to 110V, calculate the steady speed of the motor?

3) Decrease the flux by 10%, calculate the steady speed of the motor?

Solution:

According to the voltage equation:

$$C_e\Phi_N = (U_N - I_N R_a)/n_N = (220 - 41.1 \times 0.4)/1500 = 0.136$$

1) Inserting a resistance $R_{ad} = 1.65\Omega$ series with the armature circuit, the steady speed of the motor is

$$n = (U_N - I_N R_a - I_N R_{ad})/\ C_e\Phi_N$$
$$= (220 - 41.1 \times 0.4 - 41.1 \times 1.65)/0.136\text{r/min} = 1000\text{r/min}$$

2) Decrease the voltage to 110V, the steady speed of the motor is

$$n = (U - I_N R_a)/\ C_e\Phi_N = (110 - 41.1 \times 0.4)/0.136\text{r/min} = 687\text{r/min}$$

3) Decrease the flux by 10%, the steady armature current of the motor is

$$I_a = \Phi_N I_N/\Phi = 41.1/0.9\text{A} = 45.7\text{A}$$

The steady speed of the motor is

$$n = (U_N - I_a R_a)/\ C_e\Phi = (220 - 45.7 \times 0.4)/(0.136 \times 0.9)\text{r/min} = 1650\text{r/min}$$

1.8.2 The operational characteristics of the series DC motor

The construction process of a series DC motor is identical to the construction with a shunt motor except for the field. The field is connected in series with the armature and therefore it must carry the full armature current. This series field is composed with a few turns of wire, where there should be a cross section, that is sufficiently large to carry the current.

Although their constructions are similar, the properties of a series motor are completely different from those of a shunt motor. In a shunt motor, the flux per pole Φ is constant at all loads because the shunt field is connected to a line. However, in a series motor the flux per pole depends upon the armature current, and hence depends upon the load. When the current is large, the flux is large and vice versa.

The equivalent circuit of a series DC motor is shown in Fig. 1-43. In a series DC motor, the armature current, field current, and line current are all the same, the Kirchhoff's voltage law equation for this motor is $U = E_a + I_a(R_a + R_f)$ (Eq. 1-35); and the induced torque in this motor is given by Eq. 1-20, $T_e = C_T\Phi I_a$.

The flux in this motor is proportional to its armature current I_a directly. Therefore, the flux in this motor can be given by

$$\Phi = K_f I_a \tag{1-66}$$

where K_f is a constant.

The induced torque in this motor is thus given by

$$T_e = C_T \Phi I_a = C_T K_f I_a^2 \tag{1-67}$$

In other words, the torque in this motor is proportional to the square of its armature current. As a result of this relationship, it is easy to conclude that a series DC motor gives more torque per ampere than any other DC motor. Series DC motor is therefore used in applications which require high torques, for example the starter motors in cars, elevator motors, or tractor motors in locomotives.

On the other hand, if the load of the motor is lighter than a full load, the armature current and the flux will smaller than when they are in normal condition. The weakened field causes the speed to increase, and this is the same for a shunt motor with a weakened shunt field. For instance, if the load current of a series motor drops to half of its normal value, the flux will be diminished by half and hence the speed will double. Obviously, if the load is light, the speed may increase to high dangerous values. For this reason a series motor to operate at no-load is never permitted. It tends to turn out of the operation point, and the resulting centrifugal forces could split the windings out of the armature, thereby destroying the machine.

In aseries DC motor, the following characteristic curves will be considered.

1) Speed characteristic curve $n = f(I_a)$.

2) Torque characteristic curve.

3) Efficiency characteristic curve $\eta = f(I_a)$.

4) Terminal characteristic curve $n = f(T_e)$.

All equations of a series DC motor are given by

$$\left.\begin{aligned} U &= E_a + I_a R_a + I_f R_f = E_a + I_a (R_a + R_f) \\ E_a &= C_e n \Phi = C_e K_f n I_a = C'_e n I_a \\ T_e &= T_2 + T_0 \\ T_e &= C_T \Phi I_a = C_T K_f I_a^2 = C'_T I_a^2 \end{aligned}\right\} \tag{1-68}$$

where $\Phi = K_f I_f$, K_f is a constant, $C'_e = C_e K_f$, $C'_T = C_T K_f$.

1. The speed characteristic curve

The speed characteristic curve $n = f(I_a)$ of a series DC motor is a plot of the speed n as a function of armature current I_a; where the terminal voltage U is constant and there is no external resistance in the armature circuit.

From Eq. 1-68, the motor's speed is given by

$$n = \frac{U - I_a(R_a + R_f)}{C_e \Phi} = \frac{U}{C_e K_f I_a} - \frac{R_a + R_f}{C_e K_f} \tag{1-69}$$

From Eq. 1-69, the speed of the motor is inverse by proportional to the armature current I_a. The resulting speed characteristic curve of a series DC motor is shown in Fig. 1-59. While the load is increasing, the speed of a series DC motor may drop rapidly; and if the load is light or at no-load, the speed may rise to high and dangerous values or tends to run away; so a series motor to operate at no-load is never permitted.

2. The electromagnetic torque characteristic curve $T_e = f(I_a)$

The torque characteristic curve $T_e = f(I_a)$ of a series DC motor is a plot of the torque T_e as a function of armature current I_a; where the terminal voltage U is constant and there is no external resistance in the armature circuit.

According to Eq. 1-68 the torque T_e of a series DC motor is proportional to the square of its armature current I_a, i. e., $T_e = C_T K_f I_a^2$; and it can supply more torque per ampere than any other DC motor. Fig. 1-59 shows the developed torque curve for a series DC motor.

3. The efficiency characteristic curve $\eta = f(I_a)$

The efficiency characteristic curve $\eta = f(I_a)$ of a series DC motor is the same as that of a shunt DC motor as shown in Fig. 1-59.

4. The terminal characteristic curve $n = f(T_e)$

The terminal characteristic of a series DC motor is a plot of its torque T_e versus speed n; and the terminal voltage U is constant.

From Eq. 1-68, the motor's torque-speed relationship is given by

$$n = \frac{U\sqrt{C_T'}}{C_e'\sqrt{T_e}} - \frac{R}{C_e' K_f} \tag{1-70}$$

where R is the total resistance of the armature circuit.

Notice that for an unsaturated series DC motor, the speed of the motor varies as the reciprocal of the square root of the torque. This is quite an unusual relationship. This ideal torque-speed characteristic is plotted in Fig. 1-60.

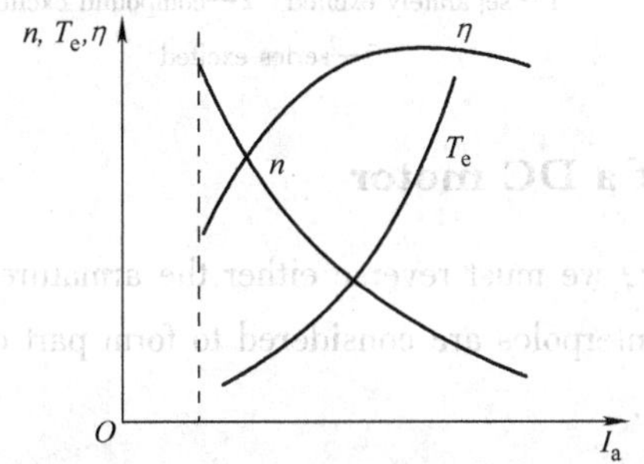

Fig. 1-59 The operation characteristic curves of a series DC motor

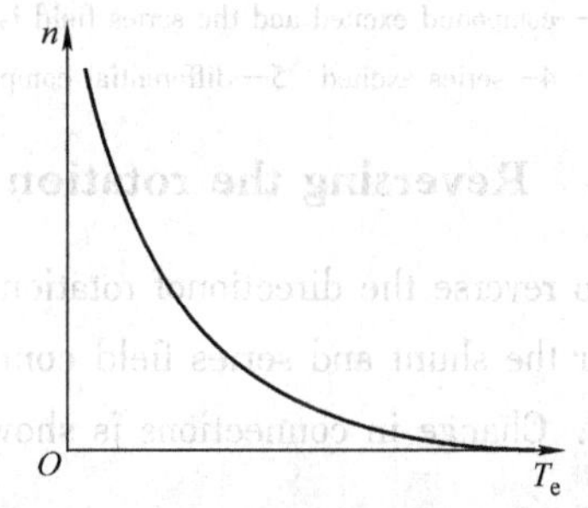

Fig. 1-60 The torque-speed characteristic of a series DC motor

1.8.3 The operational characteristics of the compound DC motor

A compound DC motor carries both a series field and a shunt field. In a cumulative compound motor, the mmf of the two fields is added. A shunt field is always stronger than a series field.

Fig. 1-44 shows the equivalent circuit of a compound motor. When the motor runs at no-load, the armature current I_a in the series winding is low and the mmf of the series field is negligible. However, the shunt field is fully excited by current I_f, hence the motor behaves like a shunt machine: there is no tendency to runway at no load.

As the load increases, the mmf of the series field increases as well; but the mmf of the shunt field remains constant. The total mmf (and the resulting flux per pole) is therefore greater than at no-load. The motor speed falls with increasing load and the speed drop from no-load to full-load is generally between 10 percent and 30 percent.

If the series field is connected, then it will oppose the shunt field. We then obtain a differential compound motor. In such a motor, the total mmf decreases with an increasing load. The speed rises as the load increases, and this may lead to instability of the motor. The differential compound motor could only apply to limited number of machines.

Fig. 1-61 shows a typical speed curve of shunt, compound, and series motors on a per-unit basis. Fig. 1-62 shows a typical torque-speed curves of shunt, compound, and series motors on a per-unit basis.

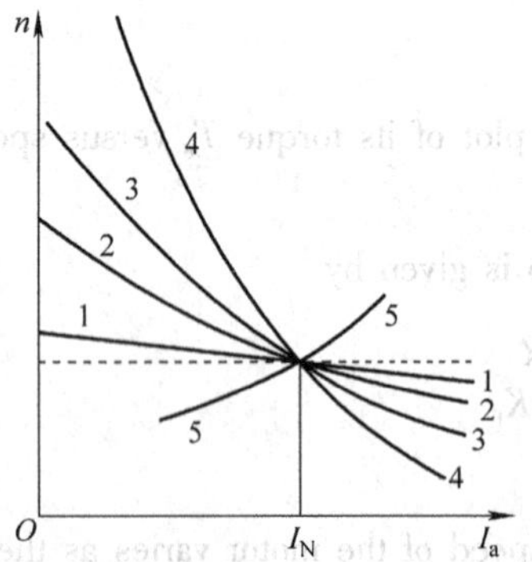

Fig. 1-61 the typical speed curves of shunt, compound, and series motors

1—shunt 2—compound excited and the shunt field is stronger 3—compound excited and the series field is stronger 4—series excited 5—differential compound

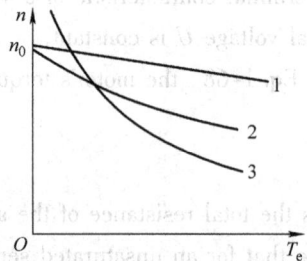

Fig. 1-62 the typical torque-speed curves of separately excited, compound, and series motors

1—separately excited 2—compound excited 3—series excited

1.8.4 Reversing the rotation direction of a DC motor

To reverse the directionof rotation of a DC motor, we must reverse either the armature connections or the shunt and series field connections. The interpoles are considered to form part of the armature. Change in connections is shown in Fig. 1-63.

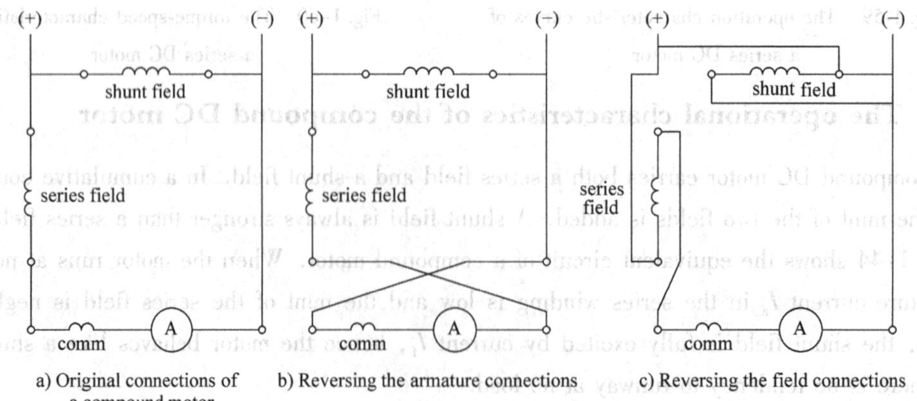

Fig. 1-63 Reverse the rotation direction of a DC motor

1.9 Speed control of shunt DC motors

Follow the speed equation of a shunt DC motor (Eq. 1-52)

$$n = \frac{U - I_a(R_a + R_{ad})}{C_e \Phi}$$

There are three methods for regulating shunt DC motror's speed, namely, by varying the voltage U (voltage control), by varying the resistance of the armature circuit (rheostatic control), and by varying the flux Φ (flux or field control).

1. Changing the resistance of the armature circuit

The operation condition of this speed control method is to insert a resistor in series with the armature circuit, but the terminal voltage U and the field current I_f are keep their rated value separately.

If a resistor is inserted in series with the armature circuit, the slope of the motor's torque-speed characteristic will drastically increase and the motor will operate slower if loaded (Fig. 1-64a).

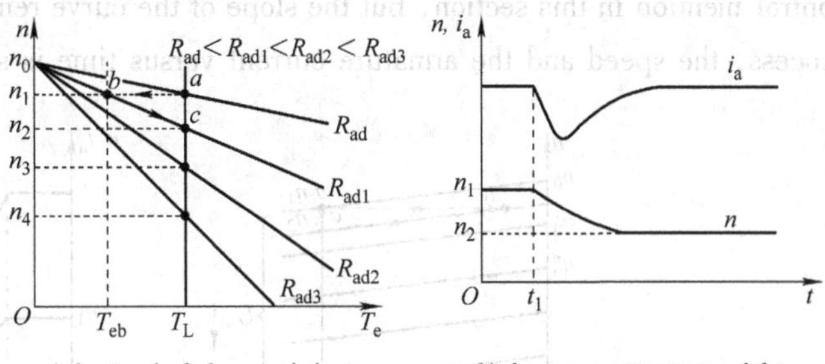

a) the terminal characteristic

b) the armature current and the speed varying with the time

Fig. 1-64 The speed control of varying the resistor of the armature circuit

The motor's terminal characteristic curves in different armature resistors are shown in Fig. 1-64a, where $R_{ad3} > R_{ad2} > R_{ad1} > R_{ad} > 0$. Assumed that the constant load torque of the DC motor is T_L, and the corresponding speeds of each terminal characteristic curves are n_1, n_2, n_3, n_4, and $n_1 > n_2 > n_3 > n_4$ (as shown in Fig. 1-64a).

The speed control is described below:

The DC motor is operating at point a. When the resistor R_{ad1} is inserted in the armature circuit, the motor speed n_1 and the internal generated voltage E_a will not change at this point. However, the DC motor will operate at point b on the terminal characteristic curve of R_{ad1}. At point b, the armature current I_a and the produced torque T_{eb} decrease, and $T_{eb} < T_L$. This causes the speed of the motor to decrease, this is the first procedure of the speed control.

The speed of the motor decrease, E_a will decrease as well. Consequently the armature current I_a and the produced torque T_e will increase, the operating point of the motor will move to point c. At point c, the speed of the motor is n_2, and $T_{ec} = T_L$, this is the second procedure of the speed control. The speed and the armature current vs time of the speed control procedures are shown in Fig. 1-64b.

The insertion of a resistor is not a very economical method for speed control, since the losses of the inserted resistor are very high. Hence, it is rarely used.

2. Changing the voltage U

The second method of speed control involves changing the voltage U that applied to the armature of the motor, at the same time without changing the voltage that applied to the field. And there is no additional resistor inserted in series with the armature circuit.

In effect, the motor must be separately excited to use armature voltage control.

If the voltage U is decreased, then the armature current I_a in the motor must decrease. As I_a decreases, the induced torque T_e decreases, causing $T_{eb} < T_L$, and thereby speed n of the motor decreases.

But as the speed n_1 decreases, the internal generated voltage E_a decreases, hence causing the armature current to increases. This increase in I_a led to the increases the induced torque T_e, causing T_e to equal T_L at a lower rotational speed n_2.

The effect of an decrease in U on the torque-speed characteristic of a separately excited motor is shown in Fig. 1-65a. Notice that the no-load speed of the motor is shifted by the method of speed control mention in this section, but the slope of the curve remains constant. During the speed control process, the speed and the armature current versus time is shown in Fig. 1-65b.

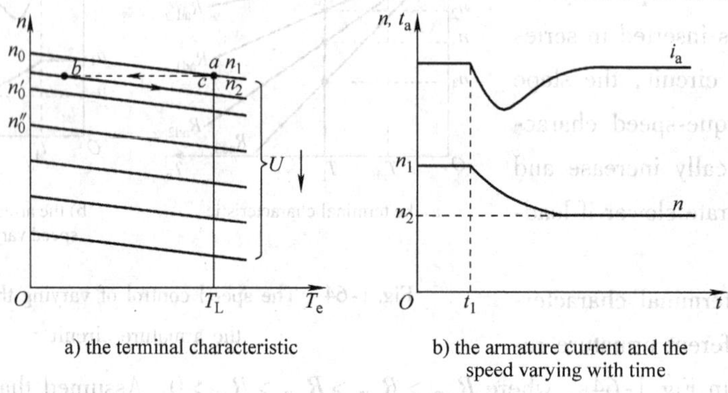

a) the terminal characteristic　　b) the armature current and the speed varying with time

Fig. 1-65　The speed control of varying the voltage applied to the armature circuit

3. Changing the flux Φ

The third method of speed control is changing the flux Φ through varying the field resistance, at the same time without changing the voltage U applied to the armature of the motor. There is no additional resistor inserted in series with the armature circuit.

The motor's terminal characteristic curves in different flux Φ are shown in Fig. 1-66a. Assumming that the load torque of the DC motor is T_L, and the corresponding speeds of each terminal characteristic curves are n_1, n_2, where $n_1 < n_2$ (as shown in Fig. 1-66a).

The speed control is described as below:

The DC motor is operating at point a, when the field resistance increases, then the field current I_f decreases, and as the field current decreases, the flux Φ decreases simultaneously. The motor speed n_1 will not change at this point, and a decrease in flux causes an instantaneous decrease in the internal generated voltage E_a. This causes a large increase in the machine's armature current $I_a = (U - E_a)/R_a$, so the produced torque T_e will increase, the operating point of the motor will then move to point b, this is the first procedure of the speed control.

At point b, the speed of the motor is n_1, and $T_{eb} > T_L$, this causes the speed of the motor to increase. Speed increase, then E_a will increase, so the armature current I_a and the produced torque T_e will decrease. The operating point of the motor will move to point c. At point c, the speed of the motor is n_2, and $T_{ec} = T_L$. This is the second procedure of the speed control.

During this procedure, the speed and the armature current versus time are shown in Fig. 1-66b. where I_{a1} is the armature current corresponding to the speed n_1 and I_{a2} is the armature current corresponding to the speed n_2. Since the inductance of the field circuit is high, the flux Φ cannot change chopping, and the produced torque varying curve is curve 3 which is shown in Fig. 1-66a.

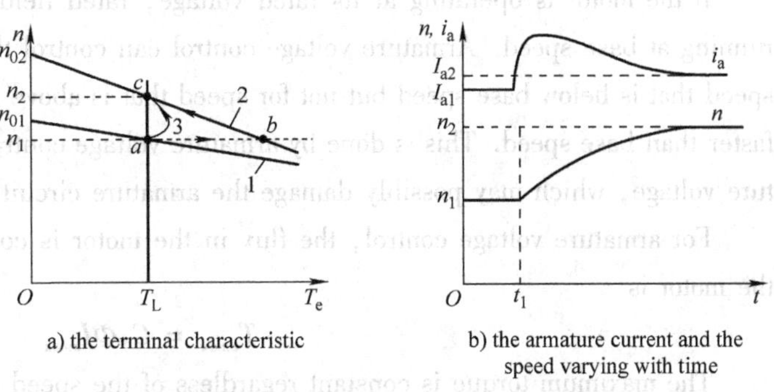

a) the terminal characteristic

b) the armature current and the speed varying with time

Fig. 1-66 The speed control of varying the flux

Notice that as the flux in the motor decreases, the no-load speed of the motor increases, while the slope of the torque-speed curve becomes steeper. This shape is a consequence of Eq. 1-58, which describes the terminal characteristic of the motor. In Eq. 1-58, the no-load speed is proportional to the reciprocal of the flux in the motor, while the slope of the curve is proportional to the reciprocal of the flux squared. Therefore, a decrease in flux causes the slope of the torque-speed curve to become steeper.

As the field resistance increased, the speed of the motor increases as well. What would happen if this effect were taken to an extreme, what if the field resistor increases? What would happen if the field circuit opens while the motor is running? As mentioned before, the flux in the motor would reduce drastically to Φ_{res}, and E_a would drop simultaneously. This would cause a really enormous increase in the armature current, and the resulting induced torque would be a little higher than the load torque on the motor. Therefore, the motor's speed starts to rise and keeps increasing; this condition is known as run away.

4. Power and torque limits for a seperately excited DC motor under the three types of speed control

The insertion of a resistor is an extravagant method of speed control, since the loss in the inserted resistor are extremely high. For this reason, it is rarely used. It will be found only in applications that the motor spends almost all its time operating at full speed or in applications that do not need a better form of speed control due to the inexpensive cost of the application.

There are two most common methods of speed control of shunt DC motor, field resistance variation and armature voltage variation, they have different safety operation ranges.

There is a significant difference between the torque and power limits on the motor, when operating these two types of speed control. The limitation in either of the cases is the heating of the armature conductors, which places an upper limit on the magnitude of the armature current I_a.

1) In armature voltage control, the lower the armature voltage on a separately excited DC motor, the slower it turns; whereas the higher the armature voltage is, the faster it turns. Since an in-

crease in armature voltage causes an increase in speed, there is always a maximum achievable speed by armature voltage control. This maximum speed occurs when the motor's armature voltage reaches its maximum permissible level.

If the motor is operating at its rated voltage, rated field current, and rated power, it will be running at base speed. Armature voltage control can control the speed of the motor. This is only for speed that is below base speed but not for speed that is above base speed. To achieve a speed that is faster than base speed. This is done by armature voltage control and it would require excessive armature voltage, which may possibly damage the armature circuit.

For armature voltage control, the flux in the motor is constant, hence the maximum torque in the motor is

$$T_{\mathrm{emax}} = C_{\mathrm{T}}\Phi I_{\mathrm{amax}} \tag{1-71}$$

The maximum torque is constant regardless of the speed of the rotation of the motor. Since the power output of the motor is given by $P = T_{\mathrm{e}}n/9.55$, the maximum power output of the motor at any speed under armature voltage control is

$$P_{\max} = T_{\mathrm{emax}}n/9.55 \tag{1-72}$$

Thus the maximum power output of the motor is directly proportional to its operating speed under armature voltage control.

2) In field resistance control or flux control, the lower the field current in a separately excited DC motor, the faster it turns; whereas the higher the field current is, the slower it turns. Since an increase in field current causes a decrease in speed, there is always a minimum achievable speed by field circuit control. This minimum speed occurs when the motor's field circuit has the maximum permissible current flowing through it.

If a motor is operating at its rated terminal voltage, rated power, and rated field current, then it will be running at rated speed, which is also known as base speed. Field resistance control can control the speed of the motor for speeds that is above base speed but not for speed that is below base speed. To achieve a speed slower than base speed, this is done by field circuit control and it would require excessive field current, which might possibly burn up the field windings.

When field resistance control is used for a DC motor, the flux will change. In this form of control, a speed increase is caused by a decrease in the motor's flux. In order to make the armature current limit not to be exceeded, the induced torque limit must decrease as the speed of the motor increases. Since the power out of the motor is given by $P = T_{\mathrm{e}}n/9.55$, and the torque limit decreases as the speed of the motor increases. The maximum power out of a DC motor under field current control is constant, while the maximum torque varies with the reciprocal of the motor's speed.

These two techniques of speed control are complementary. Armature voltage control works well for speed when below the base speed, and field resistance, field current control or flux control works well for speeds above base speed. By combining the two speed-control techniques on the same motor, it is possible to get a greater range of speed variations of up to 40 to 1 or more. Shunt and separately excited DC motors both have excellent speed control characteristics.

The seperately excited DC motor power and torque limitations for safe operation which is seen as a function of speed are shown in Fig. 1-67.

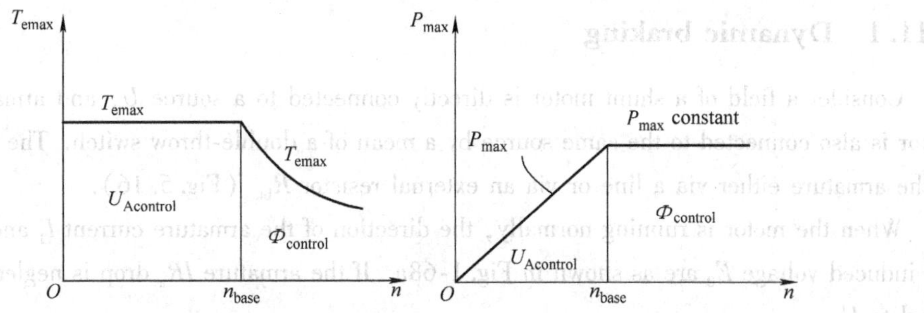

Fig. 1-67 power and torque limits as a function of speed for a separately excited motor under voltage and flux control

1.10 Starting a DC motor

If we apply full voltage to a stationary DC motor, the initial current in the armature will be very high and there will be risks of

1) Burning out the armature.

2) Damaging the commutator and brushes, due to heavy sparking.

3) Overloading the feeder.

4) Snapping off the shaft due to the mechanical shock.

5) Damaging the driven equipment because of the sudden mechanical hammerblow.

All DC motors must, therefore, be provided with a means to limit the initial current in a reasonable range of values, usually between 1.5 and twice full-load current. There are two methods to start with:

1) One of the solutions is to connect a rheostat in series with the armature. Then, the resistance is gradually reduced as the motor accelerates and it is eliminated completely, when the machine has reach the point to attain full speed.

2) Another method is to reduce the voltage applied to the armature circuit.

Nowadays, electronic methods are often used to limit the initial current and used to provide a speed control.

1.11 Braking a DC motor

Most people incline to believe that stopping the operation of DC motor is simple and trivial. Unfortunately, this is not always true. When a large DC motor is coupled to a heavy inertia load, it may take an hour or more for the system to come to a halt. Lengthy deceleration time is one of the reasons that are often unacceptable for stopping a DC motor. Thus, we must apply a braking torque to ensure a rapid stop. One way to brake the motor is by simple mechanical friction, it is the same method when we use to stop a car. A more advanced method consists of circulating a reverse current in the armature, therefore it brakes the motor electrically. Two methods are employed to create an electromechanical brake: dynamic braking and plugging.

1.11.1 Dynamic braking

Consider a field of a shunt motor is directly connected to a source U, and armature of a shunt motor is also connected to the same source by a mean of a double-throw switch. The switch connects to the armature either via a line or via an external resistor R_{bk} (Fig. 5. 16).

When the motor is running normally, the direction of the armature current I_1 and the polarity of the induced voltage E_0 are as shown in Fig. 1-68a. If the armature IR_a drop is neglected, E_0 is then equal to U.

If the switch is opened suddenly (Fig. 1-68b), the motor will continue to run, but its speed will reduce gradually due to friction and windage losses. On the other hand, since the shunt field is still excited, then the induced voltage E_0 still exist, but it is falling at the same rate as the speed. The motor is now a generator which the armature is open-circuited.

Now, let us close the switch on the second set of contacts so that the armature is connected to the external resistor suddenly (Fig. 1-68c). Voltage E_0 will immediately produce an armature current I_2. However, this current flows in the direction opposite to the original current I_1. The reverse torque, which magnitude depends upon I_2 is then developed. The reverse torque brings will bring the machine to a rapid, but very smooth break off.

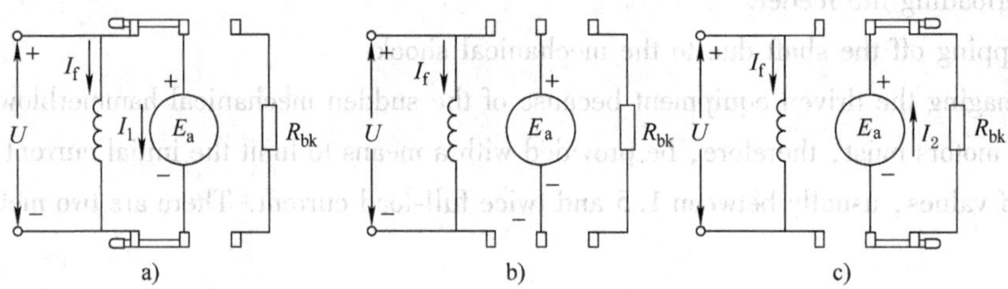

Fig. 1-68 Dynamic braking of a shunt DC motor

In practice, resistor R_{bk} is chosen so that the initial braking current is about twice the rated current. The initial braking torque is then twice the normal torque of the motor.

As the motor slows down, the gradual decrease in E_0 produces a subsequent decrease in I_2. Consequently, the braking torque reduces gradually, finally, it becomes zero when the armature ceases to turn. The speed first drops rapidly, then it drops slower later, as the armature comes to a halt.

When a DC motor is operating in dynamic braking, $U=0$, $R=R_a+R_{bk}$, where R_{bk} is the braking resistance, by substituting these expression into the equation $n=\dfrac{U}{C_e\Phi}-\dfrac{R}{C_eC_T\Phi^2}T_e$, the torque-speed characteristic can be rewritten as

$$n=-\frac{R_a+R_{bk}}{C_eC_T\Phi^2}T_e \tag{1-73}$$

Fig. 1-69 The torque-speed curve of a DC motor operating at dynamic braking

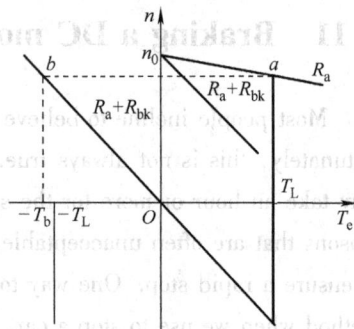

The torque-speed curve of a DC motor when it is operating at dynamic braking is shown in Fig. 1-69. From

Fig. 1-69, we can find that the torque-speed curve is a straight line passing point O; which also passes through the second and the fourth quadrant. The slope of the curve is $\beta = \frac{R_a + R_{bk}}{C_e C_T \Phi^2}$.

1.11.2 Plugging

We can stop the motor in a more rapid by using a method called plugging. It consists of sudden reverse of the armature current by reversing the terminals of the source (Fig. 1-70a).

Under normal conditions of the motor, armature current I_1 is given by

$$I_1 = (U - E_0)/R_a$$

where R_a is the armature resistance. If we reverse the terminals of the source suddenly, the net voltage which is acting on the armature circuit, then becomes $(E_0 + U)$. The so-called counter-emf E_0 of the armature is no longer counter to anything but actually adds to the supply voltage U. This net voltage would produce a powerful reverse current, perhaps 50 times greater than the full-load armature current. This current would initiate an arc around the commutator, the arc then destroy the segments, brushes and supports, even before the line circuit breakers could open.

To avoid such a catastrophe, we must limit the reverse current by introducing a resistor R_{bk} in series with the reversing circuit (Fig. 1-70b). In dynamic braking, the resistor is designed to limit the initial braking current I_2 to about twice full-load current is. In this plugging circuit, a reverse torque is developed even though when the armature has come to a stop. In effect, at zero speed, $E_0 = 0$, but $I_2 = U/(R_a + R_{bk})$, which is about one-half of the initial value. As soon as the motor stops, we must immediately open the armature circuit; otherwise it will begin to run in a different direction. Circuit interruption is usually controlled by an automatic null-speed device which mounted on the motor shaft.

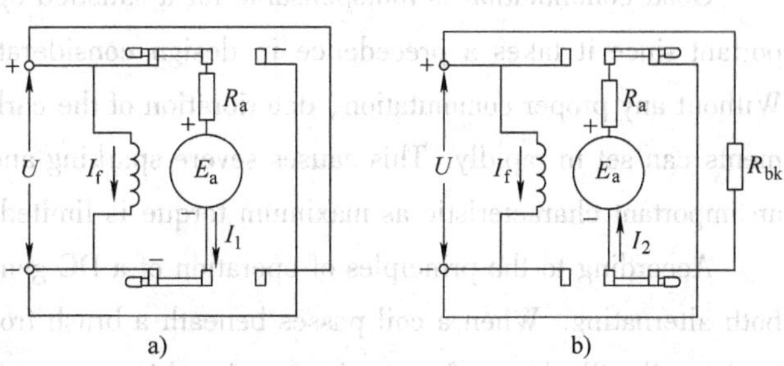

Fig. 1-70 Plugging of a shunt DC motor

When a DC motor is operating at plugging, the torque-speed characteristic can be rewritten as

$$n = -n_0 - \frac{R_a + R_{bk}}{C_e C_T \Phi^2} T_e \tag{1-74}$$

The torque-speed curve of an operation of a DC motor at plugging is the straight line bc that is shown in Fig. 1-71.

Fig. 1-72 show a series of speed versus time curves comparing a motor which is equipped with various braking method. The curves of Fig. 1-72 enable us to compare plugging and dynamic with the same initial braking current. Note that plugging stops the motor completely after an interval $2t_0$. On the other hand, if dynamic braking is used, the speed is still 25 percent of its original value at this time. Nevertheless, the comparative simplicity of dynamic braking is much more popular in most applications.

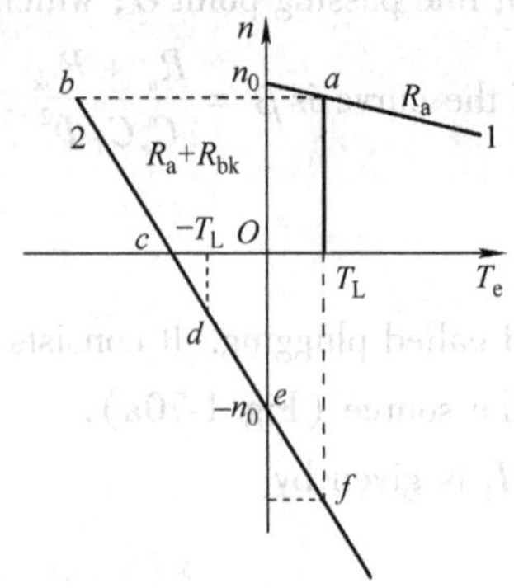

Fig. 1-71 The torque-speed curve of operation of a DC motor at plugging

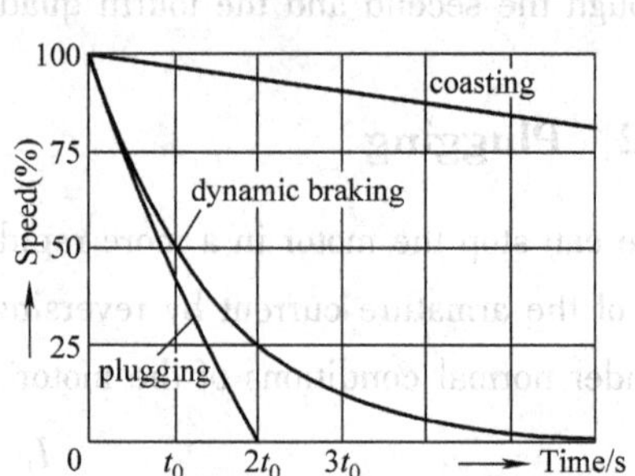

Fig. 1-72 Speed versus time curves for various braking methods

1.12 Commutation

Good commutation is indispensable for a satisfied operation of DC machines. This matter is important since it takes a precedence in design considerations over the problem of heat dissipation. Without any proper commutation, deterioration of the carbon brusher and the copper commutator segments can set in rapidly. This causes severe sparking and thereby renders the machine useless. It is an important characteristic as maximum torque is limited by commutation rather than by heating.

According to the principles of operation of a DC generator, the current and the emf in a coil are both alternating. When a coil passes beneath a brush from one pole to an adjacent pole, the current in the coil will change from $+i_a$ to $-i_a$, this process is called commutation. Fig. 1-73 shows the commutation process.

1. Electromagnetic characteristic of the Commutation

Assume that the width of the brushes and the commutator segments are the same, brushes are stationary, and the commutation segments are moving from right to left. Fig. 1-73a shows while the

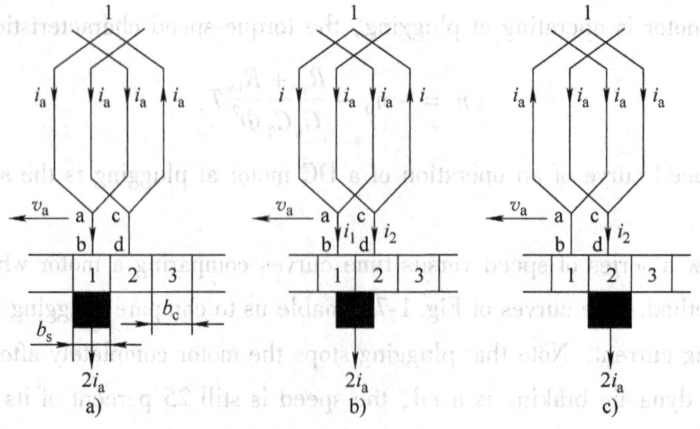

Fig. 1-73 Commutation of a DC machine

brush makes contact with the commutator segment 1, coil 1 belongs to the right circuit of the brush, and the current in coil 1 is a value of $+i_a$. When the brush makes contact with commutator segment 1and 2, coil 1 is actually short-circuited by the brush through segment 1 and 2 as shown in Fig. 1-73b, the current in coil 1 is a value of i. In Fig. 1-73c, the brush makes contact with commutator segment 2, coil 1 belongs to the left circuit of the brush, and the current in coil 1is a value of $-i_a$. During this commutation process, the current in the coil 1 will have reversed completely from $+i_a$ to $-i_a$, and the commutation period T_c is very short.

The distinction between linear commutation and voltage commutation is depicted by various commutation curve showing in Fig. 1-74.

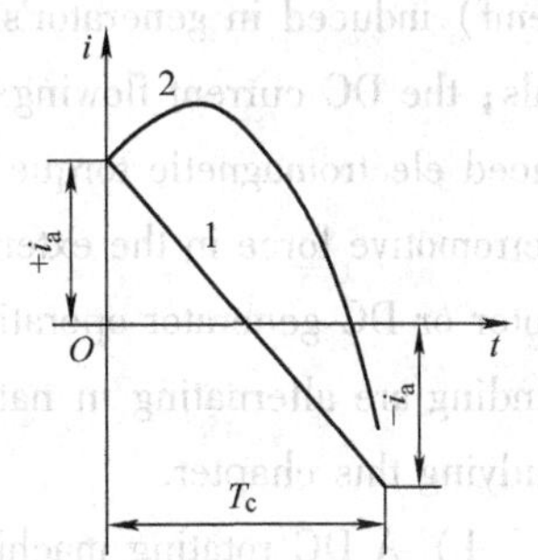

Fig. 1-74　Commutation curve of coil 1 in Fig. 1-74

Curve 1 represents a linear commutation, and it occurs only when the brush contact resistance which is the only factor influence the commutation. Under these conditions the current in the coil changes lineary during the commutation period T_c.

The commutation curve denoted as 2 in Fig. 1-74 is described as voltage commutation. As shown in curve 2, the coil current reversal is delayed either by the action of the emf of self-induction in the short-circuited coil, or by the action of the *Blv* voltage produced by the armature mmf or by the action of a combination of both. This kind of decelerated commutation is characterized by high current density at the tip of the trailing brush. The tip promotes a deterioration of the brush material and brings undesirable chemical changes in the copper segments of the commutator.

2. The methods of improving commutation

A scheme that is frequently used to neutralize this emf of self-induction and the effect of the armature mmf is the inclusion of small poles. Those small poles are located on the brush axis and excited by a coil carrying the armature current. Because of their location and function, these poles are called interpoles or commutating poles. Use of armature current excitation is dictated by the fact that both the self-induction of emf in the short-circuited coil and the flux in the brush axis. They are produced by the armature mmf, which is proportional to armature current. For generator action, the polarity of the interpole must be the same as that of the field pole into which the coil is moving. Fig. 1-75 depicts the situation for a two-pole machine.

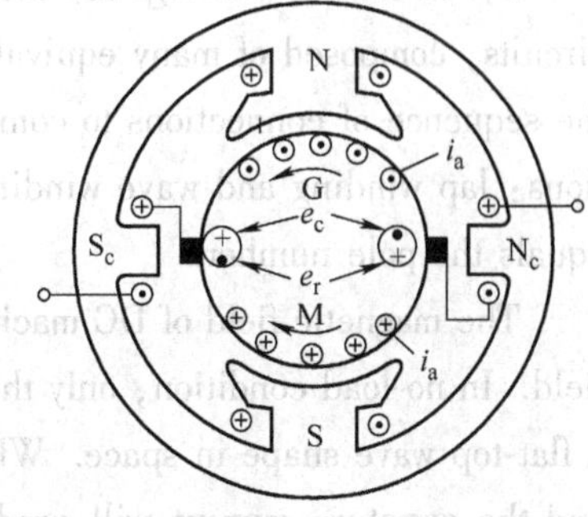

Fig. 1-75　Influence of interpoles of a DC machine

3. Compensating winding

The most effective way to prevent this flashover condition is to neutralize the armature reaction and this causes the severe peaking of air-gap flux density which will lead to breakdown. This is achieved the best by use of a *compensating winding* which is embedded in slots that are distributed along the pole faces. By allowing armature current to flow through this pole-face winding with a polarity opposite to that of the armature winding and a suitable number of ampere-turns, the armature reaction could be completely cancelled. The major disadvantage of eliminating or reducing the commutation problem is high expense on pole-face windings.

Summary

The operation process of a DC machine is based on the electromagnetic induction law and the electromagnetic force law. With the brushes and commutator, the alternating electromotive force (emf) induced in generator's armature winding is converted to the DC voltage on the output terminals; the DC current flowing through the brushes is introduced to the armature winding thus the induced electromagnetic torque is consistent in the same direction. In other words, the current and electromotive force in the exterior circuit of the machine are all DC quantities, regardless if in the DC motor or DC generator operation. However the current and emf in the interior circuit of the armature winding are alternating in nature. Students should pay more attention to the following points when studying this chapter.

1) A DC rotating machine consists of one stationary stator, one rotating rotor and an air gap between the rotor and the stator. The main poles on the stator are for the purpose of establishing the main magnetic field. Armatures are used to produce electromagnetic torque and induce electromotive force. Both electromagnetic torque and induced electromotive force (voltage) are also used to accomplish electromechanical energy conversion.

The nameplate data of a DC machine include the rated power, rated voltage, rated current, rated speed and rated excitation current.

DC machines can be classified into separately excited and self-excited machines according to their excitation type. Self-excitation can be shunt, series or compounded. When the excitation type is different; the terminal characteristics of DC machines are also different.

2) Armature windings are the main circuit of DC machines. The armature windings are closed circuits, composed of many equivalent coil elements. Armature windings are classified according to the sequence of connections to commutator segments. There are two basic armature winding connections: lap winding and wave winding. Just like the simplex lap windings, the parallel paths number equals the pole number.

The magnetic field of DC machines is determined jointly by the main pole field and the armature field. In no-load condition, only the main pole field winding is excited and the air gap flux density is a flat-top wave shape in space. When the machine is loaded, current flows into armature windings and the armature current will produce an armature magnetic field. The q-axis armature magnetic field in the air gap is a saddle shaped wave in space. The two spatial magnetic field waves can be combined to build the total magnetic field. The distortion of the total flux occurring when the load current in the armature winding increases is called armature reaction.

The induced electromotive force and electromagnetic torque of a DC machine can be calculated by two equations: $E_a = C_e \Phi n$ and $T_e = C_T \Phi I_a$.

3) The basic equations of DC machines are voltage equation, torque equation and power equation.

Induced electromotive force of DC generators is larger than armature voltage on the terminal.

The DC generator operation characteristics include the no-load characteristics, terminal characteristics and regulation characteristics at load. No-load characteristics of separately excited DC generator are similar to that of the core magnetization curves. The terminal characteristics of separately excited DC generator at load are a slightly pendulous curve.

Three conditions must be satisfied to establish a voltage in a shunt DC generator:

① There is a residual magnetic flux in the generator;

② There is correct connection of the armature winding to the field excitation winding;

③ There is less the field resistance than the critical resistance.

In DC motors, the armature voltage is greater than the induced electromotive force. The DC motor operation characteristics include the operating characteristics and mechanical characteristics.

Separately excited DC motors are basically constant speed motors. Their electromagnetic torque is proportional to armature current and has little effect to the motor speed. The characteristics of shunt DC motors are similar to that of separately excited DC motors.

Although the construction of series and shunt motors is similar, the characteristics of a series motor are completely different from those of a shunt motor. In a series motor, flux per pole depends upon the strength of armature current and the load. When the current is large, the flux is large and vice versa. If the load is small, the speed of a series DC motor may go up to a risky value, resulting in very large centrifugal forces tearing the windings out of the armature and destroying the machine. For this reason, we never allow a series DC motor to operate at no-load by an automatic disconnection switch.

4) Terminal characteristics of DC motors are a plot of its torque T_e versus speed n $[n = f(T_e)]$ while the field current I_f and the terminal voltage U remain constant. There are three types of terminal characteristics of separately excited DC motors realized by purposely designed excitation schemes.

5) To start a DC motor, a strong starting torque and as small as possible starting current are needed. In order to control the starting current within a reasonable range, there are two starting methods: one is to connect a rheostat in series with the armature winding, and the other is to reduce the voltage applied to the armature circuit.

6) In order to brake a DC motor electrically, one elegant method is to introduce a reverse current in the armature. Two methods are employed to realize the electromechanical brake: dynamic braking and plugging.

7) Speed control of DC motors is very important. There are three methods for speed regulation: by adjusting the voltage U (voltage control) applied to the armature; by varying the resistance of the armature circuit (rheostatic control); by changing the magnetic flux Φ (flux or field control).

8) Commutation is a special issue of commutator machines. Without proper commutation, the carbon brushers and copper commutator segments will worn out or damaged very quickly. Improper commutation can also cause severe sparking and eventually destroy the DC machine.

The most effective way to prevent improper commutations is the use of compensating windings and interpoles.

Questions and problems

1. Determine the polarity of the brush voltage of the following DC generators.

1) The main field poles are fixed. The brushes and armature are rotated simultaneously at the same speed.

2) The armature is fixed. The brushes and the main field poles are rotated simultaneously at the same speed.

3) The brushes are fixed. The main field poles and the armature are rotated simultaneously at different speed.

2. What are the rated current I_N and rated input power P_1 of a DC motor with $P_N = 17\text{kW}$, $U_N = 220\text{V}$, $n_N = 1500\text{r/min}$, $\eta_N = 0.85$?

3. What is the rated current I_N of a DC generator with $P_N = 100\text{kW}$, $U_N = 230\text{V}$, $n_N = 1450\text{r/min}$?

4. What is armature reaction? How does it affect the operation of a DC machine? What effect does armature reaction have on the torque-speed characteristic of a shunt DC motor?

5. What types of losses are present in a DC machine?

6. What is the speed regulation of a DC motor?

7. How can the speed of a shunt DC motor be controlled?

8. What is practical difference between a separately excited and a shunt DC motor?

9. What happens in a shunt DC motor if its field circuit opens while it is running?

10. Why is a starting resistor used in DC motor circuits?

11. How can the direction of rotation of a separately excited DC motor be reversed?

12. How can the direction of rotation of a shunt excited DC motor be reversed?

13. How does the voltage buildup occur in a shunt DC generator during starting?

14. How does armature reaction affect the terminal voltage in a separately excited DC generator?

15. Explain the concept of electrical degrees. How is the electrical angle of the voltage in a rotor conductor related to the mechanical angle of the machine's shaft ?

16. A 4-pole DC generator has a simplex lap and a full pitch winding. The flux per pole Φ equals $3.5\times10^{-2}\text{Wb}$ and the conductor number N is 157. Calculate the following:

1) The no-load induced voltage E_A when the speed is 1200r/min.

2) The electromagnetism torque T_e when the path current is 50A.

17. A shunt DC motor is running at full-load conditions, its ratings are: $U_N = 220\text{V}$; $I_N = 80\text{A}$; $R_a = 0.1\Omega$; the rated excitation voltage $U_f = 220\text{V}$; the field resistance $R_f = 88.8\Omega$; The brush voltage drop is assumed to be 2V; The stray loss is 1% of the rated power; the rated efficiency $\eta_N = 0.85$. Calculate the following:

1) The rated input power of the DC motor.

2) The rated output power of the DC motor.

3) The total losses of the DC motor.

4) The copper losses in the armature circuit.

5) The stray losses of the DC motor.

6) The total iron losses and mechanical losses.

18. A shunt DC generator is running at full-load conditions, its ratings are: $P_N = 6\text{kW}$; $U_N = 230\text{V}$; $n_N = 1450\text{r/min}$; $R_a = 0.921\Omega$; the field resistance $R_f = 177\Omega$; The brush voltage drop is assumed to be 2V; The total iron losses and mechanical losses are 313.9W; the stray loss is 60W. Calculate the following:

1) The input power at rated-load.

2) The electromagnetic power in rated state.

3) The electromagnetic torque in rated state.

4) The efficiency in rated state.

19. A shunt DC generator is running at full-load conditions, its ratings are: $P_N = 90\text{kW}$; $U_N = 230\text{V}$; $R_a = 0.04\Omega$; the field resistance $R_f = 60\Omega$; the total iron losses and mechanical losses are 2kW; The stray losses is 1% of the rated power. Calculate the following:

1) The input power at rated-load.

2) The efficiency in rated state.

20. An eight-pole, 25 kW, 120V DC generator has a simplex lap-wound armature which has 64 coils with 16 turns per coil. Its rated speed is 2400r/min.

1) How much flux per pole is required to produce the rated voltage in this generator at no-load conditions?

2) What is the current per path in the armature of this generator at the rated load?

3) What is the induced torque in this machine at the rated load?

4) If the resistance of this winding is 0.011Ω per turn, what is the armature resistance of this machine?

21. A shunt DC motor is running at full-load conditions, its ratings are: $P_N = 17\text{kW}$; $U_N = 220\text{V}$, $n_N = 3000\text{r/min}$; $I_N = 89.9\text{A}$; $R_a = 0.114\Omega$; the field resistance $R_f = 181.5\Omega$; the armature reaction can be neglected. Calculate the following:

1) The rated output torque of the motor.

2) The electromagnetic torque at rated load.

3) The efficiency at rated state.

4) The ideal no-load speed when $I_a = 0$.

22. What are the speed and electromagnetic torque of a shunt DC motor with $U_N = 220\text{V}$, $R_a = 0.316\Omega$, the ideal no-load speed $n_0 = 1600\text{r/min}$ when $I_a = 50\text{A}$?

23. The ratings of a shunt DC generator are: $P_N = 23\text{kW}$; $U_N = 230\text{V}$; $n_N = 1500\text{r/min}$; $R_a = 0.1\Omega$; field resistance $R_f = 57.5\Omega$; The armature reaction and magnetic saturation can be neglected. Description: When this machine is running at motor mode, the armature voltage and field voltage are set to be 220V, and the armature current is maintained the rated value. Calculate the following:

1) The speed of the motor.

2) The electromagnetic power of the motor.

3) The electromagnetic torque of the motor.

24. The ratings of a separately excited DC motor are: $U_N = 220V$; $I_N = 41.1A$; $n_N = 1500r/min$; $R_a = 0.4\Omega$; The load torque is a constant value. Calculate the following:

1) If a 1.65Ω resistor is suddenly inserted in series with the armature circuit of the motor, what is the instantaneous armature current? What are the armature current and its speed when the motor is running at steady state?

2) If the power supply voltage is suddenly reduced to 110V, what is the instantaneous armature current? What are the armature current and its speed when the motor is running at steady state?

3) If the magnetic flux is suddenly reduced by 10% of the rated flux, what is the instantaneous armature current? What are the armature current and its speed when the motor is running at steady state?

25. A 5.5kW, 115V shunt motor has a full-load speed of 1750 rpm, and takes a line current of 57.6A, $R_a = 0.5\Omega$, $R_f = 144\Omega$. If a resistance of 0.5Ω is suddenly inserted in series with the armature, determine for the instant at which this resistance is inserted: the induced emf, the armature current, the developed torque. Explain what happens to the operation of the motor, and determine the final steady value of speed.

26. A 5.5kW, 115V shunt motor has a full-load speed of 1750 rpm, and takes a line current of 57.6A, $R_a = 0.5\Omega$, $R_f = 144\Omega$. If the field flux is suddenly reduced 15%; for the first instant after the reduction in flux, determine: the induced emf, the armature current, the developed torque. Explain what happens to the operation of the motor, and determine the final steady value of speed. Compare the action with that of Problem 25.

27. A 120V shunt motor running at 900 rpm has an armature current of 150A. The armature resistance is 0.025 Ω. What will be the speed of the motor if, at constant developed torque and constant flux, the armature voltage is

1) increased by 15%?

2) decreased by 15%?

28. A 220V shunt motor running at 1000 rpm has an armature current of 25A. The armature resistance is $r_a = 0.2\Omega$. What will be the speed of the motor if, at constant developed torque, a resistance of 0.8 Ω is inserted in the armature circuit?

29. The ratings of a separately excited DC motor are: $P_N = 2.2kW$, $U_N = 220V$, $n_N = 1500r/min$, $I_N = 11.26A$, $R_a = 0.5\Omega$. Calculate the following:

1) The ideal no-load speed.

2) The electromagnetic torque at rated load.

3) The speed of this motor when $I_a = 0.5I_N$.

4) The armature current when $n_N = 1530r/min$.

30. The ratings of a separately excited DC motor are: $P_N = 10kW$, $U_N = 220V$, $n_N = 1500r/min$, $I_N = 53.4A$, $R_a = 0.4\Omega$, the brush voltage drop can be neglected. The motor is running at rated load conditions. Calculate the following:

1) The electromagnetic torque, output torque and no-load torqueat of rated state.

2) The ideal no-load speed and real no-load speed.

3) If the armature voltage is half of the rated voltage, what are the speed and armature current when the motor is runningat steady state?

4) If the magnetic flux decreases by 20%, what are the speed and armature current when the motor is running at steady state?

5) If a 1.5Ω resistor is inserted in series with the armature circuit, what are the speed and armature current when the motor is running at steady state?

31. The ratings of a separately excited DC motor are: P_N = 22kW, U_N = 220V, n_N = 1500r/min, I_N = 115A, R_a = 0.1Ω. If the speed drops to 1000r/min and the load of the motor is $T_L = T_N$, the no-load torque is neglected:

1) What is the resistance connected to armature circuit, when the voltage is 220V?

2) How much should the power supply voltage be reduced, when the armature resistor is 0.1Ω?

3) What are the input power and output power in the above situations?

32. The ratings of a separately excited DC motor are: P_N = 30kW, U_N = 220V, n_N = 1000r/min, I_N = 158.5A, R_a = 0.1Ω, $T_L = 0.8T_N$. Calculate the following:

1) What is the speed of this motor?

2) If a 0.3Ω is inserted in series with the armature circuit, what is the speed when the motor is running at steady state?

3) If the armature voltage is reduced to 188V, what is the instantaneous armature current? What is the speed when the motor is running at steady state?

4) If the magnetic flux is weakened by 20% of the rated flux, what are the speed and armature current when the motor is running at steady state?

33. A 5.5kW, 115V shunt DC motor has a full-load speed of 1750r/min, and takes a line current of 57.6 amperes. The shunt field resistance is 144Ω and that of the armature 0.15Ω. It is desired that this motor runs at 900r/min, when developing 15N · m torque. How much resistance should be inserted in series with the armature?

34. A 19kW, 120V shunt DC motor running at rated speed of 900r/min has an armature current of 175A. The armature resistance is r_a = 0.026Ω. What is the developed torque of the motor? What resistance must be inserted in the armature circuit in order that the motor will develop half of the rated torque at 700r/min? The flux can be assumed constant.

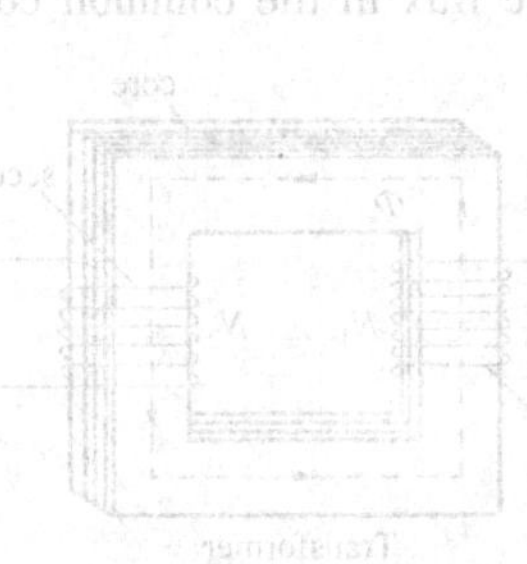

Chapter 2 Transformers

Transformers are probably one of the most useful electrical devices ever invented. A transformer can increase or reduce the voltage and current of an AC circuit. It can also isolate circuits from each other and adjust the apparent value of capacitors, inductors, or resistors. Furthermore, transformers enable us to transmit electrical energy over great distances and distribute it safely to factories and houses. All transformers are constructed based upon the laws of electromagnetic induction.

Power transformer widely used in a power system is a kind of static electrical device. It is composed of an iron core and two or more windings. For long distance transmission of electrical energy (in the power system), a step-up transformer is used to increase the voltage. Hence, the line current and the line loss will be reduced, yet the power level remains unchanged. For electricity suppliers, they make use of the step-down transformers to reduce the transmission voltage to distribution voltage in order to make it safe for domestic use.

We will study some of the basic properties of transformers in this chapter. The operating principles, the construction, the electric circuit and the magnetic fields of single-phase transformers will be introduced in the first part. The operational characteristics of single-phase transformers will also be investigated in this part. In the second part, you will be introduced to the magnetic circuits and the parallel operation of three-phase transformers. In the last part, there will be a brief introduction on autotransformers and instrument transformers.

2.1 Construction and ratings of transformers

2.1.1 Construction of transformers

As shown in Fig. 2-1, a transformer is a device that changes AC electric power at one voltage level to AC electric power at another voltage level of the same frequency based upon the laws of electromagnetic induction. It consists of two or more coils of wire wrapped around a common ferromagnetic core. These coils are usually not directly connected. The interaction between the coils is realized through the common magnetic flux in the common core.

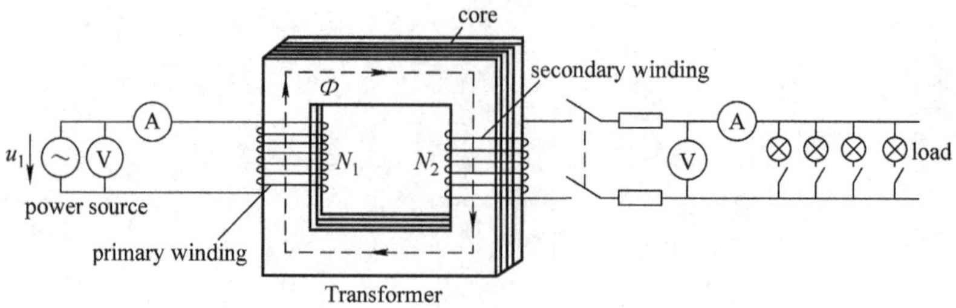

Fig. 2-1 A power transformer with core form construction

The transformer winding connected to the power source is called the primary winding (input winding with N_1 turns), and the winding connected to the load is called the secondary winding (output winding with N_2 turns). If there is a third winding on the transformer, it is called the tertiary winding.

Transformers are also used for other purposes, such as voltage sampling, current sampling and impedance transformation. However, this chapter is primarily devoted to the power transformer.

Power transformers are constructed using either the *core-form construction* or the *shell-form construction*. The former one consists of a simple rectangular laminated piece of steel with the transformer windings wrapped around two sides of the rectangle (as illustrated in Fig. 2-1 and Fig. 2-2a). The

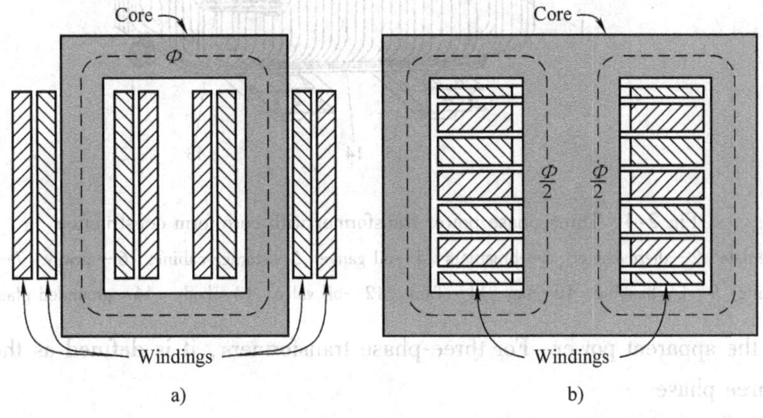

Fig. 2-2 Schematic views of core-type and shell-type transformers

latter construction consists of a three-legged laminated core with the windings wrapped around the center leg (as illustrated in Fig. 2-2b and Fig. 2-3). In order to minimize hysteresis and eddy current losses, the core is made of silicon steel laminations with high permeability. The thickness of each laminated steel ranges between 0.35mm to 0.5mm. All laminations are insulated and separated from each other. Fig. 2-4 shows a three-phase oil-immersed power transformer with core form construction.

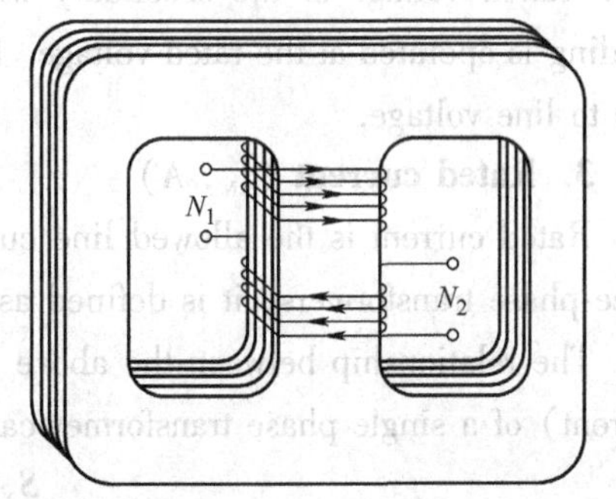

Fig. 2-3 shell form transformer construction

2.1.2 Ratings of transformers

Rated values including rated kilovolt-amperes (rated capacity), rated voltage, rated current, rated frequency and the transformer per-unit series impedance are usually marked on the nameplate. Other information such as the voltage ratings of each tap on the transformer and the wiring schematic of the transformer is also shown on the nameplate.

1. Rated capacity (S_N, kV·A or MV·A)

Rated capacity of the transformer is the guaranteed output power in the rated operation condi-

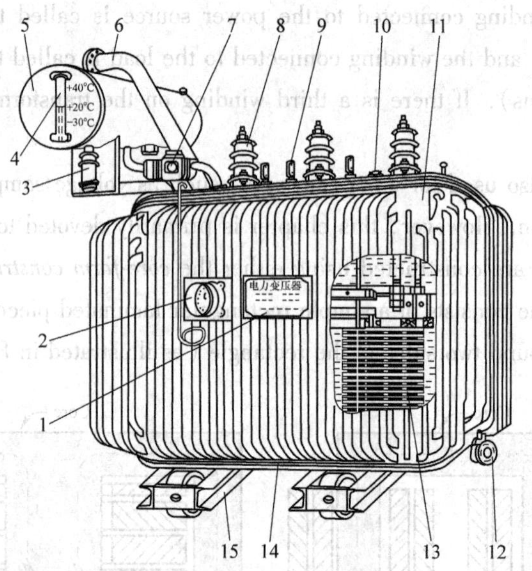

Fig. 2-4 Three-phase power transformer with core form construction

1—nameplate 2—thermometer 3—breathers 4—oil gauge 5—storage cabinet 6—airway 7—gas relay 8—HV bushing 9—LV bushing 10—tap 11—Tank 12—oil valve 13—body 14—grounded plane 15—dolly

tion, and it is the apparent power. For three-phase transformers, it is defined as the total apparent power of the three phases.

2. Rated voltage (U_N, V or kV)

Rated voltage of the primary winding is the designed voltage value under rated operation condition. Rated voltage of the secondary winding is the open-circuit voltage value when the primary winding is operated at the rated voltage. For three-phase transformers, rated voltage is defined as the line to line voltage.

3. Rated current (I_N, A)

Rated current is the allowed line current value when the transformer carries a rated load. For three-phase transformers, it is defined as the line current.

The relationship between the above rated values (i. e. rated capacity, rated voltage and rated current) of a single phase transformer can be expressed as

$$S_N = U_{1N}I_{1N} = U_{2N}I_{2N} \tag{2-1}$$

For a three-phase transformer, it becomes

$$S_N = \sqrt{3}U_{1N}I_{1N} = \sqrt{3}U_{2N}I_{2N} \tag{2-2}$$

4. Rated Frequency (f_N, Hz)

The frequency of the sinusoidal voltage is 50Hz, with reference to the industry standard in China.

Example 2-1 A 200kVA, $U_{1N}/U_{2N} = 10/6.3$kV three-phase transformer is Yd11 connected, what are the rated currents I_{1N} and I_{2N}?

Solution:

$$I_{1N} = \frac{S_N}{\sqrt{3}U_{1N}} = \frac{200}{\sqrt{3}\times 10}\text{A} = 11.5\text{A}$$

$$I_{2N}=\frac{S_N}{\sqrt{3}U_{2N}}=\frac{200}{\sqrt{3}\times 6.3}\text{A}=18.3\text{A}$$

They are both line currents, the phase currents are

$$I_{1N\phi}=I_{1N}=11.5\text{A},\ I_{2N\phi}=I_{2N}/\sqrt{3}=10.6\text{A}$$

2.2 Analysis of single phase transformers in no-load condition

2.2.1 Physical Situation of transformers in no-load condition

Fig. 2-5 shows a transformer consisting of two coils of wire wrapped around a core. The primary winding is connected to an AC power source. The secondary winding is open-circuited.

As shown in Fig. 2-5, when a sinusoidal voltage source $\dot{U}_1$ is applied to the primary winding of a transformer, an induced sinusoidal magnetizing current $\dot{I}_0$ will flow through the N_1 turns of the coil, thus produces a sinusoidal flux $\dot{\Phi}$. Since the permeability of iron core is much larger than that of the air, most of the flux remains in the core and links both windings of the transformer. The flux which

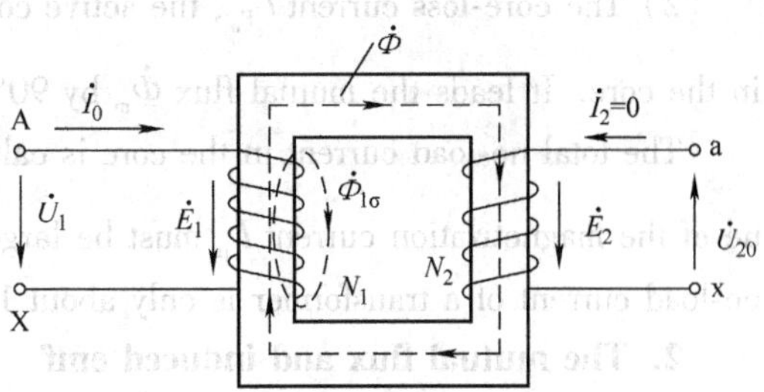

Fig. 2-5 Transformer at no-load

forms the linkage between the two windings is called the mutual flux $\dot{\Phi}$. Some of the flux leave the iron core and pass through the air or other non-ferromagnetic materials instead, the little flux that does not link the secondary winding is called the leakage flux $\dot{\Phi}_{1\sigma}$. According to Faraday's law, $\dot{E}_1$ and $\dot{E}_2$ are produced by the mutual flux $\dot{\Phi}$ in the primary and the secondary windings respectively. The leakage emf $\dot{E}_{1\sigma}$ of primary winding will be induced by the leakage flux $\dot{\Phi}_{1\sigma}$.

If the resistance of the primary winding is r_1, the voltage drop will be $\dot{I}_0 r_1$. The open-circuit voltage of the secondary winding is $\dot{U}_{20}$, which is equal to emf $\dot{E}_2$.

According to Faraday's law, the induced emfs in two windings can be expressed as:

$$e_1=-N_1\frac{\mathrm{d}\Phi}{\mathrm{d}t},\ e_2=-N_2\frac{\mathrm{d}\Phi}{\mathrm{d}t} \tag{2-3}$$

The only difference between e_1 and e_2 is brought about by the difference in the number of winding turns *i. e.* N_1 and N_2. If $N_1\neq N_2$, then $e_1\neq e_2$, which explains why transformers can change the voltage levels. The electromagnetic parameters of a transformer in no-load condition is shown in Fig. 2-6.

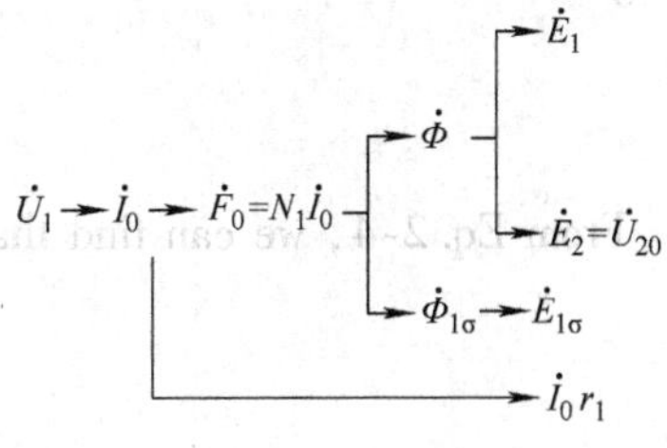

Fig. 2-6 Electromagnetic parameters of a transformer at no-load

2.2.2 The no-load current, flux and induced emf of a transformer

1. The no-load current (excitation current) $\dot{I}_0$

When an AC voltage is connected to a transformer (as shown in Fig. 2-5), current flows into the primary circuit when the secondary circuit is open circuit. This current is called no-load current which is necessary in producing magnetic flux in the transformer core. The current consists of two components:

1) The magnetization current $\dot{I}_\mu$, the reactive component contributing to the production of the total flux (mutual flux and leakage flux) in the transformer core. It is in phase with the mutual flux $\dot{\Phi}_m$, and lags $-\dot{E}_1$ by 90°.

2) The core-loss current $\dot{I}_{Fe}$, the active component leading to hysteresis and eddy current losses in the core. It leads the mutual flux $\dot{\Phi}_m$ by 90° and is in phase with $-\dot{E}_1$.

The total no-load current in the core is called the excitation current of the transformer. The value of the magnetization current $\dot{I}_\mu$ must be larger than that of the core-loss current $\dot{I}_{Fe}$. Usually, the no-load current of a transformer is only about 1% ~ 10% of the rated current.

2. The mutual flux and induced emf

If the mutual flux in the core is expressed as $\Phi = \Phi_m \sin\omega t$, ($\Phi_m$ is the magnitude of the mutual flux), according to Faraday's law, Eq. 2-3 can be expressed as:

$$\begin{cases} e_1 = -N_1\dfrac{d\Phi}{dt} = -N_1\omega\Phi_m\cos\omega t = \sqrt{2}E_1\sin(\omega t - 90°) \\ e_2 = -N_2\dfrac{d\Phi}{dt} = -N_2\omega\Phi_m\cos\omega t = \sqrt{2}E_2\sin(\omega t - 90°) \end{cases} \tag{2-4}$$

where E_1 is the rms value of the induced emf

$$\begin{cases} E_1 = \dfrac{\omega N_1\Phi_m}{\sqrt{2}} = \dfrac{2\pi f N_1\Phi_m}{\sqrt{2}} = 4.44fN_1\Phi_m \\ E_2 = \dfrac{\omega N_2\Phi_m}{\sqrt{2}} = \dfrac{2\pi f N_2\Phi_m}{\sqrt{2}} = 4.44fN_2\Phi_m \end{cases} \tag{2-5}$$

Eq. 2-5 shows that the values of E_1 and E_2 are directly proportional to the source frequency f, amplitude of the flux Φ_m and the numbers of turns of the windings (N_1 and N_2) respectively. The voltage ratio and turns ratio k of a transformer is defined as

$$k = \frac{E_1}{E_2} = \frac{N_1}{N_2} \tag{2-6}$$

From Eq. 2-4, we can find that $\dot{E}_1$ and $\dot{E}_2$ lag the mutual flux $\dot{\Phi}_m$ by 90°,

$$\dot{E}_1 = -j4.44fN_1\dot{\Phi}_m \tag{2-7}$$

$$\dot{E}_2 = -j4.44fN_2\dot{\Phi}_m \tag{2-8}$$

3. Leakage flux and related emf

According to Faraday's law, the leakage flux related emf $\dot{E}_{1\sigma}$ induced by the leakage flux $\dot{\Phi}_{1\sigma}$

has the same expression as that shown in Eq. 2-7:

$$\dot{E}_{1\sigma} = -\text{j}4.44fN_1\dot{\Phi}_{1\sigma} \tag{2-9}$$

Because the leakage flux $\dot{\Phi}_{1\sigma}$ is produced by the no-load current $\dot{I}_0$ and it passes through the air and other non-ferromagnetic materials, its magnetic flux has linear property, i. e. the value of $\dot{\Phi}_{1\sigma}$ is directly proportional to the excitation current $\dot{I}_0$ and in phase with $\dot{I}_0$. Therefore the leakage flux can be modeled by a reactance; Eq. 2-9 can be rewritten as:

$$\dot{E}_{1\sigma} = -\text{j}\dot{I}_0 x_1 \tag{2-10}$$

where x_1 is the equivalent leakage reactance of the primary winding.

According to Kirchoff's voltage law, the voltage equation of the primary winding is

$$\dot{U}_1 = -\dot{E}_1 - \dot{E}_{1\sigma} + \dot{I}_0 r_1 = -\dot{E}_1 + \text{j}\dot{I}_0 x_1 + \dot{I}_0 r_1 = -\dot{E}_1 + \dot{I}_0 z_1 \tag{2-11}$$

where z_1 is the leakage impedance of the primary winding, and $z_1 = r_1 + \text{j}x_1$. When the transformer is operated at no-load, $\dot{I}_0 z_1$ is very small and can be neglected, so Eq. 2-11 becomes

$$\dot{U}_1 \approx -\dot{E}_1 = \text{j}4.44fN_1\dot{\Phi}_\text{m} \tag{2-12}$$

In Eq. 2-12, we can find that when the leakage flux $\dot{\Phi}_{1\sigma}$ and the resistance r_1 of the primary winding is ignored, the frequency f of the power source and the turns N_1 of the winding remains unchanged, thus determining the magnitude of the mutual flux Φ_m is the only factor of the applied voltage value U_1. In other words, since the voltage of the power system is usually constant, the mutual flux Φ_m in the transformer will also be constant.

The voltage equation of the secondary winding is

$$\dot{U}_{20} = \dot{E}_2 \tag{2-13}$$

2.2.3 The equivalent circuit and phasor diagram of transformers in no-load condition

How can the core excitation effects be modeled? The no-load current has two components and one of them is the magnetization current $\dot{I}_\mu$. Since it lags $-\dot{E}_1$ by 90°, it can be modeled as a reactance x_m. The other one is the core-loss current $\dot{I}_\text{Fe}$. Since it is in phase with $-\dot{E}_1$, it can be modeled by a resistance r_m. The resistance r_m represents the iron losses and the resulting heat they produce. The magnetizing reactance x_m is a measure of the permeability of the transformer core. Thus, if the permeability is low, x_m is relatively low. The current $\dot{I}_\mu$ flowing through x_m represents the magnetizing current needed to create the mutual flux Φ in the core. Notice that both these currents are nonlinear, so the reactance x_m and the resistance r_m are only approximations of the real excitation effects. Eq. 2-7 can be expressed as:

$$-\dot{E}_1 = \dot{I}_0 z_\text{m} = \dot{I}_0(r_\text{m} + \text{j}x_\text{m}) \tag{2-14}$$

where z_m is the exciting impedance, $z_\text{m} = r_\text{m} + \text{j}x_\text{m}$.

The exciting reactance x_m models the mutual flux of magnetic circuit. The core loss is expressed by $I_0^2 r_m$ where r_m is the exciting resistance. Since the magnetic circuit of mutual flux is nonlinear, the exciting impedance changes with the saturation of the iron core. Usually the applied voltage of the transformer is at the rated value, so z_m can be treated as constant because of the constant mutual flux. The reluctance of mutual magnetic circuit is much less than that of leakage, so $x_m \gg x_1$.

The equivalent circuit of a no-load transformer is shown in Fig. 2-7a.

The phasor diagram of the no-load transformer is given in Fig. 2-7b with reference to Eq. 2-11 and Eq. 2-14.

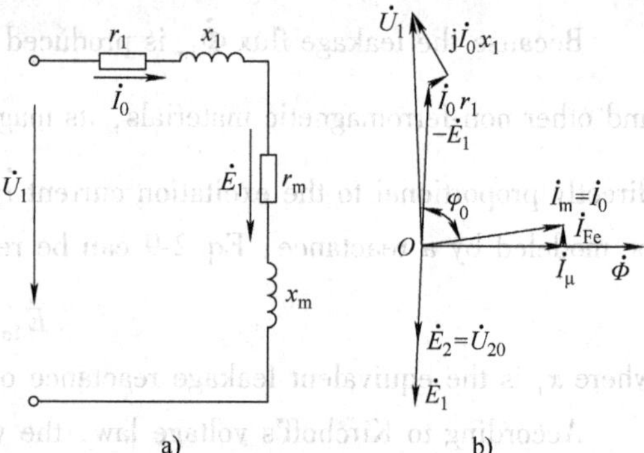

Fig. 2-7 Equivalent circuit and Phasor diagram of a transformer at no-load

In the phasor diagram, the mutual flux $\dot{\Phi}$ is fixed at an angle of 0°, induced emf $\dot{E}_1$ and $\dot{E}_2$ lag it by 90°, and $\dot{U}_{20} = \dot{E}_2$.

The no-load current $\dot{I}_0$ has two components $\dot{I}_0 = \dot{I}_m = \dot{I}_\mu + \dot{I}_{Fe}$, $\dot{I}_\mu$ is in phase with $\dot{\Phi}$ and it is much bigger than I_{Fe}. $\dot{I}_{Fe}$ leads $\dot{\Phi}$ by 90°. By the equivalent circuit and Eq. 2-11, you should be able to determine the applied voltage $\dot{U}_1$.

2.3 Operation of transformers in loaded condition

2.3.1 Physical situation of transformers in loaded condition

1. Physical relationships

Fig. 2-8 shows the situation of a transformer under load. When a load is connected to the secondary winding, a secondary current $\dot{I}_2$ will be produced. The mmf of the secondary winding is $\dot{F}_2 = N_2\dot{I}_2$. $\dot{F}_2$ tends to change the mutual flux in the core, thus changes the current of the primary winding from $\dot{I}_0$ to $\dot{I}_1$ and make the mmf of the primary winding become $\dot{F}_1 = N_1\dot{I}_1$. The mutual flux $\dot{\Phi}$ linking both windings will be produced by the primary mmf ($\dot{F}_1$) and the secondary mmf ($\dot{F}_2$). Then $\dot{E}_1$ and $\dot{E}_2$ will be induced in primary winding and secondary winding respectively. Leakage flux ($\dot{\Phi}_{1\sigma}$ and $\dot{\Phi}_{2\sigma}$) will be produced by $\dot{F}_1$ and $\dot{F}_2$ and the leakage emf $\dot{E}_{1\sigma}$ and $\dot{E}_{2\sigma}$ will be induced in each winding respectively. The resistance of each wind-

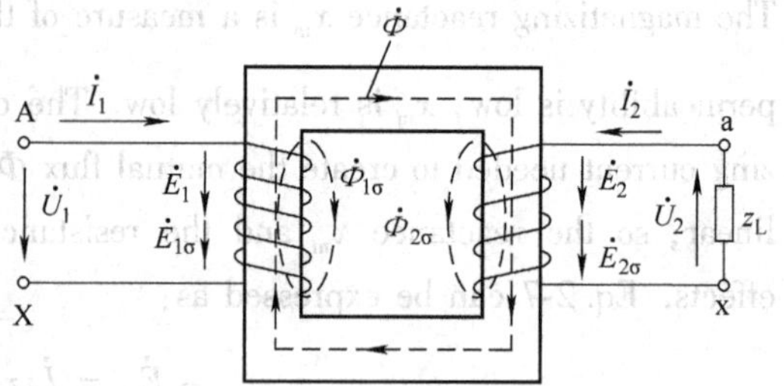

Fig. 2-8 A single-phase transformer under load

ing is r_1 and r_2. The physical relationships of a transformer under load are shown in Fig. 2-9.

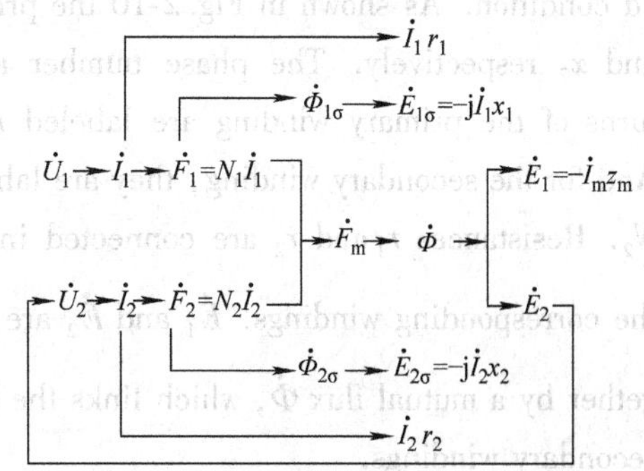

Fig. 2-9 Physical relationships of a transformer under load

2. Equation of mmf

According to Eq. 2-12, if the applied voltage U_1 is not changed, the mutual flux Φ in the transformer will be constant. In other words, since the voltage of the power system is usually constant, the mutual flux Φ in the transformer (at no-load or under load) and the mmf will be the same ($N_1 I_m$). From Fig. 2-9, the mmf equation of a transformer under load can be expressed as

$$N_1\dot{I}_m = N_1\dot{I}_1 + N_2\dot{I}_2 \tag{2-15}$$

Then

$$\dot{I}_1 = \dot{I}_m + \left(-\frac{1}{k}\dot{I}_2\right) = \dot{I}_m + \dot{I}_{1L} \tag{2-16}$$

where $\dot{I}_{1L} = -\frac{1}{k}\dot{I}_2$ is the load component of the primary current.

Eq. 2-16 shows that the primary current $\dot{I}_1$ of a transformer under load has two components, one is the magnetizing current $\dot{I}_m$ which produces the mutual flux $\dot{\Phi}$ in the core, and the other one is the load component $\dot{I}_{1L}$. It counteracts the effect of the secondary current $\dot{I}_2$.

When a transformer is loaded, the magnetizing component $\dot{I}_m$ of the primary current is smaller than $\dot{I}_1$, thus can be ignored. Eq. 2-16 becomes

$$\dot{I}_1 \approx -\frac{1}{k}\dot{I}_2 \tag{2-17}$$

It shows that the *current ratio* between primary and secondary current is approximately the inverse of the *voltage ratio* and *turns ratio* of *k*. Transformers can change not only voltage but also the current at the same time.

3. Voltage Equations

According to Kirchoff's voltage law, the voltage equations of a transformer in the loaded condition can be expressed as:

$$\dot{U}_1 = -\dot{E}_1 - \dot{E}_{1\sigma} + \dot{I}_1 r_1 = -\dot{E}_1 + j\dot{I}_1 x_1 + \dot{I}_1 r_1 = -\dot{E}_1 + \dot{I}_1 z_1 \tag{2-18}$$

$$\dot{U}_2 = \dot{E}_2 + \dot{E}_{2\sigma} - \dot{I}_2 r_2 = \dot{E}_2 - j\dot{I}_2 x_2 - \dot{I}_2 r_2 = \dot{E}_2 - \dot{I}_2 z_2 = \dot{I}_2 Z_L \tag{2-19}$$

where z_2 is the secondary leakage impedance, and $z_2 = r_2 + jx_2$; Z_L is the load impedance, and $Z_L = R_L + jX_L$.

2.3.2 The equivalent circuit and phasor diagram of a transformer in loaded condition

According to Eq. 2-18 and Eq. 2-19, we can draw a circuit diagram of a transformer in the loaded condition. As shown in Fig. 2-10 the primary and secondary leakage reactances are labeled as x_1 and x_2 respectively. The phase number and winding turns of the primary winding are labeled m_1 and N_1. And for the secondary winding, they are labeled m_2 and N_2. Resistances r_1 and r_2 are connected in series with the corresponding windings. $\dot{E}_1$ and $\dot{E}_2$ are coupled together by a mutual flux $\dot{\Phi}$, which links the primary and secondary windings.

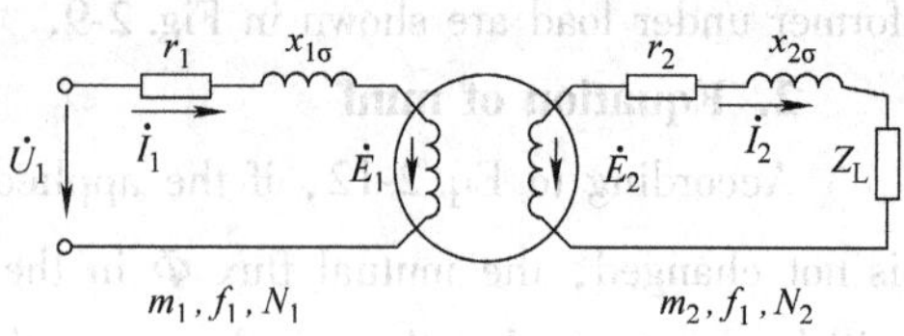

Fig. 2-10 Circuit relationships of a transformer under load

Although Fig. 2-10 is an accurate model of a transformer, it is not a very useful one for transformer analysis. To analyze practical circuits connected to transformers, it is necessary to convert the entire circuit to an equivalent circuit referring to a single voltage level.

1. Referred winding

According to Eq. 2-7 and Eq. 2-8, we can find that if $N_1 = N_2$, then $\dot{E}_1 = \dot{E}_2$. The core excitation effects is modeled in Eq. 2-14, i. e. $\dot{E}_1 = \dot{E}_2 = -\dot{I}_0 z_m = -\dot{I}_0(r_m + jx_m)$.

However, the winding turns of the two windings in a practical transformer are not equal, i. e. $N_1 \neq N_2$, so $\dot{E}_1 \neq \dot{E}_2$, and we can find that $\dot{E}_1 = k\dot{E}_2$.

In order to obtain the complete equivalent circuit of a practical transformer under load, the actual secondary winding with N_2 turns can be replaced by an equivalent winding with the same number of turns as that of the primary winding N_1. Noted that the operation of the transformer should not be affected by the winding replacement, that is, the equivalent secondary winding with N_1 turns must be able to produce the same mmf (i. e. $N_2 I_2$), copper loss, and leakage flux, and the output power must be identical to that of the original winding. The replacement process is called referring.

For the referred secondary winding with N_1 turns, the equivalent induced emf is

$$\dot{E}_2' = -j4.44fN_1\dot{\Phi}_m = \dot{E}_1 = k\dot{E}_2 \tag{2-20}$$

And the equivalent mmf of the referred secondary winding is

$$N_1\dot{I}' = N_2\dot{I}_2 \tag{2-21}$$

So the equivalent current of the referred secondary winding is

$$\dot{I}_2' = \frac{N_2}{N_1}\dot{I}_2 = \frac{1}{k}\dot{I}_2 \tag{2-22}$$

Because the copper loss and output power are not changed, i. e.

$$\dot{I}_2'^2 r_2'^2 = I_2^2 r_2,\ I_2'^2 x_2'^2 = I_2^2 x_2^2,\ I_2'^2 Z_L'^2 = I_2^2 Z_L^2 \tag{2-23}$$

we can find the equivalent impedances of the referred secondary winding using the following equations.

$$r_2' = \left(\frac{I_2}{I_2'}\right)^2 r_2 = k^2 r_2,\ x_2' = \left(\frac{I_2}{I_2'}\right)^2 x_2 = k^2 x_2,\ Z_L' = \left(\frac{I_2}{I_2'}\right)^2 Z_L = k^2 Z_L \tag{2-24}$$

Since

$$z_2' = k^2 z_2 \tag{2-25}$$

the equivalent voltage of the referred secondary winding is

$$\dot{U}_2' = \dot{I}_2'Z_L' = \frac{1}{k}\dot{I}_2(k^2 Z_L) = k\dot{U}_2 \tag{2-26}$$

In summary, when the secondary winding is referred to the primary, the voltage should be multiplied by *voltage ratio k*, all currents divided by *current ratio k*, and all impedances multiplied by k^2.

2. Equivalent circuit and phasor-diagram

After winding referring, voltage equations of practical transformers in the loaded condition become

$$\begin{cases} \dot{U}_1 = -\dot{E}_1 + \dot{I}_1 z_1 \\ \dot{U}_2' = \dot{E}_2' - \dot{I}_2' z_2' = \dot{I}_2' Z_L' \\ \dot{E}_1 = \dot{E}_2' = -\dot{I}_m z_m \\ \dot{I}_1 = \dot{I}_m + (-\dot{I}_2') \end{cases} \tag{2-27}$$

According to Eq. 2-27, the equivalent circuit of transformers referred to the primary side is shown in Fig. 2-11a. It is called the *T-type equivalent circuit* of transformers. The equivalent circuit can be referred to either its primary side or secondary side whenever it is needed in solving problems.

T-type equivalent circuit is relatively complex in practical engineering applications. It can be observed that the excitation branch of the T-type equivalent circuit adds another node to the circuit being analyzed, making the circuit analysis more complex than necessary. Actually, the excitation branch has a relatively small current I_m comparing to the load current I_1 of the transformer, i. e. $I_1 \gg I_m$. Since the voltage of primary leakage impedance is very small, the excitation branch can be moved to the front end of the transformer equivalent circuit so that the primary and secondary impedances (r_1, x_1 and r_2, x_2) are aligned in series as shown in Fig. 2-11b. This simplified equivalent circuit becomes cantilever circuit (also called *Γ-type equivalent circuit*).

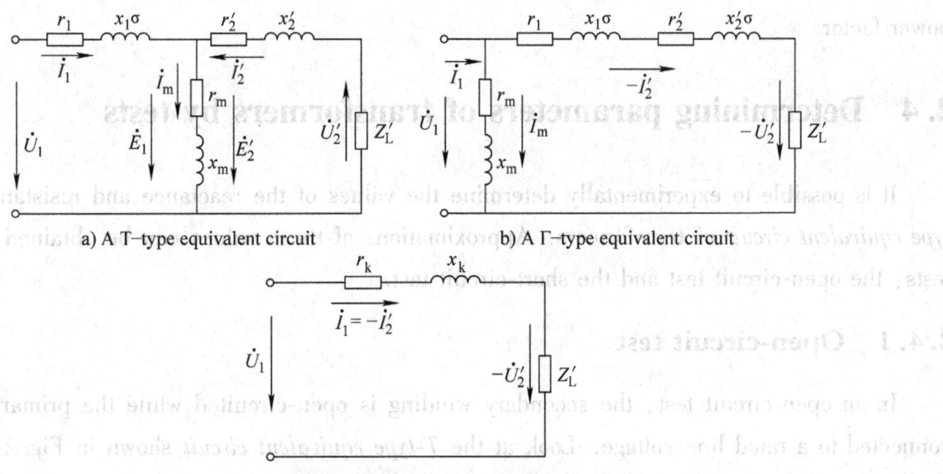

Fig. 2-11 Equivalent circuits of single-phase transformer

In some applications, the excitation branch can be totally neglected without causing serious errors. Then, we can remove the excitation branch entirely from the circuit, and get an *approximated equivalent circuit* (also called *simplified equivalent circuit*) as shown in Fig. 2-11c.

In Fig. 2-11c, r_k is the short-circuit resistance, and $r_k = r_1 + r_2'$. x_k is the short-circuit reactance, and $x_k = x_1 + x_2'$. z_k is the short-circuit impedance, and $z_k = r_k + jx_k$.

With reference to the *T-type equivalent circuit* of transformers in Fig. 2-11a, the phasor diagram of a transformer operating at a lagging power factor can be drawn as shown in Fig. 2-12.

In Fig. 2-12, the mutual flux $\dot{\Phi}$ is assumed to be at an angle 0°, and all other voltages and currents are discussed as below:

1) The emf $\dot{E}_1 = \dot{E}_2'$, and they are lagging $\dot{\Phi}$ by 90°.

2) For secondary side, because of the lagging power factor, the secondary load current $\dot{I}'_2$ lags the voltage $\dot{U}'_2$ by φ_2 angles. According to the voltage equation of equivalent secondary side i. e. $\dot{E}'_2 = \dot{U}'_2 + \dot{I}'_2(r'_2 + jx_2')$, the phasor diagram should be drawn as shown in Fig. 2-12.

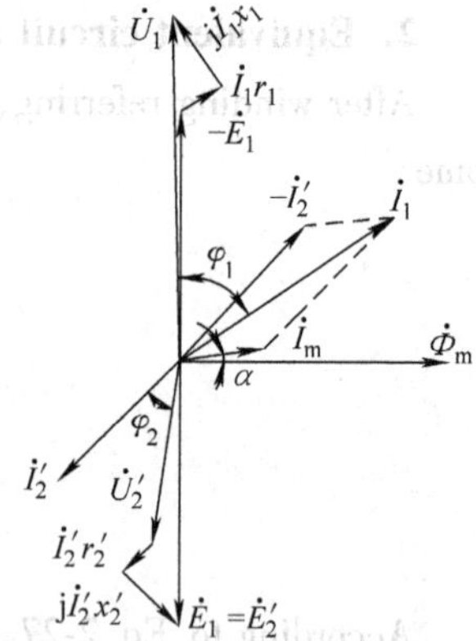

Fig. 2-12 Phasor diagram of a transformer under load

3) For primary side, according to the current equation $[\dot{I}_1 = \dot{I}_m + (-\dot{I}_2')]$ and the characteristic of magnetizing current $\dot{I}_m$, $\dot{I}_m$ leads mutual flux $\dot{\Phi}$ by a small angle α, $\alpha = \arctan\frac{r_m}{x_m}$ (as shown in Fig. 2-12), then the current $\dot{I}_1$ can be drawn out. Also, according to the primary voltage equation $(\dot{U}_1 = -\dot{E}_1 + \dot{I}_1z_1)$, the phasor diagram of the primary side can be drawn in Fig. 2-12.

In Fig. 2-12, the angle φ_1 between $\dot{U}_1$ and $\dot{I}_1$ is the power factor angle of the primary side. The angle φ_2 between $\dot{U}'_2$ and $\dot{I}'_2$ is the power factor angle of the secondary side determined by the load power factor.

2.4 Determining parameters of transformers by tests

It is possible to experimentally determine the values of the reactance and resistance in the *T-type equivalent circuit* of transformers. Approximations of these values can be obtained by only two tests: the open-circuit test and the short-circuit test.

2.4.1 Open-circuit test

In an open-circuit test, the secondary winding is open-circuited while the primary winding is connected to a rated line voltage. Look at the *T-type equivalent circuit* shown in Fig. 2-11a, in the conditions as described, all the input current will flow through the excitation branch of the transformer, since the elements r_1 and x_1 in series are very small in comparison with r_m and x_m. So eventually all the input voltage will drop across the excitation branch.

An open-circuit test connection is shown in Fig. 2-13, Fig 2-13a is the test connection for a single phase transformer and Fig. 2-13b is for a three-phase transformer.

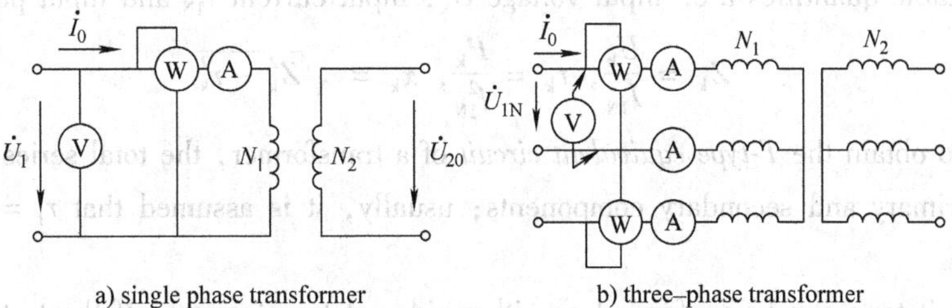

Fig. 2-13 A transformer open-circuit test connection

In open-circuit test, the rated voltage is applied on the primany side, and the secondary side is open-circuited. It is possible to determine the excitation impedance r_m and x_m, and the *turns ratio k* using 4 measured quantities i. e. input voltage U_{1N}, input current I_0 (excitation current), input power p_0 (core losses or no-load losses) and open-circuit voltage U_{20} of the secondary.

$$k = \frac{U_{1N}}{U_{20}} \tag{2-28}$$

$$Z_m = \frac{U_{1N}}{I_0}, \quad r_m = \frac{p_0}{I_0^2}, \quad x_m = \sqrt{Z_m^2 - r_m^2} \tag{2-29}$$

Usually, the open-circuit test is set on the low-voltage side and the high-voltage side is open-circuited, then the calculated excitation impedance r_m and x_m will be the values on the low-voltage side and they must be refferred to the high-voltage side, i. e. multiplied by k^2, k = hight voltage turns/low voltage turns.

2.4.2 Short-circuit test

In a short-circuit test, the secondary terminals of transformers are short-circuited while the primary terminals are connected to a fairly low-voltage source. The setup is shown in Fig. 2-14. Fig 2-14a is the test connection for a single phase transformer and Fig. 2-14b is for a three-phase transformer. Adjusting the input voltage until the current in the short-circuited windings is equal to its rated value.

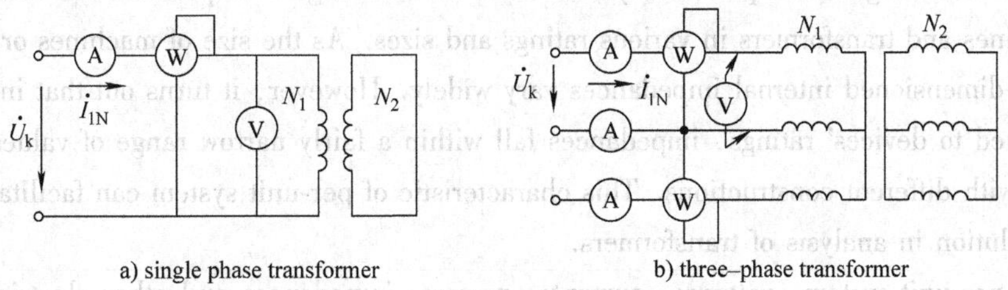

Fig. 2-14 A transformer short circuit test connection

Because the excitation impedance is very large compared to the leakage impedance, the induced current flowing through the excitation branch can be neglected during the test. Therefore, all voltage drop in the transformer can be attributed to the elements in series (leakage impedance) in the circuit (as shown in Fig. 2-11c).

It is possible to determine the total series impedance (or short-circuited impedance) r_k and x_k using 3 measurable quantities i. e. input voltage U_k, input current I_{1N} and input power P_k.

$$Z_k = \frac{U_k}{I_{1N}}, \ r_k = \frac{P_k}{I_{1N}^2}, \ x_k = \sqrt{Z_k^2 - r_k^2} \tag{2-30}$$

In order to obtain the *T-type equivalent circuit* of a transformer, the total series impedance must be split into primary and secondary components; usually, it is assumed that $r_1 = r_2' = r_k/2$, $x_1 = x_2' = x_k/2$.

Short-circuit tests can be performed on either sides of transformers. If the test is performed on the secondary side, the result will naturally generate the equivalent circuit impedances referred to the secondary side of the transformer, and vice versa. .

The resistance of windings is sensitive to the ambient temperature, and if the tested values taken are not at 75℃, the tested parameter values have to be calculated to the values at 75℃. This is because the parameter values at 75℃ are set as the standard values. If the winding is made of copper, the standard value of short-circuit resistance $r_{k75℃}$ is

$$r_{k75℃} = \frac{234.5 + 75}{234.5 + t} r_k \tag{2-31}$$

where t is the ambient temperature in testing.

If the winding is made of aluminum, the standard value of short-circuit resistance is $r_{k75℃} = \frac{228 + 75}{228 + t} r_k$.

And the standard short-circuit impedance is $Z_{k75℃} = \sqrt{r_{k75℃}^2 + x_k^2}$.

2.4.3 The per-unit system

Per-unit notation is often used when we are dealing with transformers and electrical machines. The reason is that per-unit values give us an approximate idea about the relative magnitudes of impedances, voltages, currents and powers to the rated ones. Thus, instead of dealing with ohms, amperes, volts and kilowatts, we simply work with dimensionless per-unit numbers. Consequently, we don't have to carry along units when per-unit values are used.

Another advantage of the per-unit system is the narrow range of impedance for all types of electrical machines and transformers in various ratings and sizes. As the size of machines or transformers varies, the dimensioned internal impedances vary widely. However, it turns out that in the per-unit system related to devices' ratings, impedances fall within a fairly narrow range of values for all type of devices with different constructions. This characteristic of per-unit system can facilitate the verification of solution in analysis of transformers.

In the per-unit system, voltages, currents, powers, impedances and other electrical quantities are not measured in their usual SI units (volts, amperes, watts, ohms, etc.). Instead, each electrical quantity is measured as a decimal fraction normalized by its base value. All quantities can be expressed on a per-unit basis by the following equation

$$Quantity\ per\ unit = \frac{Actual\ value}{base\ value} \tag{2-32}$$

where *actual value* is a value in volts, amperes, ohms, etc.

It is customary to select two base quantities to define a given per-unit system. The ones usually selected are voltage U and power P (or apparent power S). Once these base quantities are selected, all other base values are related to them by the usual electrical laws. For transformers, it is customary to select the rated voltage U_N and rated power P_N (or rated apparent power S_N) as the base quantities. All other base values can be computed easily by the usual electrical laws. We can find that:

1) Per-unit values of rated current, rated voltage and rated apparent power are equal to 1.

2) Per-unit values of the primary side equal that of the secondary side within the same electrical quantity. Such as

$$U_2'^* = \frac{U_2'}{U_{1N}} = \frac{kU_2}{kU_{2N}} = \frac{U_2}{U_{2N}} = U_2^*$$

3) For currents and voltages, the phase quantity and the line to line quantity have the same per-unit value. For powers, per-unit power of single-phase and three-phase is the same.

4) Some related quantities have the same per-unit value. Thus, the computation can be simplified. Such as

$$U_{kN}^* = \frac{U_{kN}}{U_{1N}} = \frac{I_{1N}Z_k}{U_{1N}} = \frac{Z_k}{U_{1N}/I_{1N}} = \frac{Z_k}{Z_{1N}} = Z_k^*$$

5) Per-unit values are related to devices' ratings, thus the impedance of machines and transformers falls within a fairly narrow range for all types of devices with different constructions. For example, the per-unit value of short-circuit impedance of a transformer ranges from 0.04 to 0.175, and the per-unit value of no-load current ranges from 0.02 to 0.1.

Example 2-2 A three-phase transformer has a rating of $S_N = 320\text{kV}\cdot\text{A}$ and voltage ratio of $U_{1N}/U_{2N} = 6300\text{V}/400\text{V}$. Its windings connection is Y-Δ. The line voltage, line current and the power used to conduct open and short-circuit tests are given. The side which the test is carried out is also provided. The following table shows the open circuit and short circuit test results.

Test	line voltage /V	line current/A	power/W	side of test
open-circuit	400	27.7	1450	low-voltage side
short-circuit	284	29.3	5700	high-voltage side

Calculate the following:

1) The per-unit values of excitation impedance and short-circuit impedances of the transformer.

2) The actual values of excitation impedance and short-circuit impedances referred to the high-voltage side.

Solution:

Method Ⅰ, Computation by actual values

(1) Find excitation impedances from an open-circuit test

Since the test is performed on the low-voltage side, rated values should be converted to phase values according to the delta connection winding.

$$U_{2N\phi} = U_{2N} = 400\text{V}$$

$$I_{0\phi} = \frac{I_0}{\sqrt{3}} = \frac{27.7}{\sqrt{3}}\text{A} = 15.99\text{A}$$

$$p_{0\phi} = \frac{p_0}{3} = \frac{1450}{3}\text{W} = 483.33\text{W}$$

The excitation impedances referred to the low-voltage side are

$$Z_m = \frac{U_{2N\phi}}{I_{0\phi}} = \frac{400}{15.99}\Omega = 25.02\Omega$$

$$r_m = \frac{p_{0\phi}}{I_{0\phi}^2} = \frac{483.33}{15.99^2}\Omega = 1.89\Omega$$

$$x_m = \sqrt{Z_m^2 - r_m^2} = \sqrt{25.02^2 - 1.89^2}\Omega = 24.95\Omega$$

(2) Calculate the turns ratio k and excitation impedances referred to the high-voltage side.

$$k = \frac{U_{1N\phi}}{U_{2N\phi}} = \frac{U_{1N}}{\sqrt{3}U_{2N}} = \frac{6300}{\sqrt{3} \times 400} = 9.09$$

$$Z'_m = k^2 Z_m = 9.09^2 \times 25.02\Omega = 2067.35\Omega$$

$$r'_m = k^2 r_m = 9.09^2 \times 1.89\Omega = 156.17\Omega$$

$$x'_m = k^2 x_m = 9.09^2 \times 24.95\Omega = 2061.57\Omega$$

(3) Find the short-circuit impedances from a short-circuit test

As the test is performed on the high-voltage side; the rated values should be converted to phase values according to the Y connection winding.

$$U_{k\phi} = \frac{U_k}{\sqrt{3}} = \frac{284}{\sqrt{3}}\text{V} = 163.97\text{V}$$

$$I_{k\phi} = I_k = 29.3\text{A}$$

$$p_{k\phi} = \frac{p_k}{3} = \frac{5700}{3}\text{W} = 1900\text{W}$$

Calculate the short-circuit impedances referred to the high-voltage side

$$Z_k = \frac{U_{k\phi}}{I_{k\phi}} = \frac{163.97}{29.3}\Omega = 5.60\Omega$$

$$r_k = \frac{p_{k\phi}}{I_{k\phi}^2} = \frac{1900}{29.3^2}\Omega = 2.21\Omega$$

$$x_k = \sqrt{Z_k^2 - r_k^2} = \sqrt{5.6^2 - 2.21^2}\Omega = 5.145\Omega$$

(4) Find the base value and per-unit value of impedances referred to the high-voltage side

$$I_{1N} = \frac{S_N}{\sqrt{3}U_{1N}} = \frac{320 \times 10^3}{\sqrt{3} \times 6300}\text{A} = 29.33\text{A}$$

$$Z_{1N} = \frac{U_{1N\phi}}{I_{1N\phi}} = \frac{U_{1N}}{\sqrt{3}I_{1N}} = \frac{6300}{\sqrt{3} \times 29.33}\Omega = 124.02\Omega$$

$$Z_m^* = \frac{Z'_m}{Z_{1N}} = \frac{2067.35}{124.02} = 16.67$$

$$r_m^* = \frac{r'_m}{Z_{1N}} = \frac{156.17}{124.02} = 1.26$$

$$x_m^* = \frac{x'_m}{Z_{1N}} = \frac{2061.57}{124.02} = 16.62$$

$$Z_k^* = \frac{Z_k}{Z_{1N}} = \frac{5.6}{124.02} = 0.0451$$

$$r_k^* = \frac{r_k}{Z_{1N}} = \frac{2.21}{124.02} = 0.0178$$

$$x_k^* = \frac{x_k}{Z_{1N}} = \frac{5.145}{124.02} = 0.0414$$

Method Ⅱ, Computation in per-unit system

(1) Find excitation impedances from an open-circuit test

The test is performed on the low-voltage side; ratings in this side are selected as the base values. The rated current is

$$I_{2N} = \frac{S_N}{\sqrt{3}U_{2N}} = \frac{320 \times 10^3}{\sqrt{3} \times 400}\text{A} = 461.89\text{A}$$

Voltage, current and power are converted to per-unit values

$$U_{2N}^* = 1$$

$$I_0^* = \frac{I_0}{I_{2N}} = \frac{27.7}{461.89} = 0.06$$

$$p_0^* = \frac{p_0}{S_N} = \frac{1450}{320 \times 10^3} = 0.004\,53$$

Per-unit values of excitation impedances are

$$Z_m^* = \frac{U_{2N}^*}{I_0^*} = \frac{1}{0.06} = 16.67$$

$$r_m^* = \frac{p_0^*}{I_0^{*2}} = \frac{0.004\,53}{0.06^2} = 1.26$$

$$x_m^* = \sqrt{Z_m^{*2} - r_m^{*2}} = \sqrt{16.67^2 - 1.26^2} = 16.62$$

(2) Find short-circuit impedances from a short-circuit test

The test isperformed on the high-voltage side, so ratings in this side are selected as the base values. The rated current is

$$I_{1N} = \frac{S_N}{\sqrt{3}U_{1N}} = \frac{320 \times 10^3}{\sqrt{3} \times 6300}\text{A} = 29.33\text{A}$$

Voltage, current and power are converted to per-unit values

$$U_k^* = \frac{U_k}{U_{1N}} = \frac{284}{6300} = 0.0451$$

$$I_k^* = \frac{I_k}{I_{1N}} = \frac{29.3}{29.33} = 0.999$$

$$p_k^* = \frac{p_k}{S_N} = \frac{5700}{320 \times 10^3} = 0.0178$$

Per-unit values of short-circuit impedances are

$$Z_k^* = \frac{U_k^*}{I_k^*} = 0.0451$$

$$r_k^* = \frac{p_k^*}{I_k^{*2}} = 0.0178$$

$$x_k^* = \sqrt{Z_k^{*2} - r_k^{*2}} = \sqrt{0.0451^2 - 0.0178^2} = 0.414$$

(3) Actual values of impedances on the high-voltage side can be calculated from their base values

$$Z_{1N} = \frac{U_{1N\phi}}{I_{1N\phi}} = \frac{U_{1N}}{\sqrt{3}I_{1N}} = \frac{6300}{\sqrt{3} \times 29.33}\Omega = 124.02\Omega$$

$$Z'_m = Z_m^* Z_{1N} = 16.67 \times 124.02\Omega = 2067.41\Omega$$

$$r'_m = r_m^* Z_{1N} = 1.26 \times 124.02\Omega = 156.27\Omega$$

$$x'_m = x_m^* Z_{1N} = 16.62 \times 124.02\Omega = 2061.21\Omega$$

$$Z_k = Z_k^* Z_{1N} = 0.0451 \times 124.02\Omega = 5.59\Omega$$

$$r_k = r_k^* Z_{1N} = 0.0178 \times 124.02\Omega = 2.21\Omega$$

$$x_k = x_k^* Z_{1N} = 0.414 \times 124.02\Omega = 5.13\Omega$$

2.5 Voltage regulation and efficiency of transformers

Because a transformer has impedances in series in its equivalent circuit, the output voltage varies with the load even if the input voltage remains constant. Voltage regulation is defined for the comparison between input and output voltage. Also, there are losses between input and output power. Hence, efficiency is defined to compare the difference between them.

2.5.1 Terminal characteristic and voltage regulation

When a transformer's secondary winding is connected to a load with a specified power factor and its primary winding is connected to the rated voltage, the relationship between the secondary voltage and the load current is called the transformer's terminal characteristic. The terminal characteristics curves of a transformer with different loads and power factor are shown in Fig. 2-15.

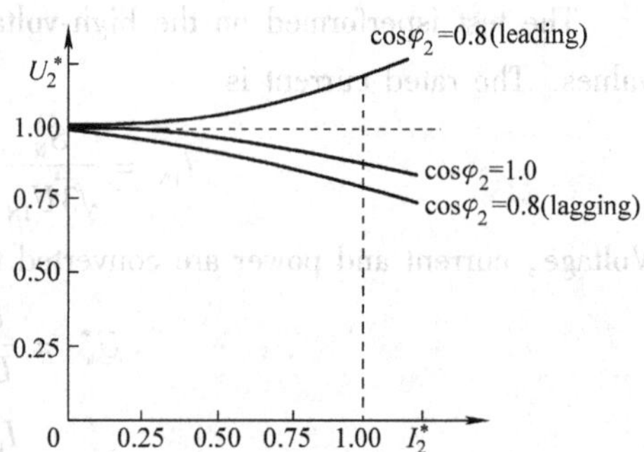

Fig. 2-15 The terminal characteristics curve of a transformer with different load

Voltage regulation is a quantity that compares the output voltage at no-load with the output voltage at a specified load, an important attribute of transformers. With the primary impressed voltage holding constant at its rated value, the voltage regulation ΔU is defined by the equations as follows:

$$\Delta U = \frac{U_{20} - U_2}{U_{20}} = \frac{U_{2N} - U_2}{U_{2N}} = 1 - U_2^* \tag{2-33}$$

where U_{2N} (or U_{20}) is the rated voltage of secondary winding; U_2 is the actual voltage of secondary winding at a specified load.

To determine the voltage regulation of a transformer, it is necessary to understand the voltage drops by the impedance in series within its equivalent circuit. Considering the simplified transformer equivalent circuit in Fig. 2-11c, the effects of excitation branch on transformer voltage regulation can be ignored, so only the series impedances are considered. Voltage regulation depends on both the magnitude of series impedances and the phase angle of the current flowing through the transformer, i. e. the power factor of the load. The easiest way to determine the effect of impedances and current phase angles on voltage regulation is to examine the phasor diagram as shown in Fig. 2-16.

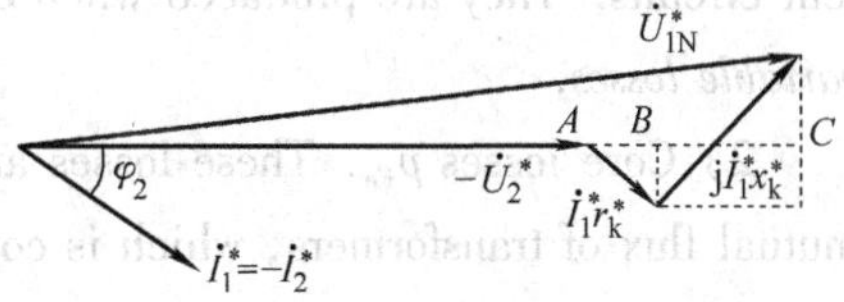

Fig. 2-16 Phasor diagram of the simplified equivalent circuit

In Fig. 2-16, the transformer is operating at load condition with lagging power factor. All quantities are per-unit values, i. e. $U_{1N}^* = 1$ and $I_2^* = I_1^*$.

Because the short-circuit impedances of the transformer are very small, the angle between $\dot{U}_{1N}^*$ and $-\dot{U}_2^*$ is very small. From Fig. 2-16, we can find that the voltage regulation is

$$\Delta U = 1 - U_2^* \approx \overline{AC} = \overline{AB} + \overline{BC}$$

where $\overline{AB} = I_1^* r_k^* \cos\varphi_2$, $\overline{BC} = I_1^* x_k^* \sin\varphi_2$, then the voltage regulation can be expressed as

$$\Delta U \approx I_1^* (r_k^* \cos\varphi_2 + x_k^* \sin\varphi_2) \tag{2-34}$$

Eq. 2-34 shows that the voltage regulation increases with the increase of the load current. It also depends on the power factor of load and short-circuit impedances of the transformer.

If the load is inductive or resistance, the voltage regulation is positive; if the load is capacitive, the voltage regulation is negative. It means that the output voltage may exceed the no-load voltage.

Example 2-3 The transformer's ratings are the same as that of Example 2-2, Find its voltage regulation with the rated load when $\cos\varphi_2 = 0.8$ (lagging), $\cos\varphi_2 = 0.8$ (leading).

Solution:

The short-circuit impedance of the transformer, $r_k^* = 0.0178$, $x_k^* = 0.0414$, $\cos\varphi_2 = 0.8$ (lag), $\sin\varphi_2 = 0.6$, so

$$\Delta U_N \approx r_k^* \cos\varphi_2 + x_k^* \sin\varphi_2 = 0.0178 \times 0.8 + 0.0414 \times 0.6 = 0.03908 \approx 3.91\%$$

$\cos\varphi_2 = 0.8$ (lead), $\sin\varphi_2 = -0.6$, so

$$\Delta U_N \approx r_k^* \cos\varphi_2 + x_k^* \sin\varphi_2$$
$$= 0.0178 \times 0.8 + 0.0414 \times (-0.6) = -0.0106 \approx -1.06\%$$

2.5.2 Efficiency

Transformers are compared and judged on their efficiencies. The efficiency of a device is the ratio between the output active power and the input active power. It is usually expressed in percentage

$$\eta = \frac{P_2}{P_1} \times 100\% \tag{2-35}$$

where P_1 is the input active power; P_2 is the output active power.

Transformer equivalent circuits make efficiency calculation easy. There are two types of losses presenting in transformers:

1) Copper (I^2R) losses p_{Cu}. These losses are accounted for by the series resistance in equivalent circuits. They are produced when current flows through windings. Copper losses are also called *variable losses*.

2) Core losses p_{Fe}. These losses are accounted for by resistor R_m. They mainly depend on the mutual flux of transformers, which is constant under a constant source voltage. Therefore, these losses are called *fixed losses* and they are approximately equal to the no-load losses of a transformer.

In Eq. 2-35, given that $P_1 = P_2 + p_{loss} = P_2 + p_{Cu} + p_{Fe}$, in order to calculate the efficiency of a transformer at a given load, we need to add up the losses from each resistor. Copper losses can be calculated by

$$p_{Cu} = I_2^2 r_k = \left(\frac{I_2}{I_{2N}}\right)^2 I_{2N}^2 r_k = I_2^{*2} p_{kN} \tag{2-36}$$

where I_2 is the load current; r_k is the short-circuit impedances; p_{kN} is the short-circuit losses; I_2^* is the per-unit value of the load current.

The total loss of the transformer is

$$\sum p = p_{Fe} + p_{Cu} = p_0 + I_2^{*2} p_{kN} \tag{2-37}$$

The output active power of the transformer is

$$P_2 = U_2 I_2 \cos\varphi_2 \approx U_{2N} I_2 \cos\varphi_2 = I_2^* S_N \cos\varphi_2 \tag{2-38}$$

where $\cos\varphi_2$ is the power factor of the load.

So the efficiency of the transformer can be calculated by

$$\eta = \frac{P_2}{P_2 + \sum p} \times 100\% = \frac{I_2^* S_N \cos\varphi_2}{I_2^* S_N \cos\varphi_2 + p_0 + I_2^{*2} p_{kN}} \times 100\% \tag{2-39}$$

Eq. 2-39 shows that the efficiency of transformers is identified by the load current and the power factor of the load. When the power factor is fixed, the curve of efficiency versus load current should be drawn as the graph shown in Fig. 2-17.

In Fig. 2-17, both the efficiency and output power of the transformer start at 0. The efficiency increases rapidly when the load current increases because the no-load losses p_0 make up the majority of the total losses. When the load current increases to a certain value, i. e. $I^*_{2(max)} = \sqrt{\frac{p_0}{p_{kN}}}$, the efficiency reaches its maximum value and at that moment, the core losses are equal to the copper losses.

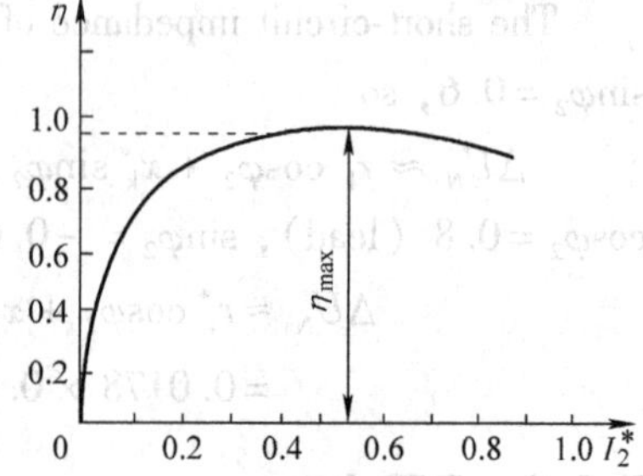

Fig. 2-17 The efficiency curve of the transformer

2.6 Three-phase transformers

Almost all major power generations and distribution systems in the world today are three-phase AC systems, so it is necessary to understand the position of transformers used in the system.

With the symmetrical operation of three-phase transformers, voltages and currents of each phase are equal in value. Thus one phase of a three-phase transformer can be treated as a single-phase transformer. Due to the limited space, only magnetic and electrical circuits of three-phase transformers will be presented in this section.

2.6.1 Magnetic circuit of three-phase transformers

Three phase transformers can be constructed in two ways. One approach is to connect three single-phase transformers in a three-phase bank as shown in Fig. 2-18. An alternative approach is to wrap a three-phase transformer consisting of three sets of windings on a common core. The construction of the magnetic circuit is shown in Fig. 2-19.

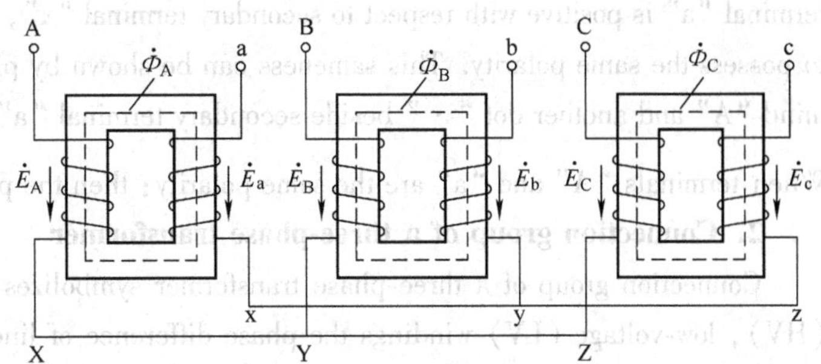

Fig. 2-18 The magnetic circuit of a three-phase transformer bank

As shown in Fig. 2-18, three single-phase transformers are connected to a three-phase AC source. Each flux passes through an independent magnetic circuit. The connection is called a three-phase transformer bank.

As shown in Fig. 2-19, if the cores of three single-phase transformers are combined as shown in Fig. 2-19a and the power source is symmetrical (i. e. Equivalent voltage magnitude and 120° phase angle difference between A, B, and C), the instantaneous flux in the middle common core will be zero. So it can be removed as shown in Fig. 2-19b. Three iron columns can be placed in a plane as shown in Fig. 2-19c. Transformers having this kind of construction are called a core-type transformer.

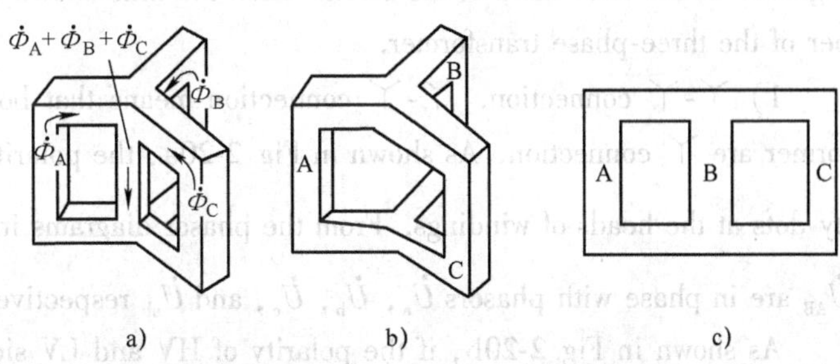

Fig. 2-19 The magnetic circuit of core-type three-phase transformer

2.6.2 Connections of three-phase transformers

A three-phase transformer consists of three transformers which are either separated or combined to one core. The primary sides and secondary sides of any three-phase transformer can be independently connected in either wye (Y) or delta (△).

1. Terminal markings and polarity

According to voltage level, there are three types of transformer winding, i. e. high-voltage, medium-voltage and low-voltage winding, and the terminal markings are designed as shown in Table 2-1.

Table 2-1 Terminal Markings

Winding type	Single-phase		Three-phase		Neutral point
	Head	End	Head	End	
high-voltage	A	X	A, B, C	X, Y, Z	N
low-voltage	a	x	a, b, c	x, y, z	n
medium-voltage	A_m	X_m	A_m, B_m, C_m	X_m, Y_m, Z_m	N_m

If the primary terminal "A" is positive with respect to primary terminal "X" and that secondary terminal "a" is positive with respect to secondary terminal "x", terminals "A" and "a" are then said to possess the same polarity. This sameness can be shown by placing a dot "·" beside primary termind "A" and another dot "·" beside secondary terminal "a". The dots are called polarity marks. When terminals "A" and "a" are the same polarity; then the phase voltage $\dot{U}_{AX}$ is in phase with $\dot{U}_{ax}$.

2. Connection group of a three-phase transformer

Connection group of a three-phase transformer symbolizes the connection between high-voltage (HV), low-voltage (LV) windings the phase difference of line voltages. It is usually explained by clock notation, i. e. the phasor of the line voltage of HV winding is regarded as the minute hand of the clock and it always fixes at 12: 00; the phasor of the corresponding line voltage of LV winding is regarded as the hour hand of the clock. Then the time shown in the clock represents the group number of the three-phase transformer.

1) Y-Y connection. Y-Y connection means that both HV and LV windings of the transformer are Y connection. As shown in Fig. 2-20a, the polarity of HV and LV windings are denoted by dots at the heads of windings. From the phasor diagrams in Fig. 2-20, phasors $\dot{U}_A$, $\dot{U}_B$, $\dot{U}_C$, and $\dot{U}_{AB}$ are in phase with phasors $\dot{U}_a$, $\dot{U}_b$, $\dot{U}_c$, and $\dot{U}_{ab}$ respectively. This connection is called Yy0.

As shown in Fig. 2-20b, if the polarity of HV and LV side are denoted by dots at the heads of HV windings and the tail of LV winding respectively, phasors $\dot{U}_A$, $\dot{U}_B$, $\dot{U}_C$, and $\dot{U}_{AB}$ are the inverse of phasors $\dot{U}_a$, $\dot{U}_b$, $\dot{U}_c$, and $\dot{U}_{ab}$ respectively. This connection is called Yy6.

From Fig. 2-20a, we can see that the polarity marks of HV and LV windings are not changed, while the phase order of LV windings are changed, i. e. 'abc' is replaced by 'bca', 'xyz' is replaced by 'yzx'. The connection of the three-phase transformer is shown in Fig. 2-21a.

In Fig. 2-21a, phasors $\dot{U}_A$, $\dot{U}_B$, and $\dot{U}_C$ are in phase with the phasors $\dot{U}_b$, $\dot{U}_c$, and $\dot{U}_a$ respectively, the phasor $\dot{U}_{ab}$ rotates 120° counter-clockwise and points to 8 o'clock, thus, the connection is called Yy8.

From Fig. 2-20a, we can see that the polarity marks of HV and LV windings are not changed,

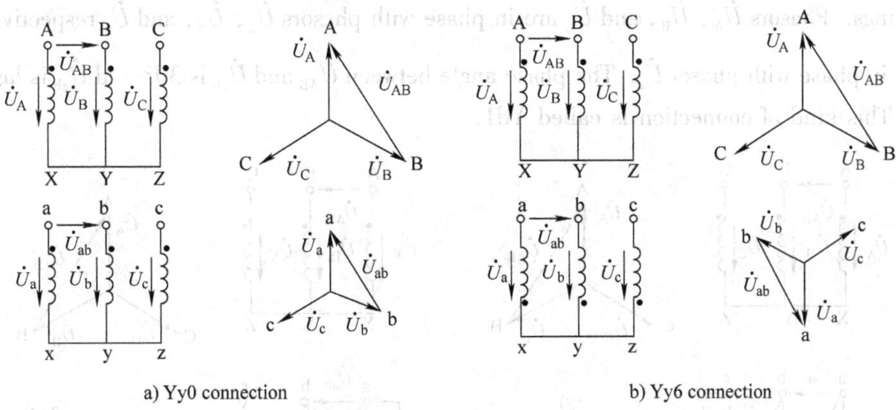

Fig. 2-20 Connections and phasor diagrams of Y-Y three-phase transformers

while the phase order of LV windings are changed, i. e. 'abc' is replaced by 'cab', 'xyz' is replaced by 'zxy'. The connection of the three-phase transformer is shown in Fig. 2-21b.

In Fig. 2-21b, phasors $\dot{U}_A$, $\dot{U}_B$, and $\dot{U}_C$ are in phase with phasors $\dot{U}_c$, $\dot{U}_a$, and $\dot{U}_b$ respectively. Phasor $\dot{U}_{ab}$ rotates 120° clockwise and points to 4 o'clock, thus, the connection is called Yy4.

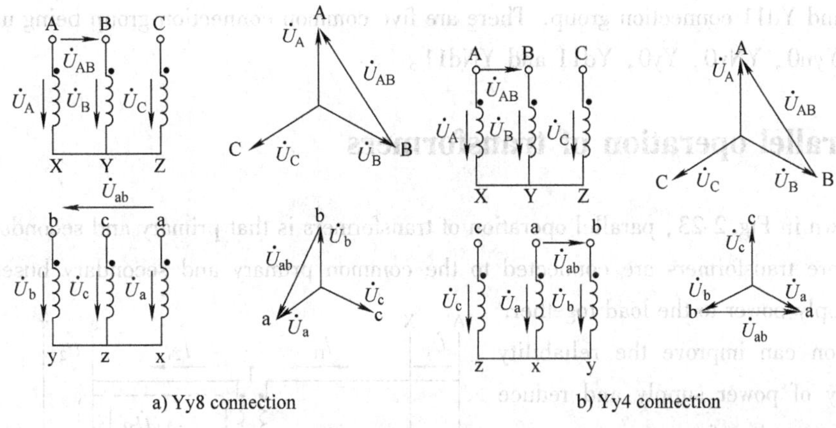

Fig. 2-21 Connections and phasor diagrams of Y-Y three-phase transformers

In summary, Y-Y connection has six kinds of even connection group, such as Yy0, Yy2, Yy4, Yy6, Yy8 and Yy10 connection group.

2) Y-Δ connection. Y-Δ connection means that HV windings are connected as Y and LV windings are connected as Δ. As shown in Fig. 2-22a, the polarity of HV and LV windings are denoted by dots at the head of the windings. Phasors $\dot{U}_A$, $\dot{U}_B$, and $\dot{U}_C$ are in phase with phasors $\dot{U}_a$, $\dot{U}_b$, and $\dot{U}_c$ respectively. Phasor $\dot{U}_{ab}$ is the inverse of phasors $\dot{U}_b$ and the phase angle between $\dot{U}_{AB}$ and $\dot{U}_{ab}$ is 30°. $\dot{U}_{ab}$ is leading $\dot{U}_{AB}$ by 30°. This connection is called Yd11.

As shown in Fig. 2-22b, the polarity of HV and LV windings are denoted by dots at the head of the windings. Phasors $\dot{U}_A$, $\dot{U}_B$, and $\dot{U}_C$ are in phase with phasors $\dot{U}_a$, $\dot{U}_b$, and $\dot{U}_c$ respectively. Phasor $\dot{U}_{ab}$ is in phase with phasor $\dot{U}_a$. The phase angle between $\dot{U}_{AB}$ and $\dot{U}_{ab}$ is 30° and $\dot{U}_{ab}$ is lagging $\dot{U}_{AB}$ by 30°. This kind of connection is called Yd1.

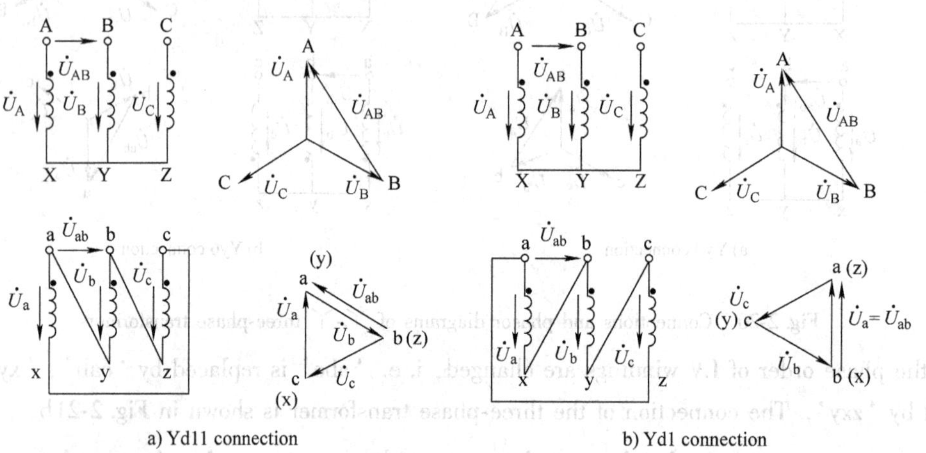

Fig. 2-22 Connections and phasor diagrams of Y-△ three-phase transformers

In summary, Y-△ connection has six kinds of odd connection group, i. e. , Yd1, Yd3, Yd5, Yd7, Yd9 and Yd11 connection group. There are five common connection group being used in power system, Yyn0, YNy0, Yy0, Yd11 and YNd11。

2.7 Parallel operation of transformers

As shown in Fig. 2-23, parallel operation of transformers is that primary and secondary windings of two or more transformers are connected to the common primary and secondary buses. The two windings supply power to the load together. This operation can improve the reliability and economy of power supply and reduce the reserved capacity of transformers.

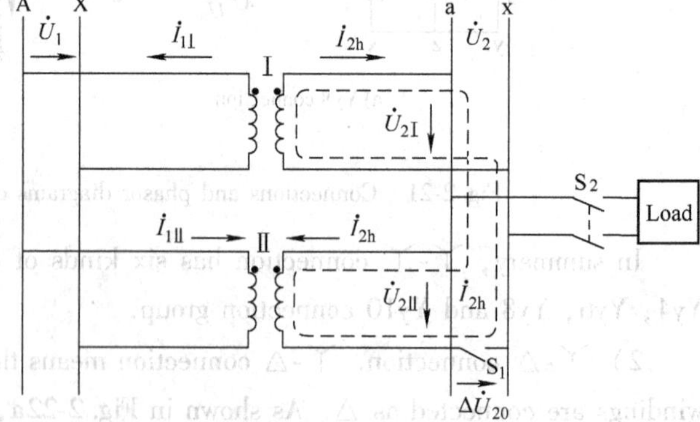

Fig. 2-23 Parallel operation of two transformers

For the parallel operation of transformers, the following three conditions must be fulfilled:

1) The rated voltage and turns ratio of the transformers must be equal.

2) Transformers must have the same connection group.

3) Transformers' per-unit values of short-circuit impedances must be equal.

1. Parallel transformers with different turns ratio

As shown in Fig. 2-23, if the turns ratio of transformer Ⅰ and Ⅱ are not equal ($k_{\mathrm{I}} < k_{\mathrm{II}}$) and their primary windings are connected to source U_1, then the secondary no-load voltages of each transformer becomes $\dot{U}_{20\,\mathrm{I}}=\dfrac{\dot{U}_1}{k_{\mathrm{I}}}$, $\dot{U}_{20\,\mathrm{II}}=\dfrac{\dot{U}_1}{k_{\mathrm{II}}}$, where $\dot{U}_{20\,\mathrm{I}}>\dot{U}_{20\,\mathrm{II}}$. There will be a voltage drop at the switch S_1, that is $\Delta\dot{U}_{20}=\dot{U}_{20\,\mathrm{I}}-\dot{U}_{20\,\mathrm{II}}$. When S_1 is closed for the parallel operation of two transformers, a loop current $\dot{I}_{2\mathrm{h}}$ will occur in the secondary windings of both transformers. It can be expressed as

$$\dot{I}_{2\mathrm{h}}=\frac{\Delta\dot{U}_{20}}{z_{\mathrm{k\,I}}+z_{\mathrm{k\,II}}} \tag{2-40}$$

where $z_{\mathrm{k\,I}}$ and $z_{\mathrm{k\,II}}$ are the short-circuit impedances referred to the secondary side of the two transformers.

Because $z_{\mathrm{k\,I}}$ and $z_{\mathrm{k\,II}}$ are very small, a small voltage drop $\Delta\dot{U}_{20}$ will cause a great loop current $\dot{I}_{2\mathrm{h}}$, thus increase the losses of transformer.

2. Parallel transformers with different connection

If two transformers have different connection, the minimum phase angle between the corresponding line voltages of the two transformers is 30°. The per-unit value of voltage drop will be $2\times\sin15°=0.518$ and a large loop current will be caused. So transformers with different connection are not allowed for parallel operation.

3. Parallel transformers with different per-unit values of short-circuit impedances

If both primary and secondary voltages of two transformers are equal, the voltage drop of their short-circuit impedances will be equal

$$\dot{I}^*_{\mathrm{I}}z^*_{\mathrm{k\,I}}=\dot{I}^*_{\mathrm{II}}z^*_{\mathrm{k\,II}} \tag{2-41}$$

where $\dot{I}^*_{\mathrm{I}}$ and $\dot{I}^*_{\mathrm{II}}$ are the per-unit values of load current of transformer I and II respectively. Then

$$\dot{I}^*_{\mathrm{I}}:\dot{I}^*_{\mathrm{II}}=\frac{1}{z^*_{\mathrm{k\,I}}}:\frac{1}{z^*_{\mathrm{k\,II}}} \tag{2-42}$$

The above equation shows that the load currents of parallel transformers are inversely proportional to the per-unit values of short-circuit impedances. If these values are not equal, the load distribution will be unbalanced, i. e. one transformer may be owe-load and the other may be over-load.

2.8 Special Transformers

Many transformers are designed to meet specific industrial applications. Although they are specially designed, they still possess the basic properties of standard transformers. As a result, the following basic properties should also be applied to the specially designed transformers.

1) The voltage induced in windings is directly proportional to the number of turns, the frequency and the flux in the core.

2) The ampere-turns of the primary winding are equal yet opposite to the ampere-turns of the secondary winding.

3) The apparent power input is equal to the apparent power output.

4) The exciting current in the primary winding can be neglected.

2.8.1 Autotransformers

An autotransformer is a single transformer winding with N_1 turns mounting on an iron core (Fig. 2-24). The winding is connected to a fixed-voltage AC source U_1 and the resulting exciting current I_0 creates an ac flux Φ_m in the core. As in any transformers, the peak value of the flux is fixed as long as U_1 is fixed.

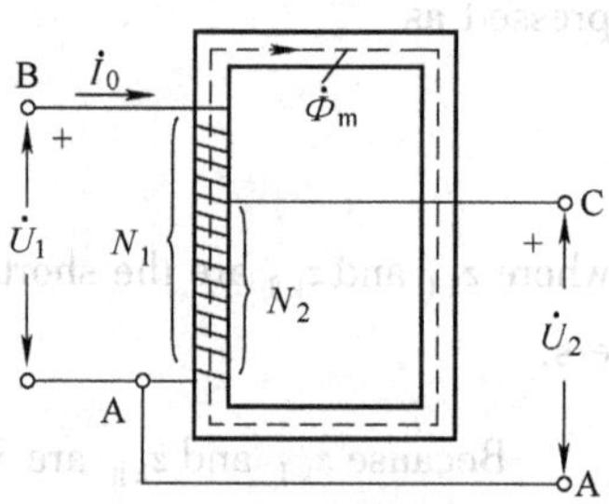

Fig. 2-24 An autotransformer

Suppose a tap C is drawn from the winding, the number of turns between A and B changes from N_1 to N_2 because the induced voltage between terminals is proportional to the number of turns, U_2 becomes

$$U_2 = (N_2/N_1) \times U_1 = U_1/k_a \tag{2-43}$$

Where k_a is called turns ratio, and $k_a = \dfrac{N_1}{N_2}$.

Clearly, this simple coil resembles transformers which have primary voltage $\dot{U}_1$ and secondary voltage U_2. However, the primary terminals (B, A) and the secondary terminals (C, A) are no longer isolated from each other because of the common terminal A.

If we connect a load to the secondary terminals CA, the resulting current I_2 will immediately cause a primary current I_1 to flow in the winding (Fig. 2-25).

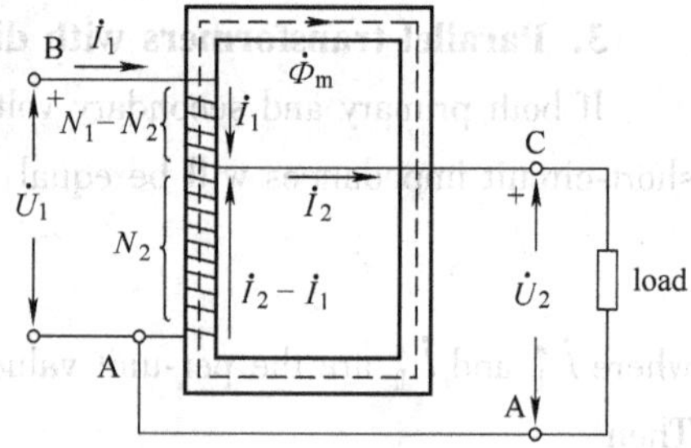

Fig. 2-25 An autotransformer under load

The BC portion of the winding carries current I_1. According to Kirchhoff's current law, the CA portion carries a current $(I_2 - I_1)$. Also, the mmf due to I_1 must equal but opposite to the mmf produced by $(I_2 - I_1)$. As a result, we have $I_1(N_1 - N_2) = (I_2 - I_1) \times N_2$. The formula can be reduced to

$$I_1 N_1 = I_2 N_2 \tag{2-44}$$

Eq. 2-44 can be expressed as

$$I_1 = I_2 N_2/N_1 = I_2/k_a \tag{2-45}$$

Finally, assuming that both the transformer losses and exciting current are negligible, the apparent power drawn by the load must equal the apparent power supplied by the source. Consequently:

$$I_1 U_1 = I_2 U_2 \tag{2-46}$$

Eq. 2-43, Eq. 2-44, Eq. 2-45 are identical to those of standard transformers which have the turns ratio N_1/N_2. However, in autotransformers, the secondary winding is actually part of the primary winding. Also, autotransformers are usually smaller, lighter and cheaper than standard transformers of equal power output. Furthermore, the absence of electrical isolation between primary and secondary windings is a serious drawback in some applications.

Autotransformers are used to start induction motors, regulate the voltage of transmission lines, and, in general, transform voltage when the primary to secondary ratio is close to 1.

2.8.2 Instrument transformers

Two types of transformers are used in power systems to do measurements for specific purposes. One is the potential transformer and the other one is the current transformer.

1. Potential transformers

Potential transformers are specially wounded transformers with high-voltage primary and low-voltage secondary (Fig. 2-26). They have very low power rating and their sole purpose is to provide a sample of the power system's voltage to the instruments monitoring it. Since the principal purpose of potential transformers is voltage sampling, they must be very accurate in order not to distort the true voltage values severely.

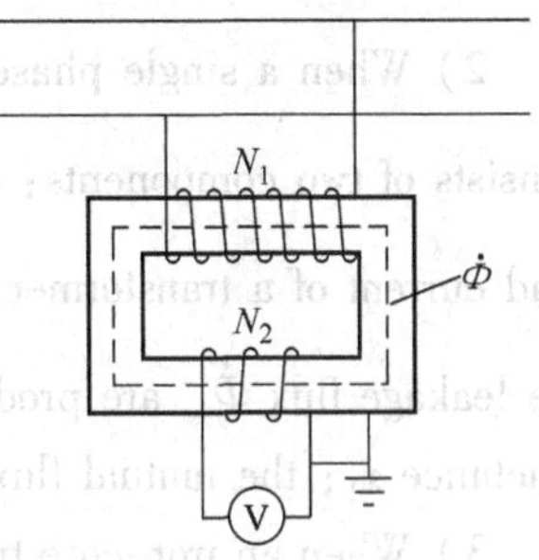

Fig. 2-26 A potential transformer

The construction of potential transformers is similar to that of conventional transformers. However, the insulation between the primary and secondary windings must be particularly good enough to withstand the full line voltage on the HV side when compared with conventional transformers.

One terminal of the secondary winding is always connected to the ground to eliminate the danger of a fatal shock when touching one of the secondary leads.

2. Current transformers

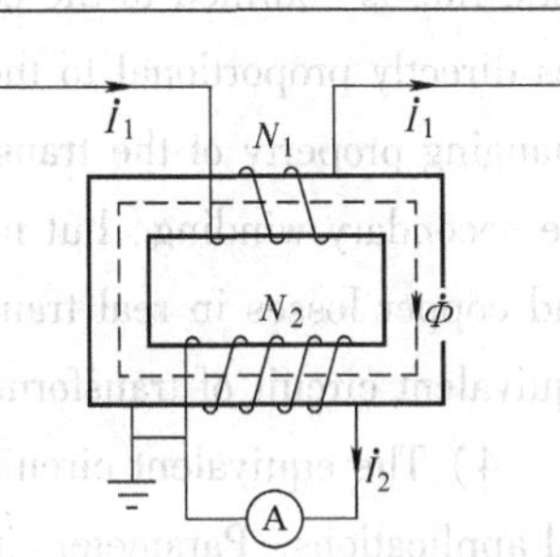

Fig. 2-27 A current transformer

Current transformers (CT) sample the current in the line and reduce it to a safe and measurable level. As shown in Fig. 2-27, the primary is connected in series with the line. The nominal secondary current is usually 5A irrespective of the primary current rating.

Because current transformers are only used for measurement and system protection, their power rating is small, generally between 15V·A and 200V·A. In the case of conventional transformers, the current ratio is inversely proportional to the number of turns of the primary and secondary windings. A current transformer having a ratio of 150A/5A has therefore 30 times more turns on the secondary than on the primary.

It is important to keep current transformers short-circuited at all times since extremely high voltage can appear across its open secondary terminals.

Summary

A transformer is a device for converting electric energy at one voltage level to at another through the action of a magnetic field. When a voltage is applied to the primary winding of a transformer, magnetic flux is produced in the core as explained by Faraday's law. The changing flux in the core then induces a voltage in the primary and secondary winding of the transformer respectively, i. e. , $\dot{E}_1 = -j4.44fN_1\dot{\Phi}_m$, $\dot{E}_2 = -j4.44fN_2\dot{\Phi}_m$. Because the transformer cores have very high permeability, the net magnetomotive force (mmf) required in the core to produce its magnetic flux is very small. Since the net mmf is very small, the primary winding's mmf must be approximately equal and opposite to the secondary winding's mmf. This fact yields the transformer current ratio. Students should pay more attention to the following points when studying this chapter.

1) The construction of a transformer consists of two or more coils and a ferromagnetic core. These coils are usually not directly connected. The interaction between the coils is realized through the common magnetic flux in the common core.

2) When a single phase transformer is operating at no-load condition, the no-load current $\dot{I}_0$ consists of two components: the magnetization current $\dot{I}_\mu$ and the core-loss current $\dot{I}_{Fe}$, and the no-load current of a transformer is only about 1% ~ 10% of the rated current. The mutual flux $\dot{\Phi}$ and the leakage flux $\dot{\Phi}_{1\sigma}$ are produced by the no-load current $\dot{I}_0$. The leakage flux can be modeled by a reactance x_1; the mutual flux can be modeled by the exciting impedance $z_m = r_m + jx_m$.

3) When an iron-core transformer is operating in loaded condition, the magnetic field is created by the combined action of the currents in both the primary and secondary windings. Most of the magnetic flux is confined to the core and links all the windings. The resultant mutual flux induces voltages directly proportional to the number of turns in the windings which is responsible for the voltage-changing property of the transformer. There is leakage flux that passes through either the primary or the secondary winding, but not both. In addition, there are hysteresis losses, eddy current losses, and copper losses in real transformers. These effects are accounted for by various resistances in the equivalent circuit of transformers.

4) The equivalent circuits of transformers are significant in both theoretical analysis and practical applications. Parameters in a transformer equivalent circuit can be measured by conducting transformer tests: open-circuit test and short-circuit test.

5) Voltage regulation and efficiency are the operating characteristics of transformers. The voltage regulation of a transformer can be calculated by the equation: $\Delta U \approx I_1^*(r_k^*\cos\varphi_2 + x_k^*\sin\varphi_2)$; and the efficiency of a transformer can be calculated by $\eta = \dfrac{I_2^* S_N\cos\varphi_2}{I_2^* S_N\cos\varphi_2 + p_0 + I_2^{*2} p_{kN}} \times 100\%$.

6) The per-unit system is a convenient method to study systems containing transformers because using per-unit system different voltage levels in the system disappear. In addition, the per-unit impedances of a transformer expressed to its own ratings base fall within a relatively narrow range, providing a convenient check for reasonableness in problem solutions.

7) The magnetic system and connection group are special contents of three-phase transformers. Three phase transformers can be constructed in two ways. One approach is to connect three single-phase transformers in a three-phase bank, and an alternative approach is to wrap a three-phase transformer consisting of three sets of windings on a common core. Connection group of a three-phase transformer symbolizes the connection between high-voltage (HV), low-voltage (LV) windings the

phase difference of line voltages.

8) The parallel operation of transformers is a common mode. For the parallel operation of transformers, the following three conditions must be fulfilled: The rated voltage and turns ratio of the transformers must be equal; Transformers must have the same connection group; Transformers' per-unit values of short-circuit impedances must be equal.

9) Many transformers are designed to meet specific industrial applications. An autotransformer is a single transformer winding with N_1 turns mounting on an iron core. *Potential transformers* are specially wounded transformers with high-voltage primary and low-voltage secondary. *Current transformers* (CT) sample the current in the line and reduce it to a safe and measurable level.

Questions and problems

1. Can a transformer get power from the power source when it is at no-load?

2. What is rated primary and secondary voltage of transformers?

3. What components compose the excitation current of a transformer? How are they modeled in the transformer's equivalent circuit?

4. What is leakage flux? How is it modeled in equivalent circuits? What is mutual flux? How is it modeled in equivalent circuits?

5. List and describe the types of losses that occur in a transformer.

6. Theratings of a three-phase transformer are: $S_N = 5000\text{kV}\cdot\text{A}$, $U_{1N}/U_{2N} = 10/6.3\text{kV}$, Y-Δ connection.

1) Find the rated voltages and rated currents (high-voltage side, low-voltage side).

2) Find the rated phase voltages and rated phase currents (high-voltage side, low-voltage side).

7. When a 60Hz transformer operates on a 50Hz system with the same voltage level, how will the following quantities change: mutual flux, no-load current, core loss, leakage and exciting reactance?

8. The ratio of a 220/110V single-phase transformer is 2, can it have the number of turns $N_1 = 2$, $N_2 = 1$? Why?

9. When the source voltage applied to a transformer increases, what will happen to the exciting impedance, exciting current and core-loss of the transformer?

10. A single-phase transformer is rated at 20kV · A, 10 000/500V and 60Hz. Can it safely supply 15kV · A to a 420V load at 50Hz? Why?

11. A single-phase transformer is rated at 18kV · A, 20 000/480V, and 60Hz. Can this transformer safely supply 15kV · A to a 415V load at 50Hz? Why or why not?

12. Why does the open-circuit test show only exciting losses but not i^2r loss? Why does the short-circuit test show only i^2r loss but not exciting losses?

13. Why does the secondary voltage of transformers change with load? Can the voltage regulation be zero with rated load?

14. Why does the power factor of loads affect the voltage regulation of transformers?

15. What quantities are related to efficiency? Under what conditions will efficiency reach the

maximum value?

16. What conditions must be met during the parallel operation of transformers? Which one is the most important?

17. What are the problems associated with the Y-Y three-phase transformer connections?

18. A 2kV·A, 220/110V single-phase transformer with parameters $r_1 = 1\Omega$, $x_1 = 6\Omega$, $r_2 = 0.25\Omega$, $x_2 = 1.5\Omega$. When the load is $Z_L = 28 + j28\Omega$, what is the primary and secondary current? What is the voltage regulation?

19. A single phase transformer has $S_N = 4.6$kV·A, $U_{1N}/U_{2N} = 380/115$V. The results of the tests are below.

Test	Voltage/V	Current/A	Power/W	Test Side
open-circuit test	115	3	60	LV side
short-circuit test	15.6	12.1	172	HV side

1) Find its excitation impedances and series impedances referred to the HV side.

2) Find the transformer's voltage regulation and efficiency at rated load and $\cos\varphi_2 = 0.8$ lagging; $\cos\varphi_2 = 0.8$ leading.

3) Find the maximum efficiency at $\cos\varphi_2 = 0.8$ lagging.

20. A three-phase transformer has $S_N = 125000$kV·A, $U_{1N}/U_{2N} = 110/11$kV, 50Hz, Y-Δ connection, The results of the tests are below.

Test	Voltage/kV	Current/A	Power/kW	Test Side
open-circuit test	11	131.22	133	LV side
short-circuit test	11.55	Rated current	600	HV side

1) Find its excitation impedances and series impedances referred to the HV side, and their per-unit values.

2) Find the transformer's voltage regulation at half rated load and $\cos\varphi_2 = 1.0$; $\cos\varphi_2 = 0.8$ (lagging); $\cos\varphi_2 = 0.8$ (leading).

3) Find the efficiency at rated load with $\cos\varphi_2 = 0.8$ (lagging) and the maximum efficiency with same power factor respectively.

21. A 20kV·A, 20 000/480V, 60Hz single-phase transformer is tested with the following results:

Open-circuit test (measured from secondary side)	Short-circuit test (measured from primary side)
$U_{OC} = 480$V	$V_{SC} = 1130$V
$I_{OC} = 1.60$A	$I_{SC} = 1.00$A
$P_{OC} = 305$W	$P_{SC} = 260$ W

Find the per-unit equivalent circuit for this transformer at 60Hz.

22. A 1000V·A, 230/115V single-phase transformer has been tested to determine its equivalent circuit. The results of the tests are below.

Open-circuit test	Short-circuit test
U_{CC} = 230V	V_{SC} = 19.1V
I_{OC} = 0.45A	I_{SC} = 8.7A
P_{OC} = 30W	P_{SC} = 42.3W

All data given were taken from the primary side of the transformer.

1) Find the equivalent circuit of this transformer referred to the low-voltage side of the transformer.

2) Find the transformer's voltage regulation at rated conditions and 0.8 PF lagging, 1.0 PF, 0.8 PF leading.

3) Determine the transformer's efficiency at rated conditions and 0.8 PF lagging.

23. A 5000kV·A 230/13.8kV single-phase power transformer has a per-unit resistance of 1 percent and a per-unit reactance of 5 percent (data taken from the transformer's nameplate). The open-circuit test performed on the low-voltage side of the transformer yielded the following data:

$$U_{OC} = 13.8\text{kV} \qquad I_{OC} = 15.1\text{A} \qquad P_{OC} = 44.9\text{kW}$$

1) Find the equivalent circuit referred to the low-voltage side of this transformer.

2) If the voltage on the secondary side is 13.8 kV and the power supplied is 4000 kW at 0.8 PF lagging, find the voltage regulation of the transformer. Find its efficiency.

24. A 750kV·A, 10/0.4kV, Yy0 three-phase transformer is tested. The following table shows the results.

Test	line Voltage /V	line Current/A	Power/W	Test Side
open-circuit	400	60	3800	LV side
short-circuit	440	43.3	10 900	HV side

Find the per-unit equivalent circuit parameters of the transformer.

25. A 5600kV·A, 10/6.3kV, Yd11 three-phase transformer is tested. The following table shows the results.

Test	line Voltage /V	line Current/A	Power/W	Test Side
open-circuit	6300	7.4	6800	LV side
short-circuit	550	323	17 900	HV side

Find the per-unit equivalent circuit parameters of the transformer.

26. A three-phase transformer bank is to handle 600kV·A and have a 34.5/13.8kV voltage ratio. Find the rating of each individual transformer in the bank (high voltage, low voltage, turns ratio, and apparent power) if the transformer bank is connected to Y-Y; Y-Δ; Δ-Y; Δ-Δ.

27. A 100 000kV·A, 230/115kV, Δ-Δ three-phase power transformer has a resistance of 0.02 pu and a reactance of 0.055 pu. The excitation branch elements are R_C = 110 pu and X_M = 20 pu.

1) If this transformer supplies a load of 80 MV·A at 0.85 PF lagging, draw the phasor diagram of one phase of the transformer.

2) What is the voltage regulation of the transformer bank under these conditions?

3) Sketch the equivalent circuit referred to the low-voltage side of one phase of this transformer. Calculate all the transformer impedances referred to the low-voltage side.

28. Three 25kV·A, 24 000/277V distribution transformers are connected in △-Y. The open-circuit test was performed on the low-voltage side of this transformer bank, and the following data were recorded:

$$U_{\text{line,OC}} = 480\text{V}, \quad I_{\text{line,OC}} = 4.10\text{A}, \quad P_{3\phi,\text{OC}} = 945\text{W}$$

The short-circuit test was performed on the high-voltage side of this transformer bank, and the following data were recorded:

$$U_{\text{line,SC}} = 1600\text{V}, \quad I_{\text{line,SC}} = 2.00\text{A}, \quad P_{3\phi,\text{SC}} = 1150\text{W}$$

1) Find the per-unit equivalent circuit of this transformer bank.

2) Find the voltage regulation of this transformer bank at the rated load and 0.9PF lagging.

3) What is the transformer bank's efficiency under these conditions?

29. We have a 100kV·A, 6000/400V, Yy0, three-phase transformer. $I_0^* = 0.065$, $p_0 = 600\text{W}$, $u_k^* = 0.05$, $p_{kN} = 1800\text{W}$.

1) Find its approximate equivalent circuit referred to the HV side.

2) Find the transformer's voltage regulation and efficiency at rated load and 0.8 PF lagging.

3) Find the maximum efficiency at 0.8PF lagging.

30. We have a 1000kV·A, 10/6.3kV, Yd11 three-phase transformer with the rated core-loss of 4.9kW and short-circuit loss of 15kW. What is the efficiency with a rated load and 0.8 PF lagging? What is the maximum efficiency at this power factor?

31. Sketch the wiring (connection) diagrams of three-phase transformers with the connection of Yy2; Yd3; Yd5; Yy10.

32. Sketch the phasor diagram and determine the connection group of the three-phase transformer, when its wiring connection is shown in following:

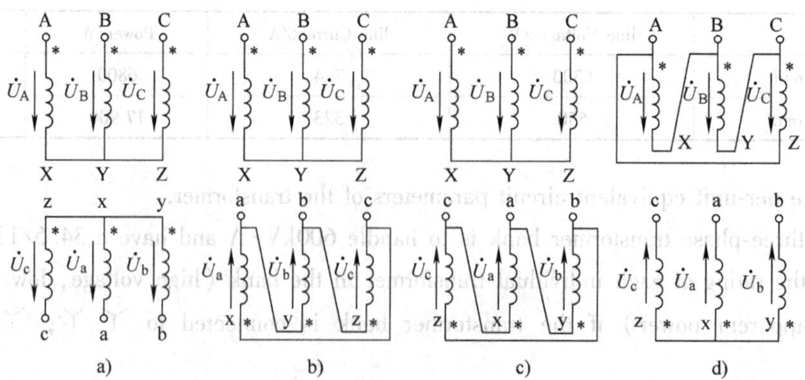

Fig. 2-28 Diagram of problem 32

Chapter 3 AC Machine Fundamentals

There are two major types of AC machines— the synchronous machine and the induction machine. A synchronous machine is motors or generators, in which there is a field winding on a rotor. The field is supplied by a separate DC voltage. The armature winding of a synchronous machine is on the stator with AC current in the stator windings. In contrast, induction machines are motors or generators, where there is AC current in a stator and rotor windings. Take the induction motor as an example, in which AC current is supplied to the stator by direct input and to the rotor by induction. The stator winding of an induction machine is the same as that of a synchronous machine. There are AC current in a stator winding, thus it is called AC winding. This chapter will first explain the configuration of AC windings, and then calculate the induction voltage in the AC windings and the mmf wave produced by the AC windings.

3.1 Operation principles of the three-phase synchronous generator

A simplified two-pole three-phase synchronous generator is shown in Fig. 3-1. The field winding on the rotor is excited by a DC voltage, which is conducted to it by the means of stationary carbon brushes bearing on slip rings or collector rings. The stator winding is a set of balanced three-phase winding with an axis of each phase displaced by 120°, and a complete winding of each phase is represented by a single coil, there are six coil ends, which are A and X, B and Y, C and Z. These coil sides are placed in slots on the stator.

Fig. 3-2 shows the radial distribution of the air-gap flux density B which is a space function of the space angle θ_s around the air-gap periphery. The flux-density wave of the air gap can be adjusted to an approximate sinusoidal distribution by properly shaping the pole faces. The rotor is driven by a prime mover at a constant speed in a counter-clockwise direction. As the rotor rotates, the flux waveform sweeps by the six coil sides. The resulting coil voltage (Fig. 3-3) is a time function which has the same waveform as the spatial distribution B. Therefore there are six sinusoidal waveforms of emf are induced in the six coil sides respectively, and they have the same magnitude. According to

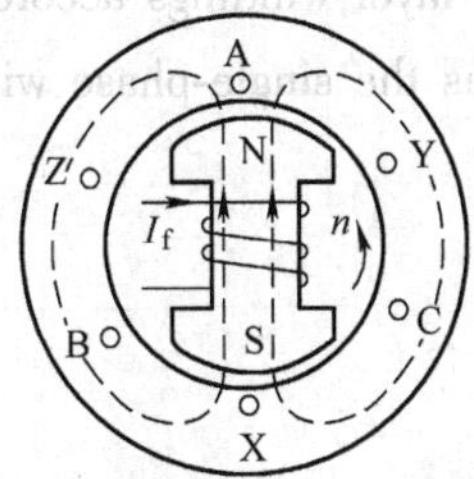

Fig. 3-1 Synchronous generator

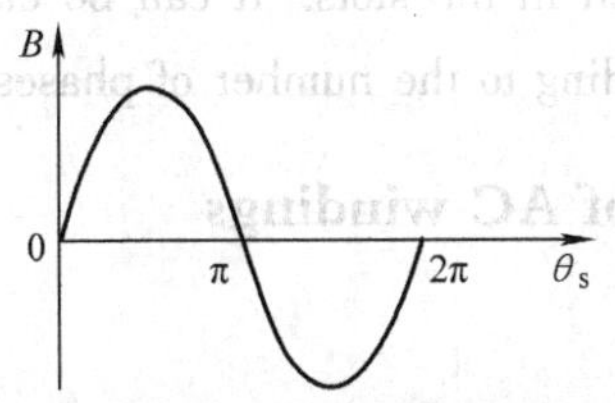

Fig. 3-2 Space distribution of flux density

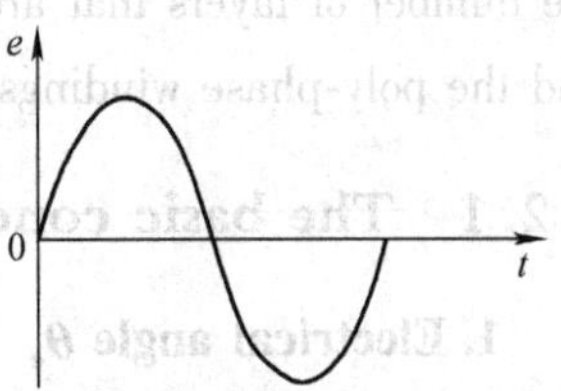

Fig. 3-3 waveform of the generated voltage

Fig. 3-1, the space phase differences between two adjacent conductors are 60°, therefore, the induced emf time phase differences between two adjacent conductors are 60°, and the induced emf phasor-diagram of the six coil sides is shown in Fig. 3-4.

Applying the Faraday's law and the right-hand rule for the field direction which is shown in Fig. 3-1, it reveals that the instantaneous voltage induced in coil sides A, Y and Z is directed out of the paper while in coil sides B, X and C it is directed into paper. Then the emf of phase A (coil sides A and X) winding will be: $\dot{E}_{AX} = \dot{E}_A - \dot{E}_X = 2\dot{E}_A$ (Fig. 3-5). The corresponding induced emfs in phase B and phase C are $\dot{E}_{BY} = 2\dot{E}_B$ and $\dot{E}_{CZ} = 2\dot{E}_C$ respectively. The induced emfs of each phase space are displaced by 120 electrical degrees as shown in Fig. 3-6.

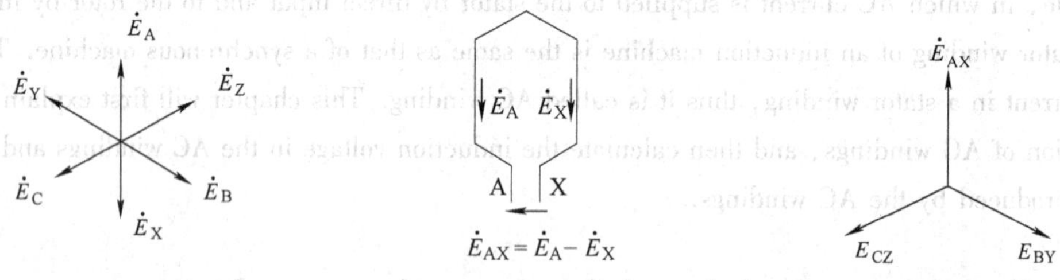

Fig. 3-4 Phasor diagram of emfs induced in conductors

Fig. 3-5 Coil emfs

Fig. 3-6 Phasor diagram of the three-phase emfs

In a three-phase synchronous generator that is shown in Fig. 3-1, a DC current is applied to a rotor winding, which will produce a rotor magnetic field. When the rotor is driven by a prime mover at a constant speed, there will be a rotating magnetic field in the machine. This rotating magnetic field induces a set of three-phase voltages within the stator windings of the generator. If the three-phase windings of the generator are connected to those three identical loads; there will be three sets of currents flowing though the windings and the loads. Eventually, the mechanical power from the prime mover is converted into an AC electric power to loads.

The set of three-phase stator windings can be connected to wye (Y) connection or delta (△) connection.

3.2 AC windings

AC windings can be classified as single-layer windings and double-layer windings according to the number of layers that are placed in the slots. It can be classified as the single-phase windings and the poly-phase windings according to the number of phases.

3.2.1 The basic concept of AC windings

1. Electrical angle θ_e

In general, if the number of magnetic poles on an AC machine stator is $2p$, then there will be p repetitions of the winding around its inner surface. When the rotor pole moves only $1/p$ distant around the stator surface, the variety of electrical quantities on the stator will in one electrical cycle. Since one electrical cycle is 360 electrical degrees, and since the mechanical motion is $(1/p) \times$ 360 mechanical degrees, therefore the relationship between the electrical angle θ_e and the mechanical angle θ_m is

$$\theta_e = p \times \theta_m \tag{3-1}$$

2. Coils

Coil is the basic element of the component of the AC windings. Each coil is consists of N_c turns wire, each turn is taped and insulated from the other turns, and each side of a turn is called *conductor*. The number of conductors on a machine's stator is given by

$$N = 2SN_c \tag{3-2}$$

where N is the number of conductors on stator; S is the number of coils on the stator which is usually equal to the number of stator slots Q for double-layer winding; and $S = Q/2$ for single-layer winding; N_c is the number of turns per coil.

Normally, a coil spans 180 electrical degrees. This means that when one side of the coil is under the axis of a given magnetic pole, the other side is under the asix of a pole of opposite polarity.

3. Pole pitch τ

The distance (in number of slots) between two adjacent poles is called *pole pitch* τ, which is also equal to the internal circumference of the stator, and it is divided by the number of the poles.

$$\tau = \frac{Q}{2p} \tag{3-3}$$

where Q is the number of slots; $2p$ is the pole number of the machine.

4. Coil pitch y_1

Thewidth of each coil is called *coil pitch*, and the distance (in number of slots) between two sides of a coil is called coil pitch y_1. If a coil spans 180 electrical degrees, i. e. $y_1 = \tau$, the voltages in the conductors on either side of coil will be exactly the same in magnitude and opposite in direction at all times, such a coil is called *full-pitch coil*. If $y_1 < \tau$, such a coil is called *fractional-pitch* coil, or short-pitch coil. If $y_1 > \tau$, the coil is called long-pitch coil.

In practice, the coil pitch lies between 80% and 100% of the pole pitch. The coil pitch is usually made less than the pole pitch in order to save some coppers and to improve the flux distribution in the air-gap.

5. Slot pitch α

The electrical degree between two adjacent slots is called *slot pitch* α, which can be expressed by

$$\alpha = \frac{p \times 360°}{Q} \tag{3-4}$$

6. The number of slots per pole per phase q

The number of slots distributed to per phase under each pole is q, it is given by

$$q = \frac{Q}{2mp} \tag{3-5}$$

where m is the number of phases.

3.2.2 Three-phase single-layer lap winding

Some AC windings are single-layer windings, this means that one side of each coil is inserted into one slot. Since a coil has two sides, the number of slots is two times the number of the coils, i. e. $Q = 2S$.

An example of the arrangement of single-layer lap windings will be illustrated as following:

Considering a stator of an AC machine, the number of slots is $Q = 36$, the number of poles is $2p = 4$, the coil number is $S = 18$, the phase number is $m = 3$, draw a developed diagram of equal-element winding with one parallel circuit ($a = 1$).

1. Slot emf diagram

As shown in Fig. 3-7, there are 36 slots in the stator and 18 coils in the 36 slots. If each of the coils has one turn, the number of conductors in the stator will be 36.

Assuming that a sinusoidal-distribution magnetic field is produced in air gap of the machine, and the rotor rotates at a constant speed in counter-clockwise direction, then sinusoidal wave emfs will be induced in the conductors, and the time phase difference between the emfs of adjacent two conductors is $\alpha = 20°$. Fig. 3-8 shows the phasor diagram of the conductor emfs, which is called slot emf diagram.

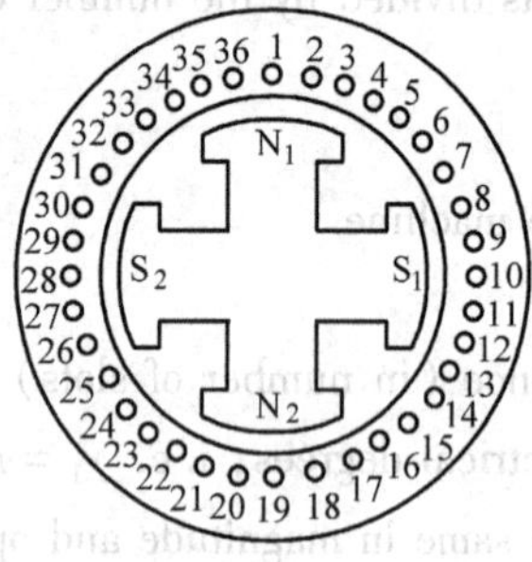

Fig. 3-7 Stator slots

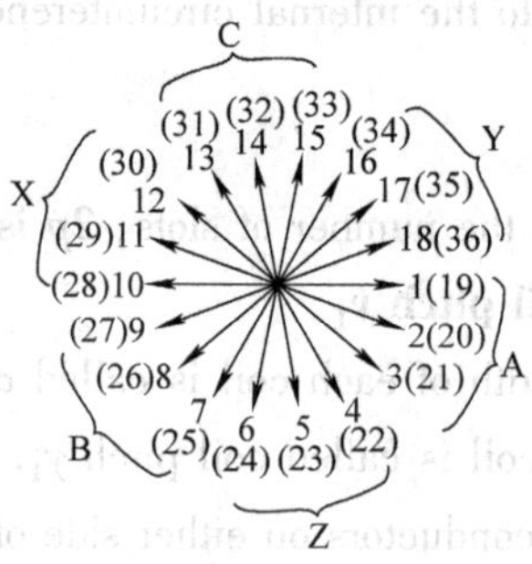

Fig. 3-8 Slot emf diagram

2. Split phase

According to Fig. 3-7, the number of slots are distributed under each pole per phase is $q = 3$. Among the range of the first pair of poles N_1 and poles S_1, conductors 1, 2, 3 and conductors 10, 11, 12 are distributed to phase A; in the range of the second pair of poles N_2 and S_2; conductors 19, 20, 21 and 28, 29, 30 are distributed to phase A. In same manner, the conductors 7, 8, 9; 16, 17, 18; 25, 26, 27 and 34, 35, 36 are distributed to phase B; the conductors 13, 14, 15; 4, 5, 6; 31, 32, 33 and 22, 23, 24 are distributed to phase C.

Then, the conductors 1, 2, 3 are connected to the conductors 10, 11, 12 respectively to create a phase group of phase A, and the conductors 19, 20, 21 are connected to the conductors 28, 29, 30 respectively to create another phase group of phase A. The two phase groups are identical and they are distributed around the circumference of the stator, they are connected in series in order to create phase A (Fig. 3-9).

Similarly, the conductors of phase B are connected to form six coils of phase B; and the conductors of phase C are connected to form six coils of phase C.

Referring back to Fig. 3-4, the phase position of the emf of the three coils is formed by conductors 1, 2, 3 and the conductors 10, 11, 12 will be same with phase position of the emf of the three conductors 1, 2, 3 respectively. The three coils are connected in series in order to form the first coil-group, and the phase position of the emf of this coil-group will be same as the phase position of the emf of the conductor 2. Then three coils are formed by conductors 19, 20, 21 and the conductors 28, 29, 30 are connected in series in order to form the second coil group, and the phase position of its emf will be the same as that of the first coil-group. Those two coil-groups are connected in series in order to form a winding of phase A with one parallel circuit, and the phase position of its emf will be the same as that of the conductor 2.

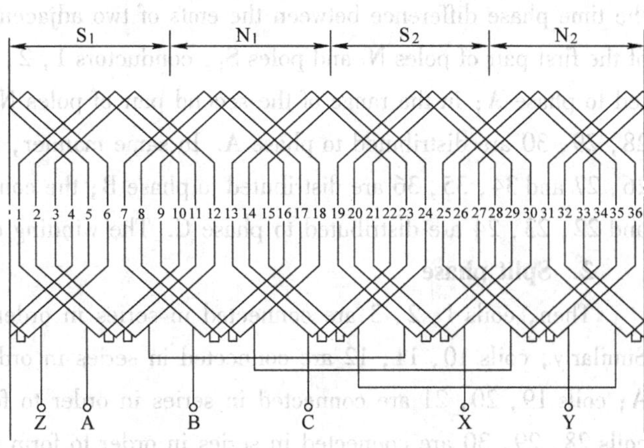

Fig. 3-9 The developed diagram of single-layer three-phase winding

We can also form a winding of phase B and phase C in the same way. The phase position of the emf of the winding of phase B is same with the phase position of emf of conductor 8, and the phase position of the emf of the winding of phase C is same with the phase position of emf of conductor 14. Moreover, the rms values of emfs of the windings of phase A, B, and C are equal. Figure 3-9 shows a developed diagram of this three-phase winding.

3.2.3 Three-phase double-layer lap winding

Except small AC machines whose power ratings are below 10kW, most AC machine's windings are double-layer windings. These improve the waveform of emf and mmf. The configuration of double-layer windings is that one side of each coil is placed in the bottom of one slot, and the other side is placed in the top of another slot; and the number of slots is equal to the number of coils, i. e. $Q = S$.

Usually, there are two types of double layer windings: lap winding and wave winding. In the following section, simplex lap winding will be introduced.

The information of a stator of an AC machineis, the number of slots is $Q = 36$, the number of poles is $2p = 4$, and the number of phases is $m = 3$; draw a developed diagram of simplex lap winding with one parallel circuit ($a = 1$).

1. Winding emf diagram

Double-layer windings are generally fractional-pitch windings, the coil pitch is $y_1 = 5/6\tau \mp \varepsilon$. In this case, $y_1 = 8$, the upper side of coil 1 is placed at the top of slot 1; and its lower side is placed at the bottom of slot 9. The upper side of coil 2 is placed at top of slot 2, and the other side is placed at the bottom of slot 10, and so on.

Referringback to Fig. 3-7, the number of slots distributed under each pole per phase is $q = 3$, the time phase difference between the emfs of two adjacent conductors is $\alpha = 20°$. Among the range of the first pair of poles N_1 and poles S_1, conductors 1, 2, 3 and conductors 10, 11, 12 are distributed to phase A; in the range of the second pair of poles N_2 and S_2; and conductors 19, 20, 21 and 28, 29, 30 are distributed to phase A. In same manner, the conductors 7, 8, 9; 16, 17, 18; 25, 26, 27 and 34, 35, 36 are distributed to phase B; the conductors 13, 14, 15; 4, 5, 6; 31, 32, 33 and 22, 23, 24 are distributed to phase C. The winding emf star diagram is shown in Fig. 3-8.

2. Split phase

Then, coils 1, 2, 3 are connected in series in order to form the first coil-group of phase A. Similarly, coils 10, 11, 12 are connected in series in order to form the second coil-group of phase A; coils 19, 20, 21 are connected in series in order to form the third coil-group of phase A; and coils 28, 29, 30 are connected in series in order to form the fourth coil-group of phase A.

3. Composition phase winding

According to the coil emf star figure, the rms values of emfs of the four coil groups are equal, and their phase positions are the same or the opposite. Therefore, we could connect four groups in series as shown in Fig. 3-10, in order to form a phase-winding of phase A with one parallel circuit.

The phase-winding of phase B and phase C with one parallel circuit can also be formed under the same manner as phase A.

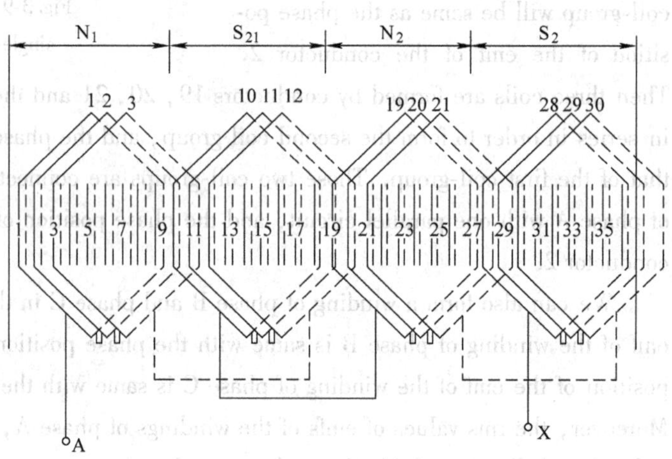

Fig. 3-10 Developed diagram of double-layer winding

3.3 Induced Emf of AC winding

3.3.1 Emf of a conductor

As shown in Fig. 3-11, a sinusoidal distribution magnetic field is formed in the air gap of the machine; the amplitude of the magnetic density is B_1. The length of a conductor in a slot is l. The linear velocity of the field cutting the conductor is v.

Therefore, a sinusoidal emf wave will be induced in the conductor, and the amplitude of the emf is $E_{m1} = B_1 lv$, its rms value is

$$E_1 = \frac{E_{m1}}{\sqrt{2}} = \frac{1}{\sqrt{2}} B_1 lv \tag{3-6}$$

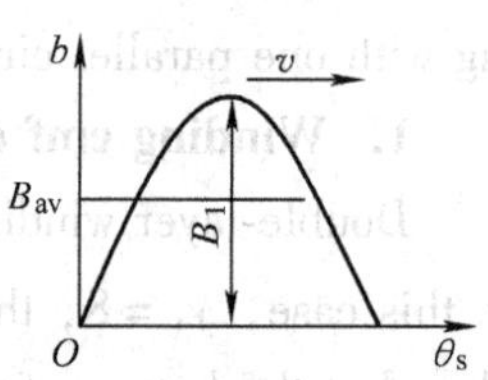

Fig. 3-11 Magnetic field and conductor

According to Fig. 3-11, the average value of flux density under one pole is $B_{av} = \frac{2}{\pi} B_1$, then the per-pole flux can be expressed as $\Phi_1 =$

$$B_{av}\tau l = \frac{2}{\pi}B_1\tau l$$

And
$$B_1 = \frac{\pi}{2\tau l}\Phi_1 \tag{3-7}$$

The linear velocity of the field can be expressed as

$$v = \frac{2p\tau n}{60} = 2\tau f \tag{3-8}$$

Substituting Eq. 3-7 and Eq. 3-8 into Eq. 3-6, then the rms value of emf of a conductor can be expressed as

$$E_1 = \frac{\pi}{\sqrt{2}}f\Phi_1 = 2.22f\Phi_1 \tag{3-9}$$

3.3.2 Emf of a coil and pitch factor

1. Full-pitch single-turn coil

As shown in Fig 3-12, a coil has two conductors in this case. When the coil pitch y_1 equals to the pole pitch τ, i. e. $y_1 = \tau$, as the solid line shown in Fig. 3-12b, the emfs of the two conductors are the same in amplitude, and opposite in phase position, therefore, the emf of the coil is,

$$\dot{E}_{c1} = \dot{E}_1 - \dot{E}_1' = 2\dot{E}_1 \tag{3-10}$$

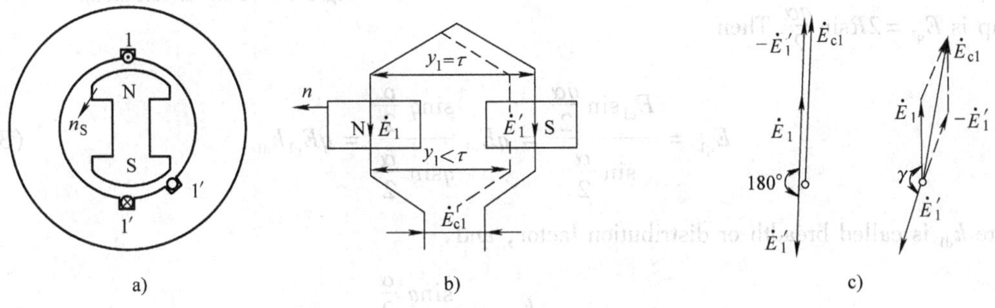

Fig. 3-12 EMF of full-pitch coil and fractional pitch coil

According to Eq. 3-9, its rms value is

$$E_{c1} = 2E_1 = 4.44f\Phi_1 \tag{3-11}$$

2. Fractional-pitch single-turn coil

When the coil pitch y_1 not equal to the pole pitch τ, i. e. $y_1 < \tau$, as the dotted line shown in Fig. 3-12b, the time phase difference of the emfs of the two conductors $\dot{E}_1$, $\dot{E}'_1$ is γ in this case, which could be expressed in electrical degree

$$\gamma = \frac{y_1}{\tau} \times 180° \tag{3-12}$$

Due to $\dot{E}_{c1} = \dot{E}_1 - \dot{E}_1'$, therefore the rms value of emf is

$$E_{c1} = 2E_1\cos\frac{180° - \gamma}{2} = 2E_1\sin\frac{\gamma}{2} = 2E_1\sin\frac{y_1}{\tau} \times 90° = 4.44fk_{p1}\Phi_1 \tag{3-13}$$

where k_{p1} is called pitch-factor, $k_{p1}=\sin\frac{y_1}{\tau}\times 90°$. For full-pitch coils, $k_{p1}=1$.

3. Multi-turns coil

If a coil has N_c turns, the emf of which is N_c times of emf of a single-turn coil, therefore, the rms value of emf of a multi-turns coil is

$$E_{c1} = 4.44fN_c k_{p1}\Phi_1 \tag{3-14}$$

3.3.3 Emf of coil group and breadth or distribution factor

A coil group has several (q) adjacent coils connected in series, so the emf of coil group is the vector sum of the emfs of each coils. In addition, the time phase difference of emfs between two adjacent coils is α. As shown in Fig. 3-13, the emf of a coil-group is $\dot{E}_{q1}=\dot{E}_{c1}+\dot{E}_{c2}+\dot{E}_{c3}$.

From Fig. 3-13, the phasors are connected end to end, and they form a portion of a regular polygon. R is the radius of the regular polygon, and $R=\frac{E_{c1}}{2\sin\frac{\alpha}{2}}$, the rms value of emf of a coil-group is $E_{q1}=2R\sin\frac{q\alpha}{2}$. Then

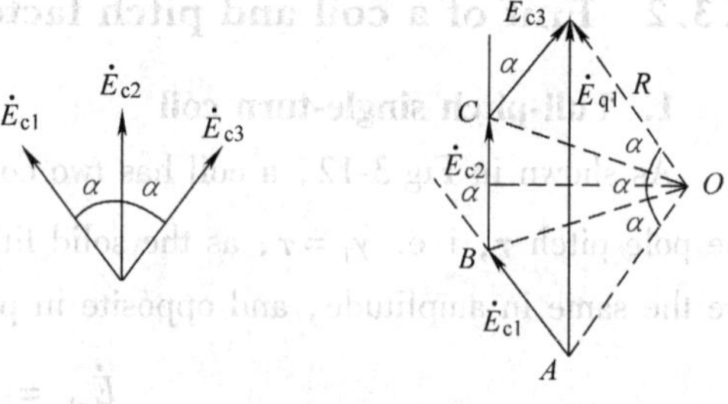

Fig. 3-13 EMF of coil-group

$$E_{q1} = \frac{E_{c1}\sin\frac{q\alpha}{2}}{\sin\frac{\alpha}{2}} = qE_{c1}\frac{\sin q\frac{\alpha}{2}}{q\sin\frac{\alpha}{2}} = qE_{c1}k_{d1} \tag{3-15}$$

where k_{d1} is called breadth or distribution factor, and

$$k_{d1} = \frac{E_{q1}}{qE_{c1}} = \frac{\sin q\frac{\alpha}{2}}{q\sin\frac{\alpha}{2}} \tag{3-16}$$

Substituting Eq. 3-14 into Eq. 3-15, then the rms value of the emf of a coil-group is

$$E_{q1} = 4.44fN_c qk_{p1}k_{d1}\Phi_1 = 4.44fN_c qk_{w1}\Phi_1 \tag{3-17}$$

where k_{w1} is called winding factor, and

$$k_{w1} = k_{p1}k_{d1} \tag{3-18}$$

3.3.4 Fundamental emf of a phase winding

Emf of each phase is the emf generated in a parallel circuit of a phase-winding. The number of coil-groups in each phase is $2p$ (double-layer windings) or p (single-layer windings). Therefore, the emf of each phase is equal to $2p/a$ or p/a times of the emf of a coil-group, the rms value of the emf of a single-layer winding per phase is

$$E_{\phi 1} = \frac{p}{a}E_{q1} = 4.44f\frac{p}{a}qN_c k_{w1}\Phi_1 = 4.44fN_1 k_{w1}\Phi_1 \tag{3-19}$$

and the rms value of the emf of a double-layer winding per phase is

$$E_{\phi 1} = \frac{2p}{a}E_{q1} = 4.44f\frac{2p}{a}qN_c k_{w1}\Phi_1 = 4.44fN_1 k_{w1}\Phi_1 \tag{3-20}$$

where N_1 is the total series number of turns per phase. For single-layer windings:

$$N_1 = \frac{p}{a}qN_c \tag{3-21}$$

For double-layer windings:

$$N_1 = \frac{2p}{a}qN_c \tag{3-22}$$

The rms voltage at the terminals of the machine depends on whether the stator winding is Y- or Δ-connection. If the stator winding is Y-connection, the terminal voltage will be $\sqrt{3}$ times $E_{\phi 1}$; on the other hand, if the stator winding is Δ-connection, the terminal voltage will just equal to $E_{\phi 1}$.

If three coils, each of N_c turns, are placed around the stator, then the voltages induced in each of them will be the same as in the magnitude but differ in phase by 120°. The resulting voltages in each of the three coils are

$$\begin{cases} e_{AA'}(t) = \sqrt{2}E_{\phi 1}\sin\omega t \\ e_{BB'}(t) = \sqrt{2}E_{\phi 1}\sin(\omega t - 120°) \\ e_{CC'}(t) = \sqrt{2}E_{\phi 1}\sin(\omega t - 240°) \end{cases} \tag{3-23}$$

3.3.5 Harmonic emf of a phase winding

If the magnetic field in the air-gap consists not only the fundamental component but also the high-order harmonic components, and there will be the high-order emf that is induced in the windings. The frequency of the ν-order emf is ν times f_1 (frequency of the fundamental emf).

According to Eq. 3-19 or Eq. 3-20, the rms of the ν-order emf can be expressed as

$$E_{\phi\nu} = 4.44f_\nu N_1 k_{w\nu}\Phi_\nu \tag{3-24}$$

where Φ_ν is the magnitude of the ν-order harmonic flux; $k_{w\nu}$ is the ν-order winding factor, and $k_{w\nu} = k_{p\nu}k_{d\nu}$; $k_{d\nu}$ is the ν-order breadth or distribution factor, and $k_{d\nu} = \dfrac{\sin q\nu\dfrac{\alpha}{2}}{q\sin\nu\dfrac{\alpha}{2}}$; $k_{p\nu}$ is the ν-order pitch-factor, $k_{p\nu} = \sin\nu\dfrac{y_1}{\tau}\times 90°$.

Example 3-1 A two-pole three-phase synchronous generator, where the number of stator slots is $Q = 48$, the stator windings is a three-phase double-layer lap winding, the coil pitch $y_1 = 20$, the total series number of turns per phase $N_1 = 16$, and the per pole flux $\Phi = 1.11$ Wb, calculate the winding factor and rms value of emf per phase.

Solution:

1) The pole pitch is $\tau = \dfrac{Q}{2p} = \dfrac{48}{2} = 24$

the slot pitch is $\alpha = \dfrac{p\times 360°}{Q} = \dfrac{1\times 360°}{48} = 7.5°$

the number of slots that are distributed under each pole per phase is $q = \dfrac{Q}{2mp} = \dfrac{48}{2\times3} = 8$

the pitch factor is $k_{p1} = \sin\dfrac{y_1}{\tau}\times90° = \sin\dfrac{20}{24}\times90° = 0.966$

the breadth factor is $k_{d1} = \dfrac{\sin\dfrac{q\alpha}{2}}{q\sin\dfrac{\alpha}{2}} = \dfrac{\sin\dfrac{8\times7.5°}{2}}{8\sin\dfrac{7.5°}{2}} = 0.956$

the winding factor is $k_{w1} = k_{p1}k_{d1} = 0.966\times0.956 = 0.923$

2) The rms of the emf per phase is

$$E_{\phi1} = 4.44fN_1k_{w1}\Phi_1 = 4.44\times50\times16\times0.923\times1.11\text{V} = 3639\text{V}$$

3.4 Operation principles of the three-phase induction motor

Fig. 3-14 shows a two-pole three-phase induction motor model, the stator of which is the same as that of the model that is shown in Fig. 3-1. Rotor conductors are placed in the rotor slots, and two ends of all conductors are connected to two end-rings respectively, so that all the rotor conductors are short circuited together.

To understand the concept of a rotating magnetic field, we will apply a set of currents to a stator of Fig. 3-14 and analyze what will happen at a specific instant of time. Assuming that the currents in the three coils are given by the equations:

Fig. 3-14 Model of an induction motor

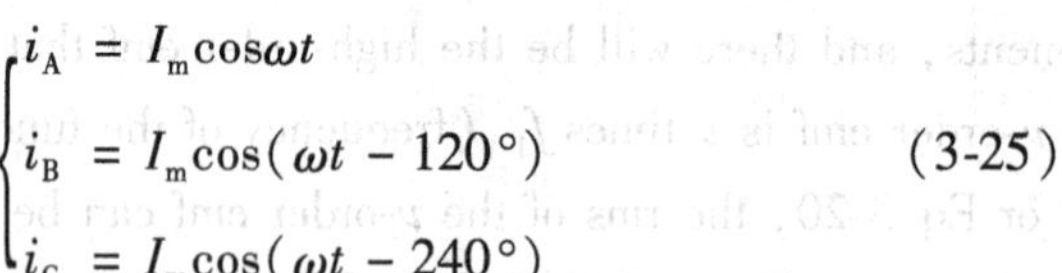

$$\begin{cases} i_A = I_m\cos\omega t \\ i_B = I_m\cos(\omega t - 120°) \\ i_C = I_m\cos(\omega t - 240°) \end{cases} \tag{3-25}$$

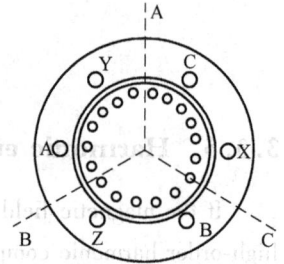

If an instantaneous value of a current is positive, the current will flow into the end (X, Y, Z) of a coil and out of the head (A, B, C) of the coil.

At time $\omega t = 0°$, the currents are $i_A = I_m$, $i_B = i_C = -\dfrac{1}{2}I_m$. Fig. 3-15a shows the direction of the current in stator conductors and a magnetic field that are produced by the currents at this moment.

Similarly, when $\omega t = 120°$ and $\omega t = 240°$, the current direction and the magnetic field at two moments are shown in Fig. 3-15b and Fig. 3-15c respectively. If the currents vary with times, the field that is formed in air gap by the currents will rotate in a counter-clockwise direction.

The fundamental principle of an AC machine operation is that if a three-phase set of currents, each of equal magnitude and differing in phase by 120°, flows in a three-phase winding, then it will produce a rotating magnetic field of a constant magnitude. The three-phase winding consists of three separate windings that are spaced 120 electric degrees apart around the surface of the machine.

Referring to Fig. 3-15, if the rotation speed of a rotor is zero, the rotor conductors will "cut" the rotating magnetic field, and emfs will be induced in them. According to the right-hand rule, the

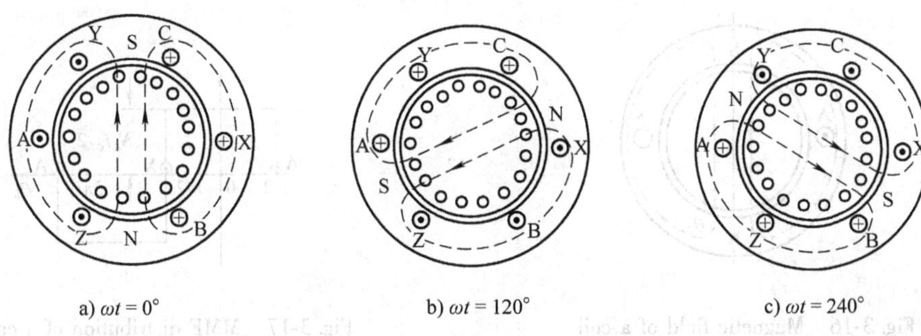

Fig. 3-15 Rotating magnetic field

direction of an emf of the conductors that are under the N pole is into paper, and the direction of an emf of the conductors which are under the S pole is out of paper. The rotor conductors are shorted by the end-rings, so that the currents will flow in the rotor conductors, and the direction of the rotor currents will be the same as the direction of the emf.

When the rotor currents interact with the rotating magnetic field in the air gap, electromagnetic force will be induced on the conductors, the direction of the force is determined by the left-hand rule. The direction of the torque is counter-clockwise; this is due to the electromagnetic forces which are called electromagnetic torque. Under the action of the electromagnetic torque, the rotor will rotate in a counter-clockwise direction, and the rotor speed will be below the speed of a rotating field.

3.5 Mmf of an AC winding

In this section, we will study the mmf that is produced by AC currents, which are flowing in the AC windings. To simplify, we assume that:

1) The currents in the windings are sinusoidal with time.
2) The currents are concentrated at center of slots.
3) Air gap is uniform, and the effect of slot-tooth is neglected.
4) The reluctance of core is zero.

3.5.1 Mmf of a single-phase winding

1. Mmf of full-pitch coil

Fig. 3-16 shows a two-pole magnetic field distribution that is produced by a current in a full-pitch coil with N_c turns. If the current is $i_c = \sqrt{2}I_c\cos\omega t$, it will produce an mmf in the air gap and the magnitude of mmf will be $N_c i_c$. Because the mmf drop of core is neglected, the air gap is uniform, and each flux line crosses the air gap twice, therefore the mmf drop across the air gap is equal to half of the total mmf, i. e. $N_c i_c/2$.

In Fig. 3- 17, the coil is shown in a developed form, the air-gap mmf distribution is shown by the step-like distribution, and the magnitude of the air-gap mmf is $N_c i_c/2$. The positive half wave means that the flux is directed from the stator into the rotor; and the negative half wave means that

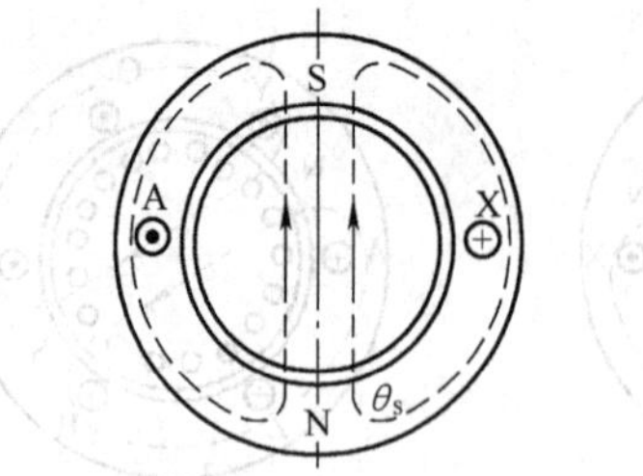

Fig. 3-16 Magnetic field of a coil

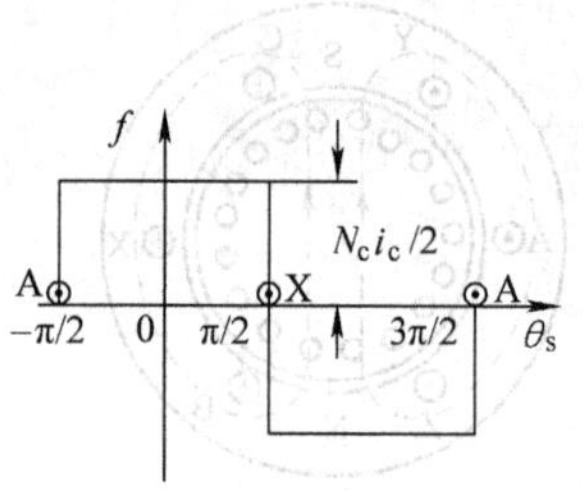

Fig. 3-17 MMF distribution of a coil

the flux is directed from the rotor into the stator. When the current varies with time, the position of the mmf wave is fixed, and the amplitude and direction of which change with current. The mmf in this case is called pulsating mmf, and it can be expressed as:

$$f_c(\theta_s,t) = \begin{cases} \dfrac{1}{2}i_c N_c = \dfrac{\sqrt{2}}{2}I_c N_c \cos\omega t & \left(-\dfrac{\pi}{2} < \theta_s < \dfrac{\pi}{2}\right) \\ -\dfrac{1}{2}i_c N_c = -\dfrac{\sqrt{2}}{2}I_c N_c \cos\omega t & \left(\dfrac{\pi}{2} < \theta_s < \dfrac{3\pi}{2}\right) \end{cases} \tag{3-26}$$

The function is represented by Eq. 3-26, which is resolved into Fourier series as following:

$$\begin{aligned} f_c(\theta_s,t) &= \frac{\sqrt{2}}{2}N_c I_c\left[\frac{4}{\pi}\left(\cos\theta_s - \frac{1}{3}\cos 3\theta_s + \frac{1}{5}\cos 5\theta_s - \cdots\right)\right]\cos\omega t \\ &= \left(F_{c1}\cos\theta_s - F_{c3}\cos 3\theta_s + F_{c5}\cos 5\theta_s - \cdots\right)\cos\omega t \end{aligned} \tag{3-27}$$

$$F_{c1} = \frac{4\sqrt{2}}{2\pi}N_c I_c = 0.9N_c I_c \tag{3-28}$$

$$F_{c\nu} = \frac{1}{\nu}F_{c1} \tag{3-29}$$

where F_{c1} is the amplitude of the fundamental component of the mmf per pole; $F_{c\nu}$ is the amplitude of the ν-order harmonic component of the mmf per pole.

Fig. 3-18 shows the fundamental and the third order components of a square wave. Notice that the amplitude of the fundamental component is at the axis of the coil.

In the design of AC machine, series efforts are made to the coils in order to minimize the higher-order harmonic components and to produce an air-gap mmf wave which consists predominantly of the fundamental component. Therefore, it is appropriate to focus on the fundamental component. From Eq. 3-27, the fundamental of the mmf of a full-pitch coil is

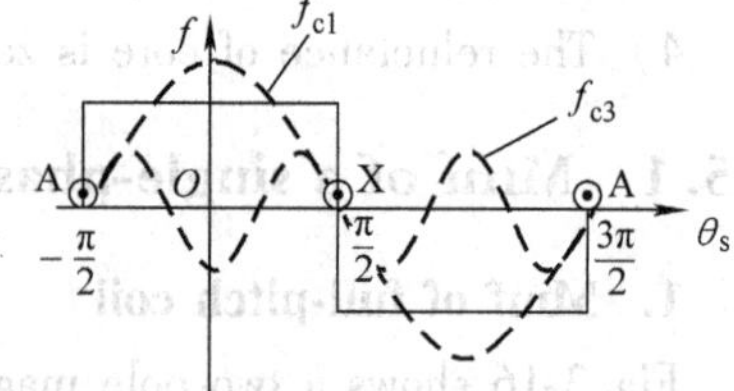

Fig. 3-18 The fundamental and higher order harmonic components of a square wave

$$f_{c1}(\theta_s,t) = F_{c1}\cos\theta_s\cos\omega t \tag{3-30}$$

2. Mmf of full-pitch coil-group

Full-pitch coil-group consists of several coils that are connected in series. Fig. 3-19 shows mmf wave of a coil-group that consists of three coils and this is a step wave. For the convenience of computation, the space-vector is introduced. Space-vector is used to represent the physical quantities that change sinusoidally in space. The length of the space-vector represents the amplitude of the quantity, and the direction of a space-vector represents the position of the positive amplitude of the quantity.

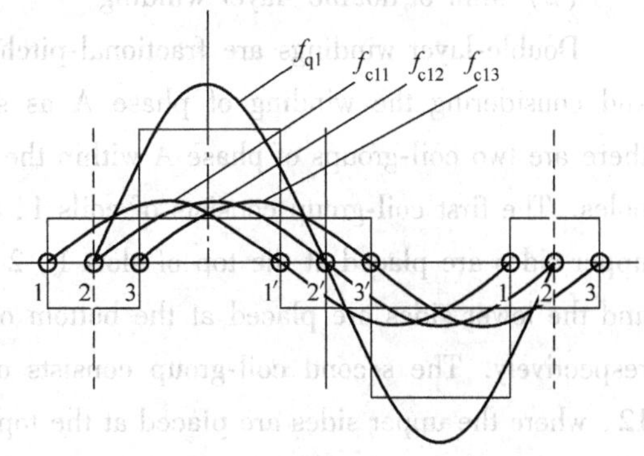

Fig. 3-19 MMF of full-pitch coil-group

As shown in Fig. 3-19, f_{c11}, f_{c12} and f_{c13} are the fundamentals of the mmf waves of three coils in a coil-group respectively. f_{q1} is the fundamental of the mmf wave of a coil-group. In Fig. 3-20, $\boldsymbol{F}_{c11}$、$\boldsymbol{F}_{c12}$、$\boldsymbol{F}_{c13}$ are the space-vectors that represent f_{c11}, f_{c12} and f_{c13}, with the phase difference of α between two adjacent space-vectors; and $\boldsymbol{F}_{q1}$ is the space-vector that represents f_{q1}.

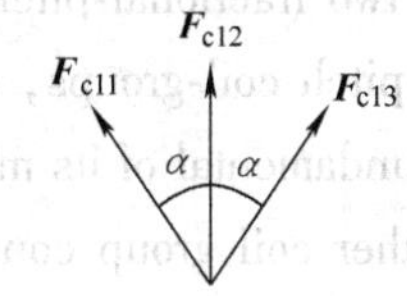

Fig. 3-20 Space vectors

Referring to the computation of the emf of a coil-group, the amplitude of fundamental of the mmf of a coil-group per pole can be expressed as:

$$F_{q1} = qF_{c1}k_{d1} = 0.9N_cI_cqk_{d1} \tag{3-31}$$

where $k_{d1} = \dfrac{\sin q\dfrac{\alpha}{2}}{q\sin\dfrac{\alpha}{2}}$ is the fundamental breadth or distribution factor.

The amplitude of the ν-order harmonic mmf of a coil-group per pole can be expressed as:

$$F_{q\nu} = qF_{c\nu}k_{d\nu} = \frac{1}{\nu}0.9N_cI_cqk_{d\nu} \tag{3-32}$$

where $k_{d\nu} = \dfrac{\sin q\nu\dfrac{\alpha}{2}}{q\sin\nu\dfrac{\alpha}{2}}$ is a ν-order breadth or distribution factor.

Notice that the amplitude of the fundamental is at the axis of a coil-group. The fundamental of the mmf of a full-pitch coil-group can be expressed as

$$f_{q1}(\theta_s,t) = F_{q1}\cos\theta_s\cos\omega t \tag{3-33}$$

3. Mmf of a single-phase winding

(1) Mmf of single-layer winding

For single-layer windings, the mmf of a single-phase winding per pole is equal to the mmf of a full-pitch coil-group per pole, therefore the amplitude of the fundamental and harmonic mmf per pole are produced by a single-phase winding, they are

$$\begin{cases} F_{\phi1} = F_{q1} = 0.9qN_cI_ck_{d1} \\ F_{\phi\nu} = F_{q\nu} = \dfrac{1}{\nu}0.9qN_cI_ck_{d\nu} \end{cases} \tag{3-34}$$

(2) Mmf of double -layer winding

Double-layer windings are fractional-pitch windings usually. Referring back to Example 3-2, and considering the winding of phase A as shown in Fig. 3-21, there are two coil-groups of phase A within the range of one pair of poles. The first coil-group consists of coils 1, 2 and 3, where the upper sides are placed at the top of slots 1, 2 and 3 respectively; and the lower sides are placed at the bottom of slot 9, 10 and 11 respectively. The second coil-group consists of coils 10, 11 and 12, where the upper sides are placed at the top of slots 10, 11 and 12 respectively; and the lower sides are placed at the bottom of slot 18, 1 and 2 respectively.

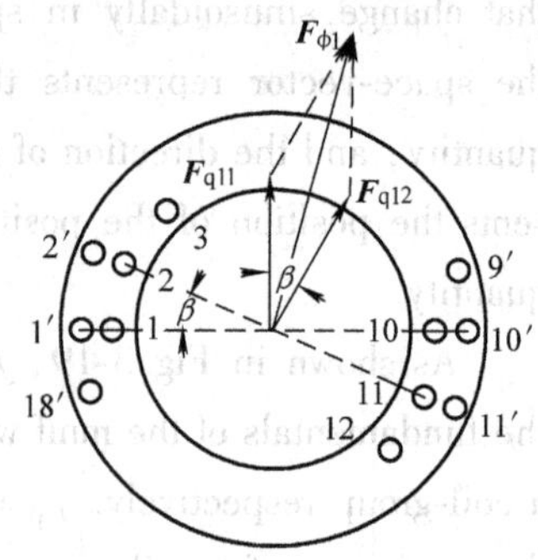

Fig. 3-21 Double-layer fractional-pitch winding within a pair of poles

Now, the two fractional-pitch coil-groups can be seen as two equivalent full-pitch coil-groups, one of them consists of the lower sides and the fundamental of its mmf is represented by a space vector $\boldsymbol{F}_{q11}$; the other coil group consists of the upper sides, and the fundamental of its mmf is represented by a space vector $\boldsymbol{F}_{q12}$. As we can see, the space phase difference between the two space-vectors is β, and $\beta=\left(1-\frac{y_1}{\tau}\right)\times 180°$. Referring to the computation of emf of a fractional pitch winding, the amplitude of fundamental of mmf per pole produced by a single-phase winding is

$$F_{\phi 1} = 2F_{q1}k_{p1} = 2\times 0.9qN_cI_ck_{p1}k_{d1} = 0.9\times 2qN_cI_ck_{w1} \tag{3-35}$$

where $k_{p1}=\sin\frac{y_1}{\tau}\times 90°$, is called pitch factor, and $k_{w1}=k_{p1}k_{d1}$ is called winding factor.

The amplitude of the ν-order harmonic mmf per pole that is produced by a single-phase winding is

$$F_{\phi\nu} = 2F_{q\nu}k_{p\nu} = \frac{1}{\nu}2\times 0.9qN_cI_ck_{p\nu}k_{d\nu} = \frac{1}{\nu}0.9\times 2qN_cI_ck_{w\nu} \tag{3-36}$$

where $k_{p\nu}=\sin\nu\frac{y_1}{\tau}\times 90°$, and $k_{w\nu}=k_{p\nu}k_{d\nu}$.

For single-layer windings:

$$qN_c = \frac{N_1 a}{p} \tag{3-37}$$

For double-layer windings:

$$qN_c = \frac{N_1 a}{2p} \tag{3-38}$$

The relationship between the phase current I_ϕ and the current I_c in a parallel circuit is

$$I_c = \frac{I_\phi}{a} \tag{3-39}$$

Both the single-layer windings and the double-layer windings, the amplitude of the fundamental mmf of the phase-winding per pole can be expressed as:

$$F_{\phi 1} = 0.9\frac{N_1 I_\phi}{p}k_{w1} \tag{3-40}$$

The amplitude of a ν-order harmonic mmf of phase-windings per pole is

$$F_{\phi\nu} = \frac{1}{\nu}0.9\frac{N_1 I_\phi}{p}k_{w\nu} \tag{3-41}$$

Notice that the amplitude of the fundamental is at the axis of the single-phase winding.

The fundamental mmf of a single-phase winding and a ν-order harmonic mmf of a single-phase winding are

$$\begin{cases} f_{\phi 1}(\theta_s, t) = F_{\phi 1}\cos\theta_s\cos\omega t \\ f_{\phi\nu}(\theta_s, t) = F_{\phi\nu}\cos\nu\theta_s\cos\omega t \end{cases} \tag{3-42}$$

3.5.2 Mmf of a three-phase winding

1. Fundamental Mmf of three-phase winding

To understand the concept of the rotating magnetic field, we will apply a set of currents to three-phase windings. Assumed that the currents in the three coils are given by the equations:

$$\begin{cases} i_A = I_m\cos\omega t \\ i_B = I_m\cos(\omega t - 120°) \\ i_C = I_m\cos(\omega t - 240°) \end{cases} \tag{3-43}$$

The three-phase windings consist of three single-phase windings, which are displaced from each other by 120 electric degree in space around the air gap of the machine; the fundamental mmfs of three single-phase windings are given by

$$\begin{cases} f_{A1}(\theta_s, t) = F_{\phi 1}\cos\theta_s\cos\omega t \\ f_{B1}(\theta_s, t) = F_{\phi 1}\cos(\theta_s - 120°)\cos(\omega t - 120°) \\ f_{C1}(\theta_s, t) = F_{\phi 1}\cos(\theta_s - 240°)\cos(\omega t - 240°) \end{cases} \tag{3-44}$$

By using a common trigonometric transformation, Eq. 3-44 can be rewritten as

$$\begin{cases} f_{A1}(\theta_s, t) = \frac{1}{2}F_{\phi 1}\cos(\theta_s - \omega t) + \frac{1}{2}F_{\phi 1}\cos(\theta_s + \omega t) \\ f_{B1}(\theta_s, t) = \frac{1}{2}F_{\phi 1}\cos(\theta_s - \omega t) + \frac{1}{2}F_{\phi 1}\cos(\theta_s + \omega t - 240°) \\ f_{C1}(\theta_s, t) = \frac{1}{2}F_{\phi 1}\cos(\theta_s - \omega t) + \frac{1}{2}F_{\phi 1}\cos(\theta_s + \omega t - 120°) \end{cases} \tag{3-45}$$

The total mmf is the sum of the contributions from each of the three phases:

$$\begin{aligned} f_1(\theta_s, t) &= f_{A1}(\theta_s, t) + f_{B1}(\theta_s, t) + f_{C1}(\theta_s, t) \\ &= \frac{3}{2}F_{\phi 1}\cos(\omega t - \theta_s) = F_1\cos(\omega t - \theta_s) \end{aligned} \tag{3-46}$$

where F_1 is the amplitude of fundamental of the mmf per pole, which is produced by a three-phase winding, and

$$F_1 = \frac{3}{2}F_{\phi 1} = \frac{3}{2}\times 0.9\frac{N_1 k_{w1}}{p}I_\phi = 1.35\frac{N_1 k_{w1}}{p}I_\phi \tag{3-47}$$

Notice that:

1) When $\omega t = 0°$, $i_A = I_m$, $i_B = i_C = -\frac{1}{2}I_m$, and $f_1\ (\theta_s,\ t)\ = F_1\cos\ (\theta_s)$. As shown in Fig. 3-22a, the amplitude of the mmf wave is F_1, and its position is at the axis of phase A.

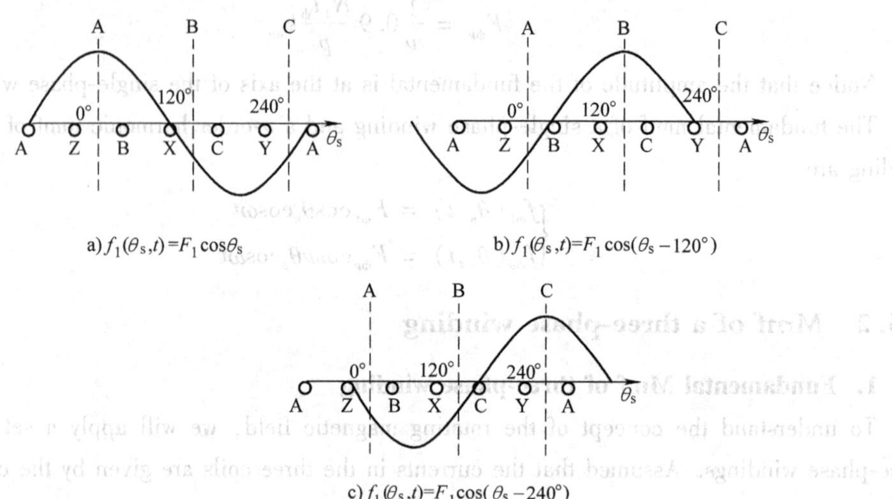

Fig. 3-22 Fundamental component of mmf

2) When $\omega t = 120°$, $i_B = I_m$, $i_A = i_C = -\frac{1}{2}I_m$, and $f_1\ (\theta_s,\ t)\ = F_1\cos\ (\theta_s - 120°)$. As shown in Fig. 3-22b, the amplitude of the mmf wave is F_1, and its position is at the axis of phase B.

Comparing this situation with 1), 120 degree has been progressed in time, and the position of the amplitude has rotated 120 electric degrees in space.

3) When $\omega t = 240°$, $i_C = I_m$, $i_A = i_B = -\frac{1}{2}I_m$, and $f_1\ (\theta_s,\ t)\ = F_1\cos\ (\theta_s - 240°)$. As shown in Fig. 3-22c, the amplitude of the mmf wave is F_1, its position is at the axis of phase C.

Comparing this situation with 2), 120 degree has been progressed in time again, and the position of the amplitude has rotated 120 electric degrees in space once more.

4) When $\omega t = 360°$, the situation will be the same as 1). Therefore, the mmf wave travels one pole-pair for one electrical cycle, and its speed [r/min] is

$$n_s = \frac{60f}{p} \tag{3-48}$$

where f is the frequency of the currents.

The characteristics of the fundamental mmf wave of a three-phase winding are illustrated as following:

1) The fundamental mmf wave of a three-phase winding is a rotating mmf wave; its speed is a synchronous speed, where $n_s = \frac{60f}{p}$.

2) When the current of any phase has reached its maximum, the positive amplitude of the mmf wave will arrive at the axis of that phase winding. Its rotating direction can therefore be reversed by reversing the phase sequence of the currents.

3) It has constant amplitude of 1.5 times of the amplitude of mmf wave which is produced by a single-phase winding alone.

2. Harmonic mmf of three-phase windings

In a same manner, the total of ν-order harmonic mmf of three-phase windings (A, B, C) can be expressed as

$$\begin{aligned} f_\nu(\theta_s,t) &= f_{A\nu}(\theta_s,t) + f_{B\nu}(\theta_s,t) + f_{C\nu}(\theta_s,t) \\ &= F_{\phi\nu}\cos\nu\theta_s\cos\omega t + F_{\phi\nu}\cos\nu(\theta_s - 120°)\cos(\omega t - 120°) + \\ &\quad F_{\phi\nu}\cos\nu(\theta_s - 240°)\cos(\omega t - 240°) \end{aligned} \tag{3-49}$$

Notice that:

1) when $\nu = 3k$ ($k = 1, 2, 3, \cdots$), i. e. $\nu = 3, 6, 9, \cdots$, the total ν-order harmonic mmf of three-phase windings (A, B, C) is not existed, and $f_\nu(\theta_s, t) = 0$.

2) when $\nu = 6k + 1$ ($k = 1, 2, 3, \cdots$), i. e. $\nu = 7, 13, \cdots$, the total ν-order harmonic mmf of three-phase windings (A, B, C) is expressed as: $f_\nu(\theta_s, t) = \frac{3}{2}F_{\phi\nu}\cos(\omega t - \nu\theta_s)$. The rotating direction of the ν-order harmonic mmf is the same as that of the fundamental mmf, its rotating speed is n_s/ν.

3) when $\nu = 6k - 1$ ($k = 1, 2, 3\cdots$), i. e. $\nu = 5, 11, \cdots$, the total ν-order harmonic mmf of three-phase windings (A, B, C) is expressed as: $f_\nu(\theta_s, t) = \frac{3}{2}F_{\phi\nu}\cos(\omega t + \nu\theta_s)$. The rotating direction of the ν-order harmonic mmf is reverse of the fundamental mmf's, and its rotating speed is n_s/ν.

Summary

This chapter introduces the configuration of AC windings, the calculations of induced emf and of mmf of AC windings. Students should pay more attention to the following points when studying this chapter.

1) AC windings are formed through several steps: distributing the conductors or coils to each phase; forming coil-groups; and connecting the coil-groups' belonging to one phase. There are some basic concepts of AC windings, such as the relationship between the electrical angle θ_e and the mechanical angle θ_m is $\theta_e = p\theta_m$; pole pitch τ; coil pitch y_1 and slot pitch α.

2) In calculating the induced emf, the pitch factor k_{p1} ($k_{p1} = \sin\frac{y_1}{\tau}\times 90°$); the breadth or distribution factor k_{d1} ($k_{d1} = \frac{E_{q1}}{qE_{c1}} = \frac{\sin q\frac{\alpha}{2}}{q\sin\frac{\alpha}{2}}$) and the winding factor k_{w1} ($k_{w1} = k_{p1}k_{d1}$) are introduced for fractional-pitch distributed windings. For full-pitch windings, the pitch factor $k_{p1} = 1$. For concentrated windings, the breadth factor $k_{d1} = 1$. For full-pitch concentrated windings, the winding factor $k_{w1} = 1$.

The rms value of the emf of a double-layer winding per phase is $E_{\phi 1} = 4.44fN_1k_{w1}\Phi_1$; the rms of the ν-order emf can be calculated by $E_{\phi\nu} = 4.44f_\nu N_1 k_{w\nu}\Phi_\nu$.

3) The characteristics of mmf of AC windings are very important. The amplitude of the fundamental mmf of a phase-winding per pole can be expressed as $F_{\phi 1}=0.9\frac{N_1 I_\phi}{p}k_{w1}$; the amplitude of a ν-order harmonic mmf of phase-winding per pole is calculated by $F_{\phi\nu}=\frac{1}{\nu}0.9\frac{N_1 I_\phi}{p}k_{w\nu}$. The amplitude of fundamental mmf produced by a three-phase winding is $F_1=1.35\frac{N_1 k_{w1}}{p}I_\phi$.

The fundamental of mmf is produced by a single-phase-winding, which is a standing wave, and the fundamental of mmf is produced by a three-phase winding which is a traveling wave. The characteristics of the fundamental mmf wave of a three-phase winding are illustrated as following:

① The fundamental component of mmf wave of a three-phase winding is a rotating mmf wave; its speed is a synchronous speed [r/min], where $n_s=\frac{60f}{p}$.

② When the current of any phase has reached its maximum, the positive amplitude of the mmf wave will arrive at the axis of that phase winding. Its rotating direction can therefore be reversed by reversing the phase sequence of the currents.

③ It has constant amplitude of 1.5 times of the amplitude of mmf wave which is produced by a single-phase winding alone.

Questions and problems

1. What are the major differences between synchronous machine and induction machine?
2. What are the natures of fundamental of mmf wave of a single-phase winding?
3. What are the natures of fundamental of mmf wave of a three-phase winding?
4. What is the relationship between the electrical frequency and magnetic field speed for an AC machine?
5. A three-phase single-layer winding, with the number of slots $Q=24$, the number of poles $2p=4$, the number of parallel circuits $a=2$. Sketch a developed diagram of the three-phase winding and calculate the winding factor?
6. A three-phase double-layer lap winding, with the number of slots $Q=24$, the number of poles $2p=2$, winding pitch $y_1=10$, and the number of parallel circuit $a=1$. Sketch the developed diagram of phase A and calculate the winding factor?
7. How to reverse the rotation direction of an induction motor?
8. Indicate the physical meanings of the pitch factor and the distribution factor?
9. If two-phase currents (the phase difference is 90°) are flowing in two-phase windings (the space difference of the two winding-axis is 90 electrical degree), examine the fundamentals of the mmf wave.
10. Indicate the physical reasons of harmonic emf induced in AC windings? How to eliminate or weaken the harmonic induced emf ?
11. Indicate the physical reasons of harmonic mmf produced in AC machines?
12. A six-pole three-phase synchronous generator, which the stator has a three-phase double-

layer lap winding, the number of slots $Q=54$, winding pitch $y_1=7$, the number of parallel circuits $a=2$, Y-connection, the turns of each coil is 10. If the flux per-pole is $\Phi_m=0.11\text{Wb}$, hence calculate the rms of the emf of the winding of one phase.

13. A four-pole three-phase induction motor, which the stator has a three-phase double-layer lap winding, with the number of slots $Q=24$, winding pitch $y_1=5$, the number of parallel circuits $a=1$, $N_c=10$ (the turns of each coil is 10) Y-connection. If three-phase currents of 50Hz and 20A (rms) flow in the windings, what is the rotation speed and amplitude of the fundamental of the mmf wave?

14. A three-phase, Y-connected, 50Hz, two-pole synchronous machine has a stator with 2000 turns of wire per phase and winding factor $k_{w1}=0.966$. What rotor flux would be required to produce a terminal (line to line) voltage of 6 kV?

15. A two-pole, three-phase, Y-connected 60Hz round-rotor synchronous generator has a armature winding with 18 turns per phase and winding factor $k_{w1}=0.933$. The fundamental flux per pole is 3.31Wb. The rotor is driven by a steam turbine at a speed of 3600 r/min. Compute the rms value of the open-circuit voltage generated in the armature.

16. A two-pole, 60Hz, three-phase, laboratory-size synchronous generator has a Y-connected armature winding, the total series number of turns per phase is $N_1=45$, the winding factor $k_{w1}=0.93$. The flux per pole is 2.1Wb, the armature current $I_a=20$ A.

1) Calculate the peak value and rms value of the induced voltage in armature-phase-winding.

2) Calculate the peak value of the fundamental mmf in armature-phase-winding.

3) Calculate the peak value of the fundamental mmf in three-phase armature windings.

17. If currents of same rms and same phase position flow in a three-phase winding, examine the fundamental of the mmf wave.

18. A three-phase induction motor has a three-phase winding of Y-connection, if one-phase of the stator windings is broken; examine the fundamentals of the mmf in the machine.

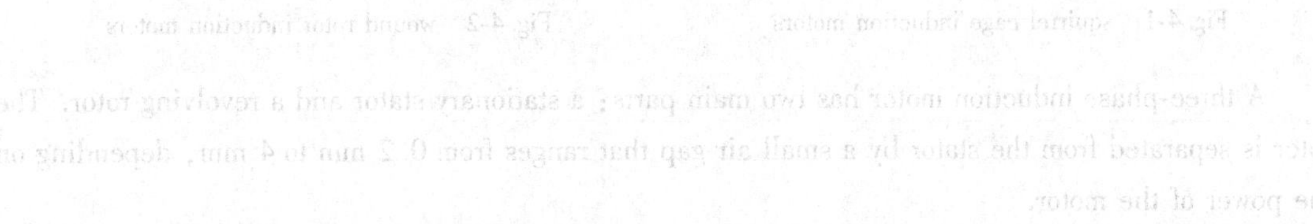

Chapter 4 Induction Machines

The induction machine is a kind of AC rotational electrical machine, which consists of motors and generators. Three-phase induction motor is widely used in the industry. It is simple, rugged, low-priced, durable and robust. At least 90 percent of drives used in the industry are induction motors. They need to run at essentially constant speed from zero to full-load. These motors are not easy to be adapted to speed control. However, variable frequency electronic drives are being used more frequently to control the speed of an induction motor.

In this chapter we are going to introduce the construction of a three-phase induction motor and how to develop fundamental equations and a equivalent circuit. Then, we are going to analyze the operation characteristics and mechanical characteristics. The start and speed control methods are also introduced.

4.1 Construction and ratings of induction motors

4.1.1 Construction of induction motors

According to the type of rotor winding, the induction motors are classified as squirrel-cage induction motors (which can also be called cage motors, Fig. 4-1) and wound-rotor induction motors (Fig. 4-2).

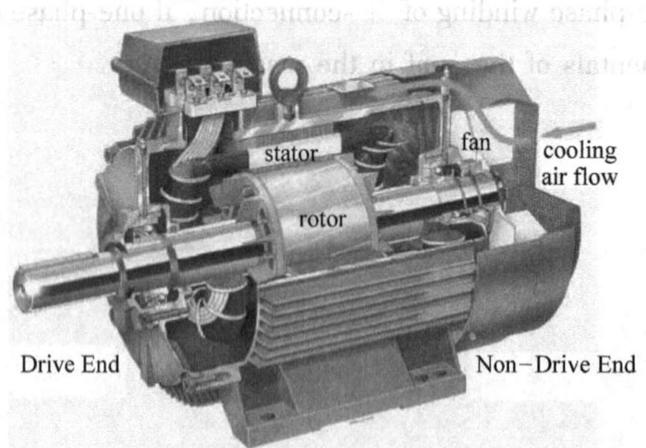

Fig. 4-1 squirrel cage induction motors

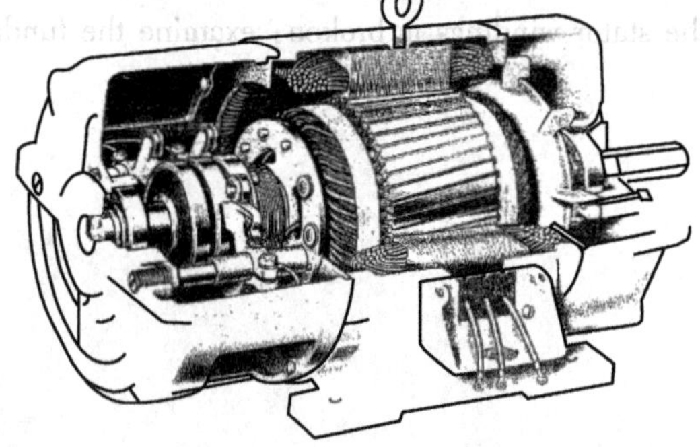

Fig. 4-2 wound rotor induction motors

A three-phase induction motor has two main parts: a stationary stator and a revolving rotor. The rotor is separated from the stator by a small air gap that ranges from 0.2 mm to 4 mm, depending on the power of the motor.

1. Stator

The stator of an induction machine consists of a stator core, a stator winding, a frame and a cover. A steel frame supports a hollow, cylindrical stacked lamination core and provides a space for the stator winding.

2. Rotor

The rotor is composed of a rotor core, a rotor winding and a shaft.

The rotor core is also composed of punched laminations. These are carefully stacked to create a series of rotor slots in order to provide space for the rotor winding. There are two types of rotor windings: squirrel-cage winding (Fig. 4-3) and conventional three-phase winding which is made of an insulated wire (i. e. wound rotor, Fig. 4-4).

Fig. 4-3 cage winding

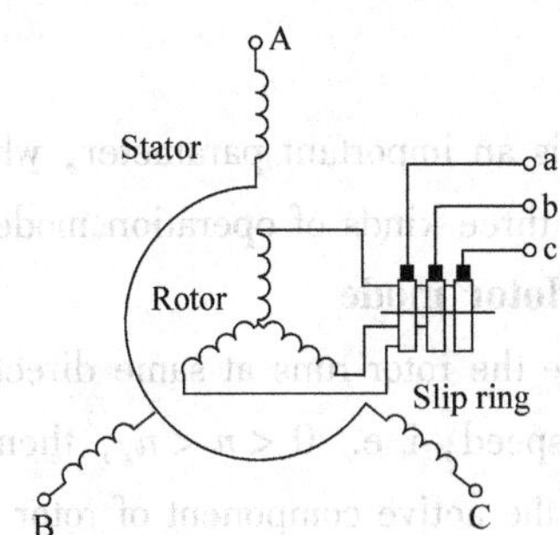

Fig. 4-4 wound rotor induction motor

A squirrel-cage rotor is composed of bare copper bars, which are slightly longer than the rotor, which are pushed into the slots. The opposite ends are welded into two copper end-rings, so that all the bars are short-circuited at common points. The entire construction (bars and end-rings) resembles a squirrel cage, from which the name is derived. In the small and medium-size motors, the bars and the end-rings are made of die-cast aluminum, which are molded to form an integral block.

A wound rotor consists a complete set of three phase windings, similar to the one on the stator. The winding is uniformly distributed in the slots and is usually connected with Wye connection (Y-connection). The terminals are connected to three slip-rings, which turns with the rotor. The revolving slip-rings and the associated stationary brushes enable us to connect the external resistors in series with the rotor windings. The external resistors are mainly used during the start-up period. Under a normal running condition, the three brushes are short-circuited.

4.1.2 Operation modes of induction machines

The operation of a three-phase induction motor is based upon the application of Faraday's law and the Lorentz force on a conductor. When the three-phase stator windings are excited from a balanced 3-phase source, it will produce a magnetic field in the air-gap by rotating at synchronous speed n_s. The synchronous speed is determined by the number of poles and the applied stator frequency f. The revolving field induces a voltage in the rotor bars, and the induced voltage creates large circulating currents which flow in the rotor bars and end-rings. The current-carrying rotor bars are immersed in the magnetic field which is created by the stator; they are therefore subjected to a strong mechanical force, and the sum of the mechanical forces on all the rotor bars produces a torque which tends to drag the rotor along the same direction as the revolving field. The rotor speed n is always less than the synchronous speed n_s so as to produce a current in the rotor bars sufficiently large enough to overcome the load torque.

Taking the rotating direction of revolving magnetic field as a reference, the difference between synchronous speed n_s and rotor speed n is named as slip speed Δn.

$$\Delta n = n_s - n \tag{4-1}$$

where Δn is the slip speed; n_s is the speed of the rotating magnetic fields (synchronous speed); n is the rotor speed.

The ratio of the slip speed and the synchronous speed is called *slip s* (per-unit slip).

$$s = \frac{n_s - n}{n_s} \tag{4-2}$$

Slip is an important parameter, which can reveal the operation mode of an induction machine. There are three kinds of operation modes, depending on the rotor speed of induction machines.

1. Motor mode

While the rotor runs at same direction as a magnetic field and the speed is lower than the synchronous speed, i. e. $0 < n < n_s$, then $0 < s < 1$. The direction of rotor induction electric motive force and the active component of rotor current is shown in Fig. 4-5b. The active component of rotor current cooperate with an air gap magnetic field, the electromagnetic torque is derived as the same direction with the air gap magnetic field. The rotor and the load rotate at the speed of n. Then the electric energy is converted into the mechanical energy and the induction machine is working as a motor.

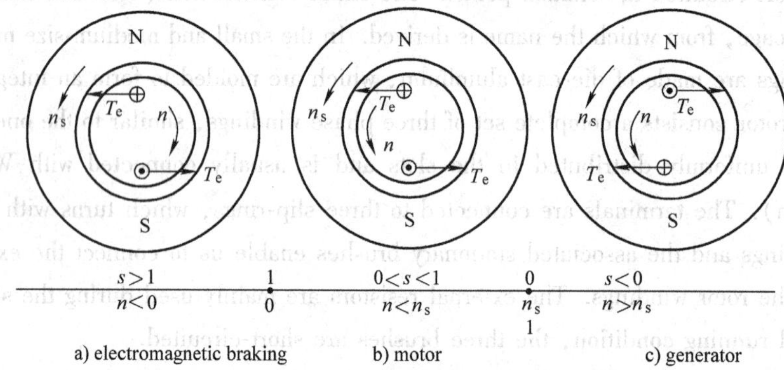

Fig. 4-5 three operation modes of the induction machine

2. Generator mode

If the rotor speed is higher than the synchronous speed, i. e. $n > n_s$, then $s < 0$ (Fig. 4-5c). Comparing to motor mode, the relative direction of the air gap magnetic field to the rotor is changed. Then, the direction of the current and the electric motive force of the rotor bar are changed. The direction of electromagnetic torque is reversed. The induction machine works as a generator and the mechanical energy is converted into electric energy.

3. Electromagnetic braking

When the direction of rotor is different from the rotation air gap field, i. e. $n < 0$, then $s > 1$

(Fig. 4-5a), the electromagnetic torque is in the same direction as the air gap magnetic field, which means that the machine is absorbing electric energy. The direction of electromagnetic torque is different from the rotor running direction, which means that the machine is absorbing mechanical energy. Both kinds of energy are converted into heat and dissipated into the electrical machine. It is working at an electromagnetic braking mode.

4.1.3 Ratings of induction motors

A nameplate of the induction motor always shows the following ratings:

(1) Rated output power P_N it means that the output power of motor is at rating status, the unit is kW.

(2) Rated voltage U_N it means that the line voltage of stator is at rating status, the unit is V.

(3) Rated current I_N it means that the line current of stator is at rating status, the unit is A.

(4) Rated speed n_N it means that the rotor speed is at rating status, the unit is r/min.

(5) Rated frequency f_N it means that frequency of power supply is at rating status, the unit is Hz.

Other than all the ratings stated above, there are some other ratings, such as efficiency η_N, power factor $\cos\varphi_N$, and temperature rise.

The input power of the induction motor is given by

$$P_1 = \sqrt{3}U_N I_N \cos\varphi_N \tag{4-3}$$

The output power of the induction motor is given by

$$P_N = P_1\eta_N = \sqrt{3}U_N I_N \eta_N \cos\varphi_N \tag{4-4}$$

4.2 Electromagnetic relations of an induction motor

4.2.1 Main flux and leakage flux of an induction motor

The induction motor is a kind of energy conversion machine that use the magnetic field as a medium. When a motor is running, the magnetic field in motor is established by both the stator current and the rotor current. The distribution of magnetic field is complex. Magnetic flux can be classified into the main flux and the leakage flux according to their function.

1. Main flux

The main flux Φ is the flux that linked simultaneously with the stator and the rotor windings. It is created by the stator and the rotor current individually or corporately. The path of the main flux is called the main magnetic path, as it is shown in Fig. 4-6. It consists of an air gap, a stator tooth, a stator yoke, a rotor tooth and a rotor yoke. Due to the main flux having inter-leakage with the stator winding and the rotor winding at same time, the induction motor generally transmits power through the main flux. The reluctance of main magnetic path is

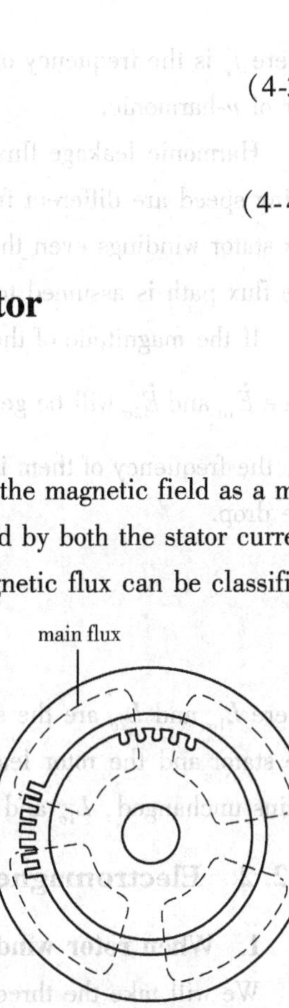

main flux

Fig. 4-6 Main flux

influenced by the saturation of iron core. The more saturated the iron core is, the greater reluctance of main magnetic path is.

2. Leakage flux

Except the main flux, all the other flux in the motor are called leakage flux. Leakage flux is classified into the stator leakage flux $\Phi_{1\sigma}$ and the rotor leakage flux $\Phi_{2\sigma}$, which are generated by the stator current and the rotor current respectively. The distribution of leakage flux is very complex. It includes the slot leakage flux, which crosses the slot (as shown in Fig. 4-7a) and the end leakage flux, which mainly surrounds the winding end as shown in Fig. 4-7b; and the harmonic leakage flux which is caused by the high-order harmonic magnetic motive force. The frequency of electric motive force in the winding that is induced by the harmonic leakage flux is the same as that is induced by the main flux, as

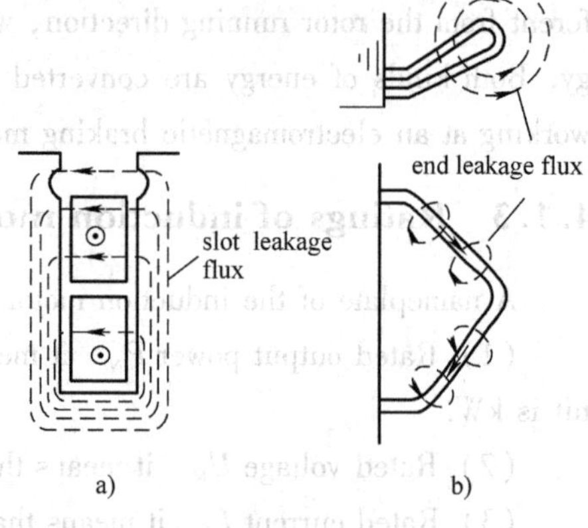

Fig. 4-7 Leakage flux

$$f_\nu = \frac{p_\nu n_\nu}{60} = \frac{\nu p \frac{n_s}{\nu}}{60} = \frac{pn_s}{60} = f_1 \tag{4-5}$$

where f_ν is the frequency of ν-harmonic, n_ν is the speed of ν-harmonic, and p_ν is the pole-pair number of ν-harmonic.

Harmonic leakage flux belongs to the leakage flux, this is because its pole-pairs number and rotating speed are different from the fundamental wave, and this can cause reactance voltage drop in the stator windings even though it goes into the rotor through the air gap. Generally speaking, leakage flux path is assumed to be unsaturated and the reluctance of leakage flux keeps constant.

If the magnitude of the stator and the rotor leakage flux varies as frequency f_1, the electromotive force $\dot{E}_{1\sigma}$ and $\dot{E}_{2\sigma}$ will be generated by the leakage flux in the stator and the rotor windings respectively, the frequency of them is also f_1. $\dot{E}_{1\sigma}$ and $\dot{E}_{2\sigma}$ can be expressed as negative leakage reactance voltage drop.

$$\begin{aligned} \dot{E}_{1\sigma} &= -\mathrm{j}\dot{I}_1 2\pi f_1 L_{1\sigma} = -\mathrm{j}\dot{I}_1 X_{1\sigma} \\ \dot{E}_{2\sigma} &= -\mathrm{j}\dot{I}_2 2\pi f_1 L_{2\sigma} = -\mathrm{j}\dot{I}_2 X_{2\sigma} \end{aligned} \tag{4-6}$$

where $L_{1\sigma}$ and $L_{2\sigma}$ are the stator and the rotor leakage inductance respectively; and $X_{1\sigma}$ and $X_{2\sigma}$ are the stator and the rotor leakage reactance respectively. Because the reluctance of leakage flux remains unchanged, $L_{1\sigma}$ and $L_{2\sigma}$ are all constant.

4.2.2 Electromagnetic relations of an induction motor

1. When rotor windings are open circuited

We will take the three-phase wound-rotor induction motor as an example to analyze the operating condition when the motor is stationary. In our study, the variables that belong to the stator and the rotor are distinguished by subscripts "1" and "2" respectively. Assuming the phase voltage and the frequency of three-phase symmetrical power supply are U_1 and f_1 respectively, the pole-pair number of stator and rotor winding is p.

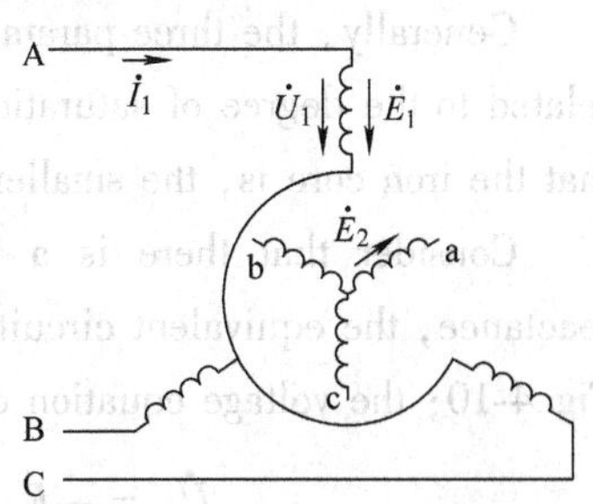

Fig. 4-8 stator and rotor circuits of wound-rotor induction motor

While the rotor winding circuit opens, the rotor current $\dot{I}_2 = 0$, the rotating speed $n = 0$, the slip $s = 1$, the situation is similar to the transformer at no-load. As shown in Fig. 4-8, Under the effect of $\dot{U}_1$, the three-phase symmetrical current $\dot{I}_1$ with the frequency f_1 flows in the stator windings, (three-phase stator current $\dot{I}_{1A}$、$\dot{I}_{1B}$、$\dot{I}_{1C}$). Positive direction of stator current $\dot{I}_1$ can be determined according to the motor convention. The speed of the fundamental rotating magnetomotive force $\boldsymbol{F}_1$ which is generated by the stator current is $n_s = \dfrac{60f_1}{p}$ and the direction of rotating is counter-clockwise. Due to $\dot{I}_2 = 0$, rotating magnetic field in the air gap is built up by $\boldsymbol{F}_1$ only. $\dot{I}_1$ is the pure magnetizing current $\dot{I}_m$ and $\boldsymbol{F}_1$ is the pure magnetizing magnetomotive force $\boldsymbol{F}_m$ as shown in Fig. 4-9. Due to the influence of magnetic hysteresis and eddy current, the flux density wave will lag the magnetomotive force wave spatially by an angle of iron loss α_{Fe}.

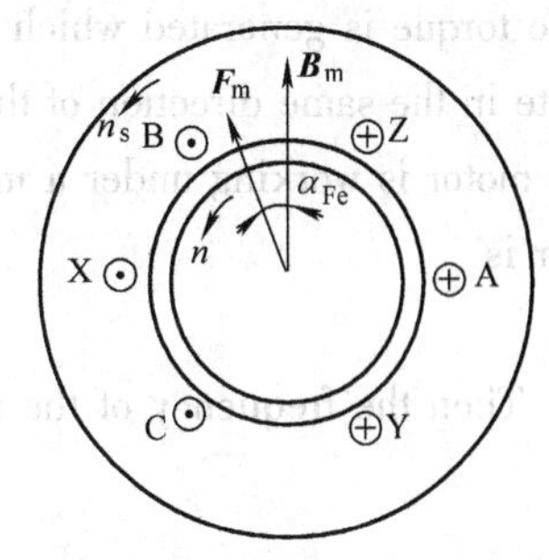

Fig. 4-9 Magnetomotive force and air-gap magnetic field

$\dot{E}_1$ and $\dot{E}_2$ are the three-phase symmetric inductive electromotive forces in the stator and the rotor winding respectively, their frequency is $f = \dfrac{pn_s}{60} = f_1$, that is the same as the power supply, the phase sequence is $A \to B \to C$ and $a \to b \to c$ respectively.

The rms value of $\dot{E}_1$ and $\dot{E}_2$ are

$$\begin{aligned} E_1 &= 4.44 f_1 N_1 k_{w1} \Phi_m \\ E_2 &= 4.44 f_1 N_2 k_{w2} \Phi_m \end{aligned} \tag{4-7}$$

where Φ_m is the magnitude of the main flux of each pole.

$$\frac{E_1}{E_2} = \frac{N_1 k_{w1}}{N_2 k_{w2}} = k_e \tag{4-8}$$

where k_e is the voltage ratio.

$\dot{E}_1$ can be represented by the negative voltage drop of impedance:

$$\dot{E}_1 = -\dot{I}_m Z_m = -\dot{I}_m (R_m + jX_m) \tag{4-9}$$

where Z_m is the magnetizing impedance, which is a comprehensive parameter of the representing magnetization properties and the iron loss; R_m is the magnetizing resistance, which is an equivalent parameter representing the iron loss; X_m is the magnetizing reactance, which is an equivalent parameter representing the magnetic properties of the main flux.

Generally, the three parameters are called magnetizing parameters collectively, their values are related to the degree of saturation of the iron core, the more saturated that the iron core is, the smaller magnetizing reactance will be.

Consider that there is a stator resistance and a stator leakage reactance, the equivalent circuit of one phase of the stator is shown in Fig. 4-10; the voltage equation can be expressed as following:

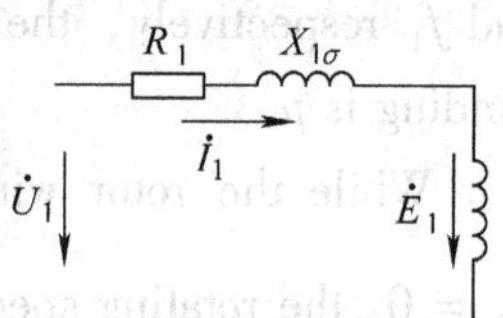

Fig. 4-10 one phase circuit of stator winding

$$\dot{U}_1 = -\dot{E}_1 + \dot{I}_1(R_1 + jX_{1\sigma}) \qquad (4\text{-}10)$$

where R_1 is the resistance of each phase of a stator winding.

2. When rotor windings are short-circuited

The rotor current is appeared when the rotor windings are short-circuited; and an electromagnetic torque is generated which is in the same direction as a rotating magnetic field. The rotor will rotate in the same direction of the rotating magnetic field with speed n, where $0 < n < n_s$. The induction motor is working under a motor mode. The speed of rotating magnetic field with respect to the rotor is

$$\Delta n = n_s - n = sn_s \qquad (4\text{-}11)$$

Then the frequency of the rotor electromotive force and the current is

$$f_2 = \frac{psn_s}{60} = sf_1 \qquad (4\text{-}12)$$

where f_2 is called slip frequency.

The relative speed of magnetic motive force to rotor is

$$n_2 = \frac{60f_2}{p} = \frac{60sf_1}{p} = sn_s = \Delta n \qquad (4\text{-}13)$$

Considering the relative speed of rotor to stator, the relative speed of rotor magnetic motive force to stator is

$$n_2 + n = \Delta n + n = (n_s - n) + n = n_s \qquad (4\text{-}14)$$

The magnetic motive force speed of the rotor and the stator is the same and remains relatively stationary, as shown in Fig. 4-11.

The speed of rotor and the stator magnetic motive force keep relative stationary, which is a necessary process to generate a constant electromagnetic torque in the induction machine.

The magnetic field of the motor is established by a stator MMF $\boldsymbol{F}_1$ and a rotor MMF $\boldsymbol{F}_2$. The excitation magnetic motive force is

$$\boldsymbol{F}_m = \boldsymbol{F}_1 + \boldsymbol{F}_2 \qquad (4\text{-}15)$$

Fig. 4-11 the speed of stator and rotor magnetic motive force

The equation above can be written as

$$\boldsymbol{F}_1 = \boldsymbol{F}_m + (-\boldsymbol{F}_2) = \boldsymbol{F}_m + \boldsymbol{F}_{1L} \qquad (4\text{-}16)$$

$\boldsymbol{F}_1$ is composed of two components, one is the excitation component, which is used to generate the main magnetic flux; another is the load component, which is used to compensate for the rotor magnetomative force.

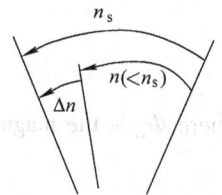

In addition, $\dot{I}_1$ consists of two components, namely

$$\dot{I}_1 = \dot{I}_{\mathrm{m}} + \dot{I}_{1\mathrm{L}} \tag{4-17}$$

where $\dot{I}_{\mathrm{m}}$ is the excitation component, which generates excitation component $\boldsymbol{F}_{\mathrm{m}}$; $\dot{I}_{1\mathrm{L}}$ is the load component, which generates load component $\boldsymbol{F}_{1\mathrm{L}}$,

$$F_{\mathrm{m}} = \frac{m_1}{2}0.9\frac{N_1 k_{\mathrm{w1}} I_{\mathrm{m}}}{p} \tag{4-18}$$

$$F_{1\mathrm{L}} = \frac{m_1}{2}0.9\frac{N_1 k_{\mathrm{w1}} I_{1\mathrm{L}}}{p} \tag{4-19}$$

Where m_1 are the phase number of stator and rotor respectively.

Then, the rms value of rotor emf is induced by the main flux.

$$E_{2\mathrm{s}} = 4.44 f_2 N_2 k_{\mathrm{w2}} \Phi_{\mathrm{m}} = 4.44 s f_1 N_2 k_{\mathrm{w2}} \Phi_{\mathrm{m}} = sE_2 \tag{4-20}$$

where subscript "s" of $E_{2\mathrm{s}}$ means that the frequency of the electromotive force of rotor is slip frequency.

The leakage reactance of the rotor is

$$X_{2\sigma\mathrm{s}} = 2\pi f_2 L_{2\sigma} = sX_{2\sigma} \tag{4-21}$$

Fig. 4-12 shows the circuit of rotor and stator when the motor is rotating. the frequency of rotor circuit is slip frequency. Voltage equation of stator and rotor can be listed as follows:

$$\dot{U}_1 = -\dot{E}_1 + \dot{I}_1(R_1 + \mathrm{j}X_{1\sigma}) \tag{4-22}$$

$$\dot{E}_{2\mathrm{s}} = \dot{I}_{2\mathrm{s}}(R_2 + \mathrm{j}X_{2\sigma\mathrm{s}}) \tag{4-23}$$

where R_2 is one-phase resistance of the rotor winding.

The electromagnetic relationship between the rotor and the stator can be summarized as Fig. 4-13.

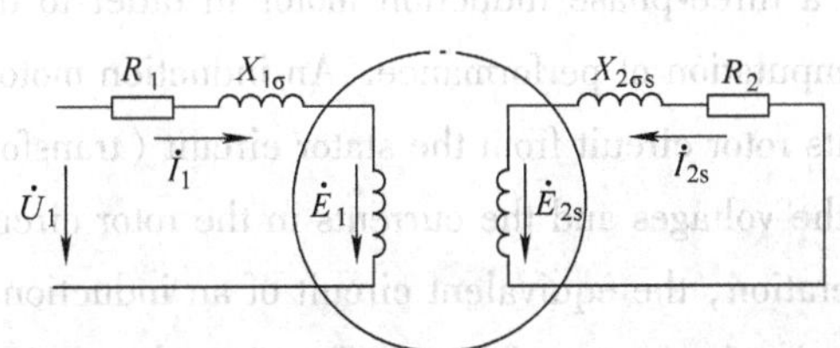

Fig. 4-12 the circuit of rotor and stator when the motor is rotating

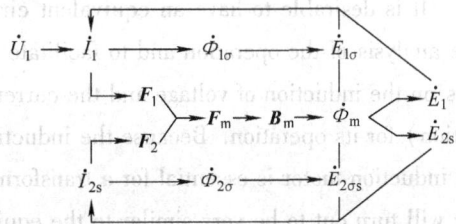

$\dot{U}_1$ → $\dot{I}_1$ → $\dot{\Phi}_{1\sigma}$ → $\dot{E}_{1\sigma}$
F_1, F_2 → F_{m} → B_{m} → $\dot{\Phi}_{\mathrm{m}}$ → $\dot{E}_1$, $\dot{E}_{2s}$
$\dot{I}_{2s}$ → $\dot{\Phi}_{2\sigma}$ → $\dot{E}_{2\sigma s}$

Fig. 4-13 The electromagnetic relationship between rotor and stator

Similar to the transformer, the stator and the rotor windings are coupling through the air-gap magnetic field and there is no direct link in the circuit. From the analysis mentioned above, the influence of the rotor, with the rotor magnetomotive force $\boldsymbol{F}_2$, on the stator depends on the amplitude and the spatial phase of $\boldsymbol{F}_2$. The amplitude of the $\boldsymbol{F}_2$ depends on the rotor current. The analysis of the spatial phase of $\boldsymbol{F}_2$ is shown in Fig. 4-14.

Fig. 4-14 shows different cases of three-phase windings of the rotor in the air-gap magnetic field. The leakage reactance is $X_{2\sigma\mathrm{s}} = 0$, this situation is shown in Fig. 4-14a. The figure shows the direction of the electromotive force in the rotor coil and the electromotive force of the rotor phase a is positive maximum.

Since the impedance angle of rotor is 0°, there is no reactive component in rotor current and the rotor current is in the same phase as the rotor electromotive force. Namely, the induced current of rotor phase a is the positive maximum.

The peak position of fundamental wave of magnetomotive force $\boldsymbol{F}_2$ is generated by rotor current which differs 90 electrical degrees from the air-gap magnetic field $\boldsymbol{B}_m$, which means that $\boldsymbol{F}_2$ is lagging $\boldsymbol{B}_m$ by 90° in the space phase.

Fig. 4-14b shows a situation under the condition the leakage reactance $X_{2\sigma s} \neq 0$.

The electromotive force of the rotor phase a is the positive maximum, and the direction of the electromotive force is the same as Fig. 4-14a in the rotor coil side, but the current of the rotor is different from Fig. 4-14a. Assuming that the impedance angle of the rotor is 60°, the value of the rotor phase b current is the negative maximum. This can be confirmed from the figure that $\boldsymbol{F}_2$ is lagging $\boldsymbol{B}_m$ by 90° + 60° in the space phase.

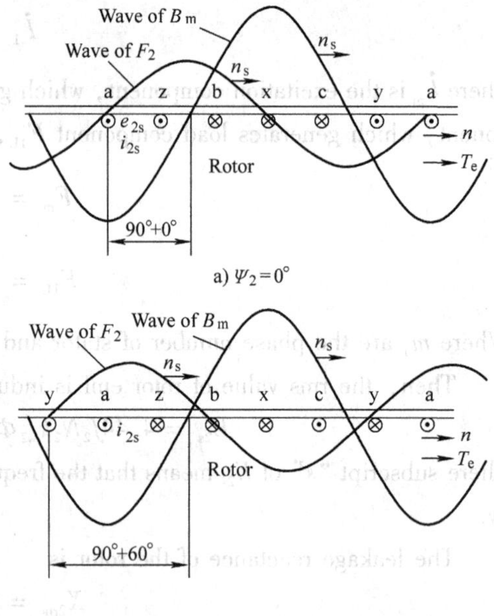

Fig. 4-14 the relative position of magnetic motive force and air-gap magnetic field of rotor

4.3 Equivalent circuit of an induction motor

It is desirable to have an equivalent circuit of a three-phase induction motor in order to direct the analysis of the operation and to facilitate the computation of performance. An induction motor relies on the induction of voltage and the currents in its rotor circuit from the stator circuit (transformer action) for its operation. Because the induction of the voltages and the currents in the rotor circuit of an induction motor is essential for a transformer operation, the equivalent circuit of an induction motor will turn out to be very similar to the equivalent circuit of a transformer. Therefore the equivalent circuit of an induction motor will be developed by starting the transformer model and by deciding how to take the variable rotor frequency f_2 and other similar induction motor effects into account. It is necessary to perform the frequency conversion and the winding conversion.

4.3.1 Frequency conversion

The purpose of the frequency conversion is to solve the problem of the different frequencies between the stator and the rotor.

Transforming the Eq. 4-23 to

$$\dot{E}_{2s} e^{j\omega_2 t} = \dot{I}_{2s} e^{j\omega_2 t} (R_2 + jX_{2\sigma s}) \tag{4-24}$$

Multiply $\frac{1}{s}\mathrm{e}^{j(\omega_1-\omega_2)t}$ on the both sides of Eq. 4-24, then

$$\dot{E}_2\mathrm{e}^{\mathrm{j}\omega_1 t} = \dot{I}_2\mathrm{e}^{\mathrm{j}\omega_1 t}\left(\frac{R_2}{s} + \mathrm{j}X_{2\sigma}\right) \tag{4-25}$$

Or

$$\dot{E}_2 = \dot{I}_2\left(\frac{R_2}{s} + \mathrm{j}X_{2\sigma}\right) \tag{4-26}$$

From Eq. 4-26, we can find that the rotor frequency is changed from $f_2 = sf_1$ to f_1. The rotor electromotive force and the leakage reactance are changed into the value at the stationary state. The rotor resistance is changed from R_2 to $\frac{R_2}{s}$. The rotor current is written in the form of $\dot{I}_2$ where the frequency is stator frequency, and its rms value remains unchanged, i. e. $I_2 = I_{2s}$. The impedance angle of the rotor circuit remains unchanged, i. e. $\psi_2 = \arctan\frac{X_{2\sigma s}}{R_2} = \arctan\frac{X_{2\sigma}}{R_2/s}$. The transformation stated above is called frequency conversion.

Fig. 4-15 is the stator and the rotor circuit after a frequency conversion. The physical meaning of a frequency conversion is changing the rotor resistance from actual resistance R_2 in the rotating state to the equivalent resistance $\frac{R_2}{s}$ at a stationary state. Because the impedance angle and the rms value of the rotor current remain unchanged after this transformation, therefore the value, the speed and the space position of the rotor magnetomotive force remain unchanged, and the impact of the rotor to the stator remains unchanged, and all the physical quantities of the stator and the power transition between the stator and rotor remain invariant.

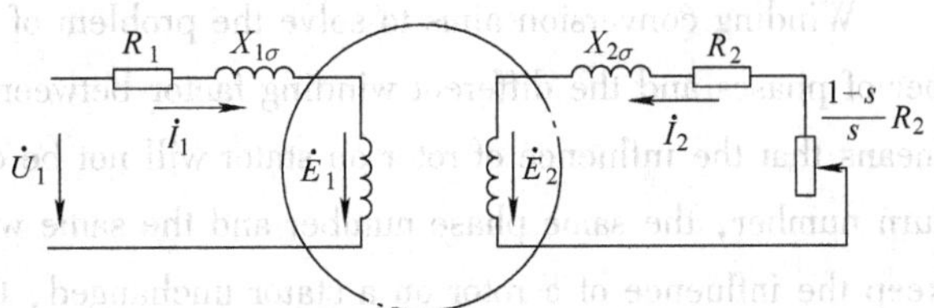

Fig. 4-15 the schematic after frequency conversion

The rotor resistance after the frequency conversion becomes

$$\frac{R_2}{s} = R_2 + \frac{1-s}{s}R_2 \tag{4-27}$$

The power consumption on the resistance of the left side of Eq. 4-27 is the electromagnetic power that is delivered from the stator to the rotor through the air gap magnetic field. Compared with the state before the frequency conversion, the electromagnetic power remains the same. The first item on the right side of Eq. 4-27 is the resistance of the rotor and its power consumption is the rotor copper loss, which is unchanged comparing with that before frequency conversion. The second item on the right side of Eq. 4-27 is the resistance that corresponds to the generated mechanical power while the rotor is rotating; its power consumption is equal to the total mechanical power that is generated on the rotor side.

To sum up, there are two functions derived from the frequency conversion. Firstly, refer the rotor frequency to the stator frequency and make both frequencies at the same rate; secondly derive the equivalent resistance that represents the mechanical power.

From Chapter 3, the mmf in stator and rotor is decided by following equations:

$$F_1 = \frac{m_1}{2}0.9\frac{N_1 k_{w1} I_1}{p}, F_2 = \frac{m_2}{2}0.9\frac{N_2 k_{w2} I_2}{p}, F_m = \frac{m_1}{2}0.9\frac{N_1 k_{w1} I_m}{p}$$

and $\boldsymbol{F}_1 = \boldsymbol{F}_m + (-\boldsymbol{F}_2)$

Where m_2 is the phase number of rotor winding.

According to the space-time vector diagram, when the time phase axis coincides with the space phase axis, the current phasor coincides with the fundamental wave of magnetic motive force vector generated by a three-phase current system. The phase relations of $\boldsymbol{F}_1$, $\boldsymbol{F}_2$ and $\boldsymbol{F}_m$ in space are the same as the phase relations of $\dot{I}_1$, $\dot{I}_2$ and $\dot{I}_m$ in time respectively, then we can get the magnetomotive force equation from the current form.

$$\dot{I}_1 = \dot{I}_m + \left(-\frac{\dot{I}_2}{k_i}\right) \tag{4-28}$$

where k_i is the current ratio, and $k_i = \frac{m_1 N_1 k_{w1}}{m_2 N_2 k_{w2}}$.

4.3.2 Winding conversion

Winding conversion aims to solve the problem of the different winding turns, the different number of phases and the different winding factor between the stator and the rotor. Winding conversion means that the influence of rotor on stator will not be changed, where the rotor winding has the same turn number, the same phase number and the same winding factor as the stator winding. In order to keep the influence of a rotor on a stator unchanged, the magnitude and the space phase of the rotor magnetomotive forced should be invariant. We need to add " ′ " to the variables of the rotor side after winding conversion.

$$m_2' = m_1, N_2' = N_1, k_{w2}' = k_{w1} \tag{4-29}$$

1. Rotor current conversion

In order to make the magnitude and the space phase of the rotor magnetomotive force invariant,

$$\frac{m_2'}{2}0.9\frac{N_2' k_{w2}' \dot{I}_2'}{p} = \frac{m_1}{2}0.9\frac{N_1 k_{w1} \dot{I}_2'}{p} = \frac{m_2}{2}0.9\frac{N_2 k_{w2} \dot{I}_2}{p} \tag{4-30}$$

Then

$$\dot{I}_2' = \frac{m_2 N_2 k_{w2}}{m_1 N_1 k_{w1}}\dot{I}_2 = \frac{\dot{I}_2}{k_i} \tag{4-31}$$

2. Rotor emf conversion

Because the magnitude and the space phase of $\boldsymbol{F}_2$ are invariant, the main magnetic flux $\boldsymbol{\Phi}_m$ remains unchanged, and then the rotor electromotive force after the conversion will be

$$\dot{E}_2' = 4.44 f_1 N_2' k_{w2}' \dot{\Phi}_m = 4.44 f_1 N_1 k_{w1} \dot{\Phi}_m = k_e \dot{E}_2 = \dot{E}_1 \tag{4-32}$$

3. Rotor impedance conversion

After the winding conversion, the active and the reactive power that are consumed in the rotor should remain unchanged, so

$$m_2' I_2'^2 \frac{R_2'}{s} = m_1 I_2'^2 \frac{R_2'}{s} = m_2 I_2^2 \frac{R_2}{s}$$

$$m_2' I_2'^2 X_{2\sigma}' = m_1 I_2'^2 X_{2\sigma}' = m_2 I_2^2 X_{2\sigma} \tag{4-33}$$

Therefore, after conversion, the rotor resistance and leakage reactance are

$$R_2' = \frac{m_2}{m_1}\left(\frac{I_2}{I_2'}\right)^2 R_2 = k_e k_i R_2,\ X_{2\sigma}' = \frac{m_2}{m_1}\left(\frac{I_2}{I_2'}\right)^2 X_{2\sigma} = k_e k_i X_{2\sigma} \tag{4-34}$$

Fig. 4-16 shows the stator and the rotor circuit of an induction motor after the frequency conversion and the winding conversion.

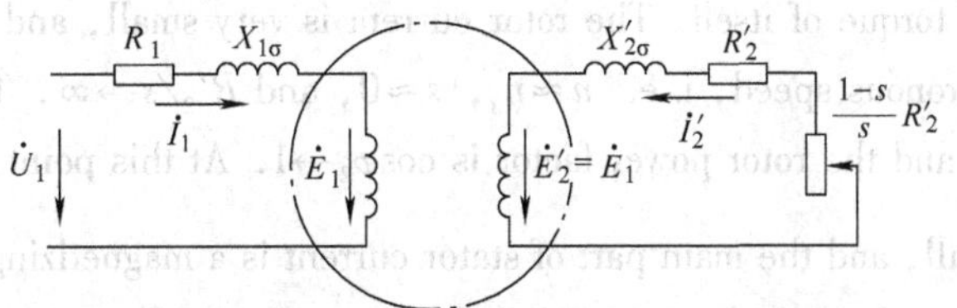

Fig. 4-16 The stator and rotor circuit after the frequency and the winding conversion

The basic equations of an induction motor can be acquired.

$$\begin{cases} \dot{U}_1 = -\dot{E}_1 + \dot{I}_1(R_1 + jX_{1\sigma}) \\ \dot{E}_2' = \dot{I}_2'(R_2'/s + jX_{2\sigma}') \\ \dot{I}_1 = \dot{I}_m + (-\dot{I}_2') \\ \dot{E}_1 = -\dot{I}_m Z_m \\ \dot{E}_1 = \dot{E}_2' \end{cases} \tag{4-35}$$

4.3.3 Equivalent circuit and phasor diagram

In Fig. 4-16, the voltage $\dot{E}_1$ is same as the voltage $\dot{E}_2$; hence these terminals may be joined to yield a complete equivalent circuit (T-shape equivalent circuit) as shown in Fig. 4-17. The phase diagram of an induction motor is shown in Fig. 4-18.

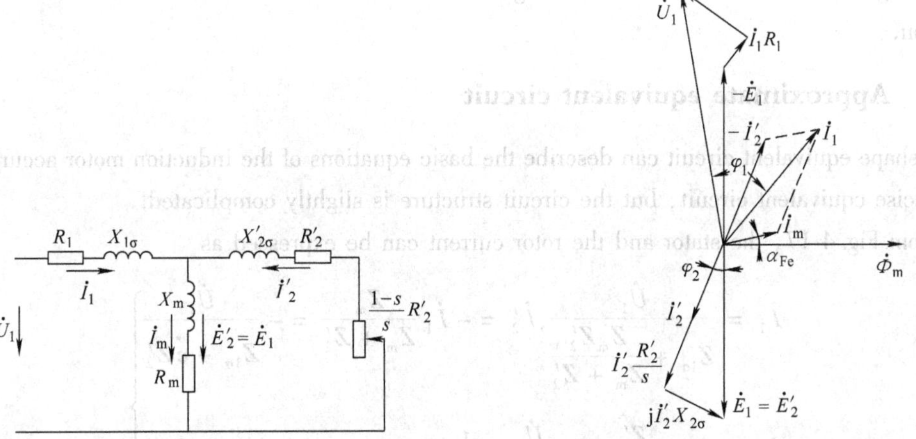

Fig. 4-17 The T shape equivalent circuit

Fig. 4-18 Phasor diagram of induction motors

It should be noted that in the calculation of the equivalent circuit, the voltage and the current should be a phase voltage and a phase current. The basic equations, an equivalent circuit and a phasor diagram reflect the physical electromagnetic relations of an induction motor in three different ways.

4.3.4 Analysis of induction motor when the load changed

When the induction motor operates at no-load, it only needs to overcome a very small no-load brake torque of itself. The rotor current is very small, and the speed of the rotor is very close to the synchronous speed, i. e. $n \approx n_s$, $s \approx 0$, and $R'_2/s \to \infty$. The rotor circuit is similar to an open circuit, and the rotor power factor is $\cos\varphi_2 \to 1$. At this point, the load component of the stator current is small, and the main part of stator current is a magnetizing current, i. e. $\dot{I}_1 \approx \dot{I}_m$. The stator power factor angle φ_1 is large, so that the stator power factor is very low. In addition, the stator leakage impedance voltage drop $\dot{I}_1(R_1 + jX_{1\sigma})$ is small, then $U_1 \approx E_1$.

If the load torque increases, the rotor current will increase, the speed will decrease, the slip will increase, R'_2/s will decrease, and the rotor power factor will decrease. Compared with no-load condition, the load component of the stator current increases, and then the stator current will increase. The voltage drop of the stator leakage impedance $\dot{I}_1(R_1 + jX_{1\sigma})$ increases, so E_1 decreases, and the main magnetic flux will reduce. At the same time, the active component of the stator current will increase, and the stator power factor will increase as well. However, if the load is large, the rotor current will also increase when the load increases continually. Due to the significant decrease is the rotor speed, the rotor frequency will increase significantly. Then, the angle φ_2 increases, so that the angle φ_1 will also increase, but the stator power factor will decrease.

When the load torque is increasing rapidly and the rotor is stopped, i. e. the rotor speed $n = 0$, $s = 1$. At this time, the stator current and the rotor current will be very large, the rotor power factor is very low, which leads to a lower stator power factor. The stator leakage impedance voltage drop is very large and the value of the main magnetic flux is about 50% ~ 60% of that at the no-load condition.

4.3.5 Approximate equivalent circuit

T-shape equivalent circuit can describe the basic equations of the induction motor accurately. It is a precise equivalent circuit, but the circuit structure is slightly complicated.

From Fig. 4-17, the stator and the rotor current can be expressed as

$$\left.\begin{aligned}\dot{I}_1 &= \frac{\dot{U}_1}{Z_{1\sigma} + \dfrac{Z_m Z'_2}{Z_m + Z'_2}}, \dot{I}'_2 = -\dot{I}_1 \frac{Z_m}{Z_m + Z'_2} = -\frac{\dot{U}_1}{Z_{1\sigma} + \dot{c} Z'_2} \\ \dot{I}_m &= \dot{I}_1 \frac{Z'_2}{Z_m + Z'_2} = \frac{\dot{U}_1}{Z_m} \frac{1}{c + \dfrac{Z_{1\sigma}}{Z'_2}}\end{aligned}\right\} \tag{4-36}$$

where $Z_{1\sigma}$ is the stator leakage impedance, and $Z_{1\sigma} = R_1 + jX_{1\sigma}$; Z'_2 is the equivalent impedance of rotor, and $Z'_2 = \frac{R'_2}{s} + jX'_{2\sigma}$; c is a complex coefficient, and $c = 1 + \frac{Z_{1\sigma}}{Z_m} \approx 1 + \frac{X_{1\sigma}}{X_m}$.

When the induction motor is operating normally, $|Z_{1\sigma}| \ll |Z_2'|$ that is $\frac{Z_{1\sigma}}{Z_2'} \approx 0$, then in Eq. 4-36:

$$\dot{I}_m \approx \frac{\dot{U}_1}{cZ_m} = \frac{\dot{U}_1}{Z_{1\sigma} + Z_m} \tag{4-37}$$

$$-\dot{I}_2' = \frac{\dot{U}_1'}{Z_{1\sigma} + cZ_2'}, \dot{I}_1 = \dot{I}_m + (-\dot{I}_2') \tag{4-38}$$

The approximate equivalent circuit can be derived as shown in Fig. 4-19:

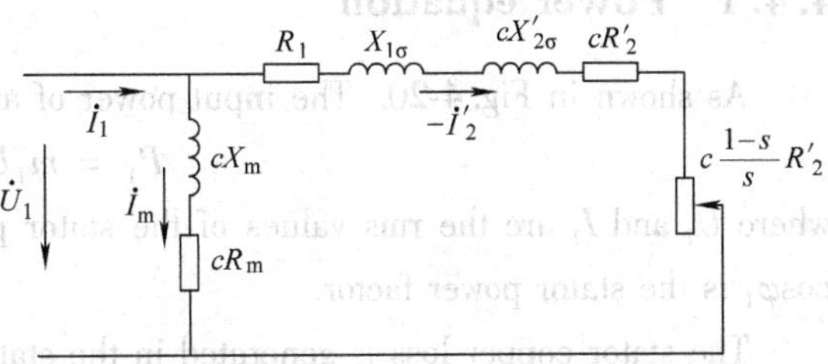

Fig. 4-19 Approximate equivalent circuit of the induction motor

Example 4-1 A three-phase induction motor, where $U_N = 380\text{V}$, $f_N = 50\text{Hz}$, $n_N = 1455\text{r/min}$; the stator windings is connected in delta (Δ-connection), where $R_1 = 1.42\Omega$, $X_{1\sigma} = 5.41\Omega$, $R_2' = 1.25\Omega$, $X_{2\sigma}' = 7.75\Omega$, $X_m = 90.4\Omega$, and $R_m = 10.6\Omega$,

1) Find the rotor frequency.

2) Find the stator current, the stator power factor and the input power at rated speed through T shape equivalent circuit.

Solution:

1) The rotor frequency: For the reason that the rated speed of induction motor is close to the synchronous speed, so that the synchronous speed should be 1500r/min, hence rated slip is

$$s_N = \frac{n_s - n_N}{n_s} = \frac{1500 - 1455}{1500} = 0.03$$

The rotor frequency is

$$f_2 = sf_1 = 0.03 \times 50\text{Hz} = 1.5\text{Hz}$$

2) At rated speed, the equivalent impedance of the rotor is

$$Z_2' = \frac{R_2'}{s_N} + \text{j}X_{2\sigma}' = \left(\frac{1.26}{0.03} + \text{j}7.75\right)\Omega = (42 + \text{j}7.75)\Omega = 42.71\angle 10.45°\Omega$$

The magnetizing impedance is

$$Z_m = R_m + \text{j}X_m = (10.6 + \text{j}90.4)\Omega = 91.02\angle 83.31°\Omega$$

Set the stator voltage phasor as $\dot{U}_1 = 380\angle 0°$, the stator current is

$$\dot{I}_1 = \frac{\dot{U}_1}{Z_1 + \frac{Z_m Z_2'}{Z_m + Z_2'}} = \frac{380\angle 0°}{(1.42 + \text{j}4.51) + \frac{91.02\angle 83.31° \times 42.71\angle 10.45°}{(42 + \text{j}7.75) + (10.6 + \text{j}90.4)}}\text{A}$$

$$= 9.84\angle -36.51°\text{A}$$

The stator power factor is

$$\cos\varphi_1 = \cos 36.51° = 0.804$$

The input power is

$$P_1 = 3U_1I_1\cos\varphi_1 = 3 \times 380 \times 9.84 \times 0.804\text{W} = 9019\text{W}$$

4.4 Power and torque equation of an induction motor

In the induction motor, the electric power input is converted into a mechanical power output. According to the process of power delivering, the power and the loss in the induction motor will be explained in following, and then the power and the torque equation will be derived by using the equivalent circuit.

4.4.1 Power equation

As shown in Fig. 4-20. The input power of an induction motor is

$$P_1 = m_1 U_1 I_1 \cos\varphi_1 \tag{4-39}$$

where U_1 and I_1 are the rms values of the stator phase voltage and the phase current respectively; $\cos\varphi_1$ is the stator power factor.

The stator copper loss is generated in the stator windings resistance and the iron loss is generated in the core, they are

$$p_{Cu1} = m_1 I_1{}^2 R_1, p_{Fe} = m_1 I_m{}^2 R_m \tag{4-40}$$

When the induction motor operates in normal condition, the speed of magnetic field that is relative to the rotor is very low, and the rotor iron loss is also very small. The main part of iron loss is the stator iron loss.

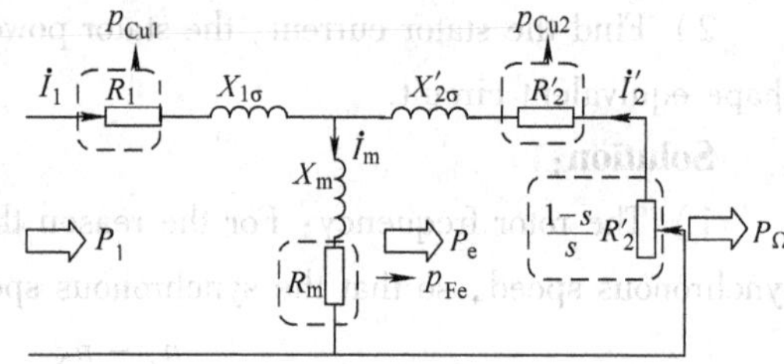

Fig. 4-20 Power and loss of the induction motor

Subtracting the stator copper loss and the iron loss from the input power, the rest power is delivered to the rotor through the air gap magnetic field. This part of the power is known as the electromagnetic power P_e:

$$P_e = P_1 - p_{Cu1} - p_{Fe} \tag{4-41}$$

According to the equivalent circuit, the electromagnetic power can be expressed as

$$P_e = m_1 E_2' I_2' \cos\psi_2 = m_1 I_2{}^2 \frac{R_2'}{s} \tag{4-42}$$

Subtracting the rotor copper loss p_{Cu2} from the electromagnetic power, the total mechanical power P_Ω is known as the conversion power and that is expressed as

$$p_{Cu2} = m_1 I_2'^2 R_2' \tag{4-43}$$

$$P_\Omega = P_e - p_{Cu2} = m_1 I_2'^2 \frac{1-s}{s} R_2' \tag{4-44}$$

By Eq. 4-42, Eq. 4-43, Eq. 4-44, the following relationship can be obtained:

$$p_{Cu2} = sP_e, \ P_\Omega = (1-s)P_e \tag{4-45}$$

Notice that the electromagnetic power P_e is delivered to the rotor, sP_e is converted into the rotor copper loss, $(1-s)P_e$ is converted to the total mechanical power. The rated slip s_N of an induction motor is very small, it is generally around 2% ~ 5%.

Then subtracting the mechanical loss p_Ω and the stray loss p_Δ from the total mechanical power P_Ω, therefore the output mechanical power P_2 will be

$$P_2 = P_\Omega - p_\Omega - p_\Delta \tag{4-46}$$

In a small cage induction motor, the stray loss is usually around 1% ~ 3% of the output power. In a large induction motor, it is usually 0.5% of the output power at full load.

Fig. 4-21 shows the power flow of an induction motor. When the load varies, the stator and the rotor current will change, but the main flux and the speed of the motor is nearly unchanged. Therefore, the mechanical losses and the iron loss are considered as the fixed losses, the stator copper loss, the rotor copper loss and the stray loss are considered as the variable losses.

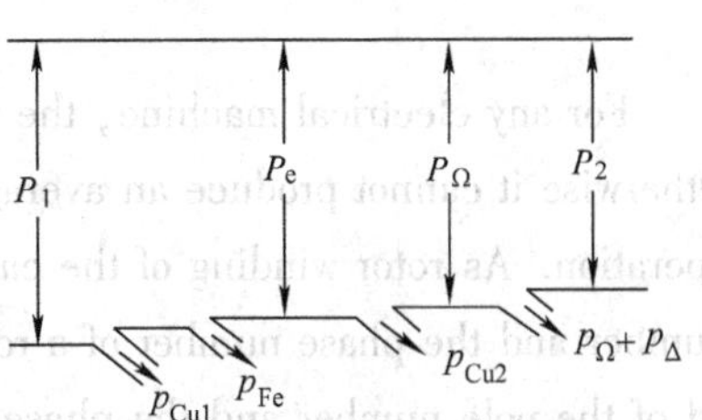

Fig. 4-21 The power flow chart of the induction motor

4.4.2 Torque equation

When the induction motor is running, there are three kinds of torque acting on the rotor, the electromagnetic torque T_e, the load torque T_2, and the no-load torque T_0. When the induction motor operates at a normal state, the electromagnetic torque is the driving torque; the load torque and no-load torque are the braking torque.

$$T_e = T_2 + T_0 \tag{4-47}$$

The relationship between the torque, the power and the losses are showed as following:

$$T_e = \frac{P_\Omega}{\Omega},\ T_0 = \frac{p_\Omega + p_\Delta}{\Omega},\ T_2 = \frac{P_2}{\Omega} \tag{4-48}$$

where Ω is the rotor mechanical angular speed.

4.4.3 Electromagnetic torque

Due to $\Omega = (1-s)\Omega_s$, in Eq. 4-48, the electromagnetic torque can be expressed as

$$T_e = \frac{P_\Omega}{\Omega} = \frac{(1-s)P_e}{(1-s)\Omega_s} = \frac{P_e}{\Omega_s} \tag{4-49}$$

Either using the total mechanical power which is divided by the mechanical angular speed, or using the electromagnetic power which is divided by the synchronous angular speed, the electromagnetic torque can be calculated as following.

Substituting $P_e = m_1 E_2' I_2' \cos\psi_2$, $E_2' = \sqrt{2}\pi f_1 N_1 k_{w1} \Phi_m$ and $\Omega_s = \dfrac{2\pi f_1}{p}$ into Eq. 4-49, we can get

$$T_e = \left(\frac{pm_1 N_1 k_{w1}}{\sqrt{2}}\right)\Phi_m I_2' \cos\psi_2 = \left(\frac{pm_2 N_2 k_{w2}}{\sqrt{2}}\right)\Phi_m I_2 \cos\psi_2 = C_T' \Phi_m I_2 \cos\psi_2 \tag{4-50}$$

where G_T' is called torque coefficient, and $C_T' = \dfrac{pm_2 N_2 k_{w2}}{\sqrt{2}}$.

Eq. 4-50 shows that the electromagnetic torque is proportional to the main flux Φ_m and the active component of the rotor current $I_2\cos\psi_2$.

4.5 The pole and phase number of the cage rotor

For any electrical machine, the pole pair number of the stator and the rotor should be the same. Otherwise it cannot produce an average electromagnetic torque and will lead to a failure in the motor operation. As rotor winding of the cage rotor motor is different from the wound rotor motor, the pole number and the phase number of a rotor winding cannot be determined easily. The calculation method of the pole number and the phase number of a cage rotor will be discussed in this section.

4.5.1 The phase number of the cage rotor

The structure of the cage rotor is symmetric and the rotor bars are uniformly distributed on the surface of therotor core. Assuming the number of rotor slots is Q_2, which is the same as the number of bar. When the magnetic field in the air-gap is rotating, the emf will be induced into the rotor bars. The rms value of the emf in each bar is equal and the phase difference of the induced emf between two adjacent bars is

$$\alpha_2 = \frac{p \times 360°}{Q_2} \tag{4-51}$$

where p is the pole pair number of the magnetic field.

Since the currents in one phase AC winding are the same phase, therefore, the phase number of cage rotor is equal to the number of bars wherein the currents are in different phases. As shown in Fig. 4-22, if Q_2/p is equal to an integer, then Q_2/p bars can produce Q_2/p emfs which are distributed symmetrically. It also can be found that the currents in Q_2/p bars are distributed symmetrically. Therefore, the cage rotor windings can be seen as a Q_2/p symmetric phase windings, which means the phase number $m_2 = Q_2/p$. If Q_2/p is not equal to an integer, then the phase number $m_2 = Q_2$, and the pole number is equal to 2.

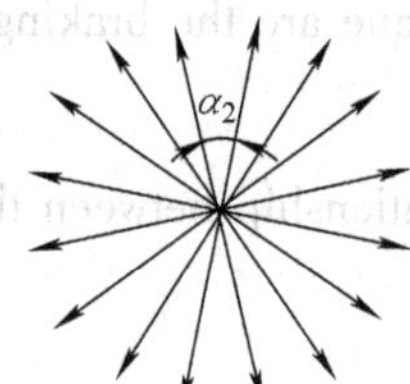

Fig. 4-22 The star map of the rotor slot emf under a pair of pole

In other words, one phase winding of per pole includes one conducting bar only, and the turns of each phase winding is

$$N_2 = \frac{1}{2} \tag{4-52}$$

Then, the cage rotor winding is not short-pitched and is not in distributed form, so the winding factor is

$$k_{w2} = 1 \tag{4-53}$$

4.5.2 Pole number of the cage rotor

In Fig. 4-23, a 16-bar squirrel-cage rotor that is placed in a 2-pole field is shown in a developed form. In Fig. 4-23a, the sinusoidal flux density wave induces a voltage in each bar which has an instantaneous value that is indicated by the solid vertical lines.

Owing to the leakage reactance, the current in the rotor bar will lag the induced emf by angle φ_2 in time. That is to say, even the emf in one bar reaches its maximum value; the current will not reach the maximum simultaneously. At a later instant of time φ_2, the bar currents assure the instantaneous values that is indicated by the solid vertical lines in Fig. 4-23b. In this time interval, the flux-density wave has traveled in its direction of rotation with the respect to the rotor through a space angle φ_2 and finally is in the position shown in Fig. 4-23b.

The corresponding rotor-mmf wave is shown by the step wave in Fig. 4-23c. The fundamental component is shown by the dashed sinusoid and the flux-density wave is shown by the solid sinusoid. Studying these figures confirms the general principle of that the number of rotor poles in a squirrel-rotor is determined by the inducing flux wave. Namely, if the pole number of magnetic field in air-gap is 2, then the pole number of the squirrel-rotor fundamental wave mmf that is produced by the rotor current will be 2. Similarly, if the pole number of the air-gap magnetic field is 4, the rotor currents will create a 4-pole fundamental wave mmf. Therefore, the pole number of the cage rotor will automatically equal to the pole number of the stator windings and it is independent from the number of the rotor bars. In addition, the rotor and the stator mmf are rotating at the same speed that always equals to the synchronous speed.

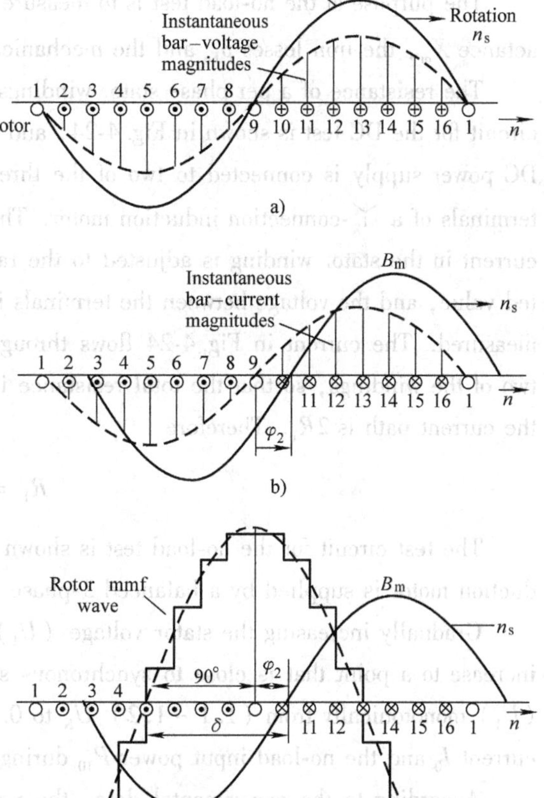

Fig. 4-23 Reactions of a squirrel-cage rotor in a 2-pole field

4.6 Determining circuit model parameters

The performance computations for a 3-phase induction motor require knowledge about the parameters of equivalent circuit. This information may be available either from the design parameters or from the appropriate tests. When the design parameters are not available, information about the magnetizing branch can be obtained from a no-load test that is performed through applying a balanced 3-phase voltage with a rated frequency to a motor which is uncoupled from its load. Information about the winding resistance and leakage reactance can be obtained from a blocked-rotor test (short-circuit test). It requires the rotor to be blocked in order to prevent rotation and the rotor winding to be short-circuited in an usual fashion.

4.6.1 No-load test

The purpose of the no-load test is to measure the magnetizing resistance R_m, the magnetizing reactance X_m, the iron losses p_{Fe} and the mechanical losses p_Ω of an induction motor.

The resistance of a per phase stator windings (R_1) could be measured by a DC test. The basic circuit for the DC test is shown in Fig. 4-24, and a DC power supply is connected to two of the three terminals of a Y-connection induction motor. The current in the stator winding is adjusted to the rated value, and the voltage between the terminals is measured. The current in Fig. 4-24 flows through two of the windings, so that the total resistance in the current path is $2R_1$. Therefore

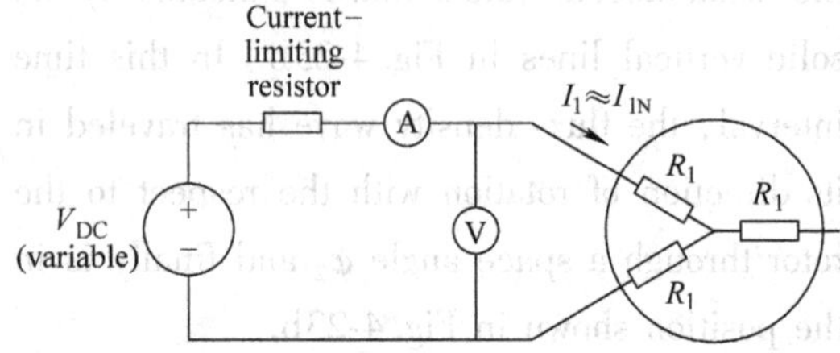

Fig. 4-24 Test circuit for a DC resistance test

$$R_1 = \frac{V_{DC}}{2I_1} \tag{4-54}$$

The test circuit for the no-load test is shown in Fig. 4-25a. In the no-load test, the 3-phase induction motor is supplied by a balanced 3-phase voltage with the rated frequency.

Gradually increasing the stator voltage (U_1) from 0V to the rated value, the rotor speed will increase to a point that is close to synchronous speed. Then decreasing the stator terminal voltage (U_1) monotonically from $(1.1 \sim 1.2)\ U_N$ to $0.3U_N$, measure the stator voltage U_1, the no-load current I_0 and the no-load input power P_{10} during the test, and 8 ~ 10 groups of data are needed.

According to the experimental data, the no-load curves of an induction motor (I_{10}, $P_{10} = f(U_1)$) can be obtained (as shown in Fig. 4-25b).

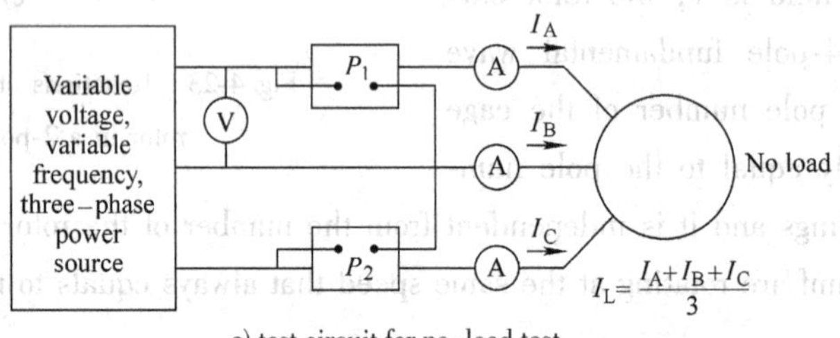

a) test circuit for no–load test

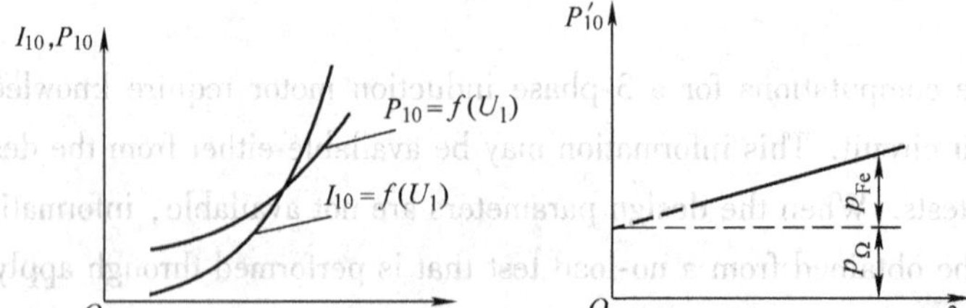

b) no–load characteristics

c) separate iron loss and mechanical loss

Fig. 4-25 The no-load test of an induction motor

When an induction motor runs at no-load, the slip is exceedingly small (possibly as small as 0.001 or less), and the rotor speed is approaching towards the synchronous speed. The resistance represents the mechanical power converted, $R_2\ (1-s)\ /s$, is much larger than the resistance corresponding to the rotor copper losses R_2 and much larger than the rotor reactance $X_{2\sigma}$, i. e. $s \to 0$, and $R_2\ (1-s)\ /s \to \infty$. When the rotor path is opened, the equivalent circuit of this motor at no-load is shown in Fig. 4-26.

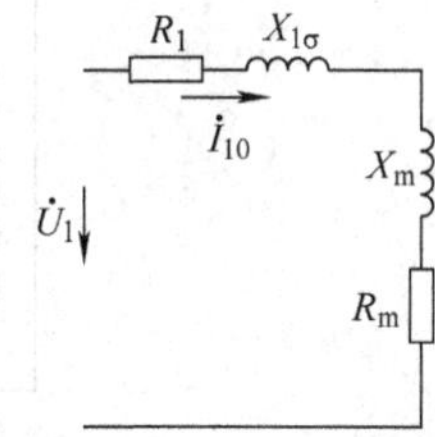

Fig. 4-26 The equivalent circuit at no-load

When this motor is at no-load conditions, the only load on the motor is friction and windage losses, so that the input power measured by the meters must be equal to the losses in the motor. The rotor copper losses are negligible. Since the current I_2 is extremely small [because $R_2\ (1-s)\ /s \to \infty$], so they may be neglected. The stator copper losses are given by $p_{Cu1} = 3I_{10}^2 R_1$.

So the input power must equal

$$P_{10} = p_{Cu1} + p_{Fe} + p_{\Omega} \tag{4-55}$$

And $P'_{10} = P_{10} - p_{Cu1} = P_{10} - m_1 I_{10}^2 R_1 = p_{Fe} + p_{\Omega}$.

Due to the mechanical losses p_{Ω} is not changing with voltage U_1, and the iron losses p_{Fe} is proportional to U_1^2, the relationship between P'_{10} and U_1^2 is approximately a straight line as shown in Fig. 4-25c. From Fig. 4-25c, the iron losses p_{Fe} can be obtained.

Magnetizing resistance R_m can be calculated according to the iron losses p_{Fe}.

$$R_m = \frac{p_{Fe}}{3I_{10}^2} \tag{4-56}$$

According to the equivalent circuit at no-load, the no-load impedance Z_0 is

$$|Z_0| = \frac{U_1}{I_{10}} \tag{4-57}$$

The no-load resistance R_0 consists of R_1 and R_m, i. e. $R_0 = R_1 + R_m$.

The no-load reactance X_0 is

$$X_0 = \sqrt{|Z_0|^2 - R_0^2} \tag{4-58}$$

The magnetizing reactance X_m is

$$X_m = X_0 - X_{1\sigma} \tag{4-59}$$

The stator leakage reactance $X_{1\sigma}$ can be gotten from the locked-rotor test.

4.6.2 Locked-rotor test

The locked rotor test is also called short-circuit test. The purpose of the locked rotor test is to obtain the short-circuit impedance Z_k, the short-circuit reactance X_k and the short-circuit resistance R_k.

In the locked rotor test, the rotor is either locked or blocked, so that it cannot move. A low voltage with a rated frequency is applied to the motor and the resulting voltage, the current, and the power are measured.

Fig. 4-27a shows the connections for the locked-rotor test. To perform a locked-rotor test, an AC voltage is applied to the stator; the voltage is then decreasing from $0.4U_N$ to 0V, meanwhile, stator current I_{1K}, stator voltage U_1 and input power P_{1k} should be measured. According to the test data, the short-circuit characteristics of an induction motor can be obtained (Fig. 4-27b).

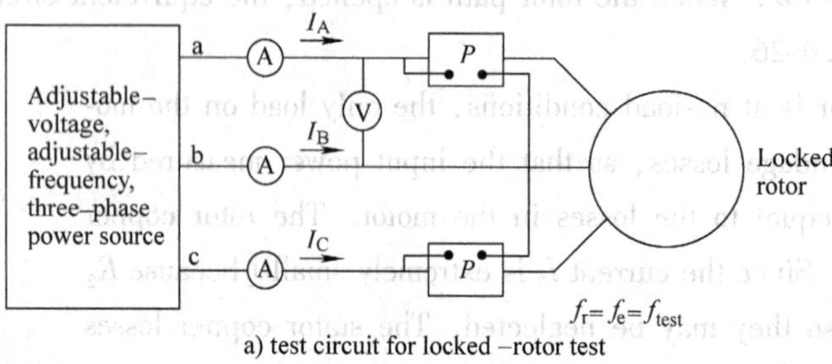

a) test circuit for locked -rotor test

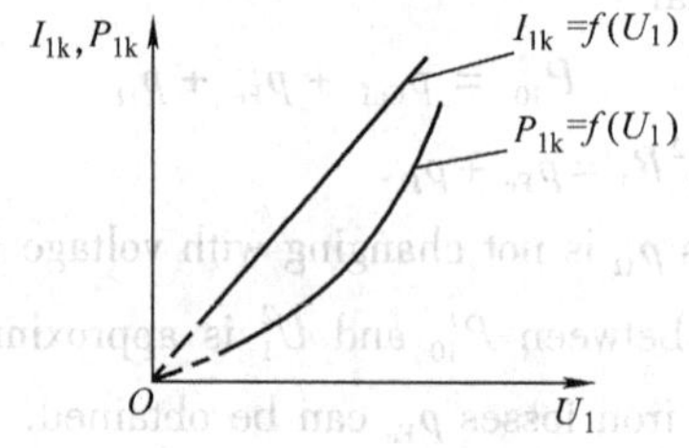

b) Short-circuit characteristics

Fig. 4-27 The locked-rotor test of an induction motor

The equivalent circuit for this test is shown in Fig. 4-28. Notice that since the rotor is not moving, the slip $s = 1$, the resistance $R_2(1-s)/s$ is equal to zero, and the rotor resistance R_2/s is equal to R_2. Since R_2 and $X_{2\sigma}$ are very small, almost all the input current will flow through them instead through the magnetizing reactance X_M which is much larger. Therefore the circuit under these conditions looks like a series combination of R_1, $X_{1\sigma}$, R_2 and $X_{2\sigma}$.

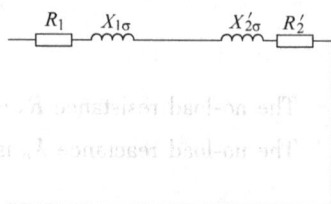

Fig. 4-28 The equivalent circuit of locked rotor test

Because the voltage is low and the speed is at zero, the input power P_k is equal to the stator copper losses $p_{Cu1} = 3I_k^2R_1$ and the rotor copper losses $p_{Cu2} = 3I_k^2R_2$, the related parameters, such as short-circuit impedance Z_k, short-circuit resistance R_k, short-circuit reactance X_k, can be therefore obtained

$$|Z_k| = \frac{U_1}{I_k},\ R_k = \frac{p_k}{3I_k^2},\ X_k = \sqrt{|Z_k|^2 - R_k^2} \tag{4-60}$$

The rotor resistance that is referred to the stator R_2' can be calculated.

$$R_2' = R_k - R_1 \tag{4-61}$$

It is hard to separate the stator leakage reactance from the rotor leakage reactance. Usually we consider that

$$X_{1\sigma} \approx X_{2\sigma} \approx \frac{1}{2}X_k \tag{4-62}$$

4.7 Operational characteristics of induction motors

The following characteristic curves of an induction motor will be discussed in this session.

1) Speed characteristic curve, $n=f(P_2)$, $U=U_N$, $f_1=f_N$.

2) Stator current characteristics curve, $I_1=f(P_2)$, $U=U_N$, $f_1=f_N$.

3) Power factor characteristics curve, $\cos\varphi_1=f(P_2)$, $U=U_N$, $f_1=f_N$.

4) Electromagnetic torque characteristics curve, $T_e=f(P_2)$, $U=U_N$, $f_1=f_N$.

5) Efficiency characteristics curve, $\eta=f(P_2)$, $U=U_N$, $f_1=f_N$.

where n is the speed of motor; T_e is the electromagnetic torque; I_1 is the stator current; $\cos\varphi_1$ is the power factor of stator; η is the efficiency of the motor; and P_2 is the output power of the motor. The characteristic curves mentioned above are shown in Fig. 4-29.

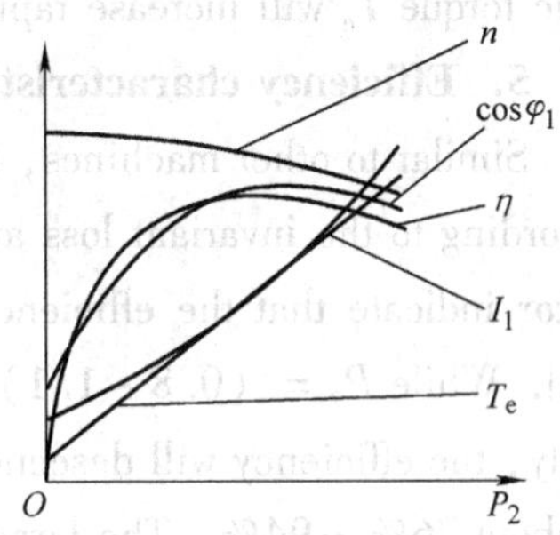

Fig. 4-29 Operational characteristics of induction motor

1. Speed characteristic curve $n=f(P_2)$

The speed characteristic curve is shown in Fig. 4-29. When the motor is running at no-load, where the output power P_2 is zero, the slip s of the motor is very small, and the speed n is close to the synchronous speed n_s. If the mechanical load increases, the motor will begin to slow down until the motor torque is equal exactly to the load torque. During this process, the output power P_2 will increase, the rotor current will increase, slip s will also increase. When P_2 is equal to P_N, the rated slip s_N is about 2% ~5%, and $n_N=(1-s_N)n_s=(95\%\sim98\%)n_s$.

2. Stator current characteristics curve $I_1=f(P_2)$

The stator current characteristic curve is shown in Fig. 4-29. When the motor is running at no-load, where the output power P_2 is zero, and the stator current I_1 is approximately equal to magnetizing current I_m.

If the mechanical load is increasing, the motor will begin to slow down until the motor torque is exactly equal to the load torque. During this process, the output power P_2 will increase and the rotor current I_2 will increase. According to this current equation $\dot{I}_1=\dot{I}_m+(-\dot{I}'_2)$, the stator current will increase rapidly .

3. Power factor characteristics curve $\cos\varphi_1=f(P_2)$

Induction motors are the inductive loads to the power grid, which needs to be absorbed inductive reactive power from the grid. The power factor of an induction motor is lagging and is always less than 1.

The power factor characteristic curve is shown in Fig. 4-29. When the motor is running at no-load, where the output power P_2 is zero, the stator current I_1 equals to I_m approximately, and the stator power factor $\cos\varphi_1$ is very low, usually equals to 0. 2 approximately.

When P_2 begins to increase, the value of s becomes smaller and the power factor of the rotor becomes bigger at this moment. While the active component of stator current increases, $\cos\varphi_1$ will increase. When P_2 increases to a certain value, while slip increasing at the same time, the power factor of rotor will decrease and $\cos\varphi_1$ will fall. Usually $\cos\varphi_1$ reaches maximum value before P_2 reaches P_N.

4. Electromagnetic torque characteristics curve $T_e = f\ (P_2)$

The electromagnetic torque characteristic curve is shown in Fig. 4-29. When the motor is running at no-load, the output power P_2 is zero, the stator current I_1 equals to I_m approximately, and the torque T_e is small, it equals to the no-load torque.

When P_2 begins to increase, according to torque equation $T_e = T_0 + T_2 = T_0 + \frac{P_2}{\Omega}$, electromagnetic torque T_e will increase rapidly.

5. Efficiency characteristics curve $\eta = f\ (P_2)$

Similar to other machines, the efficiency characteristics of the induction motor can be analyzed according to the invariant loss and the variant loss. The efficiency characteristics of the induction motor indicate that the efficiency increases rapidly when the output power P_2 increases with the load. While $P_2 = (0.8 \sim 1.1)\ P_N$, the maximum efficiency will occur. If P_2 increases continuously, the efficiency will descend as shown in Fig. 4-29. The rated efficiency of an induction motor is about 76% ~94%. The larger the induction motor's capacity is, the higher the rated efficiency will be.

4.8 Torque speed characteristics of induction motors

The torque is developed by a motor which depends upon its speed, but the relationship between those two variables cannot be expressed by a simple equation. The torque speed characteristic is also called mechanical characteristic, which is denoted by $n = f\ (T_e)$, when the voltage, the frequency and the motor parameters are constant.

According to Eq. 4-49 and Eq. 4-42, the relationship between the torque and the rotor current is

$$T_e = \frac{P_e}{\Omega_s} = \frac{m_1}{\Omega_s} I'^2_2 \frac{R'_2}{s} = \frac{m_1 I'^2_2 \frac{R'_2}{s}}{\frac{2\pi n_s}{60}} = \frac{m_1 I'^2_2 \frac{R'_2}{s}}{\frac{2\pi f_1}{p}} \tag{4-63}$$

In the T-shape equivalent circuit of an induction motor, since the magnetizing impedance Z_m is much bigger than the stator and the rotor leakage impedances Z_1 and Z'_2, the magnetizing impedance branch can be opened. Then,

$$I'_2 \approx \frac{U_1}{\sqrt{\left(R_1 + \frac{R'_2}{s}\right)^2 + (X_{1\sigma} + X'_{2\sigma})^2}} \tag{4-64}$$

Substitute Eq. 4-64 to Eq. 4-63:

$$T_e = \frac{m_1 p U_1^2 \frac{R'_2}{s}}{2\pi f_1 \left[\left(R_1 + \frac{R'_2}{s} \right)^2 + (X_{1\sigma} + X'_{2\sigma})^2 \right]} \tag{4-65}$$

Eq. 4-65 is the parameter expressions of the mechanical characteristics of an induction motor. A typical induction motor torque-speed characteristics curve is plotted in Fig. 4-30 and Fig. 4-31. Some important information about the operation of an induction motor is summarized as following:

1) The induced torque of the motor is zero at the synchronous speed.

2) The torque-speed curve is nearly linear between no load and full load condition. In this range, the rotor resistance is much larger than the rotor reactance, so the rotor current, the rotor magnetic field, and the induced torque increase linearly with the increasing slip.

3) There is a maximum possible torque that cannot be exceeded. This torque, which is called the *pullout torque* or *breakdown torque*, is 2 to 3 times the rated full-load torque of the motor.

4) The starting torque on the motor is slightly larger than its full-load torque, so this motor with start with full load.

5) The torque on the motor with given slip varies with the square of the applied voltage.

6) If the rotor of the induction motor is driven faster than the synchronous speed, then the direction of the induced torque in the machine reverses and the machine becomes a *generator*, converting the mechanical power to the electric power.

7) If the motor is turning backward that is relative to the direction of the magnetic fields, the induced torque in the machine will stop the machine rapidly and will try to rotate it in the other direction. The machine is operating at a *braking* state. Since reversing the direction of magnetic field rotation is simply a matter of switching any two of the stator phases, in fact this can be used as a way to stop an induction motor rapidly. The act of switching two phases in order to stop the motor rapidly is called *plugging*.

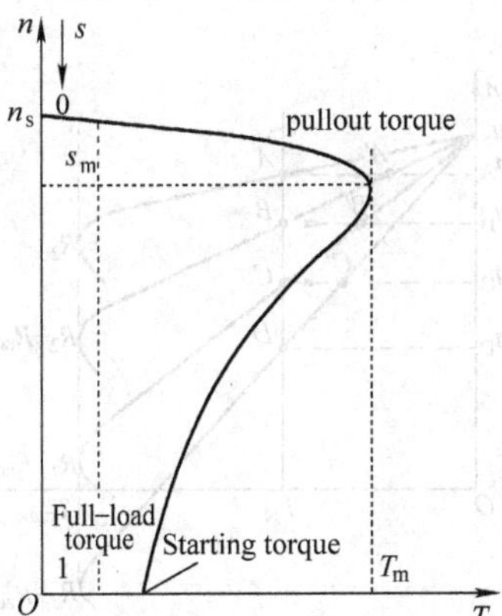

Fig. 4-30 A typical induction motor torque-speed characteristic curve

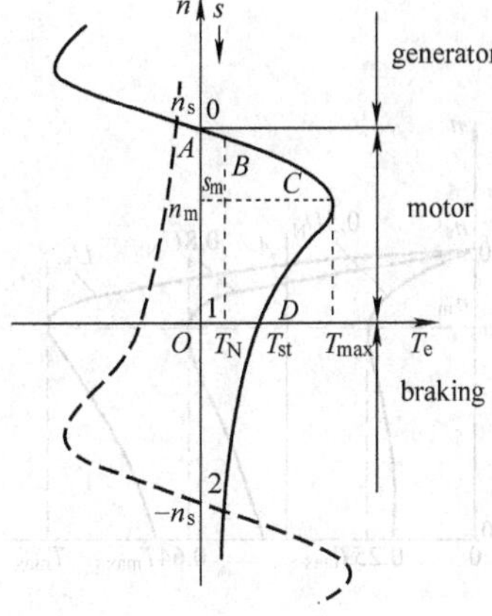

Fig. 4-31 Torque-slip curve of induction motors

1. Maximum torque (Pullout torque)

In Fig. 4-30, there is the pullout torque. According to Eq. 4-65, when $\frac{dT_e}{ds}=0$, the critical slip s_m and the maximum torque T_{max} can be calculated as following

$$s_m = \pm \frac{R'_2}{\sqrt{R_1^2 + (X_{1\sigma} + X'_{2\sigma})^2}} \tag{4-66}$$

Substitute Eq. 4-66 into Eq. 4-65, then

$$T_{max} = \pm \frac{m_1 p}{2\pi f_1} \frac{U_1^2}{2[\pm R_1 + \sqrt{R_1^2 + (X_{1\sigma} + X'_{2\sigma})^2}]} \tag{4-67}$$

In Eq. 4-66 and Eq. 4-67, the positive sign corresponds to the motor mode, the negative sign corresponds to the generator mode. Because of $R_1 \ll X_{1\sigma} + X'_{2\sigma}$, R_1 can be neglected, and then Eq. 4-66 and Eq. 4-67 are rewritten as

$$s_m \approx \pm \frac{R'_2}{X_{1\sigma} + X'_{2\sigma}} \tag{4-68}$$

$$T_{max} \approx \pm \frac{m_1 p U_1^2}{4\pi f_1 (X_{1\sigma} + X'_{2\sigma})} \tag{4-69}$$

We can get some conclusions from Eq. 4-68 and Eq. 4-69:

1) The maximum torque T_{max} of the induction machine is proportional to the square of the voltage (Fig. 4-32), and is inversely proportional to frequency and the sum of the stator and the rotor leakage reactance. It is independent of the rotor resistance.

2) The critical slip s_m is proportional to the rotor resistance and is inversely proportional to the sum of the stator and the rotor leakage reactance. s_m is independent of the voltage (Fig. 4-32).

3) The value of the maximum torque T_{max} is independent of the rotor resistance, but the critical slip s_m is directly proportional to the rotor resistance. When the rotor resistance increases, s_m increases, but T_{max} remains unchanged. Then the maximum value of n-T_e curve will move down (Fig. 4-33).

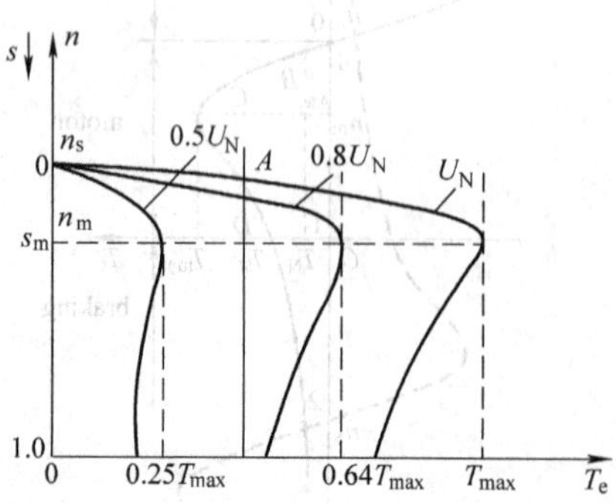

Fig 4-32 the n-T_e curve while the voltage changed

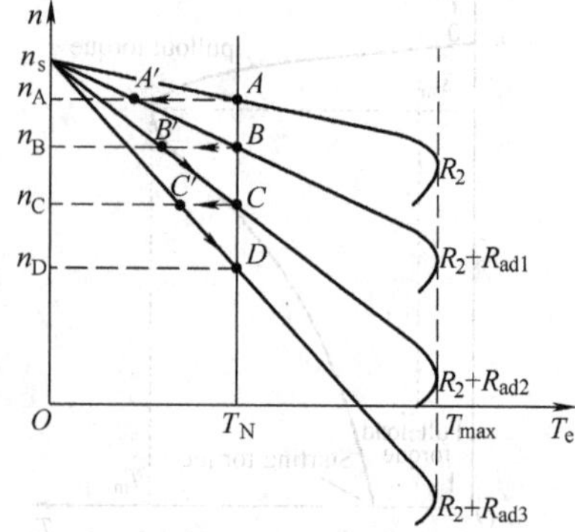

Fig 4-33 the n-T_e curve while the rotor resistance changed

The ratio of the maximum torque T_{max} and the rated torque T_N is called overload capacity λ_m

$$\lambda_m = \frac{T_{max}}{T_N} \tag{4-70}$$

If the load torque is bigger than the maximum electromagnetic torque, then the motor will stall, usually $\lambda_m = 1.6 \sim 2.5$.

2. Starting torque

When slip $s = 1$, the induced torque is called starting torque, and Eq. 4-65 can be expressed as

$$T_{st} = \frac{m_1}{2\pi f_1} \times \frac{pU_1^2 R'_2}{(R_1 + R'_2)^2 + (X_{1\sigma} + X'_{2\sigma})^2} \tag{4-71}$$

Some conclusions can be obtained from Eq. 4-71,

1) The Starting torque T_{st} is proportional to the square of voltage (Fig. 4-32).

2) If $(R_1 + R'_2) \ll (X_{1\sigma} + X'_{2\sigma})$, Eq. 4-71 will become $T_{st} \propto \frac{R'_2}{(X_{1\sigma} + X'_{2\sigma})^2}$. For the wound rotor motor, an external resistor R_{st} could be connected in series with the rotor circuit in order to increase the starting torque (Fig. 4-33).

If the maximum torque needs to be appeared at the starting point, i.e. $s_m = 1$, according to Eq. 4-66 and Eq. 4-68, the following relationship can be obtained.

$$R'_2 + R'_{st} = \sqrt{R_1^2 + (X_{1\sigma} + X'_{2\sigma})^2} \approx X_{1\sigma} + X'_{2\sigma} \tag{4-72}$$

$$R'_{st} = X_{1\sigma} + X'_{2\sigma} - R'_2 \qquad R'_{st} = R_{st} k_e k_i \tag{4-73}$$

where R'_{st} is the series resistance of the rotor which is referred to the stator side; k_e and k_i are the voltage ratio and the current ratio respectively.

Starting capacity K_{st} is the ratio of the starting torque and the rated torque.

$$K_{st} = T_{st}/T_N \tag{4-74}$$

Usually K_{st} is about $0.8 \sim 1.2$.

Example 4-2 A six-pole three-phase squirrel cage induction motor: the stator winding is Y connected, where the rated voltage $U_N = 380\text{V}$, the rated speed $n_N = 965\text{r/min}$, the power frequency $f_1 = 50\text{Hz}$, the stator resistance $R_1 = 2.1\Omega$, the stator leakage reactance $X_{1\sigma} = 3.08\Omega$, the rotor resistance $R'_2 = 1.48\Omega$, and the rotor leakage reactance $X'_{2\sigma} = 4.2\Omega$. Calculate:

1) the rated electromagnetic torque;

2) the maximum electromagnetic torque and the overload capacity;

3) the critical slip;

4) the starting torque and starting capacity.

Solution:

The synchronous speed of the motor n_s

$$n_s = \frac{60 f_1}{p} = \frac{60 \times 50}{3}\text{r/min} = 1000\text{r/min}$$

The rated slip s_N

$$s_N = \frac{n_s - n_N}{n_s} = \frac{1000 - 965}{1000} = 0.035$$

Rated phase voltage of the stator U_1

$$U_1 = \frac{380}{\sqrt{3}}\text{V} = 220\text{V}$$

1) Rated torque T_N is

$$T_N = \frac{m_1 p U_1^2 \frac{R'_2}{s_N}}{2\pi f_1\left[\left(R_1 + \frac{R'_2}{s_N}\right)^2 + (X_{1\sigma} + X'_{2\sigma})^2\right]}$$

$$= \frac{3 \times 3 \times 220^2 \times \frac{1.48}{0.035}}{2\pi \times 50 \times \left[\left(2.1 + \frac{1.48}{0.035}\right)^2 + (3.08 + 4.2)^2\right]}\text{N} \cdot \text{m}$$

$$= 35\text{N} \cdot \text{m}$$

2) The maximum electromagnetic torque T_{max}

$$T_{max} = \frac{m_1 p U_1^2}{4\pi f_1 (X_{1\sigma} + X'_{2\sigma})} = \frac{3 \times 3 \times 220^2}{4\pi \times 50 \times (3.08 + 4.2)}\text{N} \cdot \text{m} = 95.23\text{N} \cdot \text{m}$$

Overload capacity λ_m

$$\lambda_m = \frac{T_{max}}{T_N} = \frac{95.23}{35} = 2.72$$

3) Critical slip s_m

$$s_m = \frac{R'_2}{X_{1\sigma} + X'_{2\sigma}} = \frac{1.48}{3.08 + 4.2} = 0.203$$

4) Starting torque T_{st}

$$T_{st} = \frac{m_1}{2\pi f_1} \frac{p U_1^2 R'_2}{(R_1 + R'_2)^2 + (X_{1\sigma} + X'_{2\sigma})^2}$$

$$= \frac{3 \times 3 \times 220^2 \times 1.48}{2\pi \times 50 \times [(2.1 + 1.48)^2 + (3.08 + 4.2)^2]}\text{N} \cdot \text{m} = 31.18\text{N} \cdot \text{m}$$

Starting capacity K_{st}

$$K_{st} = \frac{T_{st}}{T_N} = \frac{31.18}{35} = 0.89$$

4.9 Starting the three-phase induction motor

The starting process of the motor is that the motor speed varies from zero to rated speed. The starting capacity of an induction motor is defined as $K_{st} = T_{st}/T_N$, the starting current ratio is defined as $K_I = I_{st}/I_N$. We usually hope that the starting torque is large enough, but the starting current is not large enough. the starting equipment is simple, economic and reliable.

If we connect the stator of an induction motor to a 3-phase source, rotor speed $n = 0$, and slip $s = 1$ at the starting point. According to the equivalent circuit, the starting stator current I_{st} is only restricted by the sum of the leakage impedance of the stator and the rotor, that is $Z_1 + Z'_2$. Because the leakage impedance is small, the starting current is very large, generally can reach 4 to 7 times of the rated current.

At the starting moment, the starting torque T_{st} is small. The reason for this is: As T_{st} is restricted by the electromagnetic torque $T_e = C_T\Phi_m I'_2\cos\varphi_2$, the rotor power factor angle $\varphi_2 = \arctan(sX'_2/R'_2)$ closes to 90° ($s=1$, $x'_2 \gg r'_2$), and then the rotor power factor $\cos\varphi_2$ is very small at the starting moment. Although the starting current I_{st} is very large, but the active component of the rotor current $I'_2\cos\varphi_2$ is very small.

Usually when the starting torque ratio $K_{st} = T_{st}/T_N \geqslant 1.1$, the motor can therefore be started. Fig. 4-34 shows the mechanical characteristics and the stator current characteristics when the three-phase induction motor is started directly. I_1 is the stator phase current.

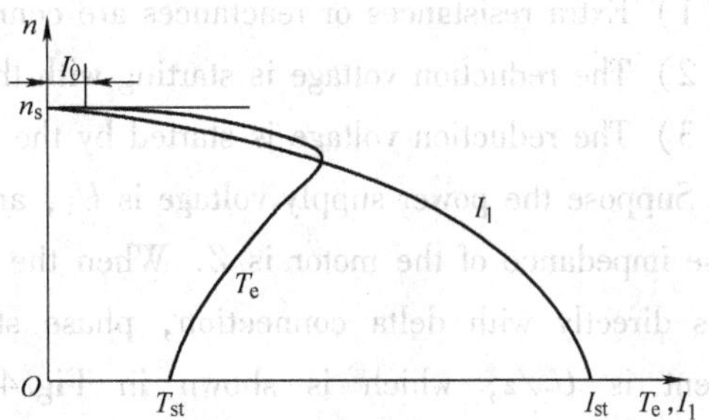

Fig. 4-34 The mechanical characteristics and the stator current characteristics under starting directly.

4.9.1 The starting methods of the three-phase induction motors

1. Direct starting

Direct starting means the stator winding is supplied under the rated voltage while the motor starting, this is also known as the full voltage starting. The advantages of direct starting are simple operation and do not need any complex starting equipments. The shortcoming of this method is that the starting current is large. Therefore, usually only the small capacity motor suitable for the directly starting. No matter the large capacity motor can be directly started or not, this depends on the grid capacity and the motor starting current. It is usually determined by the following experience equation.

$$K_I = \frac{I_{st}}{I_N} \leqslant \frac{1}{4}\left[3 + \frac{\text{total capacity of the power supply (kV·A)}}{\text{capacity of the induction motor (kV·A)}}\right] \tag{4-75}$$

The relationship between the capacity of the motor and the rated data is:

$$\text{(Starting motor capacity kV·A)} = \frac{P_N}{\eta_N\cos\varphi_{1N}} = \sqrt{3}U_{1N}I_{1N} \tag{4-76}$$

2. Reduction voltage starting

In the simplified equivalent circuit of an induction motor, if magnetizing current I_m is neglected, then

$$I_1 \approx I'_2 = \frac{U_1}{\sqrt{(R_1 + R'_2/s)^2 + (X_{1\sigma} + X'_{2\sigma})^2}} \tag{4-77}$$

When the motor is starting, by substituting $s=1$ into Eq. 4-77, the starting current is

$$I_{st} \approx \frac{U_1}{\sqrt{(R_1 + R'_{2\sigma})^2 + (X_{1\sigma} + X'_{2\sigma})^2}} \tag{4-78}$$

Eq. 4-78 shows that the starting current is proportional to the power supply voltage; therefore if we reduce the voltage in the stator windings of the motor, the starting current will decrease. But the starting torque will decrease proportionally to the square of the voltage. Therefore the reduction voltage starting method is only suitable for no load or light load condition.

There are several approaches used to reduce the voltage of the stator circuit.

1) Extra resistances or reactances are connected in series with the stator circuit.

2) The reduction voltage is starting with the autotransformer.

3) The reduction voltage is started by the switching of wye/delta connection.

Suppose the power supply voltage is U_1, and per phase impedance of the motor is Z. When the motor starts directly with delta connection, phase starting current is U_1/z, which is shown in Fig. 4-35a. Because the line current is $\sqrt{3}$ times of the phase current in the delta connection, the starting line current is $I_{stD}=\sqrt{3}U_1/z$. When the motor is starting in the wye connection (Fig. 4-35b), the phase voltage is $U_1/\sqrt{3}$, the starting current of the each phase is $U_1/(\sqrt{3}z)$. Because the line current is equal to the phase current in the wye connection, the starting current is $I_{stY}=U_1/(\sqrt{3}z)$. Then

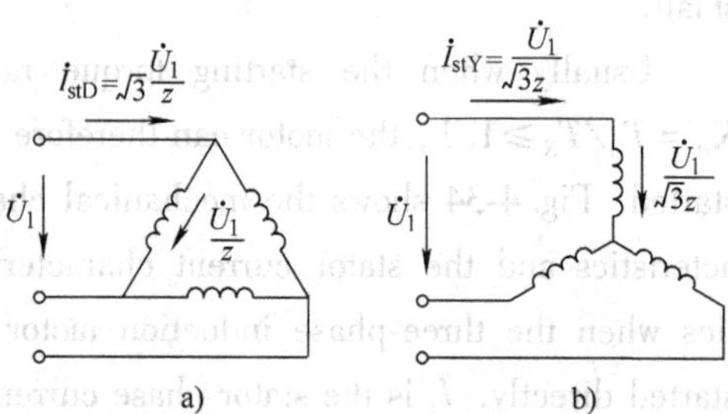

Fig.4-35 The starting current of the reduction voltage is started by switching wye/delta connection

$$\frac{I_{stY}}{I_{stD}}=\frac{U_1/(\sqrt{3}z)}{\sqrt{3}U_1/z}=\frac{1}{3} \tag{4-79}$$

Because the starting torque is proportional to the square of the phase voltage, then

$$\frac{T_{stY}}{T_{stD}}=\left(\frac{U_1/\sqrt{3}}{U_1}\right)^2=\frac{1}{3} \tag{4-80}$$

4) For a wound rotor induction motor, starting can be achieved at relatively low currents by inserting an extra resistance in the rotor circuit during starting. This extra resistance not only increases the starting torque, but also reduces the starting current.

5) For a wound rotor induction motor, starting can also be achieved at low currents and high torque by connecting the frequency-sensitive rheostat in series with the rotor circuit during starting.

Frequency-sensitive rheostat looks like the three-phase core transformer with only the primary winding connected in wye connection. The windings are connected in series with the rotor circuits of an induction motor.

When the motor is started, where rotor speed $n=0$, slip $s=1$, and the frequency of rotor current equals to the frequency of the stator voltage, i. e. $f_2=sf_1=f_1$. Both of the iron loss in the frequency-sensitive rheostat and the excitation resistance are large. Therefore there are equivalent large resistances connected in rotor circuit, and the starting current is restricted, while the starting torque is increasing.

With the speed n rising, the slip s decreases gradually, and the rotor frequency f_2 is reduced correspondingly. The magnetizing resistance R_m of the frequency-sensitive rheostat will also continue to decrease. This is equivalent to the case that the starting resistance is automatically cut off.

4.9.2 Cage rotor induction motors with high starting torque

Due to the simple structure, low cost and convenient maintenance of the cage rotor induction motors, the motors are therefore widely used in industry. However it is only suitable at no-load or light-load starting condition, due to the small starting torque. From the Eq. 4-71, increasing the rotor resistance will increase the starting torque and decrease the starting current. The cage rotor induction motor with high starting torque can be achieved by a cage rotor design.

1. Small bar cage rotor induction motors

The reactance $X_{2\sigma}$ in an induction motor equivalent circuit represents the referred form of the rotor's leakage reactance. In general, the further away from the stator of a rotor bar or part of a bar is, the greater its leakage reactance will be. Therefore, if the bars of a cage rotor are placed near the surface of the rotor (Fig. 4-36a), they will have only a small leakage flux and the reactance $X_{2\sigma}$ will also be small in the equivalent circuit. On the other hand, if the rotor bars are placed deeper into the rotor surface (Fig. 4-36b, c), there will be more leakage flux and the rotor reactance $X_{2\sigma}$ will be larger.

Fig. 4-36a shows the cross section of an induction motor rotor with small bars that are placed near the surface of the rotor. Since the cross-sectional area of the bars is small, the rotor resistance is relatively high. Since the bars are located near the stator, the rotor leakage reactance is still small. This motor is very much like a wound-rotor induction motor with an extra resistance inserted into the rotor. Because of the large rotor resistance, this motor has a pullout torque occurring at a high slip, and its starting torque is quite high as well. But the increase of the rotor resistance leads to the increase of the rotor copper loss, the operation efficiency of the motor will decrease.

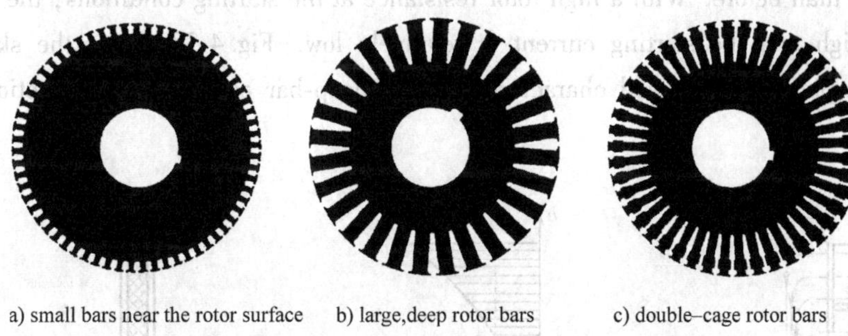

a) small bars near the rotor surface b) large,deep rotor bars c) double–cage rotor bars

Fig. 4-36 Laminations of cage induction motor rotors

2. Deep-bar and double-cage rotor induction motors

An ingenious and simple way to obtain a rotor resistance, which automatically varies with speed, is making use of the fact that the rotor frequency and the stator frequency are the same at standstill; as the motor accelerates, the rotor frequency decreases to a very low value—perhaps around 2 or 3 Hz at full load. With some suitable shapes and arrangements for the rotor bars, the squirrel-cage rotors can be designed, so that their effective resistance at 50Hz is several times of the resistance at 2 or 3 Hz.

It is possible to produce a variable rotor resistance by using the deep rotor bars (Fig. 4-36b) or the double-cage rotors (Fig. 4-36c). Fig. 4-37a shows a current flowing through the upper part of a deep rotor bar. Since current flowing in that area is tightly coupled to the stator, the leakage inductance is small for this region. Fig. 4-37b shows a current flowing deeper in the bar. Here, the leakage inductance is higher. Since all parts of the rotor bar are in parallel electrically, the bar essentially represents a series of parallel electric circuits; the upper ones have a smaller inductance and the lower one have a larger inductance.

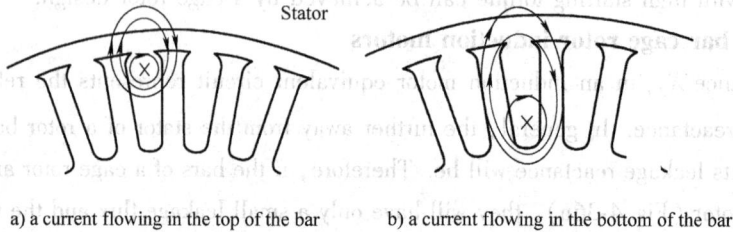

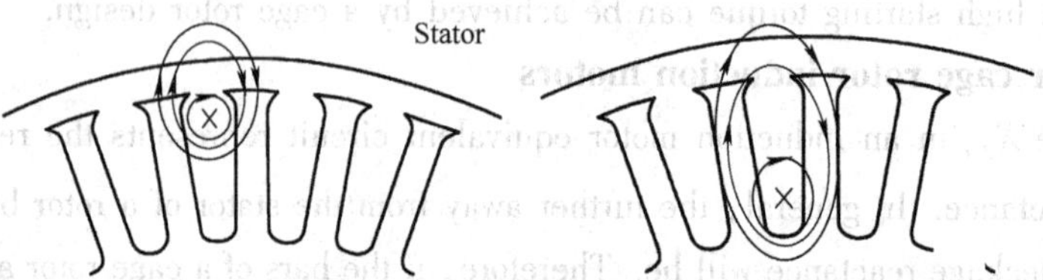

a) a current flowing in the top of the bar   b) a current flowing in the bottom of the bar

Fig. 4-37 Flux leakage in a deep-bar rotor

At low slip, the rotor's frequency is very small, and the reactances of all the parallel paths through the bar are small compared to their resistances. The impedances of all parts of the bar are approximately equal, so the current flows through all parts of the bar equally. The resulting large cross-sectional area makes the rotor resistance quite small, resulting in a large efficiency at low slips.

At high slip (starting conditions), the reactances are large compared to the resistances in the rotor bars, so all the currents are forced to flow into the low-reactance part of the bar near the stator. This phenomenon is called *skin effect*. Since the effective cross section is smaller, the rotor resistance is higher than before. With a high rotor resistance at the starting conditions, the starting torque is relatively high and the starting current is relatively low. Fig. 4-38 shows the skin effect in a deep-bar. A typical torque-speed characteristic for a deep-bar cage rotor construction is shown in Fig. 4-39.

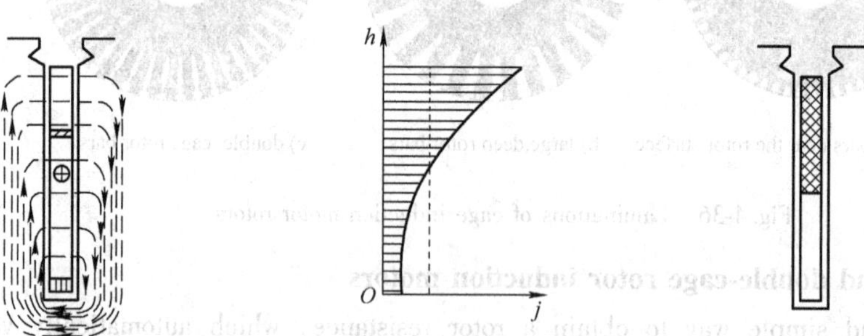

a) The distribution of the bar leakage flux   b) The distribution of the current density   c) The effective cross-sectional area of bar

Fig. 4-38 Skin Effect in a deep bar

A cross-sectional view of a double-cage rotor is shown in Fig. 4-36c. It consists of a large, low-resistance set of bars that is buried deeply in the rotor and it also consists of a small, high-resistance set of bars at the rotor surface. Double-cage rotor is similar to the deep-bar rotor, except that the difference between the low-slip and the high-slip operation is even more exaggerated.

At starting condition, only the small bar is effective, and the rotor resistance is quite high. This high resistance results in a large starting torque. However, at a normal operating speed, both bars are effective, and the resistance is almost as low as in a deep-bar rotor.

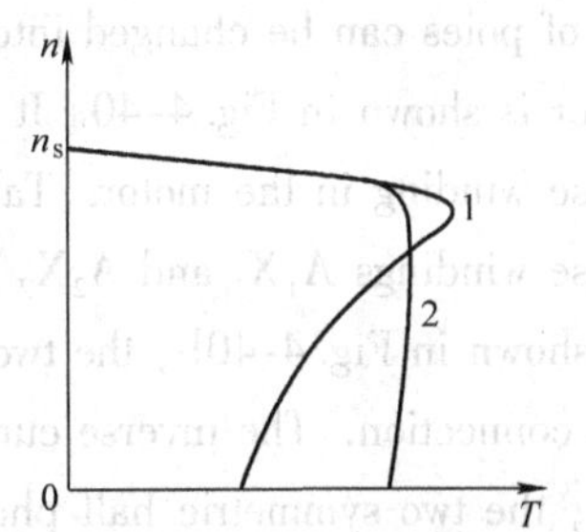

Fig. 4-39 Typical torque-speed curves for different rotor designs

1—The common cage rotor induction motor

2—The deep bar cage rotor induction motor

Notice that since the effective resistance and the leakage inductance of a double-cage and a deep-bar rotor vary with frequency, the parameters and the representation of the referred effects of the rotor resistance R_2 and the leakage inductance $X_{2\sigma}$ as viewed from the stator are not constant.

The disadvantage of double-cage rotors is that they are more expensive than the other types of cage rotors, but they are cheaper than wound-rotor motors. They demonstrate some of the features in wound-rotor motors (a high starting torque with a low starting current and a good efficiency at normal operating conditions) at lower cost and without the need of maintaining the slip rings and the brushes.

4.10 Speed control of induction motors

The simple induction motor fulfills admirably the requirements of the continuously constant-speed drives. Many motor applications, however, require several speeds, or even a continuously adjustable range of speeds. There are only two techniques by which the speed of an induction motor can be controlled. One is to vary the synchronous speed n_s, which is the speed of the stator and the rotor magnetic fields, since the rotor speed always remains near n_s. The other technique is to vary the slip s of the motor for a given load. The speed of an induction motor is given by

$$n = n_s(1 - s) = \frac{60f_1}{p}(1 - s) \tag{4-81}$$

So the synchronous speed n_s of an induction motor can be changed by

① Varying the pole number of the stator winding.

② Varying the power frequency f_1.

The slip s can be changed by

① Varying the line voltage;

② Varying the rotor resistance;

③ Inserting voltages of the appropriate frequency in the rotor circuits.

4.10.1 Speed control by changing pole number

There are two major approaches to change the number of poles in an induction motor: the method of consequent poles and the multiple stator windings.

The stator winding can be designed so that by a simple change in the coil connections the number of poles can be changed into a ratio 2∶1. The pole-changing schematic diagram of the induction motor is shown in Fig. 4-40. It is assumed that the two symmetric half phase windings make up the phase winding in the motor. Take phase A as an example. As shown in Fig. 4-40a, the two half phase windings A_1X_1 and A_2X_2 make the forward series connection to form a 4-pole magnetic field. As shown in Fig. 4-40b, the two symmetric half phase windings A_1X_1 and A_2X_2 make a reversed series connection. The inverse current in A_2X_2 leads to the 2-pole magnetic field. As shown in Fig. 4-40c, the two symmetric half phase windings A_1X_1 and A_2X_2 make a reversed parallel connection. The inverse current in A_2X_2 leads to the 2-pole magnetic field.

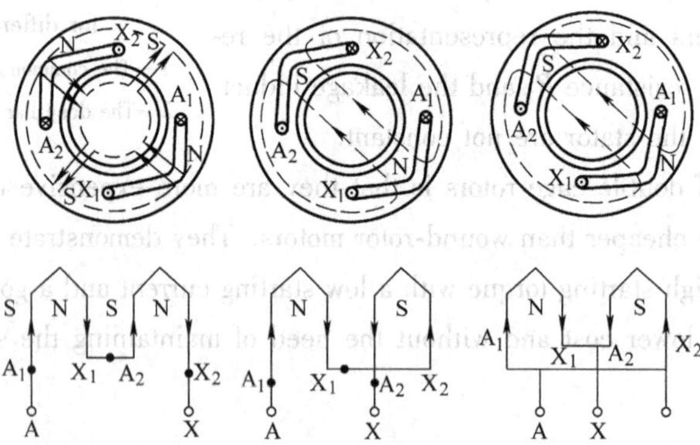

a) Forward Series Connection b) Reversed Series Connection c) Reversed Parallel Connection

Fig. 4-40 Pole-changing Schematic Diagram

The stator winding varies from the forward series connection to the reversed series connection or the reversed parallel connection to make the current of one half phase windings inversed and half the pole-pair number, so that the synchronous speed is changed twice (or half) as many times as before, and the pole number of the multiple-speed motor is usually 2/4 pole, 4/8 pole or 6/12 pole.

The rotor in such a motor adopted the cage design, since a cage rotor always has as many poles induced in it as there are in the stator and thus it is adapted when the number of stator poles changes.

The major disadvantage of the consequent-pole method of changing the speed is that the speeds must be in a ratio of 2∶1. A traditional approach to overcome this limitation is to employ a multiple stator winding with different numbers of poles and to energize one set at a time. Unfortunately, the multiple stator windings would increase the expense of the motor, therefore it is used only when it is absolutely necessary.

4.10.2 Speed control by changing frequency

The motor's synchronous speed and the rotor speed can be controlled by varying the frequency. The synchronous speed of the motor at rated conditions is known as the *base speed*, and the rated frequency is called *base frequency*. By using variable frequency control, it is possible to adjust the speed of the motor either above or below base speed. A proper designed variable-frequency induction motor drive can be very flexible. However, it is important to maintain voltage at certain level and the torque limits on the motor as the frequency is varied to ensure safe operation.

1. Speed control below base speed

When the motor is running at a speed that is below the based speed, it is necessary to reduce the terminal voltage that applied to the stator for a proper operation. The terminal voltage that is applied to the stator should decrease linearly with the decreasing stator frequency. This process is called *derating*. If it is not done, the steel in the core of the induction motor will be saturated and an excessive magnetization current will flow into the machine.

In an induction motor, if the voltage drop of the stator leakage impedance is ignored, then $U_1 \approx E_1 = 4.44 f_1 N_1 k_{w1} \Phi_m$. Therefore, in order to keep the main flux Φ_m of the motor constant, the stator terminal voltage should be adjusted proportional to the frequency, i. e.

$$\frac{U_1}{f_1} = \text{constant} \tag{4-82}$$

The torque is

$$T_e = \frac{m_1 p}{2\pi}\left(\frac{U_1}{f_1}\right)^2 \frac{f_1 \dfrac{R'_2}{s}}{\left(R_1 + \dfrac{R'_2}{s}\right)^2 + (X_{1\sigma} + X'_{2\sigma})^2} \tag{4-83}$$

The maximum torque is

$$\begin{aligned} T_{max} &= \frac{1}{2} \frac{m_1 p U_1^2}{2\pi f_1 \left[R_1 + \sqrt{R_1^2 + (X_{1\sigma} + X'_{2\sigma})^2}\right]} \\ &= \frac{1}{2} \frac{m_1 p}{2\pi}\left(\frac{U_1}{f_1}\right)^2 \frac{f_1}{R_1 + \sqrt{R_1^2 + (X_{1\sigma} + X'_{2\sigma})^2}} \end{aligned} \tag{4-84}$$

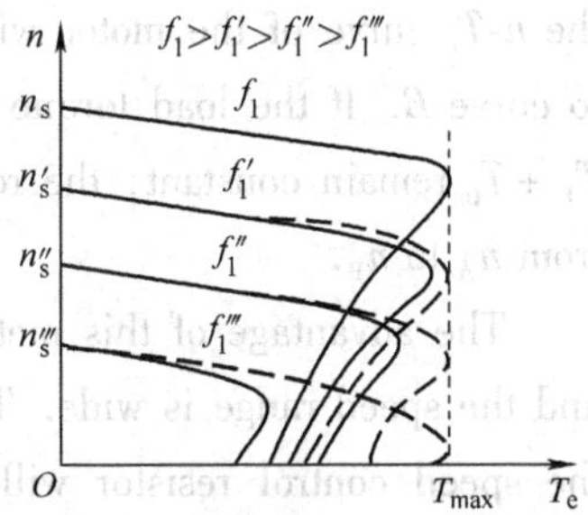

Fig. 4-41 n-T_e curves for speeds below base speed while U_1/f_1 = constant

Because $X_{1\sigma} + X'_{2\sigma}$ is proportional to the stator frequency. When the frequency is close to the base frequency, $R_1 \ll (X_{1\sigma} + X'_{2\sigma})$, and R_1 can be neglected, then the maximum torque will be nearly constant while the frequency is changed nearby the rated frequency. While the frequency is very low, R_1 cannot be neglected, the maximum torque T_{max} will decrease.

Fig. 4-41 shows n-T_e curve while the stator frequency and the voltage are below their rated values, and $\frac{U_1}{f_1}$ = constant.

2. Speed control above base speed

When the stator frequency rise from its rated value, the stator voltage is held constant at the rated value, the main flux of the motor will decrease with the increase of the stator frequency, so the maximum torque will also decrease. During the speed control above the base speed, the stator voltage is constant and the main flux is weak. Fig. 4-42 is the torque-speed curves for speed above base speed while U_1 is constant.

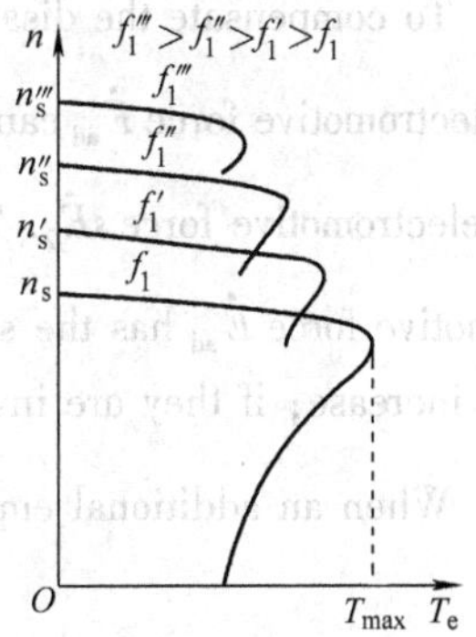

Fig. 4-42 n-T_e curves for speeds above base speed while U_1 is constant

4.10.3 Speed control by changing the stator voltage

The torque speed curve of an induction motor under different voltage is a group of curve with the same synchronous speed and the same critical slip. The maximum torque is proportional to the square of the applied voltage, as shown in Fig. 4-43. If the load belongs to the constant torque load, as T_{L1}; under different voltage condition A, B, C, the working point is n_A, n_B and n_C respectively. The speed decreases with the voltage reduction and the speed of the motor may be controlled by varying the voltage over a limited range. This kind of speed control is sometimes used on a small motor driving fan.

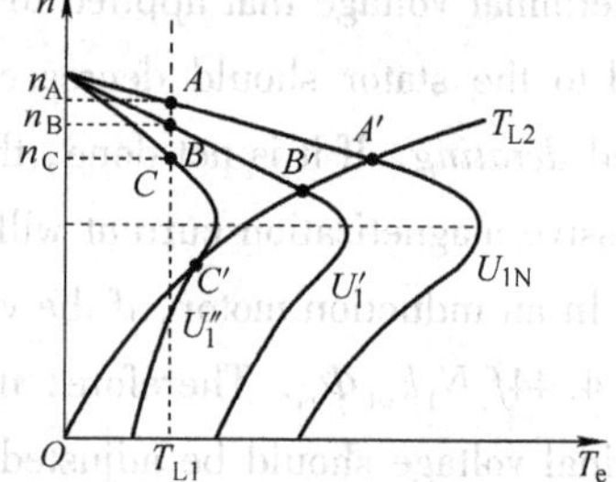

Fig. 4-43 Variable voltage speed control in an induction motor

4.10.4 Speed control by changing rotor resistance

This method is only suitable for a wound rotor induction motor.

This can be understood from Fig. 4-44, when the resistance is inserted in series with the rotor circuit, the n-T_e curve of the motor will change from curve A to curve B. If the load torque and the no-load torque $T_L + T_0$ remain constant, the rotor speed will decrease from n_A to n_B.

The advantage of this method is that it is simple and the speed range is wide. The disadvantage is that the speed control resistor will consume power. This method is mainly used in a medium and a small capacity induction motor, such as a bridge crane motor.

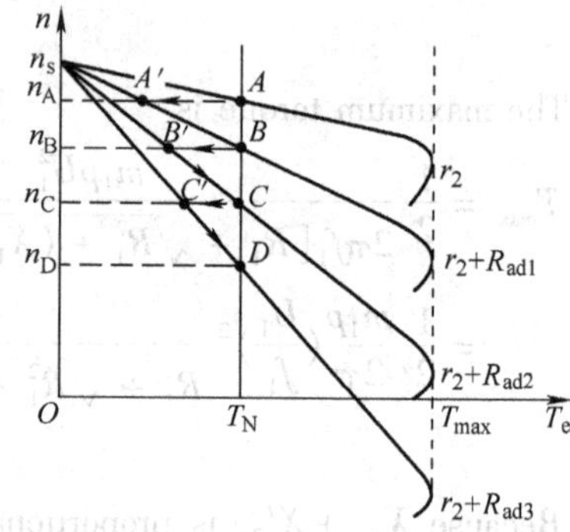

Fig. 4-44 Speed control by varying the rotor resistance in a wound rotor induction motor

4.10.5 Wound-rotor induction motor cascade speed control

To compensate the disadvantages of speed control by changing the rotor resistance, an additional electromotive force $\dot{E}_{ad}$ can be cascaded in the rotor circuit, where the frequency is equal to the rotor electromotive force $s\dot{E}_2$. This is so-called cascade speed control. If the cascading additional electromotive force $\dot{E}_{ad}$ has the same phase as the rotor electromotive force $s\dot{E}_2$, then the speed of motor will increase; if they are inversed phase, the speed of motor will decrease.

When an additional emf $\dot{E}_{ad}$ is not added to the rotor circuit, the rotor current is

$$I_2 = \frac{sE_2}{\sqrt{R_2^2 + (sX_{2\sigma})^2}} \tag{4-85}$$

When a reversed-phase additional emf $\dot{E}_{ad}$ is added to the rotor circuit, the rotor current becomes

$$I_2 = \frac{sE_2 - E_{ad}}{\sqrt{R_2^2 + (sX_{2\sigma})^2}} \quad (4\text{-}86)$$

Under this condition, the rotor current I_2 will decrease, and then the electromagnetic torque T_e will decrease, the speed of motor will also decrease whereas the slip s will increase.

According to Eq. 4-86, when slip s begins to increase, the rotor current I_2 will increase. Then the electromagnetic torque T_e will increase until T_e equals to T_L. The motor operates at low speed.

The mechanical characteristic of cascading a reversed-phase additional emf $\dot{E}_{ad}$ is shown in Fig. 4-45. Curve 1 is the natural mechanical characteristics. Curve 2 is the mechanical characteristics with a reversed-phase additional emf $\dot{E}_{ad}$. After cascading a reversed-phase additional emf $\dot{E}_{ad}$, the ideal no-load speed n'_s will be lower than the synchronous speed n_s, so it is called *low synchronous speed cascade control*. The larger the reversed-phase additional emf $\dot{E}_{ad}$ is, the lower the ideal no-load speed will be.

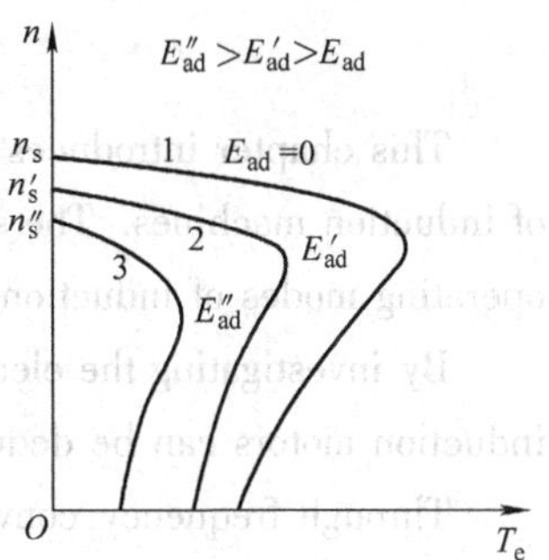

Fig. 4-45 The mechanical characteristics of cascade synchronous speed control

When a same phase additional emf $\dot{E}_{ad}$ is added in the rotor circuit, then the rotor current becomes

$$I_2 = \frac{sE_2 + E_{ad}}{\sqrt{R_2^2 + (sX_{2\sigma})^2}} \quad (4\text{-}87)$$

Under this condition, the rotor current I_2 will increase, and then the electromagnetic torque T_e will increase, the speed of motor will also increase, whereas the slip s will decrease.

According to Eq. 4-87, when s decrease, the rotor current I_2 begins to decrease, and then the electromagnetic torque T_e will decrease until T_e equal to T_L. The motor operates at high speed.

The mechanical characteristic of cascadingthe same-phase additional electromotive force $\dot{E}_{ad}$ is shown in Fig. 4-46. Curve 1 is the natural mechanical characteristics. Curve 2 is the mechanical characteristics with the same-phase additional emf $\dot{E}_{ad}$. It is called *super synchronous speed cascade control*.

During the process of the low cascade synchronous speed control, the device is producing an additional emf $\dot{E}_{ad}$ that will absorb energy from the rotor circuit and feed back to the power grid. During the process of super cascade synchronous speed control, the device is producing an additional emf $\dot{E}_{ad}$ that outputs energy to the rotor circuit and the power grid outputs energy to the stator circuit. Therefore, it is called the doubly-fed operation of the motor.

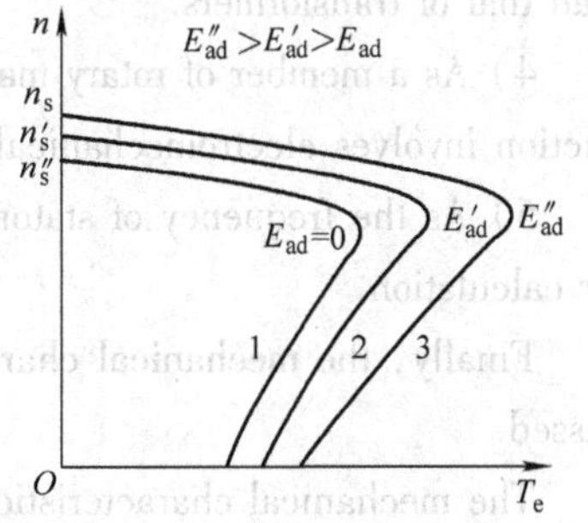

Fig. 4-46 The mechanical characteristics of low super synchronous cascade speed control

The merit of the cascade speed control is that the mechanical characteristics curve is rigid and speed can be controlled smoothly. Stepless speed regulation can be adopted and the slip loss can be feedback to the power grid. Its efficiency is increased, but the device that is producing the additional emf is complex and the cost is high. The lower the speed is, the lower the overload capacity will be. The cascade speed control is therefore suitable in the situation that the speed range is not large, it is usually around 2 ~ 4, which is similar to the fan, the rolling mill and the mine hoist.

Summary

This chapter introduces the basic construction, operation principles and three operating modes of induction machines. The slip "s" is the basic variable of induction motors which reflects different operating modes of induction machines.

By investigating the electromagnetic internal relationship of induction motors, basic equations of induction motors can be deduced.

Through frequency conversion and winding conversion, equivalent circuit of induction motors and phasor diagram can be derived. Either frequency conversion or winding conversion operates under the principle of maintaining the same magnitude and spatial phase of MMF of the rotor flux, such as the influence of rotor on stator invariable.

Basic equations, equivalent circuits and phasor diagrams are three different representations to analyze the induction motor. Among them, equivalent circuits are the most frequently used one in calculations.

Based on the equivalent circuit, power equations and operational performance of the induction motor can be derived.

By comparing analysis methods and derivation process, we can see that there are a lot of similarities between induction motors and transformers; however, the differences between them must be noted, they are as follows

1) The main flux of induction motors is a rotating magnetic field, while for transformers, it is a pulsating magnetic field.

2) The windings of induction motors are short-pitched and they are distributed windings, while the windings of transformers are full-pitch and they are concentrated windings.

3) Because of the existence of the air gap, magnetizing current of induction motors is larger than that of transformers.

4) As a member of rotary machine, induction motors produce electromagnetic torque. The production involves electromechanical energy conversion.

5) As the frequency of stators and rotors is different, frequency conversion must be conducted for calculation.

Finally, the mechanical characteristic, starting and speed control of induction motors are discussed.

The mechanical characteristic of induction motors is a nonlinear curve. The maximum torque of an induction machine is proportional to the square of voltage. However, it is independent of rotor resistance. Notice that rotor resistance is directly proportional to slip s_m. Therefore, when rotor resistance increases, s_m increases, yet T_{max} remains unchanged.

There are two main starting methods for cage induction motors: direct starting and reduced voltage starting. When a cage induction motor is started with direct starting, the starting current can be very large. But the starting torque is not comparably large. In order to reduce starting voltage, the common approach is wye-delta starting method and autotransformer starting method. For a wound-rotor induction motor, starting can be achieved at relatively low currents by inserting an extra resistance into the rotor circuit or connecting a frequency-sensitive rheostat to it during starting. The extra resistance not only increases the starting torque, but also reduces the starting current.

Speed control of induction motors can be accomplished by

1) changing the number of poles.

2) changing the applied electrical frequency.

3) changing the applied terminal voltage.

4) changing the rotor resistance and cascading additional EMF in a rotor circuit of a wound rotor induction motor.

Questions and problems

1. What are slip and slip speed in an induction motor?

2. The stator winding of a wound rotor induction motor is short-circuited. Three phase alternating current with a frequency f_1 is flowing through the rotor winding, which leads to the generation of a rotating magnetic field with a speed of $n_s = 60f_1/p$ [r/min] relative to the rotor in counter-clockwise direction. What direction does the rotor rotate in? How to calculate the slip?

3. Comparing a stationary rotor with a rotating rotor of an induction motor, how do the variable and parameters in the rotor side change?

4. When the rotor speed of an induction motor is decreasing, what happens to the rotor frequency?

5. If the frequency of an induction motor's power supply is increasing while keep the voltage constant, what happens to the main flux of the motor?

6. Why must the main flux be kept constant when an induction motor runs from no load to full load?

7. When the speed of induction motor changes, there is no relative motion between stator mmf and rotor mmf, why?

8. The voltage ratio is different from the current ratio in a three-phase induction motor, why?

9. Describe the energy transformation process of an induction motor.

10. Why is the slip very small when an induction motor is under normal operation?

11. Why is the stator current large when an induction motor is starting at rated voltage ($n = 0$)? What can be done to limit the stator current?

12. Why does the number of rotor poles automatically equal the number of stator poles in a cage rotor induction motor?

13. During a no-load test of an induction motor, when the rated voltage is dropping, the no-load current begins to drop. But when the voltage drops to a certain value, the no-load current will become to increase, why?

14. During a locked-rotor test of an induction motor, when the stator current reaches its rated value, will the electromagnetic torque be at its rated value?

15. How does an induction motor develop torque?

16. Why is it impossible for an induction motor to operate at synchronous speed?

17. What information is learned in a no-load test?

18. What information is learned in a locked-rotor test?

19. Sketch and explain the shape of a typical induction motor torque-speed characteristic curve.

20. What is a deep-bar cage rotor? Why is it used?

21. Why is the efficiency of an induction motor (wound-rotor or cage rotor) so poor at high slips?

22. Name and describe four means of controlling the speed of induction motors.

23. Why is it necessary to reduce the voltage applied to an induction motor as electrical frequency is reduced?

24. An induction motor is running at rated conditions. If the shaft load is now increased, how do the following quantities change?

① Mechanical speed; ② Slip; ③ Rotor induced voltage; ④ Rotor current; ⑤ Rotor frequency; ⑥ Synchronous speed; ⑦ Rotor copper losses.

25. A 220V, three-phase, two-pole, 50Hz induction motor is running at a slip of 5%. Find:

1) The speed of the magnetic fields in revolutions per minute.

2) The speed of rotor in revolutions per minute.

3) The slip speed of the rotor.

4) The rotor frequency in hertz.

26. A 480V, three-phase, four-pole, 60Hz induction motor is running at a slip of 0.035. Find:

1) The speed of the magnetic fields in revolutions per minute.

2) The speed of rotor in revolutions per minute.

3) The slip speed of the rotor.

4) The rotor frequency in hertz.

27. A three-phase, 60Hz induction motor runs at 890r/min at no load and at 840r/min at full load. Find:

1) How many poles does this motor have?

2) What is the slip at rated load?

3) What is the speed at one-quarter of the rated load?

4) What is the rotor's electrical frequency at one-quarter of the rated load?

28. A three-phrase induction motor runs at almost 1198 r/min at no load and 1112 r/min at full load when supplied from a 60Hz, three-phase source.

1) How many poles does this motor have?

2) What is the slip in percent at full load?

3) What is the corresponding frequency of the rotor currents?

4) What is the corresponding speed of the rotor field with respect to the rotor? With respect to the stator?

29. A 50kW, 440V, 50Hz, six-pole induction motor has a slip of 6% when operating at full-

load conditions. The friction and windage losses are 300W, and the core losses are 600W. Find the following values for the full-load conditions:

1) The shaft speed.

2) The output power.

3) The load torque (output torque).

4) The induced torque.

5) The rotor frequency.

30. A three-phase induction motor is running on full-load, its ratings are: $P_N = 7.5\text{kW}$, $U_N = 380\text{V}$, $I_N = 15.6\text{A}$, $n_N = 1426\text{r/min}$, $f_N = 50\text{Hz}$, $\cos\varphi_N = 0.87$. Find:

1) How many poles does this motor have?

2) What is the slip at rated load?

3) What is the efficiency at rated load?

31. The nameplate on a 460V, 37kW, 60Hz, four-pole induction motor indicated that its speed at rated load is 1755r/min. Assume the motor to be operating at rated load.

1) What is the slip of the rotor?

2) What is the frequency of the rotor currents?

3) What is the angular velocity of the stator-produced air-gap flux wave with respect to the stator? With respect to the rotor?

4) What is the angular velocity of the rotor-produced air-gap flux wave with respect to the stator? With respect to the rotor?

32. A three-phase induction motor is operating on full-load, $P_N = 150\text{kW}$, $f_N = 50\text{Hz}$, $2p = 4$, rotor copper loss $p_{Cu2} = 4.8\text{kW}$, mechanical loss $p_\Omega = 700\text{W}$, stray loss $p_\Delta = 1500\text{W}$. Find the electromagnetic torque and output torque.

33. A three-phase induction motor with 380V, 50Hz, four-pole, delta connection, runs at 1426r/min at rated load. Its equivalent circuit components are $R_1 = 2.865\Omega$, $X_{1\sigma} = 7.71\Omega$, $R_2' = 2.82\Omega$, $X_{2\sigma}' = 11.75\Omega$, $X_m = 202\Omega$, and R_m is neglected. Try to use the T-shape equivalent circuit to calculate I_1, P_1, $\cos\varphi_1$ and I'_2 on rated load.

34. A three-phase delta-connected induction motor with 7.5kW, 380V, 50Hz, operates with a speed of 962r/min at rated load, power factor $\cos\varphi_N = 0.827$, the stator copper loss is 470W, the iron loss is 234W, the mechanical loss is 45W, and the stray loss is 80W. Try to calculate the following quantities at rated load:

① Slip s; ② Frequency of the rotor current; ③ Rotor copper loss; ④ Efficiency; ⑤ Stator current.

35. A three-phase delta-connected induction motor with 17kW, 380V, 50Hz, is operating at rated load. Its equivalent circuit components are $R_1 = 0.715\Omega$, $X_{1\sigma} = 1.74\Omega$, $R_2' = 0.416\Omega$, $X_{2\sigma}' = 3.03\Omega$, $X_m = 75\Omega$, $R_m = 6.2\Omega$, and the mechanical loss p_Ω is 139W, the stray loss is 320W. Try to calculate the followings:

① Slip; ② Stator current; ③ Stator power factor; ④ Electromagnetic torque; ⑤ Output torque; ⑥ Efficiency.

36. A three-phase, Y-connected induction motor is running in the following conditions: $P_N =$

10kW, $U_{1N}=380V$, $f_N=50Hz$, $I_N=19.8A$, $R_1=0.5\Omega$. The no-load test results are: $U_N=380V$, $I_{10}=5.4A$, $P_{10}=425W$, $p_\Omega=170W$. The short-circuit test results are as follows,

U_{1k} (*line*) /V	200	160	120	80	40
I_{1k}/A	36	27	18.1	10.5	4
P_{1k}/W	3680	2080	920	290	40

Try to calculate:

1) X_m, R_m, $X_{1\sigma}$, $X'_{2\sigma}$ and R'_2 (Assuming $X_{1\sigma}=X'_{2\sigma}$).

2) Use T-shape equivalent circuit to determine the rated current and rated power factor (Assuming $p_\Delta=1\% P_N$).

37. A 208V, two-pole, 60Hz, Y-connected wound-rotor induction motor is rated at 11kW. Its equivalent circuit components are: $R_1=0.2\Omega$, $R_2=0.120\Omega$, $X_M=15.0\Omega$, $X_1=0.410\Omega$, $X_2=0.410\Omega$; $P_{mech}=250W$, $P_{misc}\approx 0$, $P_{core}=180W$. For a slip of 0.05, find:

The stator line current; The stator copper losses; The air-gap power; The power converted from electrical to mechanical form; The induced torque; The load torque; The overall machine efficiency; The slip at the pullout torque; The pullout torque of this motor.

38. A three-phase, Y-connected 6 poles cage rotor induction motor is running in the following conditions: $U_N=380V$, $f_N=50Hz$, $n_N=957r/min$, $R_1=2.08\Omega$. $R'_2=1.53\Omega$. $X_{1\sigma}=3.12\Omega$. $X'_{2\sigma}=4.25\Omega$. Try to calculate:

1) Rated electromagnetic torque.

2) Maximum torque and overload capacity.

3) Critical slip.

39. A three-phase, 8 poles cage rotor induction motor is running in the following conditions $P_N=260kW$, $U_N=380V$, $f_N=50Hz$, $n_N=722r/min$, overload capacity $\lambda_m=2.13$. Calculate:

1) Slip at rated load.

2) Rated electromagnetic torque.

3) The maximum torque.

4) The critical slip.

5) The electromagnetic torque whens $=0.02$.

40. A three-phase, Y connected cage rotor induction motor is running in the following conditions: $P_N=320kW$, $U_N=380V$, $f_N=50Hz$, $n_N=740r/min$, $I_N=40A$, $\cos\varphi_N=0.83$, overload capacity $\lambda_m=2.2$, starting capacity $K_T=1.93$, starting current ratio $K_I=5.04$. Find:

1) What is the starting current and starting torque when the motor is starting with rated voltage?

2) What should the resistance be connected in series with a stator winding in each phase when the starting current is limited to160A? What is the starting torque?

41. A three-phase, 2 poles cage rotor induction motor has the following ratings: $P_N=11kW$, $U_N=380V$, $f_N=50Hz$, $n_N=2930r/min$, $I_N=21.8A$, $\lambda_m=2.2$. The constant torque load $T_L=0.8T_N$ is driven by a variable frequency system, when $U_1/f_1=$ constant, the no load torque can be neglected. What is the speed of motor when $f=0.8f_N$?

Chapter 5 Synchronous Machines

A synchronous machine is an AC machine, whose speed under the steady-state condition is proportional to the frequency f of the current in its armature. The magnetic field created by the armature currents is rotating at the same speed as the one created by the field current on the rotor (which is rotating at the synchronous speed $n = n_s = \frac{60f}{p}$).

A synchronous machine can work as a generator or a motor. Three-phase synchronous generators are the primary source of all the electrical energy we consume. These machines are the largest energy converters in the world. Synchronous motors are used to convert an electrical power to a mechanical power. One of the advantages is that their power factor can be regulated.

The operation principles and the electromagnetic relation of the synchronous generator will be developed in this chapter. Then the voltage equation, the phasor diagram and the equivalent circuit of a synchronous generator are derived. The operational characteristics of a synchronous generator that are operating alone and operating together with other generators (in parallel with large power systems) will be analyzed. Finally, synchronous motors are introduced briefly at the end of this chapter.

5.1 Construction, operation modes and ratings of a synchronous machine

5.1.1 Construction of a synchronous machine

Commercial synchronous machines are built with either a stationary or a rotating DC magnetic field.

A *stationary-field* synchronous machine has the same outward appearance as a DC machine. The salient poles create a DC field, which is cut by a revolving armature. The armature possesses a 3-phase winding, of which the terminals are connected to three slip-rings that is mounted on the shaft. A set of brushes, which are sliding on the slip-rings, enable the armature connected to the external 3-phase circuits. For a synchronous generator, the armature is driven by a gasoline engine, or some other source of motive power. The stationary-field synchronous generators are used, when the power output is less than 5kV · A.

A *revolving-field* synchronous machine has a stationary armature, which is called a stator. The 3-phase stator winding is directly connected to the external 3-phase circuits, without going through those large, unreliable slip-rings and brushes. The field windings are on the rotor and are excited by a DC voltage. The revolving-field synchronous generators are usually used as the synchronous machines with a high voltage and large capacity.

According to the shape of the magnetic poles on the rotor, the revolving-field synchronous machines are classified into two types. The first is a non-salient (cylindrical) construction, as shown in Fig. 5-1a. The field windings are *distributed windings* placed in the slots on the surface of the cylindrical rotor, so the air gap of a non-salient pole machine is uniform. This generator is usually called the turbogenerator, which uses the steam turbine as the prime mover and has 2 or 4 poles number. This generator also operates the best at relatively high speeds. The second is a salient construction (Fig. 5-1b), which is classified into the low-speed type (engine- or water-driven). This is characterized physically by having salient poles, a large diameter, and a small axial length. Its air gap is not uniform; the field windings are concentrated windings. This generator is usually called hydroelectric generator, which has relatively large number of poles.

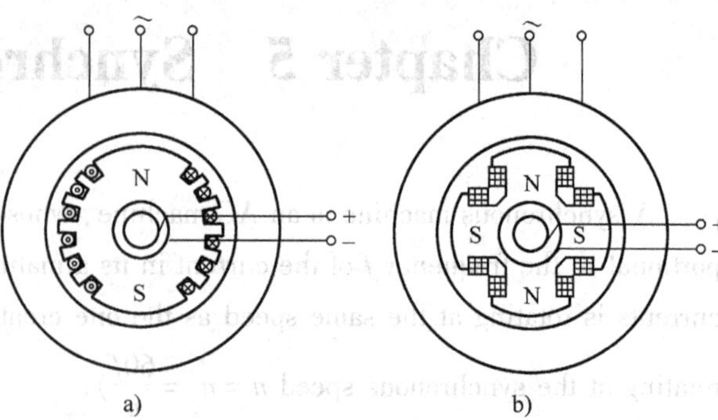

Fig. 5-1 two basic types of a revolving-field synchronous machine

There are two terms used commonly to describe the windings on a machine, one is the *field-winding*, the other is the *armature winding*. In general, the term "field windings" applies to the windings that produce a main magnetic field in a machine; and the term "armature windings" applies to the windings where the main voltage is induced. For a synchronous machine, the field windings are on the rotor, so the terms "rotor windings" and "field winding" are used interchangeably. Similarly, the terms "stator windings" and "armature windings" are used interchangeably as well.

As shown in Fig. 5-1, a three-phase synchronous machine has two main parts: a stationary stator and a revolving rotor. The rotor is separated from the stator by a uniform air gap (non-salient pole machine) or a non-uniform air gap (salient pole machine).

1. Stator

The stator of a synchronous machine consists of stator core, stator winding, frame and cover. The stator core is built up from the toothed segments of the high-quality silicon-iron steel laminations (0.5mm thick) and is covered with an insulating varnish. From an electrical standpoint, the stator of a synchronous machine is identical to that of a 3-phase induction motor. It is composed of a cylindrical laminated core, containing a set of slots that carries a 3-phase lap winding. The winding is always connected in a wye and the neutral is connected to a ground.

Fig. 5-2 shows the stator of a 200-MW turbine generator and Fig. 5-3 is a parts stator of a large hydroelectric generator.

2. Rotor

The rotor of a turbine generator is composed of rotor core, field winding, protecting ring, fan, and shaft. The rotor core of a large turbine generator is shown in Fig. 5-4.

The rotor of a hydroelectric generator is built up with field pole, field winding, field yoke, shaft, damper winding, and so on. The rotor of a 320MW hydroelectric generator is shown in Fig. 5-5.

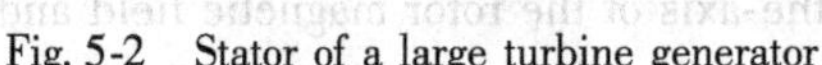

Fig. 5-2 Stator of a large turbine generator

Fig. 5-3 Parts of a stator of a large hydroelectric generator

Fig. 5-4 The Rotor core of a large turbine generator

Fig. 5-5 Rotor of a large hydroelectric generator

Fig. 5-6 shows the salient pole and the windings on the rotor of a hydroelectric generator. The salient poles of the rotor are composed of iron laminations that are much thicker (1 ~ 2mm). These laminations are not insulated, because the DC flux that they carry dose not vary. The 6 round holes in the face of the salient pole carry the bars of a squirrel-cage winding, and all the copper bars are shorted with two end rings. The squirrel-cage winding is called a *damper winding* sometimes.

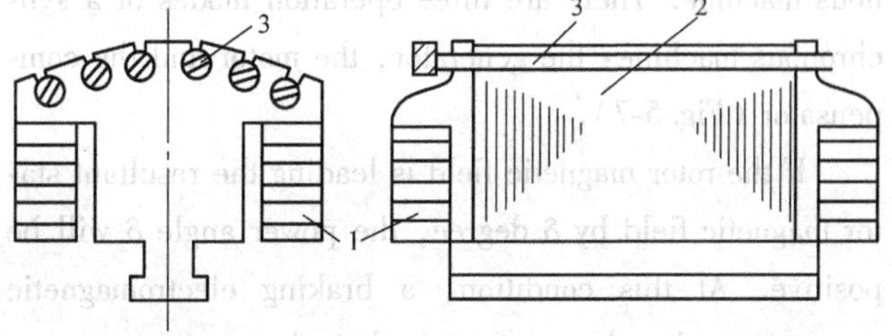

Fig. 5-6 Salient pole of a hydroelectric generator
1—Field winding 2—Field pole 3—Damper winding

Under normal condition, the damper winding does not carry any current, because the rotor turns at a synchronous speed. However, when the rotor speed begins to fluctuate for a certain reasons, a voltage will be induced in the damper winding, causing a large current to flow therein. The current reacts with the magnetic field of the stator, producing a force which damps the oscillation of the rotor. The damper winding also tends to maintain at a balanced 3-phase voltages between the lines, even when the line currents are unequal due to the unbalanced load conditions.

5.1.2 Operation mode of a synchronous machine

When a synchronous machine operates in loaded condition at a steady state, the stator (armature) three-phase windings carry the balanced three-phase currents, producing a stator magnetic field which rotates at synchronous speed. Meanwhile, a DC excitation current is supplied to the field winding on the rotor, producing a rotor magnetic field which revolves at the same speed and in the same direction as the stator magnetic field, the magnetic fields of the stator and the rotor are therefore stationary with respect to each other, producing a resultant air-gap magnetic field which is constant in amplitude and rotates at a synchronous speed.

Power angle δ is defined that it is the angle between the-axis of the rotor magnetic field and the-axis of the resultant stator magnetic field (Fig. 5-7).

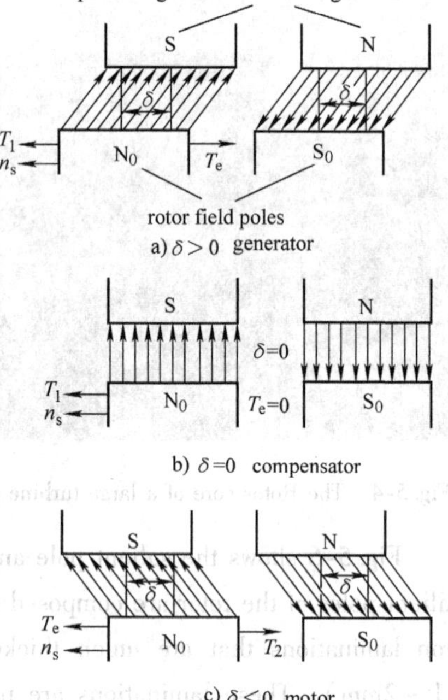

Fig. 5-7 Three operating modes of a synchronous machine

Notice that if two magnetic fields are present in a machine, then a torque will be generated and it will tend to line up the two magnetic fields. There are two magnetic fields in a synchronous machine, one magnetic field is produced by the stator currents and the other one is produced by the rotor currents. Then, a torque will be induced in the rotor, it will cause the rotor to turn and align itself with the stator magnetic field, and the electromechanical energy conversion can be achieved in the machine.

The slip s is an important variable to reflect operation modes of an induction machine. Similarly, the power angle δ is an important variable to reflect that of a synchronous machine. There are three operation modes of a synchronous machine: the generator, the motor and the compensator (Fig. 5-7).

If the rotor magnetic field is leading the resultant stator magnetic field by δ degree, the power angle δ will be positive. At this condition, a braking electromagnetic torque is produced on the rotor; the prime mover must provide a driving torque to overcome the braking electromagnetic torque, so the mechanical energy is supplied from the prime mover to the rotor. The electric energy is delivered from the stator to the power systems, and this synchronous machine works as a generator (Fig. 5-7a).

If the axis of the rotor magnetic field and the resultant stator magnetic field are aligned, the power angle δ will be zero. At this condition, no electromagnetic torque is produced on the rotor, and then no electromechanical energy conversion will be achieved. But the synchronous machine can absorb or deliver a reactive power on the power systems, and it can regulate the power factor of power systems; it works as a compensator (Fig. 5-7b).

If the rotor magnetic field rotates behind the resultant stator magnetic field, the power angle δ

will be negative. At this condition, a driving electromagnetic torque is produced on the rotor; this torque pulls a load that is connected to the shaft. The rotor delivers a mechanical energy to the load, and this synchronous machine works as a motor (Fig. 5-7c).

5.1.3 Ratings of a synchronous machine

The ratings of synchronous machine are as followings.

1) Rated capacity S_N (or rated power P_N). That is the power output of a synchronous machine at rated load. It can be expressed as an apparent power output S_N [kV·A] or an active power output P_N [kW] of a synchronous generator. It is a mechanical power output P_N [kW] from a shaft of a synchronous motor. It can be expressed as a maximum reactive power output S_N [kV·A] of a synchronous compensator.

2) Rated voltage U_N [V]. That is a line-to-line terminal voltage of a stator winding at rated load.

3) Rated current I_N [A]. That is a line current of a stator winding at rated load.

4) Rated power factor $\cos\varphi_N$. That is a power factor of synchronous machine at rated load.

5) Rated frequency f_N. That is a frequency of stator voltage at rated load, and the standard industrial frequency in China is 50Hz.

6) Rated speed n_N [r/min]. That is a speed of synchronous machine at rated load, such as synchronous speed n_s.

Other than the ratings stated above, there are some other ratings, such as the insulation class, the maximum limit of temperature rise, rated field voltage U_{fN} and current I_{fN}, and so on.

The DC field excitation of a large synchronous generator is an important part of its overall design. The reason is that the field must ensure not only a stable AC terminal voltage, but it must also respond to the sudden load changes in order to maintain stability of the system. The system that supplies DC field current is called excitation system. It includes a DC generator excitation system and a rectifier excitation system.

In a DC generator excitation system, two DC generators are used: a *main exciter* and a *pilot exciter*. The main exciter feeds the exciting current to the field of the synchronous generator by using the brushes and the slip rings. Two DC generators and synchronous generator are often mounted on a main shaft.

The rectifier excitation system includes a static rectifier system and a rotating rectifier excitation system.

In a static rectifier excitation system, the field current is supplied through the slip rings from an AC main exciter that is mounted on the main shaft of the synchronous machine and the solid-state rectifiers (either simple diode bridges or phase-controlled rectifiers).

In a rotating rectifier excitation system, the AC current is supplied from an AC main exciter (3-phase, stationary-field generator) that is mounted on the same shaft as the synchronous machine. The AC output is rectified by a group of rectifiers, then the DC output from the rectifiers is fed directly into the field winding without the need of brushes and slip rings, such a system is called a brushless excitation system (Fig. 5-8). A cutaway diagram of a complete large synchronous machine is shown in Fig. 5-9. This drawing shows an eight-pole salient-pole rotor, a stator with distributed double-layer windings, and a brushless exciter.

Fig. 5-8 A synchronous machine rotor with a brushless exciter mounted on the same shaft

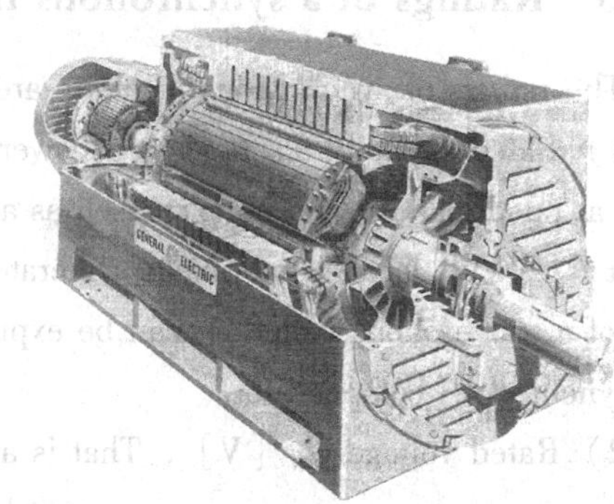

Fig. 5-9 A cutaway diagram of a large salient-pole synchronous machine with a brushless exciter

5.2 Operation of synchronous generators

In a synchronous generator, a DC current is applied to the rotor winding, which produces a rotor magnetic field. The rotor of the generator is then turned by a prime mover, producing a rotating magnetic field within the machine. This rotating magnetic field induces a three-phase set of voltages within the stator windings of the generator.

The magnetomotive force (mmf) waves and the magnetic fields in a synchronous generator that are operating at no-load and under loaded will be explored. The no-load characteristics and the armature reaction under loaded will be developed.

5.2.1 No-load operation of synchronous generators

The rotor of a synchronous generator is turned by a prime mover at the synchronous speed, the DC excitation is supplied to the field winding, the armature-winding terminals are open-circuited; and the armature current is zero, such a condition is called the no-load operation of a synchronous generator.

Fig. 5-10 shows the no-load magnetic circuit of a four-pole synchronous machine, the main flux Φ_0 is the magnetic flux that crosses the air gap and links the armature windings; the rotor leakage flux $\Phi_{f\sigma}$ is the magnetic flux that does not cross the air gap and that only links to the field winding itself. The main magnetic circuit of this machine consists of air gap, armature teeth, armature yoke, rotor pole and rotor yoke.

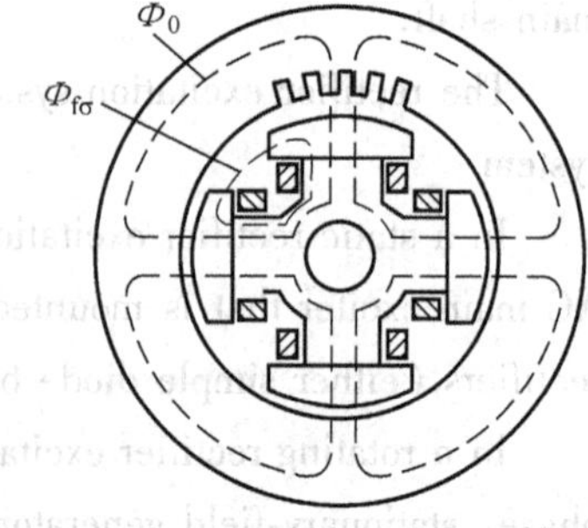

Fig. 5-10 The magnetic circuit of a synchronous machine at no-load

1. MMF wave and magnetic field at no-load

When a synchronous generator operates at no-load, the field current I_f produces the main pole mmf F_f and the rotor of the generator is driven by a prime mover rotating at synchronous speed. Then, the main pole mmf F_f becomes a rotating mmf wave $\boldsymbol{F}_f$. The space-fundamental component $\boldsymbol{F}_{f1}$ of the mmf $\boldsymbol{F}_f$ will produce a rotating magnetic field $\boldsymbol{B}_0$ in the air gap. The stator conductors cut the rotating air-gap magnetic field, and a set of symmetrical three-phase emfs will be induced in the three-phase stator windings.

Ignoring the series of the higher-order harmonics, the internal generated voltage (excitation voltage) E_0 that is produced in one phase of a synchronous generator is

$$E_0 = 4.44 f N_1 k_{w1} \Phi_0 \tag{5-1}$$

From the analysis stated above, the relationship between the no-load mmf, the magnetic field and the induced emf of a synchronous generator is

$$I_f \rightarrow \boldsymbol{F}_{f1} \rightarrow \boldsymbol{B}_0 \rightarrow \dot{\Phi}_0 \rightarrow \dot{E}_0$$

where $\dot{\Phi}_0$ is the main magnetic flux at on-load condition. it is a time phasor and its magnitude is the per pole main magnetic flux.

2. No-load (Open-circuit) characteristic

When a synchronous machine is operating at a synchronous speed n_s and the 3-phase armature-windings are open-circuited, the internal generated voltage (excitation voltage) E_0 that is produced in a one phase of a synchronous generator is related to the field current I_f as shown in Fig. 5-11. This plot is called the *magnetization curve* or the *open-circuit characteristic* of the machine.

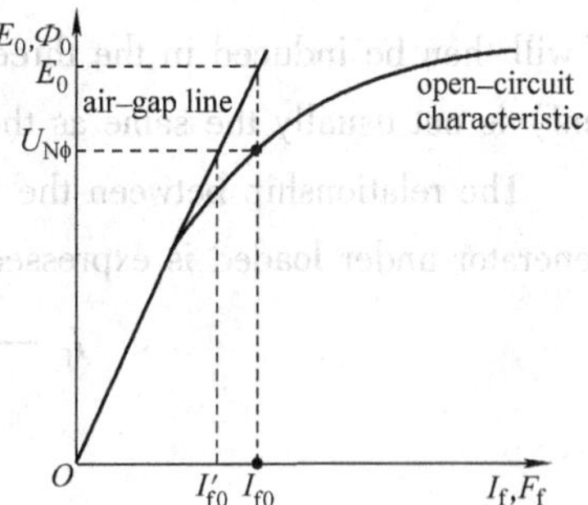

Fig. 5-11 the open-circuit characteristic of synchronous machines

Since the internal generated voltage E_0 is directly proportional to Φ_0 and to the speed, and the field current I_f is proportional to F_f; the curve of open-circuit characteristic $E_0 = f(I_f)$ is similar with the magnetization curve $\Phi_0 = f(F_f)$. In fact, the open-circuit characteristic reflects the saturation conditions of the magnetic circuit; it is an important characteristic of the synchronous machines.

If the main flux Φ_0 is small, and the magnetic circuit of the machine is unsaturated, then the open-circuit characteristic will become a straight line, which is called the air-gap line (or air-gap characteristic) (Fig. 5-11).

With the main flux Φ_0 increasing, the iron core of the machine will gradually become saturated; the open-circuit characteristic will gradually flatten out as well (Fig. 5-11). When the open-circuit voltage E_0 is equal to the rated voltage $U_{N\varphi}$, the saturation coefficient k_μ of the magnetic circuit is calculated by $k_\mu = \frac{I_{f0}}{I'_{f0}} = \frac{E_0}{U_{N\varphi}} \approx 1.1 \sim 1.25$.

Open-circuit characteristic curve can be obtained by calculating or testing. Table 5-1 shows a typical synchronous generator open-circuit characteristic.

Table 5-1 Typical synchronous generator open-circuit characteristic

$E_0{}^*$	0.58	1.0	1.21	1.33	1.40	1.46	1.51
$I_f{}^*$	0.5	1.0	1.5	2.0	2.5	3.0	3.5

5.2.2 Loaded operation of synchronous generators

1. MMF waves and magnetic fields in loaded condition

When the 3-phase stator windings of a synchronous generator are connected with the balanced three-phase loads, the DC excitation is supplied to the field winding, and the rotor is driven by the prime mover at a synchronous speed. Then, a set of balanced three-phase currents will flow into the three-phase armature windings, and produce a rotating armature fundamental mmf $\boldsymbol{F}_a$.

The rotating speed and the direction of the main pole fundamental mmf $\boldsymbol{F}_{f1}$ and the armature fundamental mmf wave $\boldsymbol{F}_a$ are the same; they are also relatively still in space, then a synthesized rotating mmf $\boldsymbol{F}$ can be produced by them, and a rotating magnetic field $\boldsymbol{B}$ will be built by $\boldsymbol{F}$ in the air gap. This means that the resultant air-gap flux $\dot{\Phi}$ in the synchronous generator depends upon the magnitude of the exciting current I_f in the field winding as well as on the value of the armature current $\dot{I}$.

The stator conductors cut the rotating magnetic field $\boldsymbol{B}$ in the air gap, a set of three-phase emfs $\dot{E}$ will then be induced in the three-phase stator windings. This induced emf $\dot{E}$ (namely synthesized emf) is not usually the same as the internal generated voltage E_0.

The relationship between the mmfs, the magnetic field and the induced emf of a synchronous generator under loaded is expressed as

$$\begin{array}{l} I_f \longrightarrow \boldsymbol{F}_{f1} \\ \qquad\qquad\qquad \searrow \\ \qquad\qquad\qquad\quad \boldsymbol{F} \longrightarrow \boldsymbol{B} \longrightarrow \dot{\Phi} \longrightarrow \dot{E} \\ \qquad\qquad\qquad \nearrow \\ \dot{I} \longrightarrow \boldsymbol{F}_a \end{array}$$

where $\dot{I}$ is the per-phase armature current; $\dot{\Phi}$ is the main flux that is produced by the synthesized mmf $\boldsymbol{F}$; $\dot{E}$ is the synthesized emf in one-phase armature winding.

2. The armature reactionin balanced load condition

When a synchronous generator's rotor is driven by the prime mover at synchronous speed and the stator windings and open-circuited, the internal generated voltage E_0 is induced in the stator windings. If a load is attached to the terminals of the generator, a current will flow in the stator windings. The 3-phase stator currents will therefore produce a magnetic field in the machine. This stator magnetic field distorts the original rotor magnetic field, hence changing the resulting phase voltage. This effect is called *armature reaction*, because the armature (stator) current affects the magnetic field which produced it.

In order to understand the armature reaction, let's introduce the *time-space vector diagram* of a synchronous generator.

The so-called *time-space vector diagram* is a diagram that the time phasor and the space phasor of the machine are both in it. For a balanced polyphase system, assumed that the time reference axis is in the same direction as the one phase winding axis (such as phase A or B or C) in the time-space vector diagram, hence it results in following relationships:

1) The mmf vector $\boldsymbol{F}$ that is produced by the balanced polyphase currents is in the same phase as the current phasor $\dot{I}$.

2) Ignoring the hysteresis and the eddy current losses; the air-gap magnetic flux density vector $\boldsymbol{B}$ that is built by the mmf $\boldsymbol{F}$ is in the same phase as the mmf vector $\boldsymbol{F}$.

3) The main flux phasor $\dot{\Phi}$ is in the same phase as the magnetic flux density vector $\boldsymbol{B}$.

According to the relationships stated above, the time-space vector diagram is shown in Fig. 5-12.

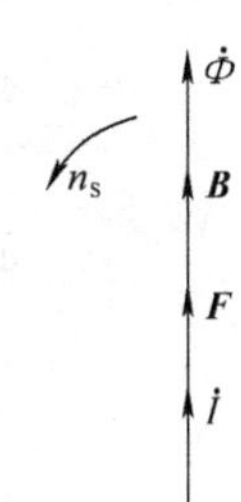

Fig. 5-12 Time-space vector diagram

The rotating angular velocity of the time phasor and the space vector are both equal to the synchronous speed n_s, so their phase relationship remains unchanged at all times. The main pole mmf $\boldsymbol{F}_{f1}$, the main pole magnetic field $\boldsymbol{B}_0$ and the main pole flux $\dot{\Phi}_0$ have the same relationships as shown in Fig. 5-12.

The kinds of the armature reaction (includes the increasing magnetization, the demagnetization and the cross magnetization) depend on the space phase angle between the armature fundamental mmf $\boldsymbol{F}_a$ and the main pole field $\boldsymbol{B}_0$ (or the main pole fundamental mmf $\boldsymbol{F}_{f1}$). The space phase angle between $\boldsymbol{F}_a$ and $\boldsymbol{B}_0$ is the same as the time phase difference between the internal generated voltage $\dot{E}_0$ and the armature current $\dot{I}$. The time phase angle between $\dot{E}_0$ and $\dot{I}$ is called *the internal power factor angle* ψ_0.

Fig. 5-13a shows a space vector diagram of a synchronous generator. In Fig. 5-13a, the rotor pole axis is defined as a direct axis or a d-axis; the axis between the two rotor poles is defined as a quadrature-axis or a q-axis. Meanwhile, the d-axis and the q-axis are both rotating at a synchronous speed counter-clockwise, and the rotor pole axis leads phase *A* winding-axis by 90° (electrical angle).

According to equation $\dot{E} = -\mathrm{j}4.44fNk_{w1}\dot{\Phi}_m$, we can find that $\dot{E}$ lags the mutual flux $\dot{\Phi}_m$ by 90° (electrical angle).

At the moment shown in Fig. 5-13a, the loop of phase A parallels to the rotor pole magnetic flux density lines, and the flux $\dot{\Phi}_{0A}$ through the loop of phase A is minimum. Then, the induced emf $\dot{E}_0$ in phase A winding is maximum, thus phasor $\dot{E}_0$ is placed in the same phase with a time-axis, and the main flux $\dot{\Phi}_0$ leads $\dot{E}_0$ by 90° electrical angle (Fig. 5-13b).

Now, supposing that the generator is connected to a lagging load, because the load is lagging, the peak current will occur at an angle behind the peak voltage, i. e. the armature current $\dot{I}$ will lag $\dot{E}_0$ an angle ψ_0 ($0° < \psi_0 < 90°$). This effect is shown in Fig. 5-13b.

In Fig. 5-13b, the time-axis and the phase A axis are placed in the same direction; the phase-axis and the time-axis are stationary. The d-axis, the q-axis, all the time phasors and all the space vectors are rotating at a synchronous speed counter-clockwise. The armature mmf $\boldsymbol{F}_a$ that is produced by three-phase currents and the armature current $\dot{I}$ are in phase; the main pole mmf $\boldsymbol{F}_{f1}$ and the main flux $\dot{\Phi}_0$ are also in phase. Then, the space phase difference between $\boldsymbol{F}_a$ and $\boldsymbol{F}_{f1}$ is equal to the time phase difference between $\dot{\Phi}_0$ and $\dot{I}$, the space phase angle (or the time phase angle) is $90° + \psi_0$, thus it is clear that the role of the armature reaction depends on the internal power factor angle ψ_0.

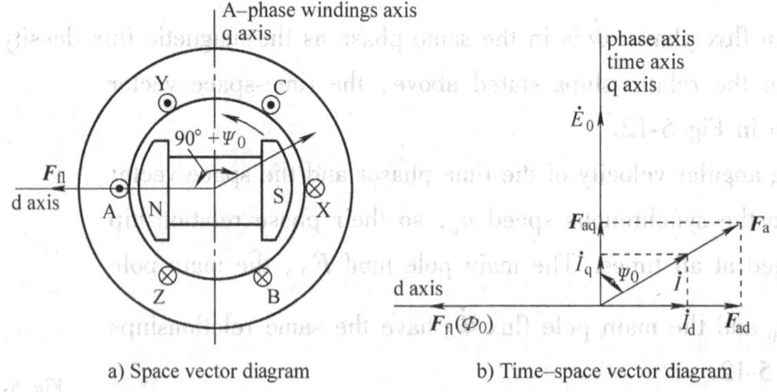

Fig. 5-13 The time-space vector diagram of a synchronous generator when $0° < \psi_0 < 90°$

1) If the internal power factor angle $\psi_0 = 0°$ (i. e. $\dot{I}$ and $\dot{E}_0$ are in phase), the time-space vector diagram is shown in Fig. 5-14.

During this condition, the armature mmf $\boldsymbol{F}_a$ is in phase with the q-axis, this armature reaction is called the quadrature-axis armature reaction.

In Fig. 5-14, the synthesized mmf $\boldsymbol{F}$ is lagging the main pole mmf $\boldsymbol{F}_{f1}$ with an angle in space. That is the main pole magnetic field, which leads to the stator synthesized magnetic field. A brake electromagnetic torque will therefore be produced on the rotor. Then, the prime mover must input more torque to overcome the brake torque, and the mechanical-electrical energy conversion can therefore be achieved.

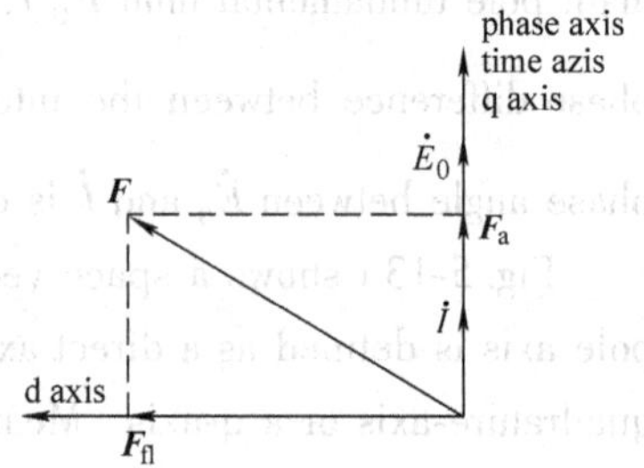

Fig. 5-14 Time-space vector of a synchronous generator when $\psi_0 = 0°$

2) If the internal power factor angle $\psi_0 = 90°$ (i. e. $\dot{I}$ lags $\dot{E}_0$ by 90°), the time-space vector diagram is shown in Fig. 5-15.

During this condition, the armature mmf $\boldsymbol{F}_a$ lags q-axis by 90°, that is $\boldsymbol{F}_a$ inversed with the d-axis. The synthesized magnetic field of the machine is weakened, and this armature reaction is called the d-axis demagnetization armature reaction, and any mechanical-electrical energy conversion can be achieved in the machine.

3) If the internal power factor angle $\psi_0 = -90°$ (i. e. $\dot{I}$ leads $\dot{E}_0$ by 90°), the time-space vector diagram is shown in Fig. 5-16.

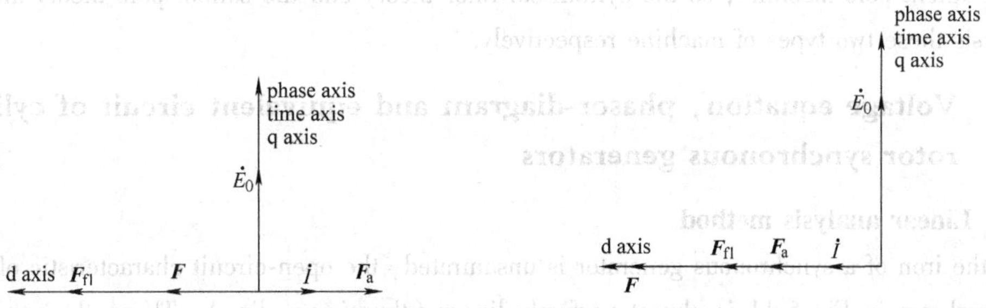

Fig. 5-15 Time-space vector of a synchronous generator when $\psi_0 = 90°$

Fig. 5-16 Time-space vector of a synchronous generator when $\psi_0 = -90°$

At this moment, the armature mmf $\boldsymbol{F}_a$ leads the q-axis by 90°, that is $\boldsymbol{F}_a$ in phase with the d-axis, and the synthesized magnetic field of the machine is enhanced; this armature reaction is called the d-axis increasing magnetic armature reaction, and any mechanical-electrical energy conversion can be achieved in the machine.

4) If the internal power factor angle $0° < \psi_0 < 90°$ (i. e. $\dot{I}$ lags $\dot{E}_0$ by a sharp angle ψ_0), the time-space vector diagram is shown in Fig. 5-13b.

At this moment, the armature mmf $\boldsymbol{F}_a$ is neither in the d-axis nor in the q-axis, it lags $\dot{E}_0$ by ψ_0.

The armature mmf $\boldsymbol{F}_a$ can be decomposed into two components, the d-axis armature mmf $\boldsymbol{F}_{ad}$ and the q-axis armature mmf $\boldsymbol{F}_{aq}$, that is

$$\boldsymbol{F}_a = \boldsymbol{F}_{ad} + \boldsymbol{F}_{aq} \tag{5-2}$$

where

$$\begin{cases} \boldsymbol{F}_{ad} = \boldsymbol{F}_a \sin\psi_0 \\ \boldsymbol{F}_{aq} = \boldsymbol{F}_a \cos\psi_0 \end{cases} \tag{5-3}$$

Accordingly, the armature current $\dot{I}$ can also be decomposed into the d-axis armature current $\dot{I}_d$ and the q-axis armature current $\dot{I}_q$, that is

$$\dot{I} = \dot{I}_d + \dot{I}_q \tag{5-4}$$

where

$$\begin{cases} I_d = I\sin\psi_0 \\ I_q = I\cos\psi_0 \end{cases} \tag{5-5}$$

During this condition, there are q-axis armature reaction and d-axis demagnetization armature reaction, and the mechanical-electrical energy conversion achieved in the machine.

5.3 Voltage equation, phasor-diagram and equivalent circuit of synchronous generators

Because the rotor structure are different between the non-salient (cylindrical) rotor machine and the salient pole machine, so the cylindrical rotor theory and the salient-pole theory are applied to analysis these two types of machine respectively.

5.3.1 Voltage equation, phasor-diagram and equivalent circuit of cylindrical rotor synchronous generators

1. Linear analysis method

If the iron of a synchronous generator is unsaturated, the open-circuit characteristic of this machine as shown in Fig. 5-11 is almost perfectly linear (the air-gap line). Then, the superposition principle can be used to analysis the magnetic field and the magnetic circuit of the synchronous generator.

When a DC current I_f is applied to the field winding of a synchronous generator, the exciting current I_f produces the rotor magnetic field. Because the rotor is driven by a prime mover at synchronous speed n_s, the rotor rotating magnetic field and the rotor mmf $\boldsymbol{F}_{f1}$ are produced. The main pole mmf (rotor mmf) $\boldsymbol{F}_{f1}$ generates the main flux $\dot{\Phi}_0$, and the internal generated voltage $\dot{E}_0$ is induced in the stator windings. Because the stator windings are connected the load, then the armature current $\dot{I}$ will flow in the stator windings. The armature mmf $\boldsymbol{F}_a$ will then produce the armature reaction flux $\dot{\Phi}_a$, and armature reaction emf $\dot{E}_a$ can be induced in the stator windings. According to the superposition principle, the per-phase synthesized emf $\dot{E}$ can be obtained by adding $\dot{E}_0$ and $\dot{E}_a$. At the same time, the armature current $\dot{I}$ can also produce the leakage flux $\dot{\Phi}_\sigma$; and the leakage emf $\dot{E}_\sigma$ can be induced in the stator windings. Then, the electromagnetic relationship between the physical quantities in the synchronous generator that mentioned above is expressed as following:

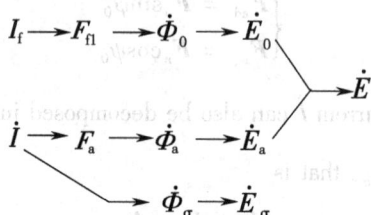

$$
\begin{array}{l}
I_f \rightarrow \boldsymbol{F}_{f1} \rightarrow \dot{\Phi}_0 \rightarrow \dot{E}_0 \searrow \\
\qquad\qquad\qquad\qquad\qquad\qquad \rightarrow \dot{E} \\
\dot{I} \rightarrow \boldsymbol{F}_a \rightarrow \dot{\Phi}_a \rightarrow \dot{E}_a \nearrow \\
\quad \searrow \dot{\Phi}_\sigma \rightarrow \dot{E}_\sigma
\end{array}
$$

(1) Voltage equation

A 3-phase synchronous generator is having terminals A, B, C, feeding a balanced 3-phase load. The machine and its load are both connected in Wye, yielding the circuit as shown in Fig. 5-17. Although the neutrals O and O_1 are not connected, they are at the same potential because the load is balanced. Consequently, we could connect them together (as indicated by the short dash line) without affecting the behavior of the voltages or the currents in the circuit.

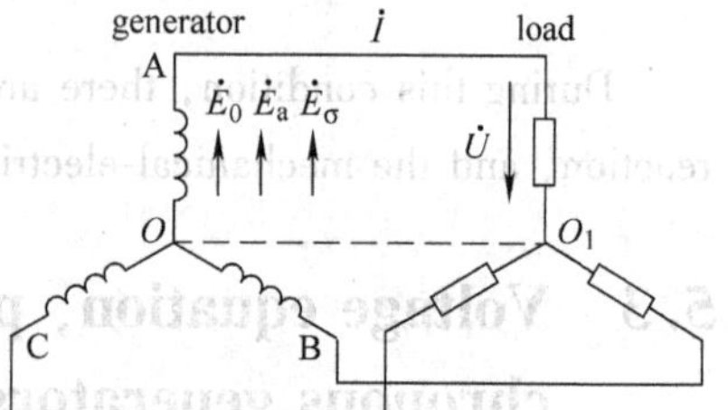

Fig. 5-17 Electric circuit of a 3-phase synchronous generator

In Fig. 5-17, the terminal voltage $\dot{U}$ of the generator is not the same as the $\dot{E}_0 + \dot{E}_a + \dot{E}_\sigma$, because there is the resistance

R_a of the armature winding and there is the resistance voltage drop $\dot{I}R_a$.

The voltage equation of the generator can be expressed as

$$\dot{E}_0 + \dot{E}_a + \dot{E}_\sigma - \dot{I}R_a = \dot{U} \tag{5-6}$$

Because the magnetic circuit of the generator is linear, the magnitude of the armature reaction emf E_a is directly proportional to the amplitude of armature reaction flux Φ_a; and the amplitude of the armature mmf F_a is proportional to the magnitude of the armature current I. Then, we can find that Φ_a is proportional to F_a as following

$$E_a \propto \Phi_a \propto F_a \propto I$$

According to equation $\dot{E} = -\mathrm{j}4.44fNk_{w1}\dot{\Phi}$, $\dot{E}_a$ lags $\dot{\Phi}_a$ by 90°, but $\dot{\Phi}_a$ and $\dot{I}$ are in the same phase, so $\dot{E}_a$ lags $\dot{I}$ by 90°. Therefore $\dot{E}_a$ can be expressed as

$$\dot{E}_a = -\mathrm{j}\dot{I}X_a \tag{5-7}$$

where X_a is called *the armature reaction reactance* (its corresponding stator self-inductance is called L_a), and $X_a = \frac{E_a}{I}$.

Similarly, the armature leakage emf E_σ can be expressed as

$$\dot{E}_\sigma = -\mathrm{j}\dot{I}X_\sigma \tag{5-8}$$

where X_σ is called the armature leakage reactance (its corresponding stator leakage-inductance is called L_σ), and $X_\sigma = \frac{E_\sigma}{I}$.

Substitute Eq. 5-7 and Eq. 5-8 into Eq. 5-6, then

$$\dot{E}_0 = \dot{U} + \dot{I}R_a + \mathrm{j}\dot{I}X_a + \mathrm{j}\dot{I}X_\sigma = \dot{U} + \dot{I}R_a + \mathrm{j}\dot{I}X_s \tag{5-9}$$

where X_s is called the *synchronous reactance*, and $X_s = X_a + X_\sigma$. The synchronous reactance X_s of a generator is internal impedance, just like its internal resistance R_a. The impedance is there, but it can neither be seen nor touched. The value of X_s is typically 10 to 100 times greater than R_a.

X_s is proportional to the square of the turns N_1 of the armature winding and to the equivalent permeance Λ_s of the armature magnetic circuit.

$$X_s \propto N_1^2\Lambda_s \propto N_1^2(\Lambda_a + \Lambda_\sigma)$$

where Λ_a is the equivalent permeance of the armature reaction magnetic circuit; Λ_σ is the equivalent magnetic permeance of the armature leakage magnetic circuit.

(2) Equivalent circuit

According to Eq. 5-9, it is possible to sketch the equivalent circuit of a 3-phase synchronous generator. The full equivalent circuit of such a generator is shown in Fig. 5-18a. It consists of the models for each phase. Each phase has an internal generated voltage with the series inductance X_s and the series resistance R_a. In effect, the three phases are identical, except that their respective voltages and currents are out of the phase by 120°.

These three phases can be either Y or Δ-connected. If they are Y-connected, then the terminal voltage U is related to the phase voltage U_ϕ by $U = \sqrt{3}U_\phi$. If the are Δ-connected, then $U = U_\phi$.

The per-phase equivalent circuit of a generator is shown in Fig. 5-18b. One important fact must be kept in mind is that when the per-phase equivalent circuit is used: the three phases have the same voltages and currents, *only* when the loads attached to them are *balanced*.

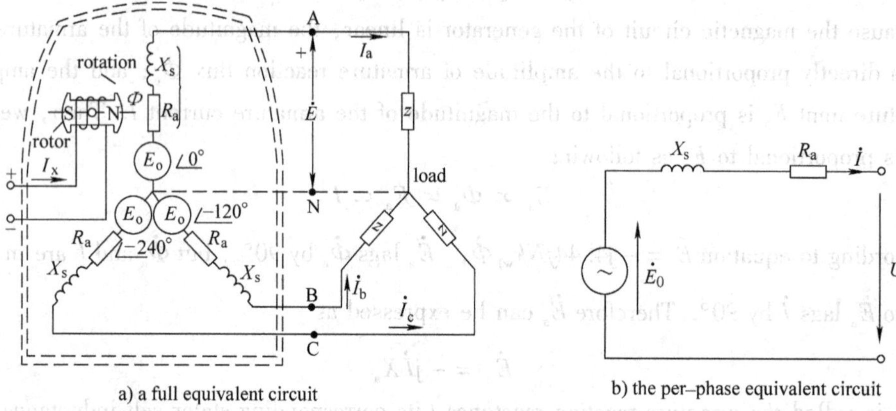

a) a full equivalent circuit b) the per–phase equivalent circuit

Fig. 5-18 The equivalent circuit of a three-phase cylindrical rotor synchronous generator

(3) Phasor diagram

Because the voltages in a synchronous generator are the AC voltages, they are usually expressed as phasors. Since the phasors have both a magnitude and an angle, the relationship between them must be expressed by a two-dimensional plot. When the voltages ($\dot{E}_0$, $\dot{U}$, j$\dot{I}X_s$, $\dot{I}R_a$) within a phase and the current $\dot{I}$ in the phase are plotted in such a fashion in order to show the relationships among them, this resulting plot is called a *phasor diagram*.

Fig. 5-19a shows these relationships, when the generator is supplying a load at unity power factor (a purely resistive load). The phase diagrams of the generators operating at lagging and leading power factors are shown in Fig. 5-19b and Fig. 5-19c respectively.

In Fig. 5-19, the power factor angle is marked by φ, it is the angle between $\dot{U}$ and $\dot{I}$; the inter-

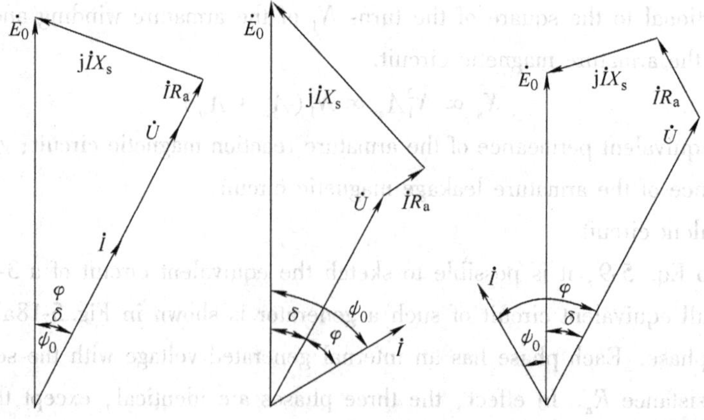

a) unity power factor b) lagging power factor c) leading power factor

Fig. 5-19 The phasor diagram of a three-phase cylindrical rotor synchronous generator

nal power factor angle is marked by ψ_0, it is the angle between $\dot{E}_0$ and $\dot{I}$; the angle between $\dot{E}_0$ and $\dot{U}$ is called the power angle δ, and $\delta = \psi_0 - \varphi$.

Notice that, for a given phase voltage and an armature current, a larger internal generated voltage $\dot{E}_0$ is needed for the lagging loads than for the leading loads. Therefore, a larger field current is needed with the lagging loads to get the same terminal voltage. Alternatively, for a given field current and a magnitude of a load current, the internal generated voltage $\dot{E}_0$ is lowered for the lagging loads and raised for the leading loads.

In the real synchronous machines, the synchronous reactance X_s is normally much larger than the winding resistance R_a, so R_a is often neglected in the qualitative study of voltage variations.

2. Nonlinear analysis method

The power output capability of the electrical machines directly depends upon the degree to which the material is worked magnetically and electrically. The higher the operating flux density in the iron parts of the machine and the current density in the copper parts are; the greater the output will be. Consequently, in most electrical machines, the level of flux density will determine if the operation is in the saturated or in the nonlinear region. Then, the open-circuit characteristic of this machine as shown in Fig. 5-11 is the actual open-circuit characteristic curve, but not the air-gap line. Therefore, the superposition principle cannot be applied to analyze the magnetic field and the magnetic circuit of the synchronous generator.

In a saturated generator, the resultant air-gap magnetic field is produced by the excitation current I_f and armature current $\dot{I}$. That is the rotor mmf $\boldsymbol{F}_{f1}$ and the armature mmf $\boldsymbol{F}_a$ are producing a resultant magnetic flux $\dot{\Phi}$ in the air-gap of the machine. The stator conductors cut the resultant magnetic flux $\dot{\Phi}$, and the induced voltage $\dot{E}$ can be generated in the stator windings. At the same time, the armature current $\dot{I}$ can also produce the leakage flux $\dot{\Phi}_\sigma$, and the leakage emf $\dot{E}_\sigma$ can be induced in the stator windings. Then, the electromagnetic relationship between the physical quantities mentioned above in the saturated synchronous generator is expressed as followings:

$$\left.\begin{array}{l} I_f \longrightarrow F_{f1} \\ \dot{I} \longrightarrow F_a \end{array}\right\} \longrightarrow F \longrightarrow \dot{\Phi} \longrightarrow \dot{E}$$

$$\dot{I} \longrightarrow \dot{\Phi}_\sigma \longrightarrow \dot{E}_\sigma \ (\dot{E}_\sigma = -\mathrm{j}\dot{I}X_\sigma)$$

Then, the stator voltage equation can be written

$$\dot{E} + \dot{E}_\sigma - \dot{I}R_a = \dot{U} \tag{5-10}$$

Substitute $\dot{E}_\sigma = -\mathrm{j}\dot{I}X_\sigma$ into Eq. 5-10, it can be rewritten by

$$\dot{E} = \dot{U} + \dot{I}(R_a + \mathrm{j}X_\sigma) \tag{5-11}$$

5.3.2 Voltage equation and phasor-diagram of salient-pole synchronous generators

In contrast to the cylindrical-pole synchronous machine, which has a uniform air gap, the salient-pole machine has a highly non-uniform air gap, because of the use of a protruding-pole structure. Consequently, the analytic methods for the cylindrical-pole machine cannot be used for the salient-pole machine with these satisfactory results. A modified analytic method is developed for the salient-pole machine in a two-reactance method (the two-reaction theory or the direct- and quadrature-axis theory).

1. Direct and quadrature-axis theory

The air gap δ of a salient pole synchronous machine under the rotor pole face is smaller (Fig. 5-20b), and the air gap δ under the inter-pole is bigger, i. e. the permeance along the polar or the direct-axis is appreciably greater than that along the interpolar or the quadrature-axis (Fig. 5-20c). Under normal operation, the armature (or the stator) winding mmf is distributed with its peak value that is located somewhere between the direct and the quadrature axises. Accordingly, it produces a significant effect in each axis in a different way; it is because of the considerable difference between the reluctance variations of each axis.

When the armature mmf is just located on the direct-axis or the quadrature-axis, the waveform of an armature magnetic field is symmetrical; and the amplitude of the fundamental magnetic field can be determined easily (Fig. 5-20b and Fig. 5-20c). If the armature mmf is located between the direct-axis or the quadrature-axis, the armature magnetic field distribution is not symmetrical, so it is difficult to directly determine the armature reaction. At this moment, we can divide armature mmf $\boldsymbol{F}_a$ into two components, one is the direct-axis component $\boldsymbol{F}_{ad}$ and the other is the quadrature-axis component $\boldsymbol{F}_{aq}$. Then calculate the armature reaction of the direct-axis and the quadrature-axis respectively. This analytic method is called *the direct-and the quadrature-axis theory.*

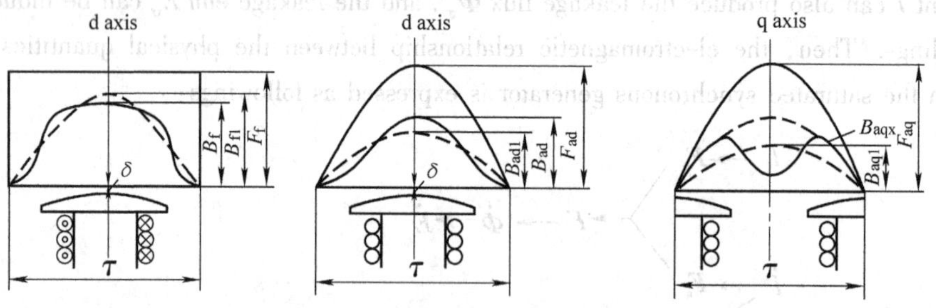

a) the main role mmf b) the armature magnetic sine wave in d-axis c) the armature magnetic sine wave in q-axis

Fig. 5-20 The air gap magnetic waveform of the salient pole synchronous generator

2. Voltage equation and Phasor diagram

If the machine's magnetic circuit is linear, the open-circuit characteristic of this machine as shown in Fig. 5-11 is almost perfectly linear (the air-gap line). Then the superposition principle can be used to analysis the magnetic field and the magnetic circuit of the synchronous generator. The electromagnetic relationship between each physical quantity is shown as followings:

$$I_f \longrightarrow F_{f1} \longrightarrow \dot{\Phi}_0 \longrightarrow \dot{E}_0$$

$$\dot{I} \begin{cases} \dot{I}_d \rightarrow F_{aq} \rightarrow \dot{\Phi}_{ad} \rightarrow \dot{E}_{ad} \\ \dot{I}_q \rightarrow F_{aq} \rightarrow \dot{\Phi}_{aq} \rightarrow \dot{E}_{aq} \\ \longrightarrow \dot{\Phi}_\sigma \rightarrow \dot{E}_\sigma\,(\dot{E}_\sigma = -\mathrm{j}\dot{I}X_\sigma) \end{cases} \qquad \dot{E}_0 + \dot{E}_{ad} + \dot{E}_{aq} \longrightarrow \dot{E}$$

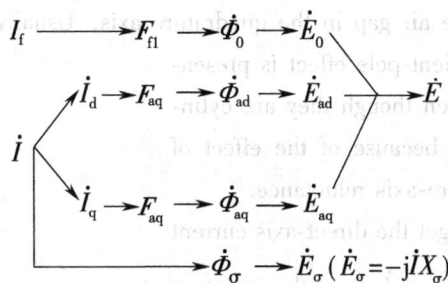

(1) Voltage equation

According to the analysis mentioned above, the stator voltage equation is

$$\dot{E}_0 + \dot{E}_{ad} + \dot{E}_{aq} + \dot{E}_\sigma - \dot{I}R_a = \dot{U} \tag{5-12}$$

where

$$E_{ad} \propto \Phi_{ad} \propto F_{ad} \propto I_d (I_d = I\sin\psi_0)$$
$$E_{aq} \propto \Phi_{aq} \propto F_{aq} \propto I_q (I_q = I\cos\psi_0)$$

And $\dot{E}_{ad}$ and $\dot{E}_{aq}$ can be expressed as

$$\begin{cases} \dot{E}_{ad} = -\mathrm{j}\dot{I}_d X_{ad} \\ \dot{E}_{aq} = -\mathrm{j}\dot{I}_q X_{aq} \end{cases} \tag{5-13}$$

where X_{ad} is called the *direct-axis armature reaction reactance*, and $X_{ad} = \dfrac{E_{ad}}{I_d}$; X_{aq} is called the *quadrature-axis armature reaction reactance*, and $X_{aq} = \dfrac{E_{aq}}{I_q}$.

Substitute Eq. 5-13 and Eq. 5-8 into Eq. 5-12, then

$$\begin{aligned} \dot{E}_0 &= \dot{U} + \dot{I}R_a + \mathrm{j}\dot{I}_d X_{ad} + \mathrm{j}\dot{I}_q X_{aq} + \mathrm{j}\dot{I}X_\sigma \\ &= \dot{U} + \dot{I}R_a + \mathrm{j}\dot{I}_d (X_{ad} + X_\sigma) + \mathrm{j}\dot{I}_q (X_{aq} + X_\sigma) \\ &= \dot{U} + \dot{I}R_a + \mathrm{j}\dot{I}_d X_d + \mathrm{j}\dot{I}_q X_q \end{aligned} \tag{5-14}$$

where X_d is called the *direct-axis synchronous reactance*, and $X_d = X_{ad} + X_\sigma$; X_q is called the *quadrature-axis synchronous reactance*, and $X_q = X_{aq} + X_\sigma$.

(2) Phasor diagram

The phasor diagram (Fig. 5-21) is drawn for the unsaturated salient-pole generator that is operating with a lagging power factor.

In Fig. 5-21, the internal generated voltage (excitation voltage) $\dot{E}_0$ equals to the phasor sum of the terminal voltage $\dot{U}$ plus the armature-resistance drop $\dot{I}R_a$ and the component synchronous-reactance drops $\mathrm{j}\dot{I}_d X_d + \mathrm{j}\dot{I}_q X_q$.

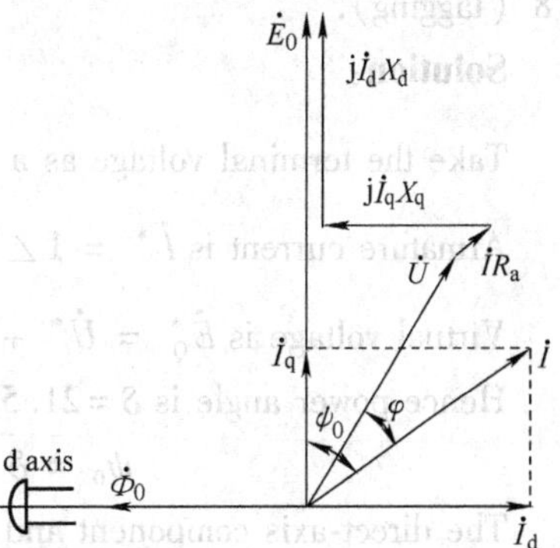

Fig. 5-21 Phasor diagram of salient-pole synchronous generator

The reactance X_q is less than the reactance X_d, because of the greater reluctance of the air gap in the quadrature axis. Usually, X_q is between 0.6 X_d and 0.7 X_d. Note that a small salient-pole effect is presented in the turboalternators, even though they are cylindrical-rotor machines, this is because of the effect of the rotor slots on the quadrature-axis reluctance.

In Fig. 5-21, in order to get the direct-axis current $\dot{I}_d$ and the quadrature-axis current $\dot{I}_q$ from the armature current $\dot{I}$, the internal power factor angle ψ_0 is necessary. In fact, the angle ψ_0 between $\dot{E}_0$ and $\dot{I}$ could not be measured, but it can be calculated in a phasor diagram in terms of geometrical method (Fig. 5-22).

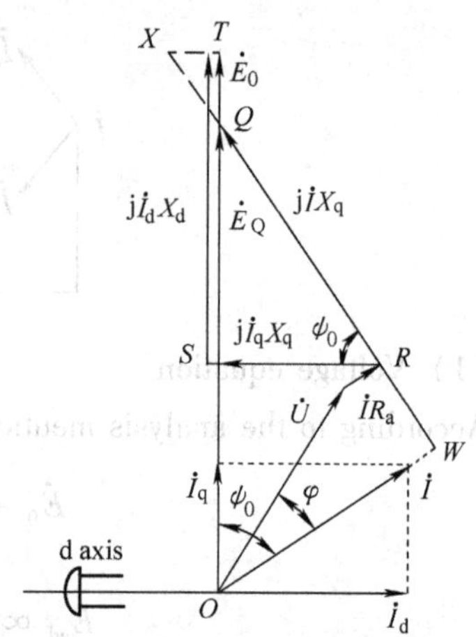

Fig. 5-22 Relations between the component voltages in a phasor diagram of a salient pole synchronous generator

In Fig. 5-22, the phasor $\overline{RQ}$ is perpendicular to the phasor $\dot{I}$, the angle between the line $\overline{RQ}$ and the phasor $j\dot{I}_qX_q$ equals to the internal power factor angle ψ_0, then the phasor $\overline{RQ}$ equals to $\dfrac{j\dot{I}_qX_q}{\cos\psi_0} = j\dot{I}X_q$, so the virtual voltage $\dot{E}_Q$ is

$$\dot{E}_Q = \dot{U} + \dot{I}R_a + j\dot{I}X_q \tag{5-15}$$

The internal power factor angle ψ_0 can be calculated by

$$\psi_0 = \arctan\frac{U\sin\varphi + IX_q}{U\cos\varphi + IR_a} \tag{5-16}$$

And

$$E_0 = E_Q + I_d(X_d - X_q) \tag{5-17}$$

Example 5-1 There is a salient pole synchronous generator, its synchronous reactance per-unit value of the direct-axis and the quadrature-axis are $X_d^* = 1.0$, $X_q^* = 0.7$ respectively, ignore the armature resistance in regardless of the magnetic saturation. Try to compute the excitation emf E_0^* and the power angle δ, when the generator is in the rated voltage, in the rated current and $\cos\varphi = 0.8$ (lagging).

Solution:

Take the terminal voltage as a reference phasor, that is $\dot{U}^* = 1\angle 0°$

Armature current is $\dot{I}^* = 1\angle -36.87°$

Virtual voltage is $\dot{E}_Q^* = \dot{U}^* + j\dot{I}^*X_q^* = 1 + j0.7\angle -36.87° = 1.526\angle 21.52°$

Hence power angle is $\delta = 21.52°$, so internal power factor angle is

$$\psi_0 = \delta + \varphi = 21.52° + 36.87° = 58.39°$$

The direct-axis component and the quadrature-axis component of the armature current are illustrated respectively

$$I_d^* = I^*\sin\psi_0 = 0.8516$$

$$I_q^* = I^* \cos\psi_0 = 0.5241$$

Excitation voltage per-unit value is

$$E_0^* = E_Q^* + I_d^*(X_d^* - X_q^*) = 1.526 + 0.8516 \times (1 - 0.7) = 1.781$$

5.4 Power and torque equations of synchronous generators

1. Power equations

The synchronous generator converts the mechanical power that is supplied by a prime-mover into electrical power that is delivered to the terminals of the armature windings by air-gap magnetic field.

Not all the mechanical power that is going into a synchronous generator becomes an electrical power that is out of the machine. The difference between the input power and the output power represents the losses of the machine. A power-flow diagram for a synchronous generator is shown in Fig. 5-23.

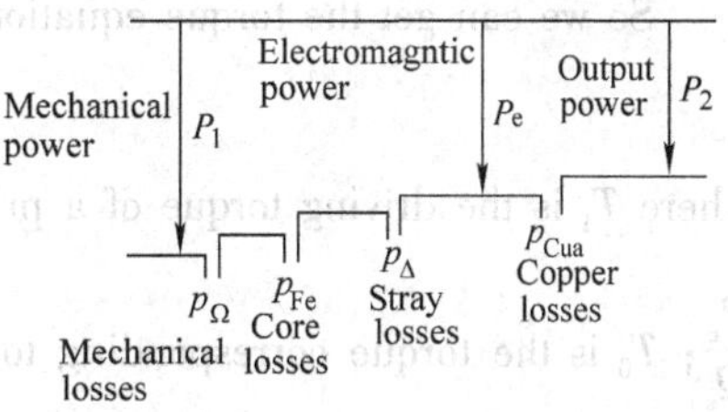

Fig. 5-23 The power-flow diagram of a synchronous generator

The input mechanical power is the shaft power in the machine P_1. The mechanical losses p_Ω (that is the windage and the friction losses), the core losses p_{Fe} and the stray losses p_Δ are subtracted from P_1 to give the electromagnetic power P_e, which is given by

$$P_e = P_1 - p_\Omega - p_{Fe} - p_\Delta \tag{5-18}$$

Usually the rotor copper losses p_{Cuf} is supplied by an independent DC power source; if the main exciter is a direct-coupling exciter, the rotor copper loss must be deducted from the input mechanical power P_1.

Stator copper losses p_{Cua} is subtracted from P_e, which gives the output electrical power of the synchronous generator P_2, that is

$$P_2 = P_e - p_{Cua} \tag{5-19}$$

where the output power $P_2 = mUI\cos\varphi$; the stator copper losses $p_{Cua} = mI^2R_a$, and m is the phase number of the stator.

2. Electromagnetic power

According to Eq. 5-19, the electromagnetic power is

$$P_e = mUI\cos\varphi + mI^2R_a = mI(U\cos\varphi + IR_a)$$

In Fig. 5-24, $U\cos\varphi + IR_a = E\cos\psi = E_Q\cos\psi_0$, so the electromagnetic power P_e can be expressed as

$$P_e = mEI\cos\psi = mE_QI\cos\psi_0 = mE_QI_q \tag{5-20}$$

where ψ is the angle between the air-gap emf ($\dot{E}$) and the armature current ($\dot{I}$).

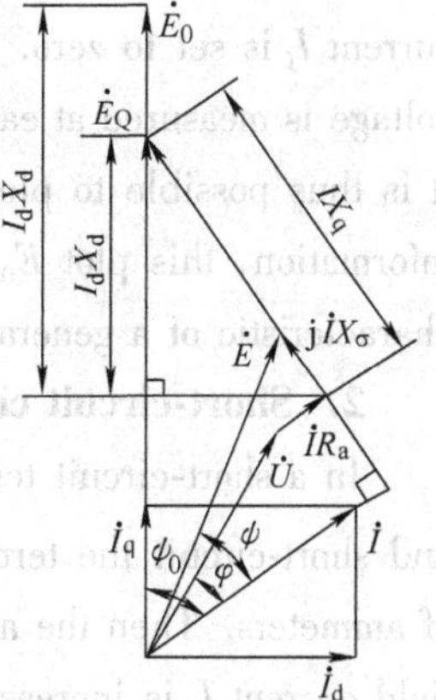

Fig. 5-24 The phasor diagram of a salient-pole synchronous generator

For the non-salient-pole synchronous machine, due to $X_d = X_q$ and $E_Q = E_0$, the electromagnetic power can be written as

$$P_e = mE_0I\cos\psi_0 = mE_0I_q \tag{5-21}$$

Form Eq. 5-20 and Eq. 5-21, notice that the more q-axis component I_q of the armature current is, the greater electromagnetic power will be. Therefore, the armature current must have a q-axis component, when synchronous motor works on the conversion of mechanical and electrical. The q-axis armature reaction becomes stronger; the electromagnetic power from the mechanical and the electrical energy conversion is more.

3. Torque equation

In Eq. 5-18, both sides are divided by the angular velocity Ω_s of the synchronous machine, thus

$$\frac{P_1}{\Omega_s} = \frac{P_e}{\Omega_s} + \frac{p_\Omega + p_{Fe} + p_\Delta}{\Omega_s}$$

So we can get the torque equation of the synchronous generator

$$T_1 = T_e + T_0 \tag{5-22}$$

where T_1 is the driving torque of a prime mover, $T_1 = \frac{P_1}{\Omega_s}$; T_e is the electromagnetic torque, $T_e = \frac{P_e}{\Omega_s}$; T_0 is the torque corresponding to the mechanical losses, the core losses and the stray losses, $T_0 = \frac{p_\Omega + p_{Fe} + p_\Delta}{\Omega_s}$. If we ignore the stray loss, the no-load torque can be expressed as $T_0 = \frac{p_\Omega + p_{Fe}}{\Omega_s}$.

5.5 Operation characteristics of synchronous generators

The operational characteristics of a synchronous generator include the open-circuit characteristics, the short-circuit characteristics, the external (or terminal) characteristics and the regulation characteristics. The open-circuit characteristics and the short-circuit characteristics can be used to calculate the synchronous reactance X_s of a synchronous machine.

1. Open-circuit characteristics

The open-circuit characteristics can be obtained from the open-circuit test. In this test, the generator is turned at the rated speed; the terminals are disconnected from all loads, and the field current I_f is set to zero. Then the field current I_f is gradually increased by steps and the terminal voltage is measured at each step along the way. With the terminals open, $I = 0$, so E_0 is equal to U. It is thus possible to plot a curve of E_0 versus I_f from these information, this plot $E_0 = f(I_f)$ is called the open-circuit characteristic of a generator (Fig. 5-25).

2. Short-circuit characteristics

In a short-circuit test, adjust the field current I_f to zero and short-circuit the terminals of the generator through a set of ammeters. Then the armature current I is measured as the field current I_f is increased. Such a plot $I = f(I_f)$ is called the short-circuit characteristic and is shown in Fig. 5-25. The phasor diagram of a generator at the short-circuit condition is shown in Fig. 5-26.

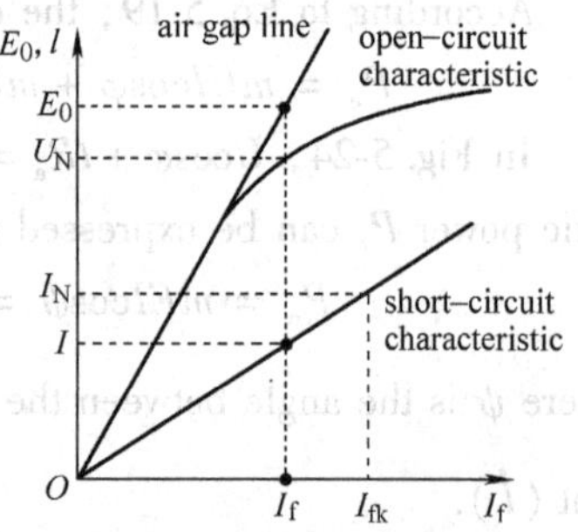

Fig. 5-25 Open-circuit characteristic curve and short-circuit characteristic curve

In Fig. 5-25, the short circuit characteristic is a straight line. The reason is that the short-circuit current only can be limited with the reactance. If the armature resistance is ignored, and $U = 0$, then the short-circuit current will be the inductive current (lagging current), i. e. $\psi_0 = 90°$, $\dot{I}_q = 0$, $\dot{I} = \dot{I}_d$, the voltage equation can be expressed as

$$\dot{E}_0 = \dot{U} + \dot{I}R_a + j\dot{I}_dX_d + j\dot{I}_qX_q \approx j\dot{I}X_d \qquad (5\text{-}23)$$

The pure lagging short-circuit current produces a direct-axis demagnetization armature mmf $\boldsymbol{F}_{ad}$, which makes the resultant mmf $\boldsymbol{F}$ decreased, so the air-gap emf E is decreased, then

$$\dot{E} = \dot{U} + \dot{I}R_a + j\dot{I}X_\sigma \approx j\dot{I}X_\sigma \qquad (5\text{-}24)$$

Fig. 5-26 phasor diagram of a generator at short-circuit condition

Form Eq. 5-24, the air-gap emf equals to the voltage drop of the armature leakage reactance. Then the magnetic circuit of the generator is unsaturated, and the short-circuit curve is linear.

In Fig. 5-25, according to Eq. 5-23, the synchronous reactance X_s can be calculated by

$$X_s = \frac{E_0}{I} \qquad (5\text{-}25)$$

where E_0 is the rated open-circuit voltage at the rated excitation current I_f (on open-circuit characteristics); I is the short-circuit current at the rated excitation current I_f (on short-circuit characteristics).

Due to the core is unsaturated in short circuit test, E_0 can be obtained from the air-gap line (Fig. 5-25), so the value of X_s is an unsaturated value.

3. The Short-circuit ratio

The *short-circuit ratio* k_c of a generator is defined as the ratio of the field current I_f, which is required for the rated voltage U_N at open-circuit, to the field current I_{fk}, which is required for the rated armature current I_N at short-circuit.

$$k_c = \frac{I_{f(U=U_N)}}{I_{fk(I=I_N)}} \qquad (5\text{-}26)$$

4. Terminal characteristics

Terminal characteristic $U = f(I)$ is the relationship between the terminal voltage U and the armature current I, when $n = n_s$, I_f = constant, $\cos\varphi$ = constant. Fig. 5-27 shows a synchronous generator terminal characteristic at different load power factors.

In Fig. 5-27, we can find that the behavior of a synchronous generator depends upon the type of load that has to supply. If the load is lagging, the terminal voltage U will be decreased by the demagnetizing effect of an armature reaction and a voltage drop of the stator leakage impedance. Thus, the terminal characteristic is decreasing (Fig. 5-27, $\cos\varphi = 1$ and $\cos\varphi = 0.8$ lagging). If the load is leading, the terminal voltage U will be increasing with increasing load (Fig. 5-27, $\cos\varphi$ = 0.8 leading).

In Fig. 5-28, the voltage regulation Δu of a generator is defined as

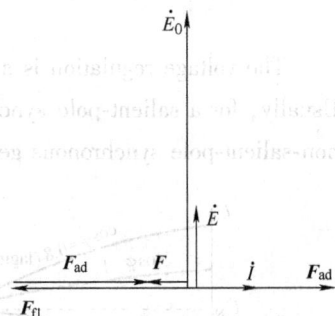

$$\Delta u = \frac{E_0 - U_{N\varphi}}{U_{N\varphi}} \times 100\% \tag{5-27}$$

The voltage regulation is an important value of the characterization of a synchronous generator. Usually, for a salient-pole synchronous generator: $\Delta u = 18\% \sim 30\%$ ($\cos\varphi = 0.8$ lagging), for a non-salient-pole synchronous generator: $\Delta u = 30\% \sim 48\%$ ($\cos\varphi = 0.8$ lagging).

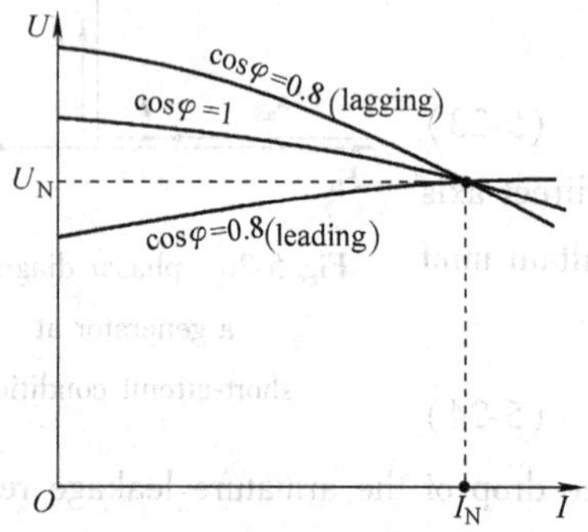

Fig. 5-27 The terminal characteristic

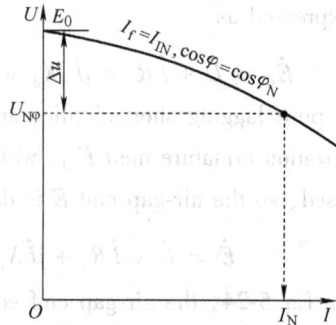

Fig. 5-28 The voltage regulation characteristic.

5. Regulation characteristics

When a single synchronous generator feeds a variable load, and the terminal voltage U must be kept constant, then we are interested in knowing how the excitation current I_f changes as a function of the load current I.

When $n = n_S$, $U = U_N$, $\cos\varphi$ = constant, the relationship between I_f and I is called the regulation characteristic $I_f = f(I)$. Fig. 5-29 shows the regulation characteristics of a synchronous generator under different load power factor.

In Fig. 5-29, when the load power factor is lagging (Fig. 5-29, $\cos\varphi = 1$ and $\cos\varphi = 0.8$ lagging), the excitation current I_f is increased with an increasing armature current I, so the regulation characteristic is rising. When the load power factor is leading (Fig. 5-29, $\cos\varphi = 0.8$ leading), the regulation characteristic is decreasing.

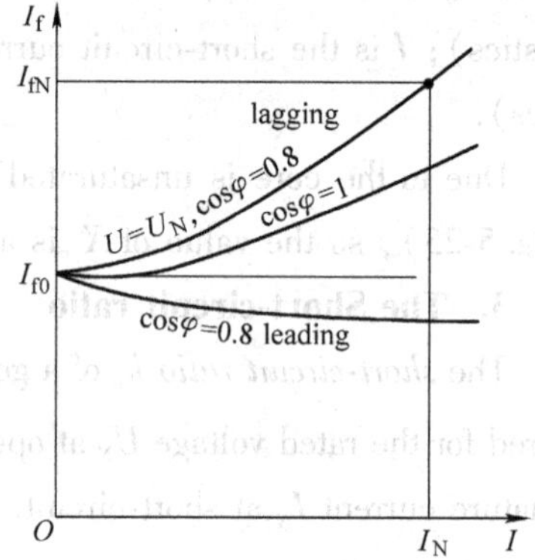

Fig. 5-29 The regulating characteristic of a synchronous generator

Example 5-2 A Y-connected stream-turbine generator, $S_N = 15000\text{kV}\cdot\text{A}$, $U_N = 6.3\text{kV}$, $\cos\varphi_N = 0.8$ (lagging load), in regardless of the magnetic saturation and the armature resistance, the following values can be obtained by an open- and a short-circuit test.

Exciting current I_f/A	102
Armature current (short-circuit characteristic) I/A	887
Line-line voltage U_L (open-circuit characteristic) /V	6300
Line-line voltage U_L (air-gap line) /V	8000

1) What are the synchronous reactance and its per-unit value?

2) What is the internal generated voltage E_n at the rated load?

Solution:

1) From the air-gap line, when $I_f = 102\text{A}$, the line-to-neutral excitation emf $E_0 = 8000/\sqrt{3} = 4618.8\text{V}$; with the same excitation current from the short-circuit characteristic, the short-circuit current can be obtained by $I = 887\text{A}$; the synchronous reactance can be expressed by

$$X_d(X_s) = \frac{E_0}{I} = \frac{4618.8}{887}\Omega = 5.207\Omega$$

Rated phase-current is

$$I_{N\phi} = \frac{S_N}{\sqrt{3}U_N} = \frac{15\,000 \times 10^3}{\sqrt{3} \times 6300}\text{A} = 1374.6\text{A}$$

Rated phase-voltage is

$$U_{N\phi} = U_N/\sqrt{3} = 6300/\sqrt{3}\text{V} = 3637.3\text{V}$$

Impedance base value is

$$Z_b = \frac{U_{N\phi}}{I_{N\phi}} = \frac{3637.3}{1374.6}\Omega = 2.646\Omega$$

The per-unit value of synchronous reactance is

$$X_d^* = \frac{X_d}{Z_b} = \frac{5.207}{2.646} = 1.968$$

Per-unit value calculation:

$$E_0^* = \frac{E_0}{U_{N\phi}} = \frac{8000/\sqrt{3}}{6300/\sqrt{3}} = 1.2698$$

$$I^* = \frac{I}{I_{N\phi}} = \frac{887}{1374.6} = 0.6453$$

$$X_d^* = \frac{E_0^*}{I^*} = \frac{1.2698}{0.6453} = 1.968$$

2) Take the terminal voltage as the reference phasor, i.e., $\dot{U}^* = 1\angle 0°$, then $\dot{I}^* = 1\angle -36.87°$

The excitation emf is

$$\dot{E}_0^* = \dot{U}^* + \text{j}\dot{I}^* X_s^* = 1 + \text{j}1.968\angle -36.87° = 2.689\angle 35.8°$$

The actual value of the excitation emf (phase emf) is

$$E_0 = E_0^* U_{N\phi} = 2.689 \times 3637.3\text{V} = 9781\text{V}$$

5.6 Parallel operation of synchronous generators

The electrical needs of the industry, the commercial establishments, and the individual consumers are supplied almost exclusively by the synchronous generators. There are virtually hundreds of such generators are operating to meet these needs. The purpose of this section is to describe some of the principles that govern the steady-state behavior of the alternators that operate in parallel.

5.6.1 The need to synchronize

First, we focus on the parallel operation of the two isolated synchronous generators of equal ratings. In Fig. 5-30, assume that the first generator G_1 is already connected to a three-phase resistive load by a closed load switch which is operating at rated frequency and rated voltage. Now, it is desirable to place a second generator parallel with the first one, as it might help to share the current load as well as to prepare for the increasing power requirements of this load in the future.

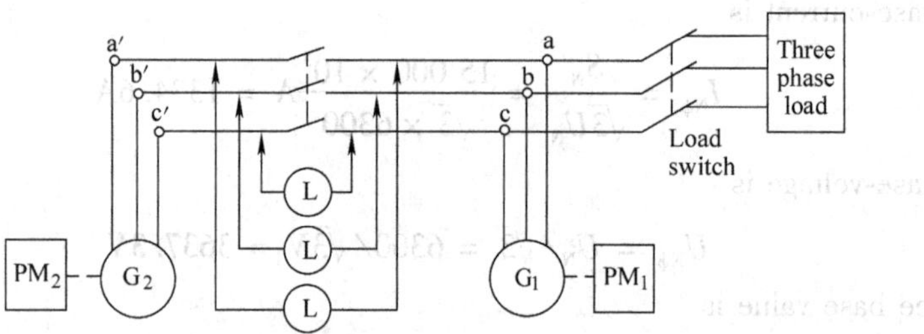

Fig. 5-30 Synchronizing two synchronous generators for parallel operation

If severe shocks on both generators are to be avoided, a procedure called *synchronizing procedure* must be followed before the switch, which parallels to two machines that can be closed. The process is started by adjusting the speed of the prime mover of generator G_2 to correspond exactly to that of generator G_1. The field current of G_2 is then adjusted to yield the same nominal voltage between lines a′, b′, and c′. A suitable bank of lamps is placed in series with lines a′-a, b′-b, c′-c (Fig. 5-30). These lamps will light up if either the voltage or the frequency of G_2 differs slightly from those of G_1.

When the two frequencies are identical but the magnitude of the line voltages are slightly different, the lamps will be on steadily. A subsequent adjustment of the field current of G_2 can then be used to cause the lamps to be darkened; the synchronizing switch can be closed as well. Thereby, both machines would be in parallel without any incident. On the other hand, if the two sets of three-phase voltages are equal in magnitude, but a bit different in frequency, the lamps will light again, but now they will flicker at a rate equal to a different frequency. At one point in a different cycle, the voltage across the lamps will reach twice the line voltage; it will be zero after half-cycle. An adjustment in the prime mover speed can serve to reduce this difference of frequency to a very small value. Then, as a dark period is approached, the line switch can be closed with a little or no disturbance.

We seldom have to connect two generators in parallel except in isolated locations. It is much more common to connect a generator to a large power system (infinite bus) that already has many alternators connected to it.

An infinite bus is a system, which is so powerful, and it imposes its own voltage and frequency upon any apparatus that is connected to its terminals. Let G_1 in Fig. 5-30 represent a large power system, and then Fig. 5-30 becomes the Fig. 5-31.

The synchronizing procedure needs to connect G_2 to the infinite bus, and this is the same as the interconnection of two isolated generators as described previously.

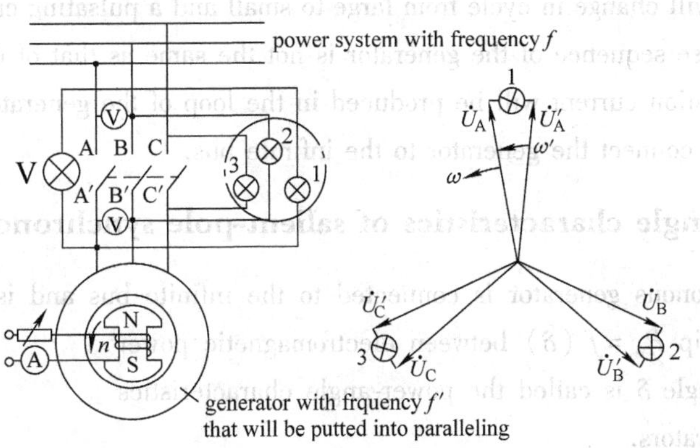

Fig. 5-31 A generator is connecting to the infinite bus

The process of generators being connected in parallel with a large system is called paralleling or synchronizing. In this process of synchronizing, a huge impact current must be prevented so as to avoid the generators being damaged and the power systems suffering from interference. Therefore, the following conditions must be met before synchronizing:

1) The phase sequence of the generator is the same as that of the power system.

2) The generator frequency f' is equal to the power system frequency f, i. e. $f' = f$.

3) The excitation emf $\dot{E}_0$ of the generator is equal to the power system voltage $\dot{U}$ (equal in magnitude and in phase), i. e. $\dot{E}_0 = \dot{U}$.

There will be a serious consequence, if it does not meet any of the conditions mentioned above, before connecting the generator in parallel with the infinite bus.

(1) If the excitation emf $\dot{E}_0$ of the generator is not equal to the power system voltage $\dot{U}$

Fig. 5-32a shows the single-phase schematic diagram of connecting a generator to the infinite bus. When the magnitudes of $\dot{E}_0$ and $\dot{U}$ are not equal (Fig. 5-32b) or the phase angles of them are not the same (Fig. 5-32c), a voltage difference $\Delta\dot{U}$ across the switch S will be produced. At this time, if the switch S is closed, there will be a transient impact current in the loop of the generator and the system. If the impact current is too large, the magnetic force that is produced by the current will be injured to the ends of the stator winding and give an impact to the generator shaft.

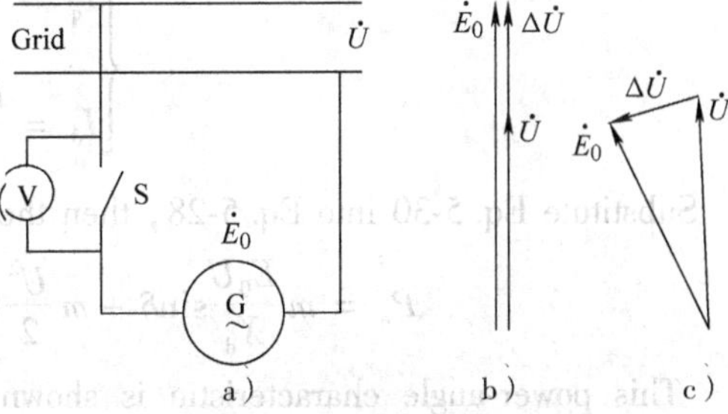

Fig. 5-32 Single-phase schematic diagram of a generator connected to the infinite bus
a) Circuit diagram
b) the magnitudes of $\dot{E}_0$ and $\dot{U}$ are not equal
c) the phases of $\dot{E}_0$ and $\dot{U}$ are not the same

(2) If the generator frequency f' is not equal to the power system frequency f

The phase difference of $\dot{E}_0$ and $\dot{U}$ will change between 0° ~ 360°. The voltage difference $\Delta\dot{U}$

across the switch S will change in cycle from large to small and a pulsating current will be produced.

(3) If the phase sequence of the generator is not the same as that of the infinite bus

A strong circulation current will be produced in the loop of the generator and the system. It is strictly prohibited to connect the generator to the infinite bus.

5.6.2 Power-angle characteristics of salient-pole synchronous generators

When a synchronous generator is connected to the infinite bus and is operating at a steady-state, the relationship $P_e = f(\delta)$ between electromagnetic power P_e and the power angle δ is called the power-angle characteristics of synchronous generators.

The phase-diagram of a salient-pole generator is shown in Fig. 5-33. The armature resistance will be neglected, because it is usually small, then the electromagnetic power P_e is equal to the output power of the machine approximately, that is

$$\begin{aligned} P_e \approx P_2 &= mUI\cos\varphi = mUI\cos(\psi_0 - \delta) \\ &= mUI(\cos\psi_0\cos\delta + \sin\psi_0\sin\delta) \qquad (5\text{-}28) \\ &= mU(I_q\cos\delta + I_d\sin\delta) \end{aligned}$$

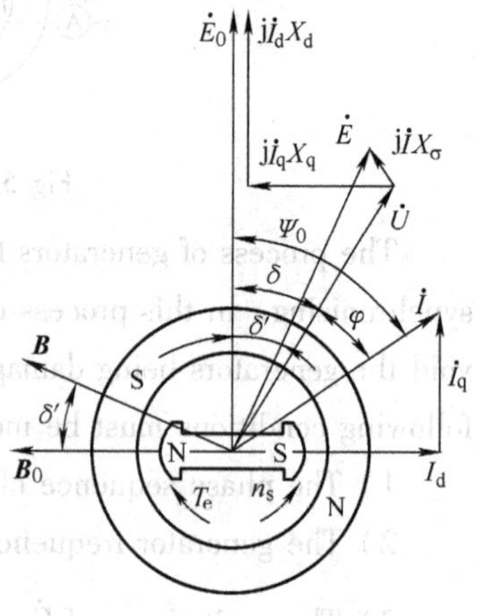

Fig. 5-33 Time-pace vector diagram of a salient-pole synchronous generator

In Fig. 5-33, we can find that

$$\begin{cases} I_qX_q = U\sin\delta \\ I_dX_d = E_0 - U\cos\delta \end{cases} \qquad (5\text{-}29)$$

Then

$$\begin{cases} I_q = \dfrac{U\sin\delta}{X_q} \\ I_d = \dfrac{E_0 - U\cos\delta}{X_d} \end{cases} \qquad (5\text{-}30)$$

Substitute Eq. 5-30 into Eq. 5-28, then the electromagnetic power P_e is expressed as

$$P_e = m\frac{E_0U}{X_d}\sin\delta + m\frac{U^2}{2}\left(\frac{1}{X_q} - \frac{1}{X_d}\right)\sin 2\delta = P_{e1} + P_{e2} \qquad (5\text{-}31)$$

This power-angle characteristic is shown in Fig. 5-34. The first term $m\dfrac{E_0U}{X_d}\sin\delta$ is the same as the expression that is obtained from a cylindrical-rotor machine; which is called the *basic electromagnetic power*, and is P_{e1} (Fig. 5-34). The second term $m\dfrac{U^2}{2}\left(\dfrac{1}{X_q} - \dfrac{1}{X_d}\right)\sin 2\delta$ introduces the effect of salient poles. It represents the fact that the air-gap flux wave creates a torque that has a tendency to align the field poles in the position of minimum reluctance. This term is called the *reluctance power*

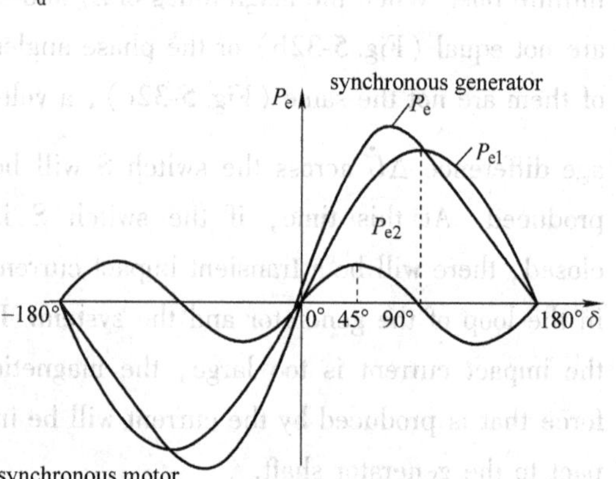

Fig. 5-34 Power-angle characteristic of a salient-pole synchronous machine

corresponding to the reluctance torque, and is denoted by P_{e2} (Fig. 5-34).

Note that the reluctance torque is independent of field excitation. Also, note that if $X_d = X_q$, as in a uniform air-gap machine, there is no preferential direction of magnetization, the reluctance torque is zero, and Eq. 5-30 reduces to the power-angle equation for a cylindrical-rotor machine, of which the synchronous reactance is X_s.

$$P_e = m\frac{E_0 U}{X_s}\sin\delta \tag{5-32}$$

When $\delta = 90°$, the maximum electromagnetic power of a cylindrical-rotor machine is

$$P_{emax} = m\frac{E_0 U}{X_s} \tag{5-33}$$

Notice that the characteristic for negative values of δ is the same except for a reversal in the sign of P_e. The generator and the motor regions are alike, if the effects of the reluctance are negligible. For generator operation state $\dot{E}_0$ leads $\dot{U}$; for motor operation state $\dot{E}_0$ lags $\dot{U}$. The Steady-state operation is stable over the range, where the slope of the power-angle characteristic is positive. Because of the reluctance torque, a salient-pole machine is stiffer than the one with a cylindrical rotor; i. e., for equal voltages and equal values of X_s. a salient-pole machine develops a given torque at a smaller value of δ, and the maximum torque that can be developed is somewhat greater.

According to Eq. 5-31 and Fig. 5-34, when $\delta = 90°$, P_{e1} achieves its maximum value, i. e., $P_{e1max} = m\frac{E_0 U}{X_d}$; when $\delta = 45°$, P_{e2} achieves its maximum value, i. e., $P_{e2max} = m\frac{U^2}{2}\left(\frac{1}{X_q} - \frac{1}{X_d}\right)$. P_e achieves its maximum value during $45° < \delta < 90°$, and $P_{emax} \neq P_{e1max} + P_{e2max}$.

5.6.3 Active power regulation and static stability

Generators supply active and reactive power to power systems, when the synchronous generators are connected in parallel with the infinite bus. How to adjust the active and reactive power will be analyzed as followings. Taking a cylindrical-rotor generator as an example, the magnetic saturation and armature resistance are ignored in this case for simplicity, the power system is known as "the infinite bus", the system voltage U = constant, and frequency f = constant.

1. Active power regulation

When a generator is connected in parallel with the infinite bus and output none of active power, i. e., $P_2 = 0$, this generator is operating at no-load condition, and it is said to float on the line. Then, the electromagnetic power $P_e = P_2 = 0$, and the power angle $\delta = 0$ (Fig. 5-35a). The excitation emf $\dot{E}_0$ is in phase with the grid voltage $\dot{U}$, and leads armature current $\dot{I}$ by 90°, in this case, the armature current is the reactive current. At this moment, the mechanical power P_1 from the prime mover is used only to overcome the no-load losses of the generator, i. e., $P_1 = p_\Omega + p_{Fe} = p_0$, and the drive torque T_1 is equal to the no-load torque T_0.

If the input power P_1 from the prime mover is increased, then the drive torque T_1 will increase and the rotor will accelerate. Thus, the main magnetic field $\boldsymbol{B}_0$ of the generator will begin to lead the air-gap magnetic field $\boldsymbol{B}$. Accordingly, the excitation emf $\dot{E}_0$ will begin to lead the terminal voltage (or grid voltage) $\dot{U}$ (Fig. 5-35b), and the power angle will increase, i. e., $\delta > 0$. Therefore, the electromagnetic power $P_e > 0$, and the generator outputs the active power to the grid.

Finally, the power angle δ will stabilize at a certain value δ_a, the electromagnetic torque T_e is equal to the remaining torque ($T_1 - T_0$), and the machine reaches a new equilibrium state (Fig. 5-35c). At this moment, the mechanical power P_1 from the prime mover deducts the no-load losses p_0, and the resultant power is equal to the electromagnetic power P_e, and it is expressed as

$$P_1 - p_0 = P_e = m\frac{E_0 U}{X_s}\sin\delta_a \tag{5-34}$$

The electromagnetic power P_e is equal to the output active power P_2 of the generator, and the generator is operating at loaded.

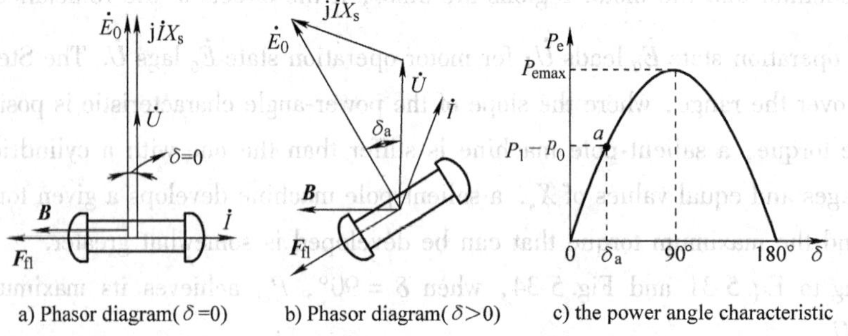

Fig. 5-35 Active power regulation of a synchronous generator in parallel with the infinite bus

The analysis mentioned above shows that increasing the input power from the prime mover, the power angle δ increases, the electromagnetic power and the output power of the generator will increase accordingly. In order to increase the output active power of the generator, the input power must increase by using the prime mover.

Note that when $\delta = 90°$, the maximum electromagnetic power $P_{emax} = m\frac{E_0 U}{X_s}$ is called the limit power of the machine. If $\delta > 90°$, the output active power of the generator will decrease, the generator cannot reach a new equilibrium state and the generator will eventually lose the synchronization.

Notice that increasing the output from the prime mover, the output active power of the generator will increase, and the power angle δ will also increase. Then, the output reactive power of the generator will also change. If the output reactive power Q of the generator is keeping constant, and the output from the prime mover is increasing to adjust the active power of the generator, the excitation emf E_0 and the power angle δ will change simultaneously.

2. Static stability

When a synchronous generator is running at a stable operation point in parallel with the grid, if some tiny disturbances from the grid or the prime mover are occurred, the operation point will be changed. If the disturbances disappear, the generator can revert to the original running state; this generator's operation is called the static stability. If the disturbances disappear, the generator cannot revert to the original running state, this operation is called unstable.

Fig. 5-36 shows the static stability of a non-salient pole synchronous generator. There are two points, A and C on the power-angle characteristic curve, the corresponding power angles are δ and δ' respectively. We want to know which the stable operation point is.

Assuming that the generatoris running at point A initially, the power angle is δ. When the output power of the prime mover increases a small ΔP_T, the power angle of the generator will increase $\Delta\delta$. Accordingly, the electromagnetic power increases by ΔP_e, eventually the generator is operating at point B, and meet the power balance condition $P_T + \Delta P_T = P_e + \Delta P_e$ (Fig. 5-36). When the disturbance disappears, because of $P_T < P_e + \Delta P_e$, the driving torque is less than the electromagnetic braking torque, so the rotor speed is decreasing. The power angle is also reduced, finally the generator's operation point will be returned to point A. Thus, point A is the static stability.

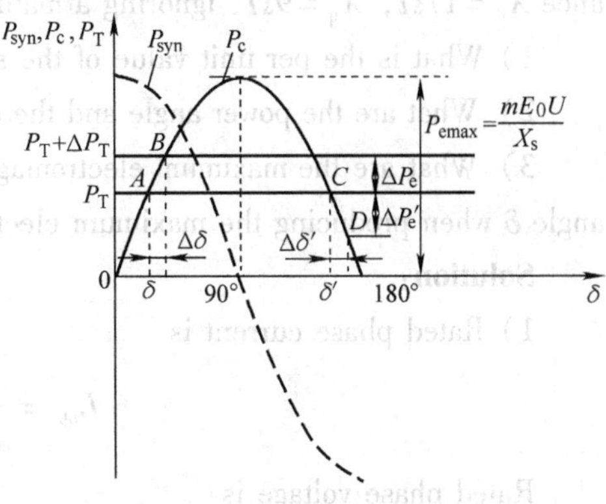

Fig. 5-36 Static stability of the synchronous generator in parallel with infinite bus

If the generator is running at point C initially, the power angle is δ'. When the output power of the prime mover increases by a small ΔP_T, the power angle of the generator will increase $\Delta\delta'$ (Fig. 5-36), but the electromagnetic power of the generator reduces the $\Delta P_e'$. And $P_T + \Delta P_T > P_e - \Delta P_e'$, the drive torque is greater than the braking torque of the generator, so the rotor speed continues to accelerate and cannot reach a new equilibrium. Even if the disturbance disappears, the output power P_T of the prime mover is greater than the electromagnetic power, and the rotor speed continues to accelerate, eventually leading to the generator out of sync, so point C is unstable.

Note that when the power angle $0° \leqslant \delta < 90°$, the generator is statically stable, at this time $\frac{dP_e}{d\delta} > 0$; if the power angle $90° < \delta \leqslant 180°$, the generator is unstable, at this time $\frac{dP_e}{d\delta} < 0$; when power angle $\delta = 90°$, the generator is the intersection of stable and unstable, at this time $\frac{dP_e}{d\delta} = 0$, this point is called the static stability limit.

The ratio synchronous power P_{syn} of a non-salient pole synchronous machine is defined as

$$P_{syn} = \frac{dP_e}{d\delta} = m\frac{E_0U}{X_s}\cos\delta \tag{5-35}$$

The curve of P_{syn} is the dotted line shown in Fig. 5-36.

The overload capacity k_p of a non-salient pole generator is defined as the ratio between the rated power and the maximum electromagnetic power, it is given by

$$k_p = \frac{P_{emax}}{P_N} \approx \frac{m\frac{E_0U}{X_s}}{m\frac{E_0U}{X_s}\sin\delta_N} = \frac{1}{\sin\delta_N} \tag{5-36}$$

where δ_N is the rated power angle of a synchronous machine. Usually, δ_N and k_p of the turbine generator are $\delta_N \approx 30° \sim 40°$, $k_p \approx 1.6 \sim 2.0$.

Example 5-3 A 8750kV·A, 11kV, 50Hz, Y-connected a three-phase turbine generator rated to run in parallel with the infinite bus, power factor 0.8 (lag), each phase synchronous reactance $X_d = 17\Omega$, $X_q = 9\Omega$, ignoring armature resistance and excluding the magnetic saturation.

1) What is the per unit value of the synchronous reactance?

2) What are the power angle and the excitation emf when the generator runs at rated conditions?

3) What are the maximum electromagnetic power P_{emax}, the overload capacity k_p and the power angle δ when producing the maximum electromagnetic power?

Solution:

1) Rated phase current is

$$I_{N\phi} = \frac{8750}{\sqrt{3} \times 11} A = 459.3A$$

Rated phase voltage is

$$U_{N\phi} = \frac{11\,000}{\sqrt{3}} V = 6350.9V$$

The per unit value of synchronous reactance is

$$X_d^* = X_d \frac{I_{N\phi}}{U_{N\phi}} = 17 \times \frac{459.3}{6350.9} = 1.229$$

$$X_q^* = X_q \frac{I_{N\phi}}{U_{N\phi}} = 9 \times \frac{459.3}{6350.9} = 0.6509$$

2) The terminal voltage as a reference phasor, calculate in unit value:

$$\dot{U}^* = 1\angle 0°, \dot{I}^* = 1\angle -36.87°$$

Virtual electromotive force is

$$E_Q^* = \dot{U}^* + j\dot{I}^* X_q^* = 1 + j0.6509\angle -36.87° = 1.485\angle 20.53°$$

So the power angle $\delta = 20.53°$, the internal power factor angle is

$$\psi_0 = \delta + \varphi = 20.53° + 36.87° = 57.4°$$

The d-axis and q-axis components of armature current are illustrated respectively as following

$$I_d^* = I^* \sin\psi_0 = \sin 57.4° = 0.8425$$

$$I_q^* = I^* \cos\psi_0 = \cos 57.4° = 0.5388$$

The per unit value of the excitation emf is

$$E_0^* = U^* \cos\delta + I_d^* X_d^* = \cos 20.53° + 0.8425 \times 1.229 = 1.972$$

The actual value of the excitation electromotive force (phase emf) is

$$E_0 = E_0^* U_{N\phi} = 1.972 \times 6350.9V = 12\,524V$$

3) The per unit value of the electromagnetic power is

$$P_e^* = \frac{E_0^* U^*}{X_d^*} \sin\delta + \frac{U^{*2}}{2}\left(\frac{1}{X_q^*} - \frac{1}{X_d^*}\right) \sin 2\delta$$

$$= \frac{1.972}{1.229} \sin\delta + \frac{1^2}{2}\left(\frac{1}{0.6509} - \frac{1}{1.229}\right) \sin 2\delta$$

$$= 1.605 \sin\delta + 0.3613 \sin 2\delta$$

Notice that in the per-unit value, the power base value is the three-phase rated apparent power $S_N = 3U_{N\phi} I_{N\phi}$.

Set $\dfrac{dP_e^*}{d\delta} = 0$, then

$$\frac{dP_e^*}{d\delta} = 1.605\cos\delta + 0.7226\cos 2\delta = 0$$

and $1.605\cos\delta + 0.7226\ (2\cos^2\delta - 1) = 0$

$$1.445\cos^2\delta + 1.605\cos\delta - 0.7226 = 0$$

So $\cos\delta = \dfrac{-1.605 + \sqrt{1.605^2 + 4 \times 1.445 \times 0.7226}}{2 \times 1.445} = \dfrac{-1.605 + 2.599}{2.89} = 0.3439$

The power angle corresponding to the maximum electromagnetic power is

$$\delta_m = \arccos 0.3439 = 69.89°$$

The per unit value corresponding to the maximum electromagnetic power is

$$P_{emax}^* = 1.605\sin 69.89° + 0.3613\sin 139.78° = 1.74$$

The maximum electromagnetic power is

$$P_{emax} = P_{emax}^* S_N = 1.74 \times 8750\text{kW} = 15\ 225\text{kW}$$

The overload capacity is

$$k_P = \frac{P_{emax}}{P_N} = \frac{1.74}{0.8} = 2.175$$

5.6.4 Reactive power regulation and V-curves

1. Reactive power regulation

When a non-salient pole generator is synchronized and connected to an infinite bus, if we want to regulate the output reactive power of the generator, and the input power P_1 of the generator is maintained constant. Then, the electromagnetic power P_e and the output active power P_2 of the generator will remain unchanged, namely

$$\begin{cases} P_e = m\dfrac{E_0 U}{X_s}\sin\delta = \text{constant} \\ P_2 = mUI\cos\varphi = \text{constant} \end{cases} \tag{5-37}$$

Because the phase number m, the terminal voltage U, and the synchronous reactance X_s are fixed values, so the equations stated can be rewritten as

$$\begin{cases} E_0\sin\delta = \text{constant} \\ I\cos\varphi = \text{constant} \end{cases} \tag{5-38}$$

If we now increase the field current I_f of the generator, and the input power P_1 of the generator is constant, i. e., Eq. 5-38 is kept unchanged. The excitation emf E_0 will increase according to Eq. 5-38, the power angle δ will increase. Because the generator is connected to an infinite bus, the generator terminal voltage $\dot{U}$ is constant, according to the voltage equation Eq. 5-9, the armature current I will increase. According to Eq. 5-38, the power factor angle φ will also become larger, then the generator output reactive power $Q = mUI\sin\varphi$ will inevitably increase. Consequently, the synchronous generator is over-excited; it supplies a reactive power to the infinite bus. The reactive power increases as the DC exciting current is raised.

Fig. 5-37 shows the phasor diagram of a generator that is excited by different field currents.

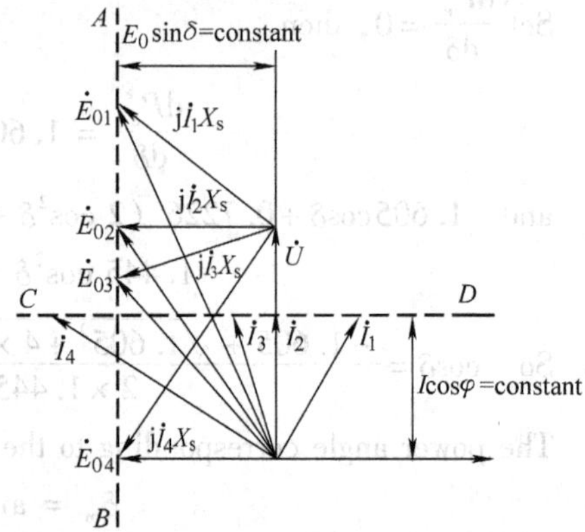

Fig. 5-37 the phasor diagram of a generator excited by different field currents

In Fig. 5-37, when the field current is I_{f2}, the excitation emf is $\dot{E}_{02}$, and the armature current is $\dot{I}_2$, the power factor of the generator $\cos\varphi = 1$, i. e., the armature current $\dot{I}_2$ is in phase with the terminal voltage $\dot{U}$. At this time, the generator is operating at a normal-excited state, it only outputs active power to the infinite bus.

If we increase the field current as I_{f1} ($I_{f1} > I_{f2}$), then the excitation emf will increase to $\dot{E}_{01}$, and the armature current is $\dot{I}_1$. $\dot{I}_1$ lags $\dot{U}$, the power factor of the generator $\cos\varphi < 1$ (lagging). At this time, the generator is operating at an over-excited state. Consequently, the generator supplies not only the active power, and also the inductive reactive power to the infinite bus.

Let us decrease the field current now as I_{f3} ($I_{f3} < I_{f2}$), the excitation emf will reduce to $\dot{E}_{03}$, and the armature current is $\dot{I}_3$, $\dot{I}_3$ leads $\dot{U}$, the power factor of the generator $\cos\varphi < 1$ (leading). At this time, the generator is operating at an under-excited state. Thus, generator supplies not only the active power, and also the capacitive reactive power to the infinite bus.

In Fig-5-37, if we continue to reduce the field current as I_{f4}, and excitation emf will reduce to $\dot{E}_{04}$, then power angle $\delta = 90°$, the generator will reach the static stability limit. If the field current is less than I_{f4}, the generator will be out of synchronism.

2. V-curves

In Fig. 5-37, notice that when the generator is operating at the normal-excited state, its armature current I_2 is at minimum value. If we increase or decrease the field current, the magnitude of the armature current will become larger. A plot of I versus I_f for a synchronous generator is shown in Fig. 5-38. Such a plot is called a *synchronous generator V curve*, because its shape looks like the letter "V".

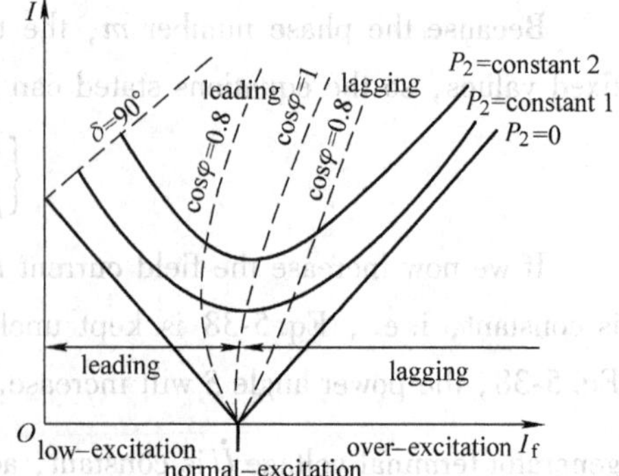

Fig. 5-38 V-curves of a synchronous generator

There are several V-curves drawn, corresponding to different active power levels. For each curve, the minimum armature current I occurs at unity power factor, when only real power is supplied to the infinite bus. At any other point on the curve, some reactive power is being supplied to or by the infinite bus.

In Fig. 5-38, notice that if the field current I_f is less than its normal excitation value, the armature current becomes leading and the generator will supply a reactive power to the power system as a capacitor would. If the field current I_f is greater than its normal excitation value, the armature current becomes lagging, and the generator will supply an inductive reactive power to the power system. Therefore, by controlling the field current of a synchronous generator, the reactive power that supplied to the power system can be controlled.

By connecting the lowest point of each V-curve, we will plot a curve of $\cos\varphi = 1$ (a dotted line in Fig. 5-38). On the right side of curve $\cos\varphi = 1$, the generator is operating at an over-excited condition, the power factor is lagging, the generator outputs the inductive reactive power to the infinite bus; on the left side of curve $\cos\varphi = 1$, the generator is operating at an under-excited condition, the power factor is leading, the generator outputs the capacitive reactive power to the power system. If the field current is very small, and the power angle $\delta > 90°$, the generator can be out of synchronism.

5.7 Synchronous motors

Synchronous machine can operate either as a generator or as a motor. When it is operating as a motor (by connecting them to a 3-phase source), they are called synchronous motor. As the name implies, the synchronous motor runs in synchronism with the revolving field. The speed of rotation is therefore tied to the frequency of the source. Because the frequency is fixed, the motor speed stays constant, irrespective of the load or voltage of the 3-phase line. Moreover, the synchronous motor can operate as a synchronous compensator machine (or phase modifier) to regulate the power factor of the power system.

5.7.1 Voltage equation, phasor diagram and equivalent circuit of synchronous motors

According to the generator conventions, the power angle δ of synchronous motors is negative, the main magnetic field lags the air gap magnetic field and the excitation emf $\dot{E}_0$ lags the terminal voltage $\dot{U}$. Therefore the electromagnetic power P_e is negative, the motor output negative power to the grid $P_2 = mUI\cos\varphi$, and the power-factor angle is a obtuse angle ($\varphi > 90°$). Apparently, it is extremely inconvenient to analyze the synchronous motor by using a negative power angle and a negative electromagnetic power.

In this section, the motor conventions are applied. The direction of the input current is the positive direction of the armature current, and the direction of the input power of the motor is also positive. the electromagnetic power P_e is positive and the power-factor angle φ is an acute angle. The phasor diagram and the equivalent circuit of a non-salient pole synchronous motor are shown in Fig. 5-39.

According to the equivalent circuit of a synchronous motor (Fig. 5-39b), the voltage equation is expressed as

$$\dot{U} = \dot{E}_0 + \dot{I}R_a + j\dot{I}X_s \tag{5-39}$$

Similarly, the voltage equation of a salient pole synchronous motor is

$$\dot{U} = \dot{E}_0 + \dot{I}R_a + j\dot{I}_d X_d + j\dot{I}_q X_q \tag{5-40}$$

The phasor diagram of a salient pole synchronous motor is shown in Fig. 5-40.

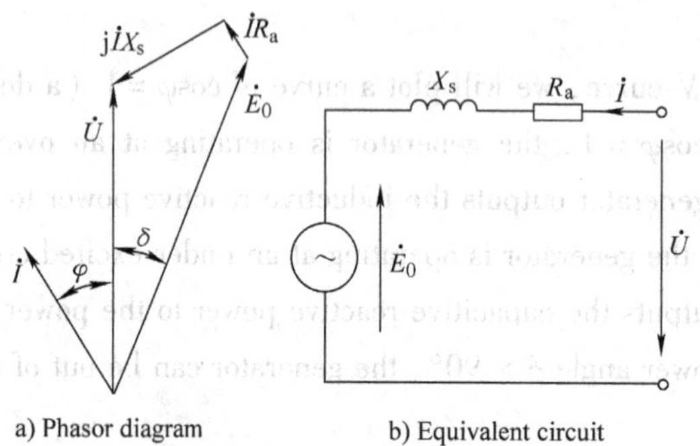

Fig. 5-39 Phasor diagram and equivalent circuit of non-salient pole synchronous motor

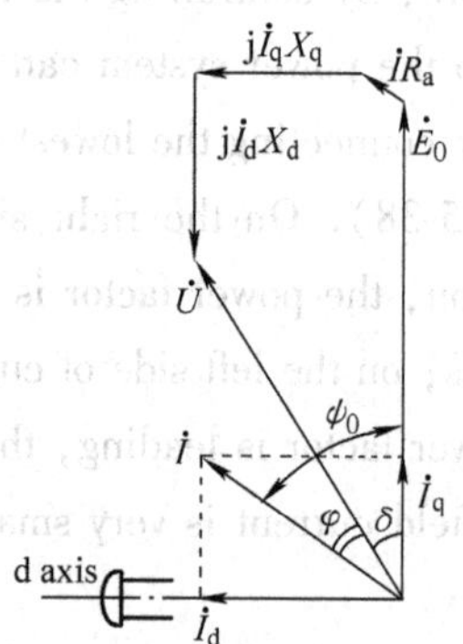

Fig. 5-40 Phasor diagram of a salient pole synchronous motor

5.7.2 Power-angle characteristic, power and torque equations of synchronous motors

The power-angle characteristic of a salient pole synchronous motor is the same as that of a synchronous generator (Eq. 5-31), that is

$$P_e = m\frac{E_0 U}{X_d}\sin\delta + m\frac{U^2}{2}\left(\frac{1}{X_q} - \frac{1}{X_d}\right)\sin 2\delta \tag{5-41}$$

When $0° \leqslant \delta \leqslant 180°$, the machine is operating as a motor and its electromagnetic power is positive.

The copper loss p_{Cua} of the armature windings is subtracted from the input electric power P_1, the rest power can convert to a mechanical power through the air-gap magnetic field. This is called the electromagnetic power P_e:

$$P_e = P_1 - p_{Cua} \tag{5-42}$$

The mechanical losses p_Ω, the stator iron losses p_{Fe} and the stray losses p_Δ are subtracted from P_e, The remaining part is called the output power P_2, which is a mechanical power

$$P_2 = P_e - p_\Omega - p_{Fe} - p_\Delta \tag{5-43}$$

Both sides of Eq. 5-43 is divided by Ω_s, the torque equation is given by

$$T_e = T_2 + T_0 \tag{5-44}$$

where T_e is electromagnetic torque, and $T_e = \frac{P_e}{\Omega_s}$; T_2 is the load torque, and $T_2 = \frac{P_2}{\Omega_s}$; T_0 is the no-load torque, and $T_0 = \frac{p_\Omega + p_{Fe}}{\Omega_s}$.

5.7.3 Reactive power regulation

When a synchronous motor is connected to a power system, it will absorb an active power and a reactive power from the grid. This section will explain how to adjust the reactive power of a synchronous motor. For simplicity, take the non-salient pole synchronous motor as an example, where the armature resistance is ignored, and the magnetic circuit is assumed to be unsaturated.

If we want to regulate the output reactive power of the motor, and the output power P_2 of the motor is maintained constant; then, the electromagnetic power P_e and the input power P_1 of the motor will remain unchanged, namely

$$P_e = m\frac{E_0 U}{X_s}\sin\delta = \text{constant}$$

$$P_1 = mUI\cos\varphi = \text{constant} \tag{5-45}$$

Because m, U and X_s are all constant, Eq. 5-45 can be written as

$$E_0\sin\delta = \text{constant}$$

$$I\cos\varphi = \text{constant} \tag{5-46}$$

Fig. 5-41 shows the phasor diagram of a motor excited by some different field currents.

In Fig. 5-41, when the field current is I_f, the excitation emf is $\dot{E}_0$, and the armature current is $\dot{I}$, the power factor of the generator $\cos\varphi = 1$, i. e., the armature current $\dot{I}$ is in phase with the terminal voltage $\dot{U}$. At this time, the motor is operating at a normal-excited condition; it only consumes active power from the infinite bus.

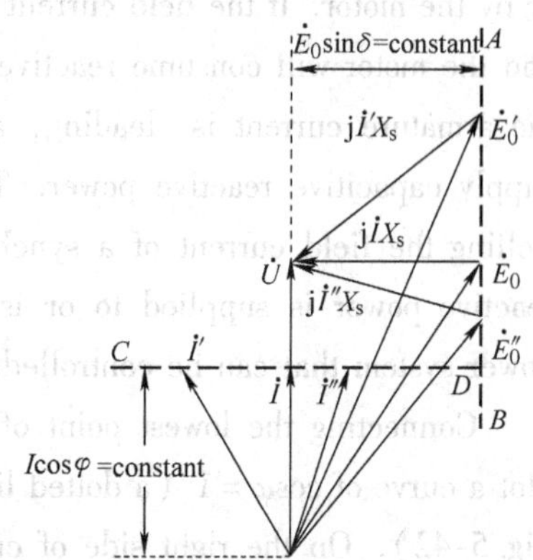

Fig. 5-41 Reactive power regulation of a synchronous motor in parallel with a infinite bus

If we increase the field current as I_f' ($I_f' > I_f$), then the excitation emf will increase to $\dot{E}_0'$, and the armature current is $\dot{I}'$. $\dot{I}'$ leads $\dot{U}$, the power factor of the motor $\cos\varphi < 1$ (leading). At this time, the motor is operating at an over-excited condition. Consequently, it consumes not only the active power, and also the capacitive reactive power from the infinite bus.

Let us now decrease the field current as I_f'' ($I_f'' < I_f$), the excitation emf will reduce to $\dot{E}_0''$, and the armature current is $\dot{I}''$, $\dot{I}''$ lags $\dot{U}$, the power factor of the motor $\cos\varphi < 1$ (lagging). At this time, the motor is operating at an under-excited condition. Thus, it consumes not only the active power, and also the inductive reactive power from the infinite bus.

Note that an increase in the field current increases the magnitude of $\dot{E}_0$, but does not affect the active power that is supplied by the motor. When the field current is increased, $\dot{E}_0$ must increase as well, but it can only do so by sliding out along the line of the constant power. This effect is shown in Fig. 5-41.

Notice that as the value of $\dot{E}_0$ increases, the magnitude of the armature current $\dot{I}$ first decreases and then increases again. At low $\dot{E}_0$, the armature current is lagging, and the motor is an inductive load. It is acting like an inductor-resistor combination, consuming the reactive power. As the field current is increased, the armature current eventually lines up with $\dot{U}$, and the motor looks purely resistive. As the field current is increased further, the armature current becomes leading, and the motor becomes a capacitive load. It is now acting like a capacitor-resistor combination, consuming the negative reactive power or, alternatively, supplying a reactive power to the system.

In Fig. 5-41, when the motor is operating at the normal-excited condition, its armature current $\dot{I}$ is at a minimum value. If we increase or decrease the field current, the magnitude of the armature current will become larger. A plot of I versus I_f for a synchronous motor is shown in Fig. 5-42. Such a plot is called *synchronous motor V-curve*, because its shape looks like the letter "V".

There are several V-curves drawn corresponding to some different active power levels. For each curve, the minimum armature current I occurs at the unity power factor, only when the active power is supplied to the motor. At any other point on the curve, some reactive power is being supplied to or by the motor. If the field current is less than its minimum value, the armature current is lagging, and the motor will consume reactive power. If the field current is greater than its minimum value, the armature current is leading, and the motor will supply capacitive reactive power. Therefore, by controlling the field current of a synchronous motor, the reactive power is supplied to or is consumed by the power system that can be controlled.

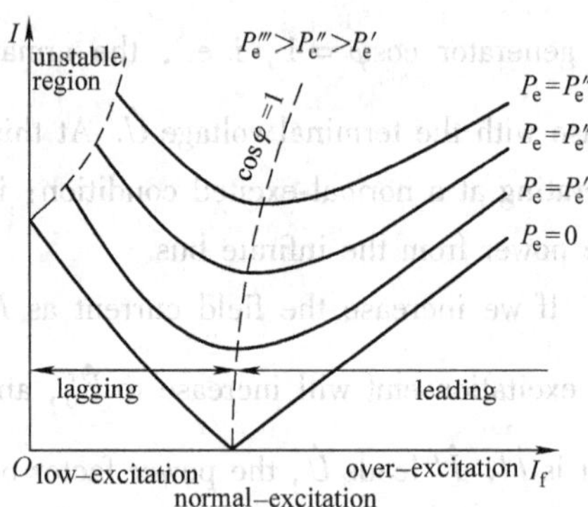

Fig. 5-42 V-curves of a synchronous motor

Connecting the lowest point of each V-curve will plot a curve of $\cos\varphi = 1$ (a dotted line that is shown in Fig. 5-42). On the right side of curve $\cos\varphi = 1$, the motor is operating at an over-excited condition and the power factor is leading, and the motor consumes a capacitive reactive power from the infinite bus, then the power factor of the power system can be improved; on the left side of curve $\cos\varphi = 1$, the motor is operating at under-excite condition. In this condition, the power factor is lagging; the motor consumes a reactive power from the power system. If the field current is very small, and the power angle $\delta > 90°$, the motor can be out of synchronism.

5.7.4 Starting synchronous motors

When the stator of the synchronous motor is connected to the 3-phase line, the stator magnetic field is rotating at the synchronous speed; the rotor of the motor is stationary, and therefore the rotor magnetic field is stationary. During an electrical cycle, the induced torque on the shaft of the rotor is first counter-clockwise, and then becomes clockwise, and the average torque over the complete cycle is zero. The motors, therefore, vibrates heavily within each electrical cycle and finally overheats. Thus, the synchronous motor cannot be started by itself.

Three basic approaches can be used to safely start a synchronous motor:

(1) Using an external prime mover

Use an external prime mover to drag the synchronous motor up to a synchronous speed, then go through the paralleling procedure, and bring the machine on the line as a generator. At last, turning off or disconnecting the prime mover will make the synchronous machine become a motor.

(2) Reducing electrical frequency

If the stator magnetic fields in a synchronous motor rotate at a speed that is low enough, the rotor will have no problem to accelerate and to lock in with the stator magnetic field. The speed of the stator magnetic fields can be increased to an operating speed by gradually increasing f up to its normal 50Hz value.

(3) Using damper windings or amortisseur windings

Damper windings are some special bars laid into the notches carved in the face of a synchronous motor's rotor and then shorted on each end by using a large shorting ring. A pole face with a set of damper windings is shown in Fig. 5-6.

If a machine has some amortisseur windings, it can be started by the following procedure:

1) Disconnect the field windings from their DC power source and short them out.

2) Apply a three-phase voltage to the stator of the motor, and let the rotor accelerate up to a near-synchronous speed. The motor should have no load on its shaft, so that its speed can approach synchronous speed n_s as closely as possible.

3) Then, connect the DC field circuit to its power source. After this is done, the motor will be locked into step at a synchronous speed, and the loads may then be added to its shaft.

Summary

Synchronous machines are usually used as generators. Turbine generators and hydroelectric generators are quite different in construction; turbine generatorsare constructed with non-salient pole rotors, which are necessary for high-speed operation driven by steam turbines or gas turbines; and hydroelectric generators are constructed with salient-pole rotors, which are crucial elements for low-speed operation driven by hydraulic turbines.

When a synchronous generator is operating in no-load condition, the fundamental wave of the main pole mmf produces a magnetic field (i. e., the no-load air-gap magnetic field). The magnetic field, which is rotating at synchronous speed, cuts the symmetrical three-phase stator windings, thus produces a set of symmetrical three-phase induced emf. The set of induced emf is called excitation emfs. The rms value of E_0 is determined by the excitation current I_f, the relationship curve between E_0 and I_f [$E_0 = f(I_f)$] is called open-circuit characteristic of synchronous machines.

When a synchronous generator is operating in a balanced load condition and there is a symmetrical three-phase current flowing into the symmetrical three-phase armature windings, an armature mmf wave rotating at synchronous speed will be produced. The influence of the fundamental wave of the armature mmf to the main pole magnetic field is called armature reaction.

The nature of armature reaction (increasing magnetic armature reaction, demagnetization armature reaction or quadrature-axis armature reaction) depends on the internal power factor angle ψ_0 (i. e., the time phase difference between the excitation emf and the armature current).

When the magnetic circuit is unsaturated, the superposition principle can be applied. The main pole mmf is produced by the main magnetic flux and the mmf produces the excitation emf in the armature windings. Similarly, the armature mmf is produced by the armature reaction flux and the mmf produces the armature reaction emf in the armature windings.

For non-salient pole machines, X_a is called armature reaction reactance; X_σ is called armature leakage reactance. X_s is called synchronous reactance, and $X_s = X_a + X_\sigma$. For salient pole machines, two-reaction theory is used to express armature mmf $\boldsymbol{F}_a$ by using d-axis component $\boldsymbol{F}_{ad}$ and q-axis component $\boldsymbol{F}_{aq}$. The corresponding armature reaction emf is $\dot{E}_{ad} = -\mathrm{j}\dot{I}_d X_{ad}$ and $\dot{E}_{aq} = -\mathrm{j}\dot{I}_q X_{aq}$. X_{ad} is called d-axis armature reaction reactance; X_{aq} is called q-axis armature reaction reactance. X_d is called the d-axis synchronous reactance, and $X_d = X_{ad} + X_\sigma$; X_q is called q-axis synchronous reactance, and $X_q = X_{aq} + X_\sigma$.

Using no-load characteristics and short-circuit characteristics, synchronous reactance X_d of synchronous machines can be calculated.

Synchronous generators are generally running in parallel with an infinite bus. Three conditions must be met for the synchronous generators. They are same phase sequence, same frequency and same terminal voltage.

Power-angle characteristic of non-salient pole synchronous generators is expressed as $P_e = m\dfrac{E_0 U}{X_s}\sin\delta$. The power-angle characteristic of salient pole synchronous generators is expressed as $P_e = m\dfrac{E_0 U}{X_d}\sin\delta + m\dfrac{U^2}{2}\left(\dfrac{1}{X_q} - \dfrac{1}{X_d}\right)\sin 2\delta$.

Power angle δ is the angle between excitation emf $\dot{E}_0$ and armature terminal voltage $\dot{U}$, it is also the space angle between the main magnetic field $\boldsymbol{B}_0$ and the air-gap magnetic field $\boldsymbol{B}$. When the main pole magnetic field leads the air gap magnetic field, δ is positive and the electromagnetic power is also positive; i. e. synchronous machine is operating at generator state. When power angle δ is zero, synchronous machines work as compensators. When power angle δ is negative, synchronous machines work as motors.

The maximum electromagnetic power of a non-salient pole machine is $P_{emax} = m\dfrac{E_0 U}{X_S}$. Notice that if the output from a prime mover increases, the output active power of the generator will increase. Also, the power angle δ will increase, and the output reactive power of the generator will also be changed. When the output reactive power Q of a generator remains unchanged, and the output of a prime mover increases to adjust the active power of the generator, the excitation emf E_0 and power angle δ will be changed simultaneously. When the field current of a synchronous generator is changed, then the output reactive power of the generator can be regulated. A plot of I versus I_f for a synchronous generator is called a *synchronous generator V-curve*.

Synchronous machines can also be run as motors. The speed of rotation is tied to the frequency of the source. If the frequency is fixed, the motor speed stays constant irrespective of the load or voltage of the 3-phase line. Moreover, synchronous motors can operate as synchronous compensator machines (or phase modifiers) to regulate the power factor of the power system.

Questions and problems

1. What is the relationship between frequency, number of pole-pairs and synchronous speed? What is the number of poles of a turbine generator with $f = 50\text{Hz}$, $n = 3000\text{r/min}$? What is the speed of a hydrogenerator with $f = 50\text{Hz}$, $2p = 100$?

2. The information of a three-phase synchronous generator is shown as follows: rated capacity is $S_N = 20\text{kV}\cdot\text{A}$, rated voltage is $U_N = 400\text{V}$ and rated power factor is $\cos\varphi_N = 0.8$ (lag). Find the rated current I_N, rated active power output P_N and reactive power output Q_N of the generator.

3. Sketch the space-time vector diagram of a non-salient pole three-phase synchronous generator operating at three-phase symmetrical lagging load, or three-phase symmetrical leading load. The armature winding resistance may be ignored, the time vector $\dot{U}$, $\dot{I}$, $\text{j}\dot{I}X_s$, $\dot{E}$ and space vector $\boldsymbol{F}_{f1}$, $\boldsymbol{F}_a$, $\boldsymbol{F}$ must be shown on the diagram. Declare the role of the armature reaction mmf $\boldsymbol{F}_a$ in both cases, and compare the magnitude of the main pole mmf $\boldsymbol{F}_{f1}$ and the synthetic magnetic force $\boldsymbol{F}$ in both cases.

4. Explain just how the synchronous impedance and armature resistance can be determined in a synchronous generator.

5. Determine the short-circuit characteristic of a synchronous generator if the motor speed is reduced from rated speed n_N to $0.5n_N$. Does the reduction in motor speed affect the result?

6. Sketch the phasor diagrams and magnetic field relationships for a synchronous generator operating at unity power factor, lagging power factor, leading power factor.

7. Why does an alternator's voltage drop sharply when it is loaded down with a lagging load?

8. Why does an alternator's voltage rise when it is loaded down with a leading load?

9. Explain, using phasor diagrams, what happens to a synchronous generator as its field current is varied. Derive a synchronous generator V-curve from the phasor diagram.

10. Why must a 60Hz generator be derated if it is to be operated at 50Hz? How much derating must be done?

11. Would you expect a 400Hz generator to be larger or smaller than a 60Hz generator of the same power and voltage rating? Why?

12. At a location factory, it is necessary to supply 300kW of 60Hz power. The only power sources available operate at 50Hz. It is decided to generate the power by means of motor-generator set consisting of a synchronous motor driving a synchronous generator. How many poles should each of the two machines have in order to convert 50Hz power to 60Hz power?

13. What are the conditions required to a synchronous generators operation in parallel with an infinite bus? If either one of the parallel conditions is not met, what will be the consequence? What means should be taken in order to meet the parallel conditions?

14. A salient pole synchronous generator is operating in parallel with an infinite bus. If the generator output active power and reactive power are adjusted to keep the output power to be constant, will the power angle δ and field current I_f be changed? What is the changing track of $\dot{I}$ and $\dot{E}_0$? (the armature resistance and the magnetic saturation are not taking into account)?

15. In a DC motor, $E_0 > U$ or $E_0 < U$ is the determining factor of the usage of the motor (to be a motor or a generator). Is this statement an accurate description for synchronous motors? Why? What are the factors to decide the state of synchronous machines (generator state or motor state) running on?

16. A synchronous machine has a synchronous reactance of 2.0Ω per phase and an armature resistance of 0.4Ω per phase. If $\dot{E}_A = 460\angle -8°V$ and $\dot{U}_\phi = 480\angle 0°V$, is this machine a motor or a generator? How much power P is this machine consuming from or supplying to the electrical system? How much reactive power Q is this machine consuming from or supplying to the electrical system?

17. When a synchronous generator is running on "over-excited condition", what is the nature of the current and power conveying to the grid? When it is running on "under-excited condition", what is the nature of the current and power conveying to the grid? What is the situation if it is a synchronous motor?

18. Why can not a synchronous motor start by itself? What techniques are available to start a synchronous motor?

19. A 25MV·A, three-phase, 13.8kV, two-pole, 60Hz, Y-connected synchronous generator was tested by the open-circuit test, and its air-gap voltage was extrapolated with the following results:

Open-circuit test					
Field current/ A	320	365	380	475	570
Line voltage/ kV	13.0	13.8	14.1	15.2	16.0
Extrapolated air-gap voltage/ kV	15.4	17.5	18.3	22.8	27.4

The short-circuit test was then performed with the following results:

Short-circuit test					
Field current/ A	320	365	380	475	570
Armature current/ A	1040	1190	1240	1550	1885

The armature resistance is 0.24Ω per phase.

1) Find the unsaturated synchronous reactance of this generator in ohms per phase and per unit.

2) Find the approximate saturated synchronous reactance X_s at a field current of 380A. Express the answer both in ohms per phase and per unit.

3) Find the approximate saturated synchronous reactance at a field current of 475A. Express the answer both in ohms per phase and per unit.

4) Find the short-circuit ratio for this generator.

20. A 1500kW three-phase hydrogenerator is operating at rated voltage $U_N = 6300V$ (Y-connection), its rated power factor $\cos\varphi_N = 0.8$ (lag), $X_d = 21.3\Omega$, $X_q = 13.7\Omega$. Ignoring the armature resistance and excluding the magnetic saturation, calculate the following:

1) The per-unit value of X_d and X_q.

2) The rated power angle δ_N and excitation emf E_0.

21. A 70000kV·A, 60000kW, 13.8kV, Y-connected three-phase turbine generator has a d-axis synchronous reactance per unit of $X_d^* = 1$ and q-axis synchronous reactance per unit of $X_q^* = 0.7$. Calculate the excitation emf E_0^*, power angle δ and voltage regulation rate Δu of this generator at rated conditions. (the armature resistance and the magnetic saturation are not taking into account).

22. A non-salient pole synchronous generator is operating in parallel with an infinite bus at rated speed. Its ratings are: $S_N = 31250\text{kV}\cdot\text{A}$, $U_N = 10.5\text{kV}$ (Y-connection), $\cos\varphi_N = 0.8$ (lagging). Its per phase synchronous reactance X_s is 7.0Ω. (the armature resistance and the magnetic saturation are not taking into account). Calculate the following:

1) The power angle δ, the electromagnetic power P_e, and the overload capacity k_p.

2) Keep the rated excitation current constant, when the output active power is reduced by half, calculate the power angle δ, the electromagnetic power P_e and the power factor angle φ. How will the output reactive power change?

3) If the excitation current increases 10%, calculate the power angle δ, the electromagnetic power P_e and the power factor angle φ. How will the output reactive power be changed?

23. A 13.8kV, 10MV·A, 0.8PF-lagging, 60Hz, two-pole, Y-connected steam-turbine generator has a synchronous reactance of 12 Ω per phase and an armature resistance of 1.5 Ω per phase. This generator is operating in parallel with a large power system (infinite bus):

1) What is the magnitude of E_A at rated conditions?

2) What is the torque angle of the generator at rated conditions?

3) If the field current is constant, what is the maximum power possible out of this generator? How much reserve power or torque does this generator have at full load?

4) At the absolute maximum power possible, how much reactive power will this generator be supplying or consuming? Sketch the corresponding phasor diagram. (Assume I_F is still unchanged.)

24. A 100MV·A, 12.5kV, 0.85PF-lagging, 50Hz, two-pole, Y-connected synchronous generator has a per-unit synchronous reactance of 1.1 and a per-unit armature resistance of 0.012.

1) What are its synchronous reactance and armature resistance in ohms?

2) What is the magnitude of the internal generated voltage E_A at the rated conditions? What is its torque angle δ at these conditions?

3) Ignoring losses in this generator, what torque must be applied to its shaft by the prime mover at full load?

25. A three-phase Y-connected synchronous generator is rated 120MV·A, 13.2kV, 0.8PF lagging, and 60Hz. Its synchronous reactance is 0.9Ω, and its resistance may be ignored.

1) What is its voltage regulation?

2) What would the voltage and apparent power rating of this generator be if it were operated at 50Hz with the same armature and field currents as it had at 60Hz?

3) What would the voltage regulation of the generator be at 50Hz?

26. A non-salient pole synchronous motor is running at rated condition, the power angle δ is 30°, and the excitation current remains constant, Declare the change of the power angle under the following situations (the armature resistance and the magnetic saturation are not taking into account).

1) Frequency of the power grid drops 5%, load torque remains unchanged.
2) Frequency of the power grid drops 5%, load power remains unchanged.
3) Both frequency and voltage of the power grid drop 5%, load torque remains unchanged.

27. A three-phase non-salient pole synchronous motor is operated with rated voltage $U_N = 380V$ (Y-connection), its synchronous reactance X_s is 5Ω, the armature resistance may be ignored. When the power angle is $\delta = 30°$ and the electromagnetic power is $P_e = 16kW$, calculate the following:

1) Excitation emf E_0 per phase.
2) If keep the exciting current unchanged, calculate the maximum electromagnetic power.

28. A three-phase synchronous motor is operated with rated power $P_N = 2000kW$, the rated voltage is $U_N = 3000V$ (Y-connection), the rated power factor is $\cos\varphi_N = 0.85$ (leading), the rated efficiency is $\eta_N = 95\%$, the pole pair number is $p = 3$, per phase stator resistance is $R_a = 0.1\Omega$, calculate the following

1) Stator input electrical power P_1 at rated operation.
2) Rated current I_N.
3) Rated electromagnetic power P_e.
4) Rated electromagnetic torque T_e.

29. A synchronous motor runs at rated voltage and absorbs rated current from the grid. Its power factor is 0.8 (leading), its d-axis and q-axis synchronous reactance per unit value are $X_d^* = 0.8$, $X_q^* = 0.5$, find the excitation emf E_0^* and the power angle δ, indicating the state of this motor (i. e. whether it is operated on the "over-excited condition" or "under-excited condition". Do not take into account of the magnetic saturation and armature resistance).

Electric Machinery

中文部分

绪　论

0.1　电机在国民经济中的作用

电能是现代能源中应用最广的二次能源，它适宜于大量生产、集中管理、远距离传输、灵活分配及自动控制，是现代最常用的一种能源。而电机是与电能的生产、变换、分配、使用和控制有关的能量转换设备，它在国民经济的各个方面都起着极其重要的作用。下面分三个方面介绍其主要作用。

1. 电机是电能的生产、传输和分配中的主要设备

电力工业的发展是以电机制造工业的发展为基础的。在发电厂中，发电机由汽轮机、水轮机、柴油机或其他动力机械带动电机的转子旋转，将一次能源（燃料燃烧的热能、水的位能、原子核裂变的原子能、风能、太阳能或潮汐能等）转化而来的机械能进一步转化为电能。显然，没有发电机就没有电能的大规模生产，所以发电机是电力系统中最关键的电气设备。发电厂一般地处偏僻区域，发电机发出的电压一般为10.5～20kV，为了把大量电能经济、远距离地输送出去，应当采用高压输电，一般输电电压为110kV、220kV、330kV、500kV或更高，此时需要采用升压变压器将发电机输出的电压升高后再进行传输。当电能到达用电地区后，为了安全用电，还需要各种等级的降压变压器将电压降低。一般电力系统及电网所需要变压器的总容量，可达到发电设备总容量的7～8倍。因此，在电力工业中，发电机和变压器是发电厂和变电站的主要电气设备。

2. 电机是各种生产机械和装备的动力设备

在工农业、交通运输及日常生活中，各种电动机被广泛用来拖动生产机械和装备。例如在机械工业中，各种工作母机都需要一台或多台不同容量的电动机来拖动和控制，磨床用电动机的转速可达每分钟数万转，甚至更高，有些机床需要多速可控电动机；冶金工业中各种高炉、转炉和平炉都需要多台电动机来驱动，大型轧钢机要用5000kW或更大功率的直流电动机；农业中的电力排灌、农副产品加工，各类企业中的鼓风、起吊、运输传送，采矿场的矿石采掘和传送，交通运输中的城市电车、铁道电力机车的牵引，以及造纸、医疗器械、家用电器等都需要各种交、直流电动机来驱动。所以，电机是各种生产机械和装备的动力设备。

3. 电机是自动控制系统中的重要部件

随着科学技术的发展，工农业和国防设施的自动化程度越来越高。各种各样的控制电机被用做执行、检测、放大和解算元件。这类电机一般功率较小、品种繁多、用途各异、精度要求较高。例如，火炮和雷达的自动定位、人造卫星发射和飞行的控制、舰船方向舵的自动控制、机床加工的自动控制和显示、电梯的自动选层与显示，以及计算机、自动记录仪表等的运行控制、监测或记录显示等，都离不开各种各样的控制电机。所以众多的各种容量的精密控制电机是整个自动控制系统中的重要部件。

总之，在电力工业中，产生电能的发电机和对电能进行变换、传输与分配的变压器是电站和变电所的主要电气设备。在机械、冶金、纺织、石油、煤炭、化工、交通运输业及其他工农业中，需要大量的电动机作为各种生产机械的动力设备。在航天、航空和国防科技等领域的自动控制技术中，需要各种各样的控制电机作为检测、随动、执行和解算元件。因此，电机在国民经济的各个方面都起着极其重要的作用。

0.2 电机的主要类别

电机是一种机电能量转换或信号转换的电磁机械装置。

按功能来分，可将电机分为不同的类型：

1）发电机，把机械能转换为电能的装置。

2）电动机，把电能转换为机械能的装置。

3）变压器、移相器、变频机、变流机，将一种形式的电能转换为另一种形式的电能的装置，分别用于改变电能的电压、相位、频率和电流。

4）控制电机，在自动控制系统中作为产生、传递信号的元件或伺服元件。

根据应用电流的种类，旋转电机可分为直流电机和交流电机两类，交流电机又可分为感应电机（也称为异步电机）、同步电机和交流换向器电机三种。

0.3 电机中所用的材料

电机中所用的材料可分为以下 4 类：

1）导电材料，用于电机中的电路系统。为减小线路损耗，要求导电材料的电阻率小。常用纯铜及铝作为导电材料。

2）导磁材料，用于电机中的磁路系统。为了在一定励磁磁动势下产生较强的磁场并降低铁损，要求导磁材料有较高的磁导率和较低的铁损系数。交流磁路中常用硅钢片、直流磁路中常用钢板和铸钢作为导磁材料。

3）绝缘材料，作为带电体之间及带电体与铁心之间的电气隔离。要求材料的介电强度高且耐热强度好。电机中所用绝缘材料，按耐热能力可分为 A、E、B、F、H 5 个等级，其最高允许工作温度分别为 105℃、120℃、130℃、155℃、180℃。

4）结构材料，支撑和连接各个部件，使各部件构成整体。要求材料的机械强度好，加工方便、重量轻，常用铸钢、铸铁、钢板、铝合金及工程塑料。

0.4 磁路

0.4.1 磁场

在实现机电能量转换的过程中，电机必须借助于磁场的媒介作用。在工程分析计算时，常把电机内的磁场问题简化为磁路问题来处理。在电机分析中，常用到以下 4 条基本原理：

1）载流导体周围存在磁场。

2）线圈与交变的磁场交链时，线圈中会产生感应电动势（这是变压器的工作原理）。

3）磁场中的载流导线会受到电磁力作用（这是电动机的工作原理）。

4）磁场中的运动导体切割磁力线会产生感应电动势（这是发电机的工作原理）。

1. 磁场和磁路分析中的常用物理量

描述磁场的物理量主要有磁感应强度（或磁通密度）$\boldsymbol{B}$、磁场强度$\boldsymbol{H}$、磁通$\boldsymbol{\Phi}$、磁动势$\boldsymbol{F}$、磁阻R_m、磁导Λ_m、磁链ψ等。

（1）磁感应强度（或磁通密度）$\boldsymbol{B}$　载流导体周围存在着磁场，描述磁场强弱和方向的物理量是磁感应强度$\boldsymbol{B}$，$\boldsymbol{B}$是矢量。磁感应强度也称为磁通密度，单位为特斯拉（T）。

（2）磁场强度$\boldsymbol{H}$　表征磁场性质的另一个基本物理量是磁场强度$\boldsymbol{H}$，$\boldsymbol{H}$是矢量，其单位为安/米（A/m）。它与磁感应强度$\boldsymbol{B}$的关系为

$$\boldsymbol{H} = \frac{\boldsymbol{B}}{\mu} \tag{0-1}$$

式中，μ为介质的磁导率。

电机中所用的介质，主要是铁磁材料和非导磁材料。空气、铜、铝和绝缘材料等为非导磁材料，它们的磁导率可认为等于真空的磁导率μ_0，$\mu_0 = 4\pi \times 10^{-7}$H/m。

磁感应强度$\boldsymbol{B}$与磁场强度$\boldsymbol{H}$的关系通常也表示为$\boldsymbol{B} = \mu_0\mu_r \boldsymbol{H}$。这里：$\mu_r$是材料的相对磁导率，为材料的磁导率与真空磁导率的比值，即

$$\mu_r = \frac{\mu}{\mu_0} \tag{0-2}$$

铁磁材料的磁导率远大于真空的磁导率，如铸钢的相对磁导率μ_r约为1000，各种硅钢片的相对磁导率μ_r为6000～7000。

（3）磁通$\boldsymbol{\Phi}$　穿过某一截面（面积为A）的磁感应强度$\boldsymbol{B}$的通量称为磁通，用符号$\boldsymbol{\Phi}$表示，即

$$\boldsymbol{\Phi} = \int_A \boldsymbol{B} \cdot \mathrm{d}\boldsymbol{A} \tag{0-3}$$

在均匀磁场中，如果截面A与$\boldsymbol{B}$垂直，则磁通$\boldsymbol{\Phi}$和磁感应强度$\boldsymbol{B}$之间的数值关系为

$$\boldsymbol{\Phi} = \boldsymbol{BA} \text{ 或 } \boldsymbol{B} = \frac{\boldsymbol{\Phi}}{\boldsymbol{A}} \tag{0-4}$$

因此，磁感应强度$\boldsymbol{B}$即单位面积上的磁通，又称为磁通密度，简称为磁密。在国际单位制中，磁通$\boldsymbol{\Phi}$的单位为韦伯（Wb），磁通密度的单位为T，$1\text{T} = 1\text{Wb/m}^2$。

（4）磁动势F　在线圈中通以电流就会产生磁场，若线圈的匝数为N，单匝线圈的电流为I，则线圈所产生的磁动势F为

$$\boldsymbol{F} = NI \tag{0-5}$$

磁动势是产生磁通的“动力”，单位为安匝（A·t）。

（5）磁阻R_m　磁阻类似于电路中的电阻，表示磁路对磁通所起的阻碍作用。磁阻与磁路的尺寸和磁路所用材料的导磁率有关。在磁路中取一段由磁导率为μ的材料构成的均匀磁路，其横截面面积为A，长度为l，则该段磁路的磁阻R_m为

$$R_m = \frac{l}{\mu A} \tag{0-6}$$

磁阻R_m的单位为亨$^{-1}$（H^{-1}）。

空气的磁导率为常数，因此一定长度气隙的磁阻是常量；铁磁材料的$\boldsymbol{B}$-$\boldsymbol{H}$曲线是非线性的，磁导率不是常数，所以铁磁材料的磁阻不是常数，而随着$\boldsymbol{B}$的变化而变化。

(6) 磁导 Λ_m 磁阻的倒数称为磁导，用 Λ_m 表示，即

$$\Lambda_m = \frac{1}{R_m} = \frac{\mu A}{l} \tag{0-7}$$

磁导的单位为亨（H）。

(7) 磁链 ψ 线圈的匝数 N 与通过线圈的磁通 Φ 的乘积，称为磁链，用 ψ 表示，即

$$\psi = N\Phi \tag{0-8}$$

2. 磁场的产生

载流导体的周围会产生磁场。当存在多根载流导体时，在磁场的分析中常采用安培环路定律，即

$$\oint \boldsymbol{H} \mathrm{d}\boldsymbol{l} = I_{net} \tag{0-9}$$

式中，$\boldsymbol{H}$ 是由电流 I_{net}（单位为 A）产生的磁场强度矢量，其单位为安匝/米（A·t/m）；$\mathrm{d}\boldsymbol{l}$ 是沿积分路径的积分单元。

图 0-1[㊀]为一个由载流线圈产生的简单铁心磁场示意图。在图 0-1 中，铁心柱上所绕线圈的匝数为 N，线圈中的电流为 i（单位为 A）；铁心的长度为 l_c（单位为 m），铁心的截面积为 A（单位为 m^2）。

依据安培环路定律，N 匝载流线圈所产生磁场的幅值与流过 N 匝线圈的总电流成正比，并且由于铁心材料为导磁性能好的铁磁材料，因此，磁场的磁通主要从铁心中通过。在图 0-1 中，安培环路定律可表示为

$$Hl_c = Ni \tag{0-10}$$

式中，H 是磁场强度矢量 $\boldsymbol{H}$ 的幅值。式（0-10）可改写为

$$H = \frac{Ni}{l_c} \tag{0-11}$$

依据式（0-1），铁心中的磁感应强度为

$$B = \mu H = \frac{\mu Ni}{l_c} = \frac{\mu_0 \mu_r Ni}{l_c} \tag{0-12}$$

于是，铁心中的磁通为

$$\Phi = BA = \frac{\mu_0 \mu_r NiA}{l_c} \tag{0-13}$$

0.4.2 磁路的概念

磁通所通过的路径称为磁路。图 0-2 为变压器磁路与四极直流电机的磁路示意图。

在电机和变压器中，当线圈中有电流通过时，线圈内部和周围就会产生磁场。由于铁心的磁导率比空气的磁导率高很多，即铁心的导磁性能好于空气的导磁性能，所以绝大部分磁通将在铁心内通过，这部分磁通称为主磁通；另外还有一小部分磁通经载流线圈和周围的空气闭合，这部分磁通称为漏磁通。主磁通和漏磁通所经过的路径分别构成主磁路和漏磁路，如图 0-2 所示。

用以激励磁路中磁通的载流线圈称为励磁线圈（或励磁绕组），励磁线圈中的电流称为励磁电流。励磁电流为直流时的磁路称为直流磁路，直流电机的磁路属于这一类。若励磁电

㊀ 本书中文部分不另提供图，请参阅英文部分对应图号的图。

流为交流（常称交流励磁电流为激磁电流），磁路中的磁通将随时间交变，这种磁路称为交流磁路，交流铁心线圈、变压器和感应电机的磁路都属于这一类。

磁路与电路有对偶关系。图 0-3 为简单电路与磁路的对偶关系。在图 0-3a 中，电压源 U 驱动产生线路中的电流 I，电流 I 流经电阻 R，它们的关系可以用电路欧姆定律表示为 $U=IR$。

根据对偶关系，在图 0-3b 中，磁动势 F 与电路中的电动势对应，由于磁路中的磁动势 $F=Ni$ 产生铁心中的磁通 Φ，F 与磁通 Φ 的关系可表示为

$$F = \Phi R_{\mathrm{m}} \tag{0-14}$$

式中，R_{m}为磁路的磁阻。

磁动势 F 的极性决定磁通 Φ 的方向，磁动势 F 的极性与线圈电流方向之间符合右手定则。若右手四指所指方向为线圈中电流的方向，则大拇指所指方向为磁动势的正方向。

根据式（0-7），磁路的磁导为磁阻的倒数，则式（0-14）可改写为

$$\Phi = F\Lambda_{\mathrm{m}} \tag{0-15}$$

根据式（0-6），当已知图 0-1 中铁心磁路的结构尺寸时，可以计算出铁心磁路的磁阻。与电路中的电阻一样，磁路中的磁阻也遵循串联与并联等效的规则。

串联磁阻的等效磁阻可通过下式计算：

$$R_{\mathrm{meq}} = R_{\mathrm{m1}} + R_{\mathrm{m2}} + R_{\mathrm{m3}} + \cdots \tag{0-16}$$

并联磁阻的等效磁阻可通过下式计算：

$$\frac{1}{R_{\mathrm{meq}}} = \frac{1}{R_{\mathrm{m1}}} + \frac{1}{R_{\mathrm{m2}}} + \frac{1}{R_{\mathrm{m3}}} + \cdots \tag{0-17}$$

利用磁路分析方法可以简化磁场的计算，但将磁场简化为磁路，必须在相应的假设条件下进行，这就使得利用磁路计算的结果存在一定的误差。产生误差的原因有以下几种情况：

1）磁路计算中，假定磁通全部在铁心中通过，没有考虑漏磁通的作用和影响。

2）在磁阻计算时铁心的长度和横截面面积均采用各自的平均值，没有考虑铁心转角的影响。

3）铁磁材料的磁导率通常是随磁场强弱的变化而变化的，而在磁路计算中，假定铁磁材料的磁导率为常值。

4）如果铁心中存在气隙，如图 0-4 所示，磁通经过气隙时，将由气隙段向外扩散，这种现象称为边缘效应。边缘效应增大了气隙的有效面积，在磁路计算中应加以考虑。气隙越大，边缘效应相应也越强。

例 0-1 如图 0-5a 所示的铁心磁路，其中有 3 段铁心的宽度相同，为 15cm，第 4 段铁心的宽度为 10cm。全部铁心的厚度为 10cm，铁心的其他尺寸均如图中所示。铁心柱上所绕线圈的匝数为 200 匝，铁心材料的相对磁导率为 2500，计算线圈中通入 1A 电流时，在铁心中产生磁通的幅值。

解：

由于 4 段铁心中有 3 段铁心的横截面面积相等，因此将铁心磁路分为两部分，各部分磁路的磁阻如图 0-5b 所示。

第一部分磁路中铁心的长度为 45cm，其横截面面积为 $10\mathrm{cm}\times 10\mathrm{cm}=100\mathrm{cm}^2$。因此，该段磁路的磁阻为

$$R_{m1}=\frac{l_1}{\mu A_1}=\frac{l_1}{\mu_r\mu_0 A_1}$$

$$=\frac{0.45\text{m}}{2500\times 4\pi\times 10^{-7}\times 0.01\text{m}^2}$$

$$=14\ 300\text{H}^{-1} \tag{0-18}$$

第二部分磁路中铁心的长度为 130cm，其横截面面积为 $15\text{cm}\times 10\text{cm}=150\text{cm}^2$。因此，该段磁路的磁阻为

$$R_{m2}=\frac{l_2}{\mu A_2}=\frac{l_2}{\mu_r\mu_0 A_2}$$

$$=\frac{1.3\text{m}}{2500\times 4\pi\times 10^{-7}\times 0.015\text{m}^2}$$

$$=27\ 600\text{H}^{-1} \tag{0-19}$$

磁路的总磁阻为

$$R_{meq}=R_{m1}+R_{m2}$$

$$=14\ 300\text{H}^{-1}+27\ 600\text{H}^{-1}$$

$$=41\ 900\text{H}^{-1} \tag{0-20}$$

总磁动势为

$$F=Ni=200\text{匝}\times 1.0\text{A}=200\text{A}\cdot\text{t} \tag{0-21}$$

因此，铁心中的总磁通为

$$\Phi=\frac{F}{R_{meq}}=\frac{200\text{A}\cdot\text{t}}{41\ 900\text{H}^{-1}}$$

$$=0.0048\text{Wb} \tag{0-22}$$

例 0-2 图 0-6 为一铁心磁路示意图，其铁心长度为 40cm，其中有一段 0.05cm 的气隙。铁心的横截面积为 12cm^2，铁心材料的相对磁导率为 4000，铁心上所绕线圈的匝数为 400 匝，设气隙的边缘效应将气隙横截面积增大 5%。计算：

1）磁路的总磁阻（包括气隙部分）。

2）若要在气隙中产生 0.5T 的磁感应强度，线圈中的电流应该为多少？

解：

图 0-6a 的等效磁路如图 0-6b 所示。

1）铁心磁路的磁阻：

$$R_c=\frac{l_c}{\mu A_c}=\frac{l_c}{\mu_r\mu_0 A_c}$$

$$=\frac{0.4\text{m}}{4000\times 4\pi\times 10^{-7}\times 0.0012\text{m}^2}$$

$$=66\ 300\text{H}^{-1} \tag{0-23}$$

气隙的有效横截面面积为 $1.05\times 12\text{cm}^2=12.6\text{cm}^2$，因此，气隙段的磁阻为

$$R_a=\frac{l_a}{\mu_0 A_a}$$

$$=\frac{0.0005\text{m}}{4\pi\times 10^{-7}\times 0.00126\text{m}^2}$$

$$=316\ 000\text{H}^{-1} \tag{0-24}$$

总磁阻为

$$
\begin{aligned}
R_{eq} &= R_c + R_a \\
&= 66\ 300\mathrm{H}^{-1} + 316\ 000\mathrm{H}^{-1} \\
&= 382\ 300\mathrm{H}^{-1}
\end{aligned} \tag{0-25}
$$

可见，气隙部分的磁阻占全部磁路磁阻的绝大部分。

2）若在气隙中的磁感应强度为0.5T，则线圈中的电流应为

$$
\begin{aligned}
i &= \frac{BAR_{eq}}{N} \\
&= \frac{0.5\mathrm{T} \times 0.00126\mathrm{m}^2 \times 383\ 200\mathrm{H}^{-1}}{400\text{ 匝}} \\
&= 0.602\mathrm{A}
\end{aligned} \tag{0-26}
$$

0.4.3　铁磁材料的性质

为了在一定的励磁磁动势作用下能激励较强的磁场，电机和变压器的主磁路常采用磁导率较高的铁磁材料制成。下面介绍铁磁材料的磁化特性。

1. 铁磁材料的磁化

铁磁材料包括铁、镍、钴等以及它们的合金。铁磁材料在外磁场中呈现很强的磁性的现象称为铁磁材料的磁化。铁磁材料内部存在着许多很小的、被称为磁畴的天然磁化区。在铁磁材料未放入磁场之前，这些磁畴杂乱无章地排列着，其磁效应相互抵消，对外部不呈现磁性；当铁磁材料放入外磁场后，在外磁场的作用下，磁畴的轴线将趋于一致，形成一个附加磁场，叠加在外磁场上，使合成磁场大大增强。由于磁畴所产生的附加磁场比非铁磁材料在同一磁场强度下所激励的磁场强得多，所以铁磁材料的磁导率要比非铁磁材料的磁导率大很多。电机中常用的铁磁材料，其磁导率 μ 一般为真空磁导率 μ_0 的 2000 ~ 6000 倍。

磁化是铁磁材料的一种特性。

2. 磁化曲线和磁滞回线

铁磁材料的磁化特性可用磁化曲线和磁滞回线来表示。

（1）起始磁化曲线　在非铁磁材料中，磁感应强度 B 和磁场强度 H 之间呈线性关系，如图 0-7 中的虚线所示，其斜率就是 μ_0。铁磁材料的 B 和 H 之间呈非线性关系，把一块尚未磁化的铁磁材料进行磁化，当磁场强度 H 由零逐渐增大时，磁感应强度 B 将随之增大，此时的曲线 $B = f(H)$ 就称为起始磁化曲线，如图 0-7 中所示的 $B = f(H)$ 曲线。

起始磁化曲线基本可以分为 4 段：开始磁化时，外磁场强度 H 较弱，磁感应强度 B 增加较慢，如图 0-7 中的 Oa 段所示。随着外磁场的增强，材料内部大量磁畴开始转向，趋向于外磁场方向，此时 B 增加较快，如图 0-7 中 ab 段所示。若继续增大外磁场 H，由于大部分磁畴已经趋向于外磁场方向，可转向的磁畴越来越少，所以 B 增加得越来越慢，如图 0-7 中 bc 段所示，这种现象称为饱和。达到饱和后，磁化曲线基本成为与非铁磁材料的 $B = \mu_0 H$ 特性相平行的直线，如图 0-7 中 cd 段所示。磁化曲线开始拐弯的点（见图 0-7 中的 b 点）称为膝点。

由于铁磁材料的磁化曲线不是一条直线，所以 $\mu_{Fe} = B/H$ 也随 H 值的变化而变化，如图 0-7 中的曲线 $\mu_{Fe} = f(H)$ 所示。

设计电机和变压器时，为使主磁路中有较大的磁通而又不过分增大励磁磁动势，通常把铁心内的工作磁感应强度选择在膝点附近。

（2）磁滞回线　若将铁磁材料进行周期性磁化，B 和 H 之间的变化关系就会变成如图 0-8 中的曲线 *Oabcdefa* 所示。由图可见，当磁场强度 H 开始从零增加到 H_m 时，B 也相应从零增加到 B_m。若此时逐渐减小 H，则 B 将沿曲线 ab 下降。当 $H=0$ 时，B 值并不等于零，而等于 B_r。这种去掉外磁场之后，铁磁材料内仍然保留的磁感应强度 B_r，称为剩余磁感应强度，简称剩磁。要使 B 从 B_r 减小到零，必须加上相应的反向磁场，此反向磁场强度称为矫顽力，用 H_c 表示。B_r 和 H_c 是铁磁材料的两个重要参数。铁磁材料所具有的这种 B 的变化滞后于 H 变化的现象称为磁滞。呈现磁滞现象的 B-H 闭合回线称为磁滞回线，如图 0-8 中的曲线 *abcdefa* 所示。

磁滞现象是铁磁材料的另一个特性。

磁滞回线窄、剩磁 B_r 和矫顽力 H_c 都小的材料，称为软磁材料，如图 0-9a 所示。常用的软磁材料有铸铁、铸钢和硅钢片等。软磁材料的磁导率较高，故用于制造电机和变压器的铁心。

磁滞回线宽、剩磁 B_r 和矫顽力 H_c 都大的材料，称为硬磁材料，如图 0-9b 所示。由于剩磁 B_r 大，可制成永久磁铁，因此硬磁材料也称为永磁材料，如铝镍钴、铁氧体、稀土钴、钕铁硼等。

（3）基本磁化曲线　对同一铁磁材料，选择不同的磁场强度 H_m 进行反复磁化，可得到一系列大小不同的磁滞回线，如图 0-10 所示。将各磁滞回线的顶点连接起来，所得到的曲线称为基本磁化曲线或平均磁化曲线。基本磁化曲线不是起始磁化曲线，但差别不大。直流磁路计算时所用的磁化曲线就是基本磁化曲线。

0.4.4 铁心损耗

铁心损耗又称铁损，包括磁滞损耗和涡流损耗。

当铁磁材料置于交变磁场中时，材料被反复交变磁化，与此同时，磁畴间相互不停地摩擦并消耗能量，从而造成损耗，这种损耗称为磁滞损耗。分析表明，磁滞损耗 p_h 与磁场的交变频率 f、铁心的体积 V 和磁滞回线所包围的面积成正比。实验证明，磁滞回线所包围的面积与最大磁感应强度 B_m 的 n 次方成正比，故磁滞损耗 p_h 可写成

$$p_h = C_h f B_m^n V \tag{0-27}$$

式中，C_h 为磁滞损耗系数，其大小取决于材料的性质。

对一般的硅钢片，$n=1.6\sim2.3$。由于硅钢片磁滞回线的面积小，故电机和变压器的铁心常采用硅钢片叠成。

由于铁心是导电的，故根据电磁感应定律，当通过铁心的磁通随时间变化时，铁心中将产生感应电动势，并引起环流。这些环流在铁心内部围绕磁通呈旋涡状流动，故称为涡流，如图 0-11 所示。涡流在铁心中引起的损耗称为涡流损耗。分析表明，磁场交变频率 f 越高，磁感应强度 B 越大，铁心中的感应电动势就越大，涡流损耗也就越大；若铁心的电阻率越大，涡流所流过的路径越长，涡流损耗就越小。对于由硅钢片叠成的铁心，经推导可知，涡流损耗 p_e 为

$$p_e = C_e \Delta^2 f^2 B_m^2 V \tag{0-28}$$

式中，C_e为涡流损耗系数，其大小取决于材料的电阻率；Δ 为硅钢片的厚度。

为了减小涡流损耗，电机和变压器的铁心都用含硅量较高的薄硅钢片（0.35～0.5mm）叠成。

铁心中的磁滞损耗和涡流损耗之和称为铁心损耗（简称铁损），用p_{Fe}表示。当磁通的交变频率为f时，有

$$p_{Fe} = p_h + p_e = C_h f B_m^n V + C_e \Delta^2 f^2 B_m^2 V \tag{0-29}$$

对于一般的硅钢片，在正常的工作磁感应强度范围内（$1T < B_m < 1.8T$），式（0-29）可近似写成

$$p_{Fe} \approx C_{Fe} f^{1.3} B_m^2 G \tag{0-30}$$

式中，C_{Fe}为铁心的损耗系数；G 为铁心的重量。

式（0-30）表明，铁心的损耗与磁通交变频率f的1.3次方、最大磁感应强度B_m的二次方和铁心的重量 G 成正比。

0.5 相关的电磁定律

1. 电磁感应定律

随时间变化的磁场会产生感应电动势，此现象称为电磁感应。

若线圈的匝数为 N，所通过的磁通为 Φ，当磁通 Φ 随时间变化时，线圈内将产生感应电动势 e。e 的大小与 N 和磁通的变化率 $\mathrm{d}\Phi/\mathrm{d}t$ 成正比。e 的实际方向由楞次定则判定如下：感应电动势的方向为阻止磁通变化的方向。于是 e 的数学表达式为

$$e = -N\frac{\mathrm{d}\Phi}{\mathrm{d}t} \tag{0-31}$$

式（0-31）中的负号表示感应电动势的方向为阻碍磁通变化的方向。

如果磁路中的磁通由多个线圈的电流产生，且每个线圈中的磁通不相等，则式（0-31）可改写为

$$e = -\sum_{i=1}^{N} e_i = -\sum_{i=1}^{N}\frac{\mathrm{d}\Phi_i}{\mathrm{d}t} = -\frac{d}{\mathrm{d}t}\left(\sum_{i=1}^{N}\Phi_i\right) = -\frac{\mathrm{d}\psi}{\mathrm{d}t} \tag{0-32}$$

式中，ψ 是线圈的磁链，且

$$\psi = \sum_{i=1}^{N}\Phi_i \tag{0-33}$$

另外，长度为 l 的直导线在均匀磁场中运动时，若导线切割磁力线的速度为 v，导线所在处的磁感应强度为 B，当导线、磁感应强度 B、导线的运动速度 v 三者互相垂直，则导线中感应电动势为

$$e = Blv \tag{0-34}$$

感应电动势 e 的方向用右手定则来确定，即把右手手掌伸开，大拇指与其他 4 指成 90°，如图 0-12 所示，让磁力线指向手心，大拇指指向导线运动方向，则 4 指所指方向就是导线中感应电动势 e 的方向。

2. 电磁力定律

载流导体在磁场中要受到力的作用，该力称为电磁力。在均匀磁场中，若载流导体与磁

感应强度 B 的方向垂直，导线长度为 l，流过的电流为 i，则载流导体所受到的电磁力 f 为

$$f = Bli \tag{0-35}$$

电磁力 f 的方向可用左手定则来确定，即把左手伸开，大拇指与其他 4 指成 90°，如图 0-13 所示，让磁力线指向手心，4 指的指向为导体中电流的方向，则大拇指的指向就是导线所受电磁力 f 的方向。

第1章 直流电机

直流电机是一种能够实现机电能量转换的装置。将直流电能转换成机械能的旋转电机称为直流电动机，而将机械能转换为直流电能的旋转电机则称为直流发电机。

直流电动机具有过载能力强、起动转矩大、制动转矩大、调速范围广、调速的平滑性好、调速方式易于控制、调速时损耗小等许多优点。因此，直流电动机被广泛地应用在电力机车、无轨电车、轧钢机、机床和各种设备中。但随着电力电子技术的发展，各种大功率电力电子器件的涌现，以及直流电机与交流电机相比具有结构复杂、成本较高等缺点，使得直流电机有逐步被交流电机取代的趋势。尽管如此，分析和研究直流电机仍有一定的理论价值和实际意义。

本章首先介绍直流电机的工作原理和基本结构，以及电枢绕组和气隙磁场，然后导出感应电动势和电磁转矩公式，以及直流电机的基本方程，最后介绍直流电机的稳态运行性能和换向问题。

1.1 直流电机的基本结构、工作原理、励磁方式和额定值

1.1.1 直流电机的基本结构

直流电机由静止的定子和旋转的转子（也称为电枢）构成，定子和转子之间存在不均匀的气隙。图 1-1 为一台直流电机的主要部件。

定子的作用是产生主磁场和作为电机的机械支撑，它包括主磁极、换向极、机座、电刷装置等。转子（电枢）的作用是感应电动势并产生电磁转矩，以实现机电能量转换，它包括电枢铁心、电枢绕组、换向器等。图 1-2 是直流电机的径向剖面图。

1. 定子

主磁极也称为主极，其作用是产生气隙主磁场。在一般的大、中型直流电机中，主磁极铁心一般用厚度为 1~1.5mm 的钢板冲片叠压紧固而成。绕制好的励磁绕组套在主磁极铁心外面，整个主磁极用螺钉固定在机座上。各主磁极上励磁绕组的连接必须使其通过励磁电流时，相邻磁极的极性呈 N 极和 S 极交替排列。主磁极铁心的下部（称为极靴）比套绕组的部分（极身）要宽，以使励磁绕组牢固地套在主磁极铁心上。图 1-3 为主磁极的结构。

换向极也称为附加极，其作用是改善换向。它装在相邻两主磁极之间，由铁心和绕组构成，如图 1-4 所示。换向极铁心一般用整块钢或钢板叠片而成。换向极绕组与电枢绕组串联。

机座通常由铸钢铸成或由厚钢板焊成，是电机的机械支撑，用来固定主磁极、换向极和端盖；同时它又是电机磁路的一部分。机座上作为磁路的部分常称为磁轭。

电刷装置是将直流电压、直流电流引入或引出电枢绕组的装置，它由电刷、刷握和铜丝辫构成，如图 1-5 所示。电刷由石墨制成，放在刷握内，用弹簧紧压在换向器表面。一般电刷装置的组数与电机的主磁极极数相等。

2. 转子

电枢铁心是电机主磁路的一部分，铁心中嵌放着电枢绕组。为减小电机中的铁损，常将

电枢铁心用厚度为0.5mm的硅钢片叠压而成，冲片圆周外缘均匀地冲有许多齿和槽，槽内嵌放电枢绕组；冲片上一般还冲有许多圆孔，以形成改善散热的轴向通风孔。图1-6为带换向器的电枢铁心，图1-7为带风扇和换向器的斜槽电枢铁心，图1-8为电枢铁心冲片。

电枢绕组由许多按一定规律连接的线圈组成，用来感应电动势和通过电流，是直流电机的主电路。线圈一般用带绝缘的圆形或矩形截面导线绕制而成，嵌放在电枢槽中，线圈的一条有效边嵌放在某个槽的上层，另一条有效边则嵌放在另一槽的下层，如图1-9所示。槽内的线圈上、下层之间以及线圈与铁心之间均有绝缘，如图1-10所示。在槽口处用槽楔压紧绕组，端部用钢丝或无纬玻璃丝带扎紧，以防止绕组被离心力甩出。

换向器由许多彼此绝缘的换向片构成，如图1-11所示。它和电刷一起能将外部通入的直流转换成绕组内的交流或反之。电枢绕组每个线圈的两端分别焊接到两个换向片上。

1.1.2 直流发电机的工作原理

直流电机分为直流发电机和直流电动机两大类。将机械能转换为直流电能的旋转电机称为直流发电机，而将直流电能转换成机械能的旋转电机称为直流电动机。

图1-12是直流发电机的工作原理示意图。N、S是固定的主磁极，线圈的两个出线端A、D分别与两个互相绝缘的换向铜片1、2连接。换向片与转轴一起转动。电刷x、y静止不动，电刷与换向片接触，将线圈与外电路接通。

用原动机拖动电枢线圈按逆时针方向以60r/min的速度连续旋转时，线圈边将切割磁场，产生感应电动势。根据右手定则，可确定电刷x为正极性，y为负极性。在外电路闭合的情况下，电刷x、y之间的感应电动势波形如图1-13所示，可见负载中流过恒定方向的直流电流。

在线圈连续旋转的过程中，可观察到：

1）由于电枢的连续旋转，线圈边中感应电动势的方向是交变的。

2）不管线圈如何旋转，同一磁极下线圈边中感应电动势的方向是固定不变的。

3）由于电刷和换向器的作用，使得电刷x、y的极性不变。

所以直流发电机线圈内的电动势和电流是交变的，而经电刷至外电路中的电流则是方向恒定的直流。

若直流发电机的电枢上只有一个线圈，电刷间的电动势是脉动电动势。实际上直流电机的电枢上装有许多隔开一定距离的线圈，这样，由电刷导出的电动势将是脉动较小的直流电动势。

直流发电机的工作原理是：当原动机拖动电枢以恒定方向旋转时，电枢绕组的线圈边切割主磁极磁场，感应出交变的电动势，经电刷和换向器的“整流”作用，在电刷间引出恒定方向的直流，将机械能转换成直流电能输出。

1.1.3 直流电动机的工作原理

图1-14是直流电动机的工作原理示意图。当直流电流从电刷x流入，经换向片1、线圈边、换向片2、电刷y流出时，电枢上的载流导体在主磁极磁场中将受到电磁力作用；根据左手定则，电磁力所形成的转矩使线圈沿逆时针方向转动。

在直流电流一直加在电刷x、y的过程中，可观察到：

1）经电刷和换向器的作用，在不同位置时，线圈边中电流的方向是交变的。

2）不管线圈运动到哪个位置，同一主磁极下线圈边中电流方向固定不变，因而同一主磁极下线圈边所受电磁力的方向亦不变，于是线圈将受到恒定方向转矩的作用，使电枢能连续旋转。

实际上，直流电动机的电枢上有许多线圈，这些线圈产生的转矩合成为总转矩，拖动负载转动。

直流电动机的工作原理是：在两个电刷端加上直流电压，经电刷和换向器作用将外电路中的直流电能引入电枢线圈中，并确保同一主磁极下线圈边中电流方向不变，使该主磁极下线圈边所受电磁力的方向亦不变，从而使电枢能连续旋转，将输入的直流电能转换成机械能输出，拖动生产机械。

可见，当改变直流电机外部的约束条件时，它既可作为电动机运行，也可作为发电机运行。在两电刷端外加直流电压，此时电机作为直流电动机运行，将电能变换为机械能；若用原动机拖动电机的电枢旋转，此时电机将作为直流发电机运行，将机械能变换为电能。这种同一台电机既能作电动机亦能作发电机运行的现象在电机理论中称为可逆原理。

1.1.4 直流电机的励磁方式

直流电机的励磁方式，是指励磁电流的供给方式。根据励磁电路与电枢电路的连接关系，可将直流电机分为他励和自励两类。直流电机的运行特性随励磁方式的不同有很大差别。

1. 他励式

他励直流电机励磁绕组的电流由外电源供给，与电枢回路没有电的联系，如图 1-15a 所示。

2. 自励式

1）并励直流电机的励磁绕组与电枢绕组并联，励磁绕组的端电压等于电枢绕组的端电压，如图 1-15b 所示。

2）串励直流电机的励磁绕组与电枢绕组串联，励磁绕组的电流与电枢绕组的电流相等，如图 1-15c 所示。

3）复励直流电机的主极上有两个励磁绕组，其中一个与电枢绕组并联，另一个和电枢绕组串联，如图 1-15d 所示。当串励绕组与并励绕组产生的磁动势方向相同时，称为积复励；当两者所产生磁动势的方向相反时，称为差复励。

1.1.5 直流电机的铭牌数据

每台直流电机的机座上都有一个铭牌，如图 1-16 所示。铭牌上标有一些额定数据，称为铭牌数据。铭牌数据是电机制造厂在设计时对电机的一些电量或机械量所规定的数据，通常有：额定电压 U_N（单位为 V）、额定电流 I_N（单位为 A），额定功率 P_N（单位为 kW 或 W）、额定转速 n_N（单位为 r/min）、额定效率 η_N、额定励磁电压 U_{fN}（单位为 V），额定励磁电流 I_f（单位为 A）等。

1. 额定功率 P_N

额定功率是指电机在额定条件下运行时的输出功率。对发电机，是指线路端点输出的电功率，有

$$P_N = U_N I_N \tag{1-1}$$

对电动机，是指轴上输出的机械功率，有

$$P_N = U_N I_N \eta_N \tag{1-2}$$

式中，η_N为电机的额定效率。

2. 额定电压 U_N

额定电压是指电机在额定条件下运行时，直流电机出线端的电压。

3. 额定电流 I_N

额定电流是指电机在额定条件下运行时，直流电机的线路电流。

4. 额定转速 n_N

额定转速是指电机在额定条件下运行时的转速。

此外，铭牌上有时还列有绝缘等级和励磁方式等。

电机在实际应用时，其电量和机械量一般不允许超过额定值，因为这会降低电机的使用寿命，甚至损坏电机；但若长期使电机在低负载下运行，一方面则表示电机没有得到充分利用，另一方面表示系统效率降低、不经济。所以应当根据实际负载情况来选用电机，使电机在多数时间内接近于额定状态运行。

例 1-1 一台直流发电机的额定数据为：额定功率 $P_N = 10\text{kW}$，额定电压 $U_N = 230\text{V}$，额定转速 $n_N = 2850\text{r/min}$，额定效率 $\eta_N = 0.85$。求电机的额定电流以及在额定负载时电机的输入功率各为多少？

解：

根据式（1-1），该直流发电机的额定电流为

$$I_N = P_N/U_N = 10 \times 1000/230\text{A} = 43.48\text{A}$$

电机在额定负载时的输入功率为

$$P_1 = P_N/\eta_N = 10 \times 1000/0.85\text{W} = 11\,765\text{W}$$

例 1-2 一台直流电动机的额定数据为：额定功率 $P_N = 17\text{kW}$，额定电压 $U_N = 220\text{V}$，额定转速 $n_N = 1500\text{r/min}$，额定效率 $\eta_N = 0.83$。求电机的额定电流以及在额定负载时电机的输入功率各为多少？

解：

根据式（1-2），电机在额定负载时的输入功率为

$$P_1 = P_N/\eta_N = 17 \times 1000/0.83\text{W} = 20\,482\text{W}$$

电机的额定电流为

$$I_N = P_1/U_N = 20\,482/220 = 93.1\text{A}$$。

1.2 直流电机的电枢绕组

电枢绕组是直流电机的主要电路，是直流电机实现机电能量转换的枢纽。设计电枢绕组时要求：

1）能通过规定的电流和产生足够大的电动势。

2）尽可能节省有色金属和绝缘材料。

3）保证换向良好。

直流电机的电枢绕组是由结构和形状相同的线圈按照一定的规律连接而成的闭合绕组。线圈有单匝、多匝之分，图 1-17 为一个两匝的叠绕和波绕元件。不论是单匝或多匝线圈，它的两个边分别安放在不同的槽中，如图 1-9 所示。在槽内能切割主磁场、感应电动势和产

生电磁转矩的线圈边，称为有效边；而处于槽外，仅起连接作用的部分称为端接。线圈的两个出线端分别称为首端和尾端。电枢绕组一般做成双层绕组，将线圈的一个有效边放在槽的上层，称为上层边（绘图时画成实线）；另一个有效边放在有一定距离的另一个槽的下层，称为下层边（画成虚线）。

描述电枢绕组的常用术语有：极数 $2p$，极距 τ，槽数 Q，元件数 S，换向片数 K，第一节距 y_1，第二节距 y_2，合成节距 y 和换向器节距 y_c。

1）极距 τ 指一个主磁极在电枢表面所跨的距离，用长度表示时为

$$\tau = \pi D_a/2p$$

式中，D_a为电枢外径，p 为电机的极对数。

用槽数表示时，极距 $\tau = Q/2p$，此时的 τ 可能不是整数。

2）电枢绕组的基本形式如图 1-18 所示。在图 1-18a 所示的叠绕组中，线圈 a 由导体 a'-a''组成，线圈 b 由导体 b'-b''组成。线圈 a 的尾端 a''与线圈 b 的首端 b'连接，它们处于同一极下，且线圈 b 的尾端 b''与线圈 c 的首端 c'相连接，以此类推，最后一个线圈的尾端与第一个线圈的首端相连接，形成一个闭合回路。请注意，同一个线圈的两个导体分别处于不同极性的主磁极下，以使得线圈的感应电势可以相加。

在图 1-18b 中，与叠绕组类似，波绕组也是一个闭合绕组。第一条支路的线圈通过换向片 1、10、19 相连接，第二条支路的线圈通过换向片 19、9、18 相连接，以此类推，最后一条支路的线圈通过换向片 12，2，11 相连接，且最后一个线圈的两个导体分别与换向片 11 和 1 相连接，最终形成闭合回路。

由于同一个线圈的首端和尾端将分别与两个不同的换向片相连接，因此，对于叠绕组和波绕组而言，绕组元件的数目 S 与换向片的数目 K 相等，即 $K=S$。

3）合成节距 y、第一节距 y_1、第二节距 y_2和换向器节距 y_c。

合成节距 y 指相串联的两个元件的对应元件边（如上元件边）在电枢表面所跨的距离，一般用槽数来表示。第一节距 y_1 指一个元件的两个元件边在电枢表面所跨的距离，用槽数表示时，y_1应当是一个整数。第二节距 y_2指相串联的两个相邻的元件中，前一个元件的下层边与后一个元件的上层边之间在电枢表面所跨的距离，一般也用槽数来表示。换向器节距 y_c 指同一元件的两出线端所接的两换向片之间的距离，一般用换向片数来表示，如图 1-18 所示。几种节距之间的关系如下：

$$y = y_1 - y_2 \quad \text{（叠绕组）} \tag{1-3}$$

$$y = y_1 + y_2 \quad \text{（波绕组）} \tag{1-4}$$

$$y = y_c \quad \text{（叠绕组和波绕组）} \tag{1-5}$$

无论是叠绕组还是波绕组，为了使元件的感应电动势最大，应使第一节距 y_1 接近于极距 τ，即

$$y_1 = Q/2p \pm \varepsilon \tag{1-6}$$

此处 ε 为使 y_1凑成整数的一个小数。若 $\varepsilon=0$，表示线圈为整距线圈；若 $\varepsilon \neq 0$，且取“+”时，表示线圈为长距线圈；若 $\varepsilon \neq 0$，且取“-”时，表示线圈为短距线圈。

对于单叠绕组，有

$$y = y_c = \pm 1 \text{（“+”表示右行，“-”表示左行）} \tag{1-7}$$

对于单波绕组，其换向器节距 y_c应满足下式：

$$y = y_c = (K \pm 1)/p \tag{1-8}$$

式中，K为换向片数。

在式（1-8）中，如取“－”，则绕行一周后，比出发时的换向片后退一片，这种绕组称为“左行”绕组；如取“＋”，则绕行一周后，比出发时的换向片前进一片，这种绕组称为“右行”绕组。一般都采用左行绕组。

1.2.1 单叠绕组

单叠绕组的连接特点是：同一个元件的两个出线端连接于相邻的两个换向片上，相邻元件依次串联，后一个元件的首端与前一个元件的尾端连在一起，并接到同一个换向片上，最后一个元件的尾端与第一个元件的首端连在一起，形成一个闭合回路。紧相串联的两个元件的端接部分紧“叠”在一起，所以形象地称为“叠绕组”。

下面以极数 $2p=4$，槽数、元件数和换向片数为 $Z=S=K=16$ 的直流电机为例，说明右行整距单叠绕组的连接规律和展开图，并说明单叠绕组支路的组成情况。具体步骤如下：

1. 节距计算

对于右行整距单叠绕组，$y=y_c=1$；根据式（1-6），第一节距 $y_1=16/4=4$；根据式（1-3），第二节距 $y_2=y_1-y=4-1=3$。

2. 绕组连接表

绕组连接的顺序可以用绕组连接表来表示，如图 1-19 所示。

从图 1-19 中可以看出，16 个元件依次和 16 个换向片连在一起，最后形成一个闭合回路。

3. 绕组展开图

假设把电枢从某一个齿槽中间沿轴向切开，并将其展开成平面的绕组连接图，就称为绕组展开图。该图可以根据绕组连接表中元件的连接顺序依次画出，如图 1-20 所示。图中上层边用实线段表示，下层边用虚线段表示。从第 1 号换向片出发，第 1 号换向片与第 1 号元件的上层边 1 相接，根据 $y_1=4$，第 1 号元件的下层边应在第 5 号槽中，且与第 2 号换向片相接；第 2 号换向片与第 2 号元件的上层边 2 相接，第 2 号元件的下层边应放在第 6 号槽中，并与第 3 号换向片相接；如此继续，最后第 16 号元件的下层边与第 1 号换向片相接，从而所有元件组成一个闭合回路。

4. 安放主磁极和电刷

取主磁极宽度为 0.75τ，交替地将 N、S 极性的 4 个主磁极均匀地放置在绕组展开图中，各主磁极在圆周上的位置也是均匀对称的。两个主磁极间的轴线称为几何中性线。

为使电刷端能获得最大电动势，被电刷短接的元件的电动势应最小。若线圈端部是对称的，电刷应放在主磁极中心线下，如图 1-20 所示。

5. 单叠绕组的瞬间电路图

在图 1-20 所示的瞬间，电刷 A_1、B_1，A_2、B_2 分别与换向片 1、2，5、6，9、10，13、14 相接触。根据元件上层边号码与所连接的换向片号码一致的原则，可画出此瞬间的绕组电路图，如图 1-21 所示。

6. 单叠绕组的并联支路数

从图 1-21 中可以清晰地看出，电枢绕组由 4 条支路并联组成，即同一主磁极下相邻的元件依次串联后构成一条支路。电机有几个主磁极，其电枢绕组就有几条支路。所以单叠绕

组的支路数等于电机的极数，即

$$2a = 2p \tag{1-9}$$

式中，p 为电机主磁极的极对数；a 为支路对数。

从图 1-21 中还可看出，单叠绕组的支路电动势由电刷引出，所以电刷数必定等于支路数，也即主磁极数。电枢端电压就是每条支路的电压。电枢电流 I_a 为每条支路电流 i_a 的总和，即 $I_a = 2ai_a$。

1.2.2 单波绕组

单波绕组的连接特点是：同一个元件的两个出线端所接的两个换向片相隔较远，其距离接近于一对极距，元件串联后形成波浪形，所以形象地称为“波绕组”，如图 1-19b 所示。

下面以极数 $2p = 4$，槽数 Z 与元件数 S 和换向片数 K 为 $Z = S = K = 15$ 的直流电机为例，绘制左行短距单波绕组展开图，并说明单波绕组的构成和支路组成情况。

1. 节距计算

由于是左行单波绕组，所以换向器节距 $y_c = (K-1)/p = (15-1)/2 = 7$；采用短距绕组，则第一节距 $y_1 = 15/4 - \varepsilon = 3$；第二节距 $y_2 = y_c - y_1 = 7 - 3 = 4$。

2. 绕组连接表

绕组连接的顺序可以用绕组连接表来表示，如图 1-22 所示。

从表中可以看出，全部元件经换向片连接在一起，最后形成一个闭合回路。

3. 绕组展开图

根据绕组连接表的顺序，画出相应的绕组展开图，如图 1-23a、b 所示。图中上层边用实线段表示，下层边用虚线段表示。由于第 1 号换向片与第 1 号元件的上层边 1 相接，根据 $y_1 = 3$，第 1 号元件的下层边应在第 4 号槽中，再由换向器节距 $y_c = 7$ 可知，第 1 号元件的下层边应与第 8 号换向片相接。接下去，第 8 号换向片与第 8 号元件的上层边 8 相接，由 $y_1 = 3$，第 8 号元件的下层边应放在第 11 号槽中，并由换向器节距 $y_c = 7$ 可知，应与第 15 号换向片相接。如此继续，最后第 9 号元件的下层边与第 1 号换向片相接，从而所有元件组成一个闭合回路。

4. 单波绕组的并联支路数

主磁极和电刷的安放原则与单叠绕组相同，如图 1-23b 所示。

在图 1-23b 所示的瞬间，根据元件上层边号码与所接换向片号码一致的原则，可画出此瞬间的绕组电路图，如图 1-24 所示。

从图 1-24 中可清晰地看出，电枢绕组由 2 条支路并联组成，即同一种极性下的所有元件串联起来，通过电刷构成一条支路。所以单波绕组的支路数恒等于 2，即

$$2a = 2 \tag{1-10}$$

另外，虽然单波绕组只有两条支路，需要一对电刷即可，但考虑到支路的对称性和电刷下电流密度不致过高的要求，一般仍取电刷数等于主磁极极数。

1.3 直流电机的磁场

直流电机中有放置于主磁极上的励磁绕组和放置于电枢铁心槽中的电枢绕组。当励磁绕组中有电流流过，而电枢绕组中电流为零时，直流电机为空载运行，此时电机内部的磁场由

励磁绕组的电流产生；当励磁绕组和电枢绕组中都有电流流过时，直流电机为负载运行，此时，电机内部的磁场由励磁绕组电流和电枢绕组电流共同产生。

1.3.1 直流电机空载时的气隙磁场

气隙磁场是直流电机机电能量转换的媒介。直流发电机空载是指电机的出线端没有电功率输出的运行状态，此时电枢电流等于零；直流电动机空载是指电动机的轴上不带负载，电动机没有机械功率输出时的运行状态，此时电枢电流很小，可忽略不计。所以，直流电机空载时的磁场是由主磁极励磁电流产生的。

图 1-25 和图 1-26 分别表示一台 2 极和一台 6 极的直流电机的磁路分布。从图中可见，每极磁通在定子轭处分成两部分，每部分磁通的磁路均经过定子轭部、极身、气隙、电枢齿、电枢铁心，然后再经过电枢齿、气隙和极身回到定子轭部，形成闭合回路，在该磁通路径中经过气隙、电枢齿和极身各两次。

图 1-27 所示为一台 4 极直流电机的空载磁场分布图。从图中可见，大部分磁力线从主磁极 N 极出发，经过气隙和电枢齿，进入电枢铁心，然后再经过电枢齿和气隙，回到相邻的主磁极 S 极，最后经过定子磁轭，回到出发的主磁极 N 极，形成闭合回路。这部分通过气隙，同时与电枢绕组和励磁绕组相交链的磁通，称为主磁通，用 Φ_0 表示。电枢旋转时，电枢绕组将切割主磁通 Φ_0 而产生感应电动势。还有一小部分不进入电枢铁心，只与励磁绕组相交链的磁通，称为主磁极漏磁通，用 Φ_σ 表示。

从图 1-27 可见，主磁通回路中的气隙较小，所以主磁路磁阻比较小；而漏磁通回路中的气隙较大，因而漏磁路磁阻较大，故主磁极漏磁通的数量比主磁通的数量要小很多。通常，主磁极漏磁通的数量只有主磁通的 2%～8%。

直流电机主磁极的极靴宽度 b_p 一般约为极距 τ 的 75% 左右，如图 1-28 所示。极靴下的气隙不均匀，在主磁极中心线附近，气隙较小且均匀，接近极尖处气隙较大。这样，在气隙较小处，磁通密度 B 较大；而在气隙较大处，磁通密度 B 较小；在两个主磁极之间的几何中性线处，磁通密度 $B=0$。不计齿、槽影响时，空载时的气隙磁通密度分布是一个在空间静止的平顶波，如图 1-28 所示。

在电机中，常将主磁极轴线称为直轴，相邻两主磁极之间的轴线称为交轴，也称为几何中性线。

1.3.2 直流电机带负载时的气隙磁场和电枢反应

当直流电机带上负载后，电枢绕组中有电流流过，电枢电流将产生电枢磁动势，因此负载时的气隙磁场将由主磁极磁动势和电枢磁动势共同建立。负载时，电枢磁动势对主极磁场的影响，称为电枢反应。为了分析电枢反应的作用，先假定励磁电流为零，分析电枢电流所产生的电枢磁场的分布情况；然后在不考虑饱和的情况下，将励磁电流所产生的主磁极磁场和电枢电流产生的电枢磁场合成，从而可以弄清电枢磁动势对主磁极磁场的影响。

1. 电枢磁动势与磁感应强度分布

图 1-29a 为一台 4 极电机，电刷位于几何中性线上。设主磁极的励磁电流为零，电枢表面光滑，电枢绕组为整距绕组，导体在电枢表面均匀分布。

电枢虽然在旋转，但由于电刷和换向器的作用，使得每个主磁极下电枢元件中电流的方向固定不变，因而电枢磁动势以及由它建立的电枢磁场在空间应静止不动，且电枢磁动势的方向与电刷轴线重合。所以，当电刷位于几何中性线上时，电枢磁动势是交轴磁动势，且电枢磁场与主磁极磁场始终保持相对静止。

在图 1-29a 中，磁力线 aa 由电枢绕组电流产生。电枢绕组磁动势关于主磁极轴线对称，并随着距离 x 的增大而增大，因此，对于均匀分布绕组，在一对极（2τ）范围内，电枢磁动势在空间分布呈三角形，如图 1-29b 所示。

电枢磁场的磁通密度 B_{ax} 与该处的磁动势 F_{ax} 成正比，与气隙长度 δ_x 成反比，即 $B_{ax}=\mu_0 F_{ax}/\delta_x$，由图 1-29 可知，在主磁极下气隙均匀，磁通密度 B_{ax} 与磁动势 F_{ax} 成正比；而在两主磁极之间，由于气隙增大，磁感应强度将被大大削弱，于是整个电枢磁场 B_{ax} 在空间呈马鞍形分布，如图 1-29c 所示。

在图 1-29 中，如果将电刷从几何中性线（交轴）处移开，图 1-29b 所示的磁动势波形形状不会改变，只是随移动的电刷位移一定的距离，其过零点将不再在磁极轴线位置，其最大值点也不再在交轴位置，因此，相应的磁通密度的波形也将不再与图 1-29c 所示的相同。

图 1-30a、b 分别为一台 2 极电机，其电刷位于几何中性线上和偏离几何中心线时，电枢绕组中电流方向的示意图。由于电刷是支路电流和电枢表面电流分布的分界线，若电枢上半圆周导体电流的方向为流入纸面，则下半圆周的导体电流方向为流出纸面。

2. 带负载时的气隙合成磁场与电枢反应

(1) 交轴电枢反应　图 1-31a 为只有主磁极励磁时的磁场分布示意图；图 1-31b 为没有主磁极励磁，只有电枢绕组电流产生的电枢磁场的分布示意图。图 1-32 则表示电机带负载时磁场分布示意图。

前面已分别对主磁极磁场和电枢磁场进行了分析，由于主磁极磁场和电枢磁场在空间位置上相对静止，所以利用叠加原理，可以将平顶波分布的主磁极磁通密度与按马鞍形分布的电枢磁通密度进行叠加，得到直流电机带负载时的气隙磁通密度分布，如图 1-33a 所示。其中 b_a 表示电枢磁密波，b_0 表示主磁极磁密波，b_δ 表示气隙合成磁密波。图 1-33b 为负载时气隙磁场示意图。

由 1-33a 可知，在每个主磁极下，主磁场的一半被削弱，另一半被增强。当电机空载时，几何中性线处的主磁极磁场为零。通常将磁感应强度为零的位置称为物理中性线，空载时物理中性线和几何中性线重合。电机带负载后，由于电枢反应的影响，使气隙磁场发生畸变，电枢表面上磁感应强度为零的位置将随之移动，使得物理中性线和几何中性线不再重合。如图 1-33 所示，对于发电机，物理中性线将顺着电枢旋转方向，从几何中性线向前移过 α 角；对电动机，则逆着电机旋转方向移过 α 角。

当磁路不饱和时，主磁场被削弱的数量与增强的数量相等，如图 1-33a 中面积 $A_1=A_2$ 所示；因此带负载时每极下的合成磁通与空载时相等，即总体上看，交轴电枢反应既无增磁、亦无去磁作用。实际电机中，常存在磁饱和现象，由于磁饱和的影响，增磁部分将使该处的饱和程度提高，使铁心的磁阻增大，从而使气隙磁感应强度比不计饱和时降低；而去磁部分的气隙磁感应强度与不计饱和时基本一致。因此，带负载时每极的磁通将比空载时略有减小，总体呈现去磁性。带负载时实际的气隙合成磁场如图 1-33a 中的虚线所示。

综上所述，当电刷在几何中性线上时，电枢反应为交轴电枢反应，其作用为：使气隙磁场发生畸变，物理中性线偏离几何中性线；考虑饱和时，每极磁通略有减小，呈现一定的去磁性。

正是由于电枢磁动势和主磁极磁场的互相作用，直流电机才能实现机电能量的转换。此外，电枢反应对电机运行特性也有较大影响：对于电动机，将影响其电磁转矩和转速；对于发电机，将影响其感应电动势和端电压。

（2）电刷偏离几何中性线时的电枢反应　若电刷从几何中性线移过 β 角（相应的电枢表面弧长为 b_β），电枢反应将包括交轴电枢反应和直轴电枢反应两部分，且直轴电枢反应可能起增磁作用，也可能起去磁作用。图 1-34a 所示的电枢磁动势可分解为两部分：一部分为 $\tau-2b_\beta$ 范围内的载流导体所产生的交轴电枢磁动势，如图 1-34b 所示；另一部分为 $2b_\beta$ 范围内的载流导体所产生的直轴电枢磁动势，如图 1-34c 所示。图 1-35 为电枢磁动势分布波形，图中曲线 2 是交轴电枢磁动势波形，曲线 3 是直轴电枢磁动势波形，曲线 1 是合成电枢磁动势波形。

1.4　直流电机的感应电动势和电磁转矩

1.4.1　直流电机的感应电动势

电枢旋转时，电枢导体切割气隙磁场，电枢绕组中将会感应出电动势。直流电机电枢的感应电动势是指电枢绕组的一对正、负电刷间的电动势，即每条支路的感应电动势。从电刷两端看，每条支路在任何瞬时所串联的导体数是相等的，而且每条支路中的导体分布在同一主磁极（或同种极性的主磁极）下的不同位置，如图 1-36 所示。所以每个导体中感应电动势的瞬时值是不同的，但任何瞬时构成支路的情况基本相同，且每个主磁极下的气隙磁场，除极性不同外，其磁感应强度的分布情况完全相同。因此，无论是何种绕组，每条支路中各导体电动势瞬时值的总和可以认为不变。所以，计算支路电动势，可先计算一根导体在一个主磁极下（即一个极距内）的平均电动势 E_{av}，再乘上一条支路的总导体数 $N/(2a)$，就可以得到电枢感应电动势 E_a。

226

在图 1-36 中，假定一个磁极下的平均气隙磁感应强度为 B_{av}，则一根导体在一个主磁极下（即一个极距内）运动时其平均感应电动势 E_{av} 为

$$E_{av}=B_{av}lv \tag{1-11}$$

式中，B_{av} 为一个主磁极内气隙磁感应强度的平均值（T）；E_{av} 为一根导体的平均感应电动势（V）；l 为电枢导体的有效长度（m）；v 为电枢导体切割气隙磁场的线速度（m/s），它与电枢旋转速度 n 的关系为 $v=2p\tau n/60$。

B_{av} 与每极磁通 Φ 的关系为

$$B_{av}=\Phi/(\tau l) \tag{1-12}$$

将式（1-12）和 $v=2p\tau n/60$ 代入式（1-11），可得

$$E_{av}=2p\Phi n/60 \tag{1-13}$$

于是电枢感应电动势 E_a 为

$$E_a=\frac{N}{2a}E_{av}=\frac{N}{2a}2p\Phi\frac{n}{60}=\frac{pN}{60a}\Phi n=C_e\Phi n \tag{1-14}$$

式中，C_e 称为电动势常数，$C_e=\dfrac{pN}{60a}$，由电机的结构参数决定。若磁通 Φ 的单位为 Wb，转速 n 的单位为 r/min，则感应电动势 E_a 的单位为 V。

式（1-14）是电枢的感应电动势公式。由此可见，感应电动势 E_a 与每极磁通 Φ 和电枢旋转速度 n 的乘积成正比。

特别地，若不计磁饱和的影响，则每极磁通 Φ 将与励磁电流 I_f 成正比，即 $\Phi = K_f I_f$。于是有

$$E_a = C_e \Phi n = K_f C_e I_f n = G_{af} I_f \Omega \tag{1-15}$$

式中，$G_{af} = \dfrac{60}{2\pi} C_e K_f = C_T K_f$ 且 $G = \dfrac{pN}{2\pi a}$；Ω 为电机的机械角速度。

由此可见，不计磁饱和时，电枢的感应电动势 E_a 与励磁电流 I_f 和电枢机械角速度 Ω 的乘积成正比。

1.4.2　直流电机的电磁转矩

当直流电机带负载时，电枢绕组中有电流流过，载流的电枢绕组在气隙磁场中将受到电磁力作用，其大小可以利用电磁力定律来计算。

假设电刷位于几何中性线上，绕组为整距，则一个主磁极下载流导体中的电流方向相同。另外，每个主磁极下的气隙磁场，除极性不同外，其磁感应强度的分布情况亦完全相同。因此，只需要计算单根导体在一个主磁极下（即一个极距内）所受到的平均电磁力和电磁转矩，然后再乘以电枢表面的总导体数，就可以得到作用在整个电枢上的电磁转矩。

在图 1-36 中，一个主磁极下的平均气隙磁感应强度为 B_{av}，假定导体中的电流为 I_c，电枢总电流为 I_a，则有

$$I_c = I_a/(2a) \tag{1-16}$$

单根导体在一个主磁极下（即一个极距内）所受到的平均电磁力为

$$f_{av} = B_{av} l I_c \tag{1-17}$$

式中，f_{av} 为平均电磁力（N）。

单根导体产生的平均电磁转矩为

$$T_{av} = \frac{D_a}{2} f_{av} = \frac{D_a}{2} B_{av} l I_c \tag{1-18}$$

式中，T_{av} 为平均电磁转矩（N · m）；D_a 为电枢直径（m），$D_a = 2p\tau/\pi$。

若电枢表面的总导体数为 N，则总的电磁转矩为

$$T_e = N B_{av} l D_a I_a/(2a \times 2) \tag{1-19}$$

将 $B_{av} = \Phi/\tau l$ 和 $D_a = 2p\tau/\pi$ 代入式（1-19），可得

$$T_e = \frac{pN}{2\pi a} \Phi I_a = C_T \Phi I_a \tag{1-20}$$

式中，C_T 称为转矩常数，$C_T = \dfrac{pN}{2\pi a}$，它由电机的结构参数决定。如果每极磁通 Φ 的单位为 Wb，电枢电流 I_a 的单位为 A，则电磁转矩 T_e 的单位为 N · m。

式（1-20）就是直流电机的电磁转矩公式。由此可见，直流电机的电磁转矩 T_e 与每极磁通 Φ 和电枢电流 I_a 的乘积成正比。

特别地，若不计磁饱和的影响，电机的每极磁通 Φ 与励磁电流 I_f 成正比，即 $\Phi = K_f I_f$，其中 K_f 为磁通与励磁电流间的比例系数。于是

$$T_e = C_T\Phi I_a = K_f I_f C_T I_a = G_{af} I_f I_a \tag{1-21}$$

式中，G_{af}为转矩与电枢电流乘积之间的比例常数。

可见，不计磁饱和时，电磁转矩 T_e与励磁电流 I_f和电枢电流 I_a的乘积成正比。

从感应电动势公式（1-14）和电磁转矩公式（1-20）可知，同一台电机的转矩常数与电动势常数之间有如下关系：

$$C_T = \frac{30}{\pi}C_e = 9.55C_e \tag{1-22}$$

例 1-3 一台直流电机的极对数 $p=3$，采用单叠绕组，电枢绕组总导体数 $N=398$，气隙每极磁通 $\Phi=2.1\times10^{-2}$Wb，当转速分别为 $n_1=1500$r/min 和 $n_2=500$r/min 时，求电枢绕组的感应电动势各为多少?

解:

由于绕组是单叠绕组，绕组的并联支路对数与电机极对数相等，即 $a=p=3$。根据式（1-14）所示感应电动势公式，当电机转速 $n_1=1500$r/min 时，电枢绕组一条支路的感应电势为

$$E_a = \frac{pN}{60a}\Phi n_1 = \frac{3\times398}{60\times3}\times2.1\times10^{-2}\times1500\text{V} = 208.95\text{V}$$

当电机转速 $n_2=500$r/min 时，电枢绕组一条支路的感应电动势为

$$E_a = \frac{pN}{60a}\Phi n_1 = \frac{3\times398}{60\times3}\times2.1\times10^{-2}\times500\text{V} = 69.65\text{V}$$

例 1-4 一台直流电机的极对数 $p=3$，采用单叠绕组，电枢绕组总导体数 $N=398$，气隙每极磁通 $\Phi=2.1\times10^{-2}$Wb。当电枢电流分别为 $I_1=10$ A 和 $I_2=15$ A 时，求电机的电磁转矩各为多少?

解:

由于绕组是单叠绕组，绕组的并联支路对数与电机极对数相等，即 $a=p=3$；

根据式（1-20）所示电磁转矩公式，当电枢电流 $I_1=10$ A 时，电机的电磁转矩为

$$T_e = \frac{pN}{2\pi a}\Phi I_a = \frac{3\times398}{2\pi\times3}\times2.1\times10^{-2}\times10\text{N}\cdot\text{m} = 13.3\text{N}\cdot\text{m}$$

当电枢电流 $I_2=15$ A 时，电机的电磁转矩为

$$T_e = \frac{pN}{2\pi a}\Phi I_a = \frac{3\times398}{2\pi\times3}\times2.1\times10^{-2}\times15\text{N}\cdot\text{m} = 20\text{N}\cdot\text{m}$$

例 1-5 一台直流发电机，其额定功率 $P_N=17$kW，额定电压 $U_N=230$V，额定转速 $n_N=1500$r/min，极对数 $p=3$，电枢绕组总导体数 $N=780$，采用单叠绕组，气隙每极磁通 $\Phi=1.3\times10^{-2}$Wb。求电机的额定电流和电枢绕组的感应电动势各为多少

解:

根据式（1-3），该直流发电机的额定电流 $I_N=P_N/U_N=17\times1000/230\text{A}=73.91\text{A}$。由于绕组是单叠绕组，绕组的并联支路对数与电机极对数相等，即 $a=p=3$。根据式（1-14），当电机转速 $n_N=1500$r/min 时，电枢绕组一条支路的感应电动势为

$$E_a = \frac{pN}{60a}\Phi n_1 = \frac{3\times780}{60\times3}\times1.3\times10^{-2}\times1500\text{V} = 253.5\text{V}$$

1.5　直流电机的等效电路

1.5.1　直流发电机的等效电路

1. 他励直流发电机

他励直流发电机的等效电路如图 1-37 所示，其中 U 表示发电机电枢两端的电压，I 表示发电机的线路电流，E_a 为发电机的感应电动势，电枢电流为 I_a。由图可见，他励直流发电机的电枢电流与线路电流相等，即

$$I_a = I \tag{1-23}$$

从等效电路可得到他励直流发电机的电压方程为

$$\begin{cases} E_a = U + I_a R_a \\ U_f = I_f R_f \end{cases} \tag{1-24}$$

2. 并励直流发电机

并励直流发电机的等效电路如图 1-38 所示，其中电枢电流等于线路电流与励磁电流之和：

$$I_a = I + I_f \tag{1-25}$$

从等效电路可得到并励直流发电机的电压方程为

$$\begin{cases} E_a = U + I_a R_a \\ U = I_f R_f \end{cases} \tag{1-26}$$

3. 串励直流发电机

串励直流发电机的等效电路如图 1-39 所示，其中电枢电流等于线路电流，也等于励磁电流：

$$I_a = I = I_f \tag{1-27}$$

从等效电路可得到并励直流发电机的电压方程为

$$E_a = U + I_a(R_a + R_f) \tag{1-28}$$

4. 复励直流发电机

复励直流发电机的等效电路如图 1-40 所示，图 1-40a 为长并励的连接，图 1-40b 为短并励的连接，其中电枢电流等于线路电流与励磁电流之和：

$$I_a = I + I_f \tag{1-29}$$

从等效电路可得到并励直流发电机的电压方程为

$$\begin{matrix} E_a = U + I_a(R_a + R_{sf}) \\ I_f = U/R_f \end{matrix} \tag{1-30}$$

1.5.2　直流电动机的等效电路

1. 他励直流电动机

他励直流电动机的等效电路如图 1-41 所示。其中，电枢电流与线路电流相等，励磁电流等于励磁电压除以励磁回路电阻。

$$\begin{cases} I_a = I \\ I_f = U_f/R_f \end{cases} \tag{1-31}$$

从等效电路可得到他励直流电动机的电压方程为

$$E_a = U - I_a R_a \tag{1-32}$$

2. 并励直流电动机

并励直流电动机的等效电路如图 1-42 所示。其中，电枢电流等于线路电流与励磁电流之差，励磁电流等于励磁电压除以励磁回路电阻。

$$\begin{cases} I_a = I - I_f \\ I_f = U/R_f \end{cases} \tag{1-33}$$

从等效电路可得到并励直流电动机的电压方程为：

$$E_a = U - I_a R_a \tag{1-34}$$

3. 串励直流电动机

串励直流电动机的等效电路如图 1-43 所示。其电压方程为

$$U = E_a + I_a(R_a + R_f) \tag{1-35}$$

4. 复励直流电动机

复励直流电动机的等效电路如图 1-44 所示，其中图 1-44a 为长并励连接，图 1-44b 为短并励连接。

1.6 直流电机的损耗与功率流程

直流电机的效率由以下公式定义：

$$\eta = \frac{P_{out}}{P_{in}} \times 100\% \tag{1-36}$$

式中，P_{out}、P_{in} 分别为电机的输出功率和输入功率。

输入功率与输出功率之间的差为电机内部的总损耗 $\sum p_{loss}$，因此也可将效率表示为

$$\eta = \frac{P_{out}}{P_{in}} \times 100\% = \frac{P_{in} - \sum p_{loss}}{P_{in}} \times 100\% \tag{1-37}$$

1.6.1 直流电机内部的损耗

直流电机中的损耗主要包括消耗于绕组电阻中的铜损、电刷的接触损耗、消耗于铁心中的铁损以及转子旋转时的机械损耗和带负载时的附加损耗等。

1. 铜损

铜损主要包括电枢回路铜损 p_{Cua} 和励磁回路铜损 p_{Cuf}。由于电枢回路铜损 p_{Cua} 与电枢电流直接相关，它们随负载电流的变化而变化，所以通常称为可变损耗。

$$p_{Cua} = I_a^2 R_a \tag{1-38}$$

$$p_{Cuf} = I_f^2 R_f = U_f I_f \tag{1-39}$$

式中，I_a 为电枢电流；I_f 为励磁电流；R_a 为电枢电阻；R_f 为励磁回路电阻。

2. 电刷接触损耗

电刷接触损耗 p_c 表示为

$$p_c = 2\Delta U_s I_a \tag{1-40}$$

式中，I_a 为电枢电流，$2\Delta U_s$ 为一对正负电刷的接触压降。

3. 铁损

铁损 p_{Fe}是电枢旋转时，气隙主磁通在电枢铁心内交变而引起的损耗。它通常与磁感应强度的二次方（B^2）成正比，还与转子转速的 1.5 次方（$n^{1.5}$）成正比。

4. 机械损耗

机械损耗 p_Ω 包括轴承的摩擦损耗、电刷的摩擦损耗及定、转子的通风损耗等。

机械损耗 p_Ω 和铁损 p_{Fe}与负载大小无关，电机空载时即存在，所以这两项之和通常称为空载损耗，即 $p_0=p_\Omega+p_{Fe}$。在运行过程中 p_0的数值几乎不变，所以空载损耗也称为不变损耗。

5. 杂散损耗

杂散损耗 p_s也称为附加损耗，大致包括：结构部件在磁场内旋转而产生的损耗；因电枢齿、槽的影响，使气隙磁通产生脉动而在主磁极铁心中和电枢铁心中产生的脉动损耗；因电枢反应使气隙磁场畸变而在电枢铁心中产生的损耗；由于电流分布不均匀而增加的电刷接触损耗以及换向电流所产生的损耗等。这些损耗难以精确计算，带额定负载时，通常按额定功率的 0.5% ~1% 来估算；带其他大小负载时，p_s随 $(I/I_N)^2$ 的变化而变化。

1.6.2　直流发电机的功率流程

直流发电机的功率流程如图 1-45 所示，其中输入的机械功率 P_1除去杂散损耗 p_s，机械损耗 p_Ω 和铁损 p_{Fe}后，其余将转化为电磁功率 P_e，也称为转换功率。该转换功率可表示为

$$P_e = T_e\Omega \tag{1-41}$$

式中，Ω 是发电机的机械角速度，$\Omega=\dfrac{2\pi n}{60}$（r/s）。

电磁功率还可表示为

$$P_e = E_aI_a \tag{1-42}$$

从电磁功率 P_e中再扣除电枢回路铜损 p_{Cua}、励磁回路铜损 p_{Cuf}和电刷的接触损耗 p_c，余下的即为发电机线端输出的电功率 P_2。

$$P_1=T_1\Omega = p_\Omega + p_{Fe} + p_s + P_e \tag{1-43}$$

$$P_e = p_{Cua} + p_{Cuf} + p_c + P_2 \tag{1-44}$$

$$P_2 = UI \tag{1-45}$$

在图 1-45 中，T_1为原动机的输入转矩，$T_1 = P_1/\Omega$，T_e为发电机的电磁转矩，$T_e=P_e/\Omega$；T_0为空载转矩，$T_0=p_0/\Omega$。

$$T_1 = T_e + T_0 \tag{1-46}$$

1.6.3　直流电动机的功率流程

直流电动机的功率流程如图 1-46 所示。

在图 1-46 中，T_2为直流电动机的输出转矩，$T_2 = P_2/\Omega$；T_e为电动机的电磁转矩，$T_e = P_e/\Omega$；T_0为空载转矩，$T_0=p_0/\Omega$。

$$T_e = T_2 + T_0 \tag{1-47}$$

1.7 直流发电机的运行特性

直流发电机的运行特性主要是指空载特性、外特性和调节特性。影响这些特性的物理量有端电压 U、负载电流 I、励磁电流 I_f 以及发电机的转速 n。直流发电机的运行特性与发电机的励磁方式有关。

在直流发电机的外特性中，定义发电机的电压变化率（VR）为

$$VR = \frac{U_{nl} - U_{fl}}{U_{fl}} \times 100\% \tag{1-48}$$

式中，U_{nl} 为发电机的空载端电压；U_{fl} 为发电机带额定负载时的端电压。

1.7.1 他励直流发电机的运行特性

1. 他励直流发电机的空载特性

空载特性是指电机的转速 $n=n_N$，电枢电流 $I_a=0$ 时，电枢的空载端电压 U_0 与励磁电流 I_f 之间的关系 $U_0=f(I_f)$。

空载运行时电枢电流 $I_a=0$，电枢回路中没有电阻压降，所以空载电压就等于电枢感应电动势，即 $U_0=E_a$。又因发电机的转速恒定，故电动势 E_a 与主磁通 Φ_0 成正比。另一方面，因为励磁绕组匝数恒定，故励磁磁动势 F_f 与励磁电流 I_f 成正比，所以空载特性曲线 $U_0=f(I_f)$ 与磁化曲线 $\Phi_0=f(F_f)$ 的形状相似。

（1）磁通与励磁电流的关系　他励直流发电机空载运行时，励磁电流的变化将引起发电机感应电动势的相应变化。逐渐增大励磁电流 I_f，发电机的磁动势将增大，即每极磁通 Φ 将增大，于是可画出每极磁通 Φ 随 I_f 变化的曲线，如图 1-47 所示。

从图 1-47 可知，当励磁电流较小时，磁通较小，发电机的铁心处于不饱和状态，此时磁通随励磁电流的增加线性增加，如 Oa 段所示。随着励磁电流继续增大，发电机铁心和主磁极均将逐渐进入饱和状态，此时，随着励磁电流的增加，磁通的增加量变小，曲线进入饱和区域的转折点 ab 段，然后进入饱和区域的 bc 段。

（2）感应电动势与励磁电流的关系　依据式（1-14），发电机的感应电动势 $E_a=C_e\Phi n$ 与电机的每极磁通 Φ 和电机转速 n 的乘积成正比。但感应电动势与励磁电流的关系是怎样的呢？假设发电机的转速恒定，则发电机的空载端电压与感应电动势相等，即 $U_0=E_0$，且它们直接与每极磁通 Φ 成正比。使励磁电流从零开始逐渐增加，读取相应的空载电压 U_0 的值，直到空载电压 $U_0=(1.1\sim1.3)U_N$ 为止。然后逐渐减小励磁电流 I_f 至零，读取相应的空载电压 U_0 的值；当 $I_f=0$ 时，空载电压 U_0 不等于零，而有很小的值，此电压称为剩磁电压。改变励磁电流 I_f 的方向，重复上述步骤，可得到磁滞回线的整个下降曲线，如图 1-48 所示。然后根据对称关系画出磁滞回线的上升曲线，并找出磁滞回线的平均曲线，如图 1-48 中的虚线所示，此虚线即为发电机的空载特性曲线。

改变发电机的转速 n 时，可得到不同的空载特性曲线。

并励和复励直流发电机的空载特性，均以他励的形式来测取。空载特性常用于确定磁路和运行点的饱和程度。

（3）感应电动势与速度的关系　依据式（1-14），当励磁电流不变时，发电机的感应电动势与转速成正比。若改变发电机的旋转方向，则感应电动势的极性也将发生改变。

2. 他励直流发电机的外特性

他励直流发电机的外特性是指 $n=n_N$，$I_f=I_{fN}$，当负载变化时，电枢端电压 U 与电枢电流 I_a 的关系曲线 $U=f(I_a)$。其中，额定励磁电流 I_{fN} 是指电枢端电压为额定电压 U_N、负载电流为额定电流 I_N 时的励磁电流。他励直流发电机的电压方程为

$$U = E_a - I_a R_a \tag{1-49}$$

依据式（1-49）可知，外特性曲线应近似为一条直线，如图1-49所示。

他励直流发电机的外特性曲线略为下垂。电压下降原因可利用电压方程式（1-49）来分析。随着负载电流的增大，一方面电枢反应的去磁作用增大，使气隙合成磁通略为减小，从而电枢感应电动势也随之减小；另一方面，随着电枢电流的增大，电枢回路的电阻压降增大。这两方面的原因都使得发电机的端电压下降。

3. 他励直流发电机的调节特性

他励直流发电机的调节特性是指 $n=n_N$，当负载变化使 $U=U_N$ 时，励磁电流 I_f 与电枢电流 I_a 之间的关系曲线 $I_f=f(I_a)$。它反映了负载变化时，如何调节励磁电流 I_f 才能维持发电机的端电压 U 不变。

他励直流发电机的调节特性曲线如图1-50所示。由图可见，调节特性曲线随着负载电流的增大而上翘，原因是负载电流增大时，要维持端电压为常数，必须增大励磁电流使感应电动势增大，以补偿电枢电阻压降和电枢反应的去磁作用。

4. 他励直流发电机的电压调节

他励直流发电机端电压的调节可以通过改变发电机的感应电动势来实现。根据电压方程 $E_a=U+I_aR_a$ 可知，当 $E_a=C_e\Phi n$ 增加时，U 将增加，反之亦然。因此，可以用以下两种方法来调节他励直流发电机的端电压：

1）改变发电机的转速。当 n 增大时，$E_a=C_e\Phi n$ 将增大，于是 $U=E_a-I_aR_a$ 也将增大。

2）改变励磁电流。当励磁回路电阻 R_f 减小时，励磁电流将增大，$I_f=U_f/R_f$，于是每极磁通 Φ 将增大。随着磁通的增加，$E_a=C_e\Phi n$ 一定增加，最终使得 $U=E_a-I_aR_a$ 增加。

1.7.2 并励直流发电机的运行特性

并励直流发电机的等效电路如图1-38所示，其空载特性和调节特性与他励直流发电机相似，这里主要分析并励直流发电机的自励和外特性。

1. 并励直流发电机的自励

并励和复励直流发电机均为自励发电机，它们不需要外部直流电源供给励磁电流。这样就会产生一个问题：发电机的电压是如何建立的？下面以并励直流发电机为例，说明空载电压的建立和条件。

当用原动机拖动并励直流发电机的电枢旋转，使发电机的转速 $n \approx n_N$。并励直流发电机要能建立空载电压，发电机内部必须要有剩磁。这样，当电枢导体旋转并切割剩磁时，将会产生不大的剩磁电动势。该剩磁电动势由电枢两端回授到励磁绕组上，将产生一个很小的励磁电流。如果此时励磁绕组的极性能使该励磁电流产生的磁动势与剩磁方向相同而形成正反馈，就能使气隙磁场得到加强，电枢感应电动势增大，从而使励磁电流和气隙磁场得到进一步加强，这样空载电压就可能建立起来。如果励磁绕组的极性不当，使该励磁电流产生的磁动势与剩磁方向相反而形成负反馈，则气隙磁场将被削弱，使电枢感应电动势减小，空载电

压就无法建立。

图 1-51 中的曲线 1 表示并励直流发电机的空载特性 $U_0=f(I_f)$，曲线 2 为励磁回路的电阻线，即励磁电压 U_f 随励磁电流 I_f 变化的伏安特性（$U_f=R_fI_f$）。当不计电枢电阻上的电压降和电枢反应的影响时，励磁绕组的端电压 U_f（也即电枢端电压 U_0）与励磁电流 I_f 之间，在磁路上满足空载特性曲线，在电路上又要满足伏安特性曲线。由此可知，这两条曲线的交点应是正反馈时空载电压建立后的运行点。发电机要自励，这两条曲线必须有交点。若励磁回路电阻较大，这两条曲线就可能没有交点，如图 1-51 中的直线 4，这时空载电压将无法建立。图中直线 3 与空载特性曲线 1 相切，此时励磁回路的电阻称为临界电阻。因此发电机要自励，励磁回路的电阻必须小于相应转速时的临界电阻。

综上所述，并励直流发电机自励建压必须满足 3 个条件：

1）发电机的主磁路必须有剩磁。

2）励磁绕组与电枢绕组连接的极性要正确，使励磁电流产生的磁动势与剩磁方向一致。

3）励磁回路的总电阻必须小于该转速下的临界电阻。

2. 并励直流发电机的外特性

并励直流发电机的外特性是指转速 $n=n_N$，励磁回路的总电阻 $R_f=r_f+R_{pf}$ 为常数时，发电机端电压 U 和负载电流 I 的关系曲线 $U=f(I)$。与他励时不同，并励的外特性只能保持励磁电阻 R_f 不变；当端电压 U 变化时，励磁电流 I_f 将随之变化，不像他励那样可以保持为常数。

并励直流发电机的外特性可通过试验法测取。当用原动机拖动电枢至额定转速 n_N，并带负载运行时，调节励磁回路的调节电阻 R_{pf} 和负载电阻 R_L，使发电机达到额定状态。然后，保持励磁回路总电阻和转速 n_N 不变，从额定值起逐步减小负载电流，一直到零（负载断开）为止。在此过程中依次读取端电压 U 和相应的负载电流 I，可绘出外特性如图 1-52 所示。

与他励时的外特性相比较，并励直流发电机外特性下降的幅度较大，即在同一负载电流 I 下，端电压 U 较低。这是由于随着负载电流 I 的增大，除电枢反应的去磁作用和电枢回路电阻压降增大外，并励时励磁电流 I_f 将随着端电压 U 的降低而减小，从而引起主磁通和电枢感应电动势的进一步下降。所以在相同负载时，并励发电机的端电压要比他励时下降得多，而且会出现“拐弯”现象。

外特性曲线出现“拐弯”现象的原因是：当端电压较高时，发电机磁路处于较饱和区，励磁电流 I_f 的变化对感应电动势的影响不大，故负载电阻 R_L 减小，负载电流 I 增大，端电压 U 将下降；当 I 增大到临界电流 I_{cr} 之后，发电机进入不饱和区，I_f 的减小使电动势 E_a 和端电压 U 下降得很快，从而导致负载电阻 R_L 减小时电流 I 反而减小的现象。一般并励直流发电机的 I_{cr} 为额定电流 I_N 的 2～3 倍。

3. 并励直流发电机的调节特性

并励直流发电机的调节特性与他励直流发电机的相似。

4. 并励直流发电机端电压的调节

与他励直流发电机相似，可以用两种方法来调节并励直流发电机的端电压。

1）改变发电机的转速 n。

2）改变发电机的励磁电阻，即改变发电机的励磁电流。

例 1-6 一台并励直流发电机，$P_N=9\text{kW}$，$U_N=115\text{V}$，$n_N=1450\text{r/min}$，$R_a=0.15\Omega$，额定运行时励磁回路的总电阻 $R_{fN}=33\Omega$，发电机的铁损为 410W，机械损耗为 101W，附加损耗按额定功率的 0.5% 估算。试求：

（1）带额定负载时该发电机的电磁转矩。

（2）带额定负载时该发电机的效率。

解：

（1）该发电机的额定电流为

$$I_N = P_N/U_N = 9000/115\text{A} = 78.26\text{A}$$

励磁电流为

$$I_f = U_N/R_{fN} = 115/33\text{A} = 3.48\text{A}$$

电枢电流为

$$I_a = I_N + I_f = 78.26 + 3.48\text{A} = 81.74\text{A}$$

该发电机的电磁功率为

$$P_e = P_N + p_{Cuf} + p_{Cua} = 9000 + 115 \times 3.48 + 81.74^2 \times 0.15\text{W} = 10\,402.4\text{W}$$

电磁转矩为

$$T_e = P_e/\Omega = 10402.4 \times 60/(2 \times 3.14 \times 1450)\text{N}\cdot\text{m} = 68.54\text{N}\cdot\text{m}$$

（2）额定负载时的效率为

$$\eta_N = P_N/(P_e + p_{Fe} + p_\Omega + p_{ad}) = 9000/(10402.4 + 410 + 101 + 0.005 \times 9000) = 82.13\%$$

1.7.3　复励直流发电机的运行特性

1. 复励直流发电机的外特性

复励直流发电机分为积复励和差复励两类。在积复励发电机中，并励绕组和串励绕组产生的磁动势相互增强，且并励绕组起主要作用，它使发电机空载时产生额定电压，串励绕组的作用是补偿负载时电枢电阻压降和电枢反应的去磁作用。图1-40a为“长并励”连接的积复励直流发电机的等效电路，图1-40b为“短并励”连接的积复励直流发电机的等效电路。根据串励绕组的补偿程度，积复励直流发电机又分为平复励、过复励和欠复励三种。若发电机在额定负载时的端电压等于空载电压，就称为平复励，此时串励绕组的磁动势恰好能补偿电枢反应的去磁作用和电枢电阻压降。若串励绕组的补偿有余，则额定负载时的端电压将高于空载电压，称为过复励。反之，补偿不足就称为欠复励。在差复励直流发电机中，并励绕组和串励绕组产生的磁动势是相互抵消的。

复励直流发电机的外特性曲线如图1-53所示。

2. 复励直流发电机端电压的调节

与并励直流发电机相似，可以用两种方法来调节复励直流发电机的端电压。

1）改变发电机的转速 n。

2）改变发电机的励磁电阻，即改变发电机的励磁电流。

1.8　直流电动机的运行特性

直流电动机的运行特性包括工作特性和机械特性（也称为转矩-转速特性）。其中，工作特性是选用直流电动机的一个重要依据；机械特性表征电动机输出的机械性能，它与转子运动方程式一起可以决定拖动系统的运行状态。直流电动机的工作特性因励磁方式不同，差别很大。

他励（并励）直流电动机的工作特性是指：当电动机的端电压为额定电压 U_N，电枢回路无外串电阻，励磁电流为额定励磁电流 I_{fN} 时，他励（并励）直流电动机的转速 n、电磁

转矩 T_e和效率 η 与输出功率 P_2之间的关系，即 n，T_e，$\eta=f(P_2)$。在电动机的实际运行中，电枢电流 I_a可测，且 I_a随负载的增大而增大，所以也可以将工作特性表示为 n，T_e，$\eta=f(I_a)$。机械特性是指：电动机电枢两端加上额定电压 U_N，励磁电流 I_f为常数时，电动机转速 n 与电磁转矩 T_e之间的关系 $n=f(T_e)$。

直流电动机的转速变化率（SR）定义为

$$SR=\frac{n_{nl}-n_{fl}}{n_{fl}}\times 100\% \tag{1-50}$$

式中，n_{nl}为电动机空载时的转速，n_{fl}为电动机带额定负载时的转速。

1.8.1 他励和并励直流电动机的运行特性

他励直流电动机的运行特性（包括工作特性与机械特性）可以通过试验来测取。给励磁回路加上励磁电压 U_f，在电枢两端加额定电压 U_N，调节磁场电阻 R_{pf}和电动机的负载，使电动机输出功率为额定功率 P_N、转速为额定转速 n_N，此时的励磁电流称为额定励磁电流 I_{fN}。保持 $U=U_N$、$I_f=I_{fN}$不变，改变电动机的负载，可测得相应的转速 n、输出转矩 T_2和电枢电流 I_a（或输出功率 P_2），可得到如图 1-54 所示的工作特性。工作特性包括转速特性、转矩特性和效率特性；机械特性是电机转速 n 与电磁转矩 T_e 之间的关系曲线 $n=f(T_e)$，且机械特性又分为固有机械特性和人为机械特性两类。

1. 转速特性 $n=f(I_a)$

转速特性是指当 $U=U_N$，$I_f=I_{fN}$，电枢回路无外串电阻时，电动机的转速 n 与电枢电流 I_a之间的关系 $n=f(I_a)$。

他励直流电动机的电压方程为

$$U=C_e n\Phi+I_aR_a \tag{1-51}$$

根据感应电压公式，可推导得到电机的转速为

$$n=\frac{E_a}{C_e\Phi}=\frac{U}{C_e\Phi}-\frac{R_a}{C_e\Phi}I_a \tag{1-52}$$

式（1-52）通常称为电动机的转速公式。若忽略电枢反应的去磁作用，则每极磁通 Φ 与电枢电流 I_a无关，为一常数，于是式（1-52）可改写成直线方程式的形式：

$$n=n_0-\beta I_a \tag{1-53}$$

式中，n_0 为理想空载转速，即 $I_a=0$ 时电动机的转速，$n_0=\frac{U}{C_e\Phi}$；β 为直线的斜率，$\beta=\frac{R_a}{C_e\Phi}$。

可见，他励直流电动机的转速特性是一条斜率为 β 的向下倾斜的直线，如图 1-54 所示。

实际上，当他励直流电动机磁路饱和时，随着电枢电流 I_a的增大，交轴电枢反应的去磁作用可使转速 n 升高，使转速特性成为上翘的直线，这将使拖动系统不稳定。为保证电动机的稳定性，必须采取措施使转速特性成为向下倾斜的直线。

2. 转矩特性 $T_e=f(I_a)$

转矩特性是指当 $U=U_N$，$I_f=I_{fN}$，电枢回路无外串电阻时，电磁转矩 T_e与电枢电流 I_a之间的关系 $T_e=f(I_a)$。

由转矩公式 $T_e=C_T\Phi I_a$ 可见，$I_f=I_{fN}$且不计磁饱和时，Φ 为常数，于是电磁转矩与电枢

电流成正比，即 $T_e=f(I_a)$ 为一条直线随电枢电流的增加，T_e 线性增加；当负载（或电枢电流）较大时，由于电枢反应使磁通 Φ 略为减小，故转矩特性偏离直线，呈略为下降趋势；当电机空载时，电枢电流 $I_a=I_{a0}$，电磁转矩 $T_e=T_0$，如图 1-54 所示。

3. 效率特性 $\boldsymbol{\eta=f(I_a)}$

效率特性是指当 $U=U_N$，$I_f=I_{fN}$，且电枢回路无外串电阻时，效率 η 与电枢电流 I_a 之间的关系 $\eta=f(I_a)$。通常将效率表示为

$$\eta=\frac{P_2}{P_1}\times100\%=\frac{P_1-\sum p_{loss}}{P_1}\times100\% \tag{1-54}$$

式中，P_2 为电动机的输出功率；P_1 为电动机的输入功率；$\sum p_{loss}$ 为电动机内部的总损耗。

他励直流电动机的效率特性曲线如图 1-54 所示。当负载较小时，电动机的效率较低；当负载增大时，效率随输出功率的增大而增大。但当负载增大到一定程度时，由于铜损随电枢电流的增大而快速增大，电动机的效率将重新开始下降。

4. 机械特性

机械特性是指直流电动机加上额定电压 U_N 和额定励磁电流 I_{fN} 时，转速 n 与电磁转矩 T_e 之间的关系 $n=f(T_e)$。

根据他励直流发电机的电压方程和感应电动势的计算公式，可得到

$$U=C_en\Phi+I_aR_a \tag{1-55}$$

由于 $T_e=C_T\Phi I_a$，于是电枢电流 I_a 可表示为

$$I_a=\frac{T_e}{C_T\Phi} \tag{1-56}$$

结合式（1-55）和式（1-56），可得到

$$U=C_en\Phi+\frac{T_e}{C_T\Phi}R_a \tag{1-57}$$

最后，得到电动机的转速与电磁转矩的关系式为

$$n=\frac{U}{C_e\Phi}-\frac{R_a}{C_eC_T\Phi^2}T_e \tag{1-58}$$

如果忽略电枢反应的影响，上述关系可表示为直线关系：

$$n=n_0-\beta' T_e \tag{1-59}$$

式中，n_0 是理想空载转速，即当 $T_e=0$ 时电机的转速，$n_0=\dfrac{U}{C_e\Phi}$；β' 是直线的斜率，$\beta'=\dfrac{R_a}{C_eC_T\Phi^2}$。

电动机的实际空载转速可由下式计算：

$$n_0'=n_0-\frac{R_a}{C_eC_T\Phi^2}T_0 \tag{1-60}$$

式中，T_0 为空载损耗转矩，$T_0=\dfrac{p_0}{\Omega}$。

他励直流电动机的机械特性为一条向下倾斜的直线，如图 1-55a 所示。由图可见，电动机带负载后，转速 n 有一些下降。

根据式（1-58）可得，电动机负载后的转速降 Δn 为

$$\Delta n = \frac{R_a}{C_e C_T \Phi^2} T_e = \beta' T_e \tag{1-61}$$

由式（1-61）可知，β 越大，转速降 Δn 越大，表明机械特性越软；反之，则机械特性越硬。影响机械特性软硬的主要因素为：电枢回路所串电阻 R 的大小和电动机每极磁通 Φ 的大小。机械特性的硬度也可用额定转速的变化率 $\Delta n_N\%$ 来表示。$\Delta n_N\%$ 定义为

$$\Delta n_N\% = \frac{n_0 - n_N}{n_N} \times 100\% \tag{1-62}$$

式中，n_N 为电动机的额定转速。

当电动机所带负载增大时，由于饱和时交轴电枢反应将产生去磁作用，使得每极磁通 Φ 略为减小，使电动机的转速 n 稍有上升，可能使机械特性在负载较大时出现上翘的现象，如图 1-55b 所示。这种上翘的机械特性曲线可能会使拖动系统不能稳定运行。为避免机械特性上翘，常在主磁极上加上匝数很少的串励绕组，让其产生的磁动势抵消电枢反应的去磁作用，上述串励绕组被称为稳定绕组。

5. 人为机械特性

他励直流电动机的机械特性又分为固有机械特性和人为机械特性。外加电压为额定电压，电动机的主磁通为额定磁通，且电枢回路无外串电阻时的机械特性称为他励直流电动机的固有机械特性。如果人为地改变电动机的某一个参数，如电枢回路电阻 R、每极磁通 Φ 或电源电压 U，就可得到人为机械特性。他励直流电动机有以下 3 种人为机械特性。

（1）改变电枢回路电阻时的人为机械特性　当 $U = U_N, \Phi = \Phi_N, R = R_a + R_S$ 时，他励直流电动机的人为机械特性表达式为

$$n = \frac{U_N}{C_e \Phi_N} - \frac{R_a + R_S}{C_e C_T \Phi_N^2} T_e \tag{1-63}$$

式中，R_S 是电枢回路外串的电阻。

依据式（1-58）可知，电枢回路外串电阻 R_S 时，理想空载转速 n_0 保持不变，但机械特性曲线的斜率 β 随外串电阻 R_S 的增大而增大，使人为机械特性变软，如图 1-56 所示。在相同的负载下，电动机在稳态运行时的转速下降 Δn，随外串电阻 R_S 的增大而增大。所以，电枢回路外串电阻的人为机械特性，是一族通过理想空载转速点 n_0 而斜率不同的直线族。

（2）改变电源电压时的人为机械特性　当 $R = R_a$，$\Phi = \Phi_N$ 时，改变外加电源电压 U，可得到电动机的人为机械特性的表达式为

$$n = \frac{U}{C_e \Phi_N} - \frac{R_a}{C_e C_T \Phi_N^2} T_e \tag{1-64}$$

从式（1-64）可知，改变电源电压 U 时，斜率 β 未变，即人为机械特性的硬度与固有机械特性的硬度相同，但其理想空载转速 n_0 随外加电压 U 的降低而减小，如图 1-57 所示。

在实际工作中，电源电压一般是从额定电压 U_N 向下调节。从图 1-57 可知，在相同的负载下，电动机的转速降 Δn 不随外加电压的变化而变化，所以改变电源电压时的人为机械特性是一族斜率相同的平行线。

（3）减小磁通时的人为机械特性　当 $U = U_N$，$R = R_a$ 时，改变励磁电流 I_f 的大小，可得到电动机的人为机械特性表达式为

$$n = \frac{U_N}{C_e \Phi} - \frac{R_a}{C_e C_T \Phi^2} T_e \tag{1-65}$$

从式（1-65）可知：改变磁通 Φ 时，人为机械特性的斜率 β 随磁通 Φ 的减弱而增大；即特性的硬度随磁通 Φ 的减弱而降低；理想空载转速 n_0 则随磁通 Φ 的减弱而增大，如图 1-58b 中所示。在实际电动机中，由于存在磁饱和现象，所以磁通一般从额定磁通向下调节。从图 1-58b 可知，在相同的负载下，电动机的转速降 Δn 随磁通 Φ 的减小而增大。所以减小磁通 Φ 时的人为机械特性，是一族理想空载转速 n_0 逐渐增大，斜率 β 也逐渐增大的直线。

在改变电枢回路电阻和改变电源电压时的人为机械特性中，因为每极磁通 Φ 不变，故 $T_e \propto I_a$，所以它们的机械特性 $n = f(T_e)$ 就代表了转速特性 $n = f(I_a)$。但是对于减小磁通的人为机械特性，由于每极磁通 Φ 是个变量，因而转速特性 $n = f(I_a)$ 曲线与机械特性 $n = f(T_e)$ 曲线是不同的，如图 1-58a、b 所示。

并励直流电动机的运行特性与他励直流电动机的相似。

例 1-7　一台并励直流电动机，额定电压 $U_N = 220V$，额定电枢电流 $I_{aN} = 75A$，额定转速 $n_N = 1000r/min$，电枢回路电阻 $R_a = 0.26\Omega$（包括电刷接触电阻），额定运行时励磁回路总电阻 $R_{fN} = 91\Omega$。额定负载时电动机的铁损为 600W，机械损耗为 1989W。试求：

（1）带额定负载时该电动机的输出转矩。

（2）带额定负载时该电动机的效率。

解：

（1）该电动机的励磁电流为

$$I_f = U_N/R_{fN} = 220/91A = 2.42A$$

线路电流为

$$I_N = I_{aN} + I_f = 75 + 2.42A = 77.42A$$

输入功率为

$$P_1 = U_N I_N = 220 \times 77.42W = 17\,032.4W$$

电枢绕组感应电动势为

$$E_a = U_N - I_{aN}R_a = 220 - 75 \times 0.26V = 200.5V$$

电磁功率为

$$P_e = E_a I_{aN} = 200.5 \times 75W = 15\,037.5W$$

输出功率为

$$P_2 = P_e - p_{Fe} - p_{\Omega} = 15\,037.5 - 600 - 1989W = 12\,448.5W$$

输出转矩为

$$T_2 = P_2/\Omega = 12\,448.5 \times 60/(2 \times 3.14 \times 1000)N \cdot m = 118.9N \cdot m$$

（2）额定负载时的效率

$$\eta_N = P_2/P_1 = 12\,448.5/17\,032.4 = 73.1\%$$

例 1-8　一台他励直流电动机，$P_N = 1325kW$，$U_N = 750V$，$I_a = 1930V$，$n_N = 200r/min$，$R_a = 0.0161\Omega$。设电动机在额定负载下以 200r/min 的转速稳态运行，若负载的总制动转矩保持不变，试求

（1）在电枢回路突然串入电阻 $R_{ad} = 0.0746\Omega$，当串入电阻的最初瞬间和达到稳态运行时电动机的电枢电流和转速。

（2）减小电动机的励磁电流，使磁通 Φ 减少 10%，当达到稳态时电动机的电枢电流和转速。

解：

从他励直流电动机电压方程式，可得到额定负载时的感应电动势 E_a 为

$$E_a = U - I_aR_a = 750 - 1930 \times 0.0161\text{V} = 719\text{V}$$

（1）当在电枢回路突然串入电阻 R_{ad} 时，因惯性作用，电动机的转速不能突变，因此感应电动势 $E_a = 719\text{V}$ 保持不变，于是串入电阻的瞬间，电枢电流为

$$I_a = (U - E_a)/(R_a + R_{ad}) = (750 - 719)/(0.0161 + 0.0743)\text{A} = 343\text{A}$$

可见，串入电阻的瞬间电枢电流显著下降，电磁转矩也随之下降，使电动机的转速开始下降。

由于总的制动转矩保持不变，当电动机达到稳态时，电磁转矩将恢复到原来的数值，电枢电流也将回到原来的数值，即 $I_a = 1930\text{A}$。此时的感应电动势 E'_a 为

$$E'_a = U - I_a(R_a + R_{ad}) = 750 - 1930 \times (0.0161 + 0.0743)\text{V} = 575\text{V}$$

稳态时，励磁电流和电枢电流与串入电阻前一样，因此磁通 Φ 未变，而感应电动势与转速成正比，所以稳态时电动机的转速 n' 为

$$n' = nE'_a/E_a = 200 \times 575/719\text{r/min} = 160\text{r/min}$$

由此可见，转速降低的程度与串入电阻 R_{ad} 的大小成正比。

（2）当磁通 Φ 降低 10%，即为原来的 0.9 倍时，由于总制动转矩不变，最后稳态时电磁转矩一定回到原来的数值。由于调速前后磁通之比 $\Phi/\Phi'' = 1/0.9$，故稳态时电机的电枢电流 I''_a 和感应电动势 E''_a 分别为

$$I''_a = I_a\Phi/\Phi'' = 1930 \times 1/0.9\text{A} = 2145\text{A}$$

$$E''_a = U - I''_aR_a = 750 - 2145 \times 0.0161\text{V} = 715\text{V}$$

于是，可得到稳态时电机的转速为

$$n'' = nE''_a\Phi/E_a\Phi'' = 200 \times 715/(719 \times 0.9)\text{r/min} = 221\text{r/min}$$

可见，当磁通减小时，电动机的转速将升高。

例 1-9 一台他励直流电动机，$U_N = 220\text{V}$，$n_N = 1500\text{r/min}$，$I_N = 41.1\text{A}$，$R_a = 0.4\Omega$，在额定负载转矩情况下：

（1）如在电枢回路串入电阻 $R_{ad} = 1.65\Omega$，求串入电阻后电动机的稳态转速。

（2）如将电源电压降低为 110V，求电动机的稳态转速。

（3）若将磁通减小 10%，求电动机的稳态转速。

解：

根据他励直流电动机的电压方程式可知：

$$C_e\Phi_N = (U_N - I_NR_a)/n_N = (220 - 41.1 \times 0.4)/1500 = 0.136$$

（1）在额定负载转矩下，电枢回路串电阻调速，达到稳态后，电枢电流将保持为额定电流不变（$I_a = I_N$），故电动机的转速将下降为

$$n = (U_N - I_NR_a - I_N R_\Omega)/ C_e\Phi_N$$

$$= (220 - 41.1 \times 0.4 - 41.1 \times 1.65)/0.136\text{r/min} = 1000\text{r/min}$$

（2）在额定负载转矩下，降低电源电压时，电枢电流将保持为额定电流不变（$I_a = I_N$），故电动机的转速将下降为

$$n = (U - I_NR_a)/ C_e\Phi_N = (110 - 41.1 \times 0.4)/0.136\text{r/min} = 687\text{r/min}$$

（3）在额定负载转矩下，通过减小磁通调速时，据 $T_e = C_e\Phi_N I_N = C_e\Phi I_a$，可得稳态电

枢电流为

$$I_a = \Phi_N I_N / \Phi = 41.1/0.9\text{A} = 45.7\text{A}$$

故电动机的转速将上升为

$$n = (U_N - I_a R_a) / C_e \Phi = (220 - 45.7 \times 0.4)/(0.136 \times 0.9)\text{r/min} = 1650\text{r/min}$$

1.8.2　串励直流电动机的运行特性

串励直流电动机的等效电路如图 1-43 所示，其特点是电枢电流、励磁电流与线路电流三者相等。电机的电压方程为

$$U = E_a + I_a(R_a + R_f) \tag{1-66}$$

电机的电磁转矩公式为

$$T_e = C_T \Phi I_a$$

串励直流电动机的每极磁通 Φ 与电枢电流 I_a 成正比，即 $\Phi = K_f I_a$，其中 K_f 是比率常数。于是电机的电磁转矩为

$$T_e = C_T \Phi I_a = C_T K_f I_a^2 \tag{1-67}$$

可见，串励直流电动机的电磁转矩与电枢电流的二次方成正比。

串励直流电动机的运行特性包括：

1）转速特性 $n = f(I_a)$。

2）转矩特性 $T_e = f(I_a)$。

3）效率特性 $\eta = f(I_a)$。

4）机械特性 $n = f(T_e)$。

由于串励直流电动机的电枢电流 I_a 与励磁电流 I_f 相等，因此气隙磁通随电枢电流 I_a 的变化而有较大变化。若不考虑磁饱和，串励电动机的电压方程和转矩方程为

$$\begin{cases} U = E_a + I_a R_a + I_f R_f + 2\Delta U_s = E_a + I_a(R_a + R_f) + 2\Delta U_s \\ E_a = C_e n \Phi = C'_e n I_a = C'_e n I_f \\ T_e = T_2 + T_0 \\ T_e = C_T \Phi I_a = C'_T I_a^2 = C'_T I_f^2 \end{cases} \tag{1-68}$$

式中，$\Phi = K_f I_f$，K_f 表示主磁通与励磁电流的比例系数；$C'_e = C_e K_f, C'_T = C_T K_f$。

串励直流电动机的工作特性是指，端电压为额定电压 U_N 时，转速 n、电磁转矩 T_e 和效率 η 与输出功率 P_2（或电枢电流 I_a）之间的关系，即 n，T_e，$\eta = f(P_2)$ 或 n，T_e，$\eta = f(I_a)$。

1. 转速特性 $n = f(I_a)$

不饱和时 $\Phi = K_f I_f$，可得串励电动机的转速为

$$n = \frac{U - I_a(R_a + R_f)}{C_e \Phi} = \frac{U}{C_e K_f I_a} - \frac{R_a + R_f}{C_e K_f} \tag{1-69}$$

可见转速 n 与电枢电流 I_a 成反比，即 $n = f(I_a)$ 为一条双曲线；所以串励直流电动机的转速 n 随电枢电流 I_a 的增大而迅速下降，如图 1-59 所示。

串励直流电动机在空载或轻载时，I_a 很小，$I_a = I_f$，因而每极磁通 Φ 也很小，由式（1-69）可知，电动机的转速将非常高，出现“飞车”现象，非常危险。因此，串励直流电动机不允许空载或轻载运行。

2. 转矩特性 $T_e = f(I_a)$

由式（1-68）可知，当串励直流电动机的磁路不饱和时，电磁转矩 $T_e = C_T\Phi I_a = C_T K_f I_a^2$，此时电磁转矩按电流二次方的比例增大。当电动机的磁路高度饱和时，$\Phi \approx$常数，电磁转矩按电流一次方的比例增大。一般情况下，串励直流电动机的电磁转矩按大于电流一次方的比例增大，如图1-59所示。与并励（他励）直流电动机相比，串励直流电动机的电磁转矩 T_e随电枢电流 I_a的增加而上升较快；转速 $n=0$ 时，串励直流电动机具有较大的起动转矩。过载时，电动机的转速会自动下降，而输出功率却变化不大；当负载减轻时，转速又会自动上升。所以串励直流电动机特别适用于电力机车等拖动场合。

3. 效率特性 $\eta = f(I_a)$

串励直流电动机的效率特性与他励直流电动机相似，如图1-59所示。

4. 机械特性 $n = f(T_e)$

由式（1-68），当串励直流电动机的磁路不饱和时，$\Phi = K_f I_f$，可得机械特性方程为

$$n = \frac{U}{C'_e}\frac{\sqrt{C'_T}}{\sqrt{T_e}} - \frac{R}{C'_e K_f} \tag{1-70}$$

式中，R 为电枢回路总电阻，包括电枢绕组电阻、励磁回路电阻、换向极绕组电阻和电刷接触电阻。

式（1-70）表明，此时的机械特性是一条双曲线，转速 n 大致与 $\sqrt{T_e}$ 成反比。电磁转矩 T_e增大，转速 n 迅速下降，特性很软，如图1-60所示。

1.8.3 复励直流电动机的运行特性

复励直流电动机中有串励和并励绕组，当串励和并励绕组的磁动势相加时，电动机为积复励，通常并励绕组的作用强于串励绕组的作用。

图1-44为复励直流电动机的等效电路。当电动机空载运行时，串励绕组中的电枢电流很小，可忽略不计，此时并励绕组起主要作用，电动机近似于并励直流电动机，不会出现空载飞车的现象。

由于复励直流电动机同时具有并励绕组和串励绕组，因此其工作特性介于并励直流电动机和串励直流电动机之间。并励绕组的磁动势起主要作用时，复励直流电动机的特性接近于并励直流电动机的特性，但当电枢反应的去磁作用较强时，仍能获得下降的转速特性，以保证电动机的稳定运行，此时的串励绕组称为“稳定绕组”。串励绕组的磁动势起主要作用时，复励直流电动机的特性接近于串励直流电动机的特性，但空载时不会出现“飞车”的危险，即复励直流电动机可以空载或轻载运行。为避免运行时产生不稳定现象，通常采用积复励。

复励直流电动机的转速特性曲线如图1-61中曲线2、3、5所示。由图可知，其转速特性介于并励和串励直流电动机的转速特性之间。

复励直流电动机的机械特性曲线如图1-62中的曲线2所示。积复励直流电动机的用途很广，常用于无轨电车的拖动。

1.8.4 直流电动机旋转方向的改变方式

直流电动机的旋转方向可以通过如下任意一种方法来改变（原始连接见图1-63a）：

1）改变电枢绕组的连接（见图1-63b）；

2）同时改变并励和串励绕组的连接（见图1-63c）。

1.9　并励直流电动机的转速调节

电力拖动系统的电气调速方法是人为改变电动机的电气参数，使电力拖动系统运行在不同的人为机械特性上，从而在相同的负载下获得不同的速度。

根据并励直流电动机的转速公式 $n=\dfrac{U-I_a(R_a+R_{ad})}{C_e\Phi}$ 可知，改变电动机的转速可以有3种方法：

1）调节电枢外串电阻 R_{ad}，使转速从 n_N 向下调节。

2）调节电枢端电压 U，使转速从 n_N 向下调节。

3）调节励磁磁通 Φ，使转速从 n_N 向上调节。

1. 电枢回路串电阻调速

电动机外加额定电压 U_N，励磁电流为额定励磁电流 I_{fN} 时，通过调节电枢回路串联的附加电阻 R_{ad} 来实现转速的改变。

图1-64表示电枢回路串入电阻的调速过程，图1-64a中 $R_{ad3}>R_{ad2}>R_{ad1}>0$。假定电动机拖动恒转矩负载 T_L，相应的转速为 n_1、n_2、n_3、n_4，且 $n_1>n_2>n_3>n_4$。当电枢回路串联附加电阻 R_{ad1} 时，由于机械惯性，转速 n_1 和感应电动势 E_a 不能突变，电动机的工作点由原来的 a 点水平跃变到 b 点，电枢电流和电磁转矩相应减小，使得 $T_{eb}<T_L$，于是系统减速，这是调速的第一阶段。随着转速 n 及感应电动势 E_a 的下降，电枢电流 I_a 及电磁转矩 T_e 将开始回升，使工作点从 b 点移向 c 点，直到转速达到 n_2，$T_{ec}=T_L$ 时，重新建立新的平衡，调速过程结束，系统将运行于 c 点，这是第二阶段。调速过程中转速 n 和电枢电流 i_a 随时间的变化曲线如图1-64b所示。

电枢串电阻调速的经济性较差，一般用于串励或复励直流电动机拖动的电车等生产机械上。

2. 降低电源电压调速

降低电源电压调速简称降压调速，需要专用的调压直流电源。降压调速时电枢回路中没有附加电阻，且励磁电流为额定励磁电流 I_{fN}，通过平滑地调节电枢外加电压，可以得到一族平行于固有机械特性的人为机械特性，如图1-65a所示。降压调速的调速过程与电枢回路串联电阻调速基本相似，调速过程中电枢电流和转速随时间的变化曲线如图1-65b所示。降压调速多用于对调速性能要求较高的生产机械上，如机床、轧钢机、造纸机等。

3. 弱磁调速

弱磁调速是在电枢外加额定电压 U_N、电枢回路中不附加电阻，通过减小励磁电流 I_f（即减小每极磁通 Φ）来实现转速调节。图1-66a为弱磁调速时的机械特性曲线，其中曲线1表示固有机械特性，曲线2表示减小磁通后的人为机械特性。图1-66b表示弱磁调速过程中电枢电流和转速随时间的变化曲线。

假设系统原来在 a 点运行，当减小励磁电流 I_f 时，每极磁通 Φ 将相应减小，由于机械惯性，转速 n_1 不突变，故感应电动势 E_a 将随 Φ 的减小而减小，从而使电枢电流 $I_a=(U-E_a)/R_a$ 增大，而且 I_a 的增大倍数比 Φ 减小的倍数大，使电磁转矩 T_e 随电枢电流 I_a 的增大而增大，工作点由 a 点移到 b 点，这是调速的第一阶段。在 b 点，出现 $T_e > T_L$，$dn/dt > 0$，系统加速，转速从 n_1上升，使感应电动势 E_a回升，从而引起电枢电流 I_a回落，电磁转矩 T_e减小，工作点从 b 点向 c 点移动，当转速到达 n_2时，$T_e = T_L$，系统达到新的稳定点 c，调速过程结束，这是调速的第二阶段。

调速过程中电枢电流和转速随时间的变化曲线如图 1-66b 所示。图中 I_{a1} 指调速前与 n_1 对应的稳态电枢电流，I_{a2} 指调速后与 n_2对应的稳态电枢电流。由于励磁回路的电感较大，因此磁通不可能突变，电磁转矩的实际变化曲线如图 1-66a 中的曲线 3 所示。

需要特别注意，弱磁调速时不允许励磁回路断线。若励磁回路开路，磁通 Φ 变为剩磁，此时电枢电流 I_a将很大，转速将飞速上升，可能将电枢损坏，因此必须对励磁电路采取相应的保护措施。为了扩大调速范围，常常把降压调速和弱磁调速结合起来，在额定转速以下采用降压调速，在额定转速以上采用弱磁调速。

4. 并励直流电动机三种调速方式中功率与转矩的变化特点

电动机调速时功率与转矩的容许输出是调速的技术指标之一。电动机在额定转速下容许输出的功率和转矩主要决定于电动机的发热，而电动机的发热主要取决于电枢电流的大小。在调速过程中，只要在不同转速下电枢电流 I_a不超过额定值 I_N，电动机就能长时间安全运行。所以，额定电流是判定电动机能否长期运行的限度（忽略电动机低速运行时散热情况变差所产生的影响）。当电枢电流保持为额定值 I_N时，电动机的容量就能充分利用。

（1）电枢回路串电阻调速和降低电源电压调速　电枢回路串电阻调速和降低电源电压调速的条件是，励磁电流保持为额定励磁电流 I_{fN}（即每极磁通为额定磁通 Φ_N）不变。若在不同转速时保持电枢电流为额定值 I_N不变，则调速过程中电动机的最大电磁转矩为

$$T_{emax} = C_T \Phi I_{amax} \tag{1-71}$$

其最大输出功率为

$$P_{max} = T_{emax} n/9.55 \tag{1-72}$$

由式（1-63）和式（1-64）可知，电枢回路串联电阻调速和降低电源电压调速时，其容许输出转矩为常数，不随转速变化，因此称为恒转矩调速方式；而容许输出功率与转速成正比，随转速下降而减小，转速范围为 $0 \sim n_{base}$，如图 1-67 所示。

（2）弱磁调速　在弱磁调速过程中，$\Phi \neq$常数，所以电磁转矩 T_e也随之变化，即 $T_e \neq$常数。若在不同转速时保持电枢电流为额定值 I_N不变，则根据他励直流电动机的电压方程式可得

$$\Phi = \frac{U_N - I_N R_a}{C_e n} = \frac{C_2}{n}$$

式中，C_2为比例常数。

弱磁调速过程中的电磁转矩为

$$T_e = C_T \Phi I_N = C_T C_2 I_N / n = C_3 / n$$

电磁功率为

$$P_e = T_e n/9.55 = C_3/9.55 = C$$

式中，C_3为比例常数。

可见，采用弱磁调速时，其容许输出功率为常数，不随转速变化，因此称为恒功率调速方式；而容许输出转矩则与转速 n 成反比，随转速的增大而减小，转速从 n_{base} 增大，如图 1-67 所示。

1.10　直流电动机的起动

电动机通入电源后，从静止到达一种稳态转速的过程称为起动过程。若直接加上额定电压起动，电动机在转速 $n=0$ 时的电枢电流，称为起动电流，用 I_{st}表示；相应的电磁转矩称为起动转矩，用 T_{st}表示。通常对电动机起动的要求是：

1）要有足够大的起动转矩 T_{st}。

2）起动电流 I_{st}必须在允许范围内。

3）起动设备简单、经济、可靠。

直流电动机常用的起动方法有三种：直接起动、电枢回路串电阻起动和降压起动。

1. 直接起动

直接起动又称为全压起动。起动时先给励磁绕组加上励磁电流，使主磁场完全建立，然后把电枢接到输出额定电压的电源上，此时电枢电流与主磁极磁场相互作用，产生电磁转矩，电动机就会起动起来。直接起动时，开始时电动机的转速 $n=0$，电枢的感应电动势 E_a 亦为0，故起动电流 I_{st}很大，可达到额定电流 I_N的十几倍。过大的起动电流 I_{st}，会造成换向困难，容易引起环火，而且还会使电网电压发生瞬时跌落，影响电网上其他设备的正常工作；同时，过大的起动电流将产生很大的起动转矩，使传动机构受到很大的冲击力，容易损坏设备，所以直接起动只限于容量较小的直流电动机。

2. 电枢回路串电阻起动

为限制起动电流，起动时可在电枢回路串入可变电阻 R_{st}（通常称为起动电阻），待转速上升后再逐步将起动电阻切除。只要 R_{st}选得合适，就可将起动电流 I_{st}限制在允许的范围内。这种起动方法的缺点是，大容量电动机采用的起动变阻器较为笨重，而且起动过程中能量损耗也很大。

3. 降压起动

当直流电动机的电枢回路由专用调压直流电源供电时，通过调节加到电枢上的电压，也可限制起动电流。

降压起动是一种比较理想的起动方法。起动过程中的损耗小，起动比较平稳，但必须有专用的调压直流电源。此方法多用于要求经常起动的场合和大中型电动机的起动，实际使用的直流伺服系统多采用降压起动方法。

1.11　直流电动机的制动

根据电动机的电磁转矩 T_e和转速 n 方向之间的关系，可以把电动机分为电动和制动两种运行状态。当电磁转矩 T_e的方向与转速 n 的方向相同时，称为电动运行状态；此时电动机吸收电能，并将它转化为机械能输出。当电磁转矩 T_e的方向与转速 n 的方向相反时，称为制动运行状态；此时电动机吸收机械能，并将它转化为电能输出；或者同时吸收机械能和电能，将其消耗在电阻中。电动机的电气制动方法有能耗制动和反接制动。

1.11.1　能耗制动

能耗制动是把正常运行的电动机的电枢从电网断开，接上外加的制动电阻 R_{bk}构成闭合回路，如图1-68所示，在制动过程中，保持励磁电流的大小和方向不变。

假设电动状态时，电机的电枢电流 I_1 和感应电动势 E_0 的方向如图1-68a所示，若忽略电枢电阻压降 I_1R_a，则感应电动势 E_0 等于外加电压 U。

在制动瞬间，即打开开关的瞬间（见图1-68b），电机将继续旋转，但其转速将逐渐降低。由于励磁电流不变，使得电机的感应电动势 E_0 会随转速的降低而减小，此时电机为电枢绕组开路的发电机，电枢绕组感应电势 E_0 的方向与电动状态时的相同。

当电枢绕组与外部电阻 R_{bk} 连接，即将开关合向外部电阻 R_{bk} 时（见图1-68c），电机将成为发电机，向 R_{bk} 送电，此时电枢电流 I_2 的方向与 I_1 的方向相反，$I_2 = -\dfrac{E_0}{R_a + R_{bk}}$，从而电磁转矩 T_e 也将改变方向，并与转速 n 方向相反，使电机处于制动状态。制动过程中，电机靠系统的动能发电，并将发出的电能消耗在电枢回路的电阻上，因此称为能耗制动。

能耗制动时，$U = 0, R = R_a + R_{bk}$（R_{bk} 为制动电阻），则能耗制动的机械特性为

$$n = -\frac{R_a + R_{bk}}{C_e C_T \Phi^2} T_e \tag{1-73}$$

由式（1-73）可知，能耗制动的机械特性是通过原点，位于第二、四象限的直线，如图1-69所示。机械特性的斜率与电枢回路串入电阻 R_{bk} 时的人为特性的斜率相同，两条特性曲线相互平行。

改变制动电阻 R_{bk} 的大小，可以改变能耗制动机械特性曲线的斜率，从而改变起始制动转矩的大小，以及下放重物位能性负载的稳定速度。制动电阻 R_{bk} 越小，起始制动转矩亦越大，可以缩短制动时间，但下放位能性负载的稳定速度越小；但制动电阻过小，将会造成制动电流过大。通常最大制动电流限制在 $2 \sim 2.5I_N$。

能耗制动具有线路简单、可靠的特点，制动过程中不需要从电网吸收电功率，比较经济，并且当转速 $n = 0$ 时，电磁转矩 $T_e = 0$。对于反抗性恒转矩负载的拖动系统可以实现准确停车，也可用于稳定下放位能性重物。

1.11.2 反接制动

反接制动是把运行于电动状态下的电枢外接电压的极性突然改变，并同时在电枢回路中串入限流制动电阻 R_{bk}，如图1-70所示。

在正常的电动状态下（见图1-70a），电机的电枢电流 $I_1 = (U - E_0)/R_a$，其中 R_a 是电枢电阻。制动时（见图1-70b），保持励磁电流的大小和方向不变，将电枢电源反接，并接入制动电阻 R_{bk}，由于惯性，转速的大小和方向不变，故电枢绕组感应电动势 E_0 的大小和方向也不变，但由于电源反接，故电枢电流变为 $I_a = \dfrac{-U - E_0}{R_a + R_{bk}}$，其方向与电动状态时电枢电流的方向相反，从而使电磁转矩 T_e 的方向改变，使之与转速 n 的方向相反，于是电动机处于制动状态。制动时，电动机的机械特性为

$$n = -n_0 - \frac{R_a + R_{bk}}{C_e C_T \Phi^2} T_e \tag{1-74}$$

机械特性曲线如图1-71所示。

从机械特性曲线可知，反接制动时，电动机的工作点从原来电动机状态的 a 点水平跃变

到 b 点，此时电磁转矩反向，使转速迅速下降，沿直线 2 到达 c 点。c 点转速为零，但电磁转矩不为零，如果要求停车，应立即断开电源，否则电动机将反向运行。若负载是反抗性恒转矩负载，并且 c 点的电磁转矩值大于负载转矩，则电动机将反向起动，沿直线 2 到 d 点，在反向电动状态下稳定运行；若在 c 点电动机的电磁转矩小于负载转矩，则电动机将堵转，此时也必须立即切断电源。若负载是位能性恒转矩负载，则电动机将反向起动，沿直线 2 到 f 点，在反向回馈制动状态下稳态运行。

反接制动过程中（图 1-71 中 bc 段），电压、电枢电流和电磁转矩均为负，而转速和感应电动势为正，输入功率 $P_1 = UI_a > 0$，说明电源仍向电动机输入电功率；电动机的输出功率 $P_2 = T_e\Omega < 0$，说明负载从轴上向电动机输入机械功率；电动机的电磁功率 $P_e = E_aI_a < 0$，说明轴上输入的机械功率转化为电枢回路的电功率。由此可见，反接制动时，电源输入的电功率和负载输入的机械功率转化成电功率后，全部消耗在电枢回路的电阻上。

图 1-72 为当初始制动电流相同时，不同制动方式下电动机的转速随制动时间变化的曲线。从图中可见，在 $2t_0$时刻，反接制动的电动机转速将减小为零，而对于能耗制动的电动机，此时的转速为初始制动转速的 25%。可见反接制动能快速使电机停车。

1.12 直流电机的换向

换向问题是装有换向器电机的一个独有问题。换向不良，会在电刷下产生有害的火花。当火花超过一定程度时，会烧坏换向器和电刷，从而影响电机的正常运行。换向过程十分复杂，有电磁、机械和电化学等各方面的因素相互交织在一起。本节主要介绍直流电机换向过程中的电磁现象以及如何改善换向。

从直流发电机的工作原理可知，电枢绕组中的电动势和电流是交变的，通过换向器和电刷的作用，在电刷间则可获得直流电压和电流。当旋转的电枢元件从一条支路经电刷而进入另一条支路时，元件中的电流要改变方向，这种元件中电流方向的改变过程称为换向。图 1-73 为单叠绕组元件中电流的换向过程。

1. 换向的电磁过程

如图 1-73 所示，假设电刷的宽度与换向片的宽度相等，电刷不动，换向器从右往左运动。在图 1-73a 中，电刷与换向片 1 接触时，元件 1 属于电刷右边的支路，其电流方向如图中箭头所示，设此时元件 1 中的电流为 $+i_a$。随着电枢的旋转，电刷与换向片 1、2 同时接触，如图 1-73b 所示，此时元件 1 被电刷短路，元件 1 进入换向过程，其中的电流为 i。随着电枢的进一步旋转，电刷与换向片 2 接触，如图 1-73c 所示，此时元件 1 将属于电刷左边的支路，此时元件 1 中电流变为 $-i_a$，换向结束。元件 1 称为换向元件。换向过程所经历的时间称为换向周期 T_c，换向周期 T_c很短，一般只有千分之几秒。

在换向期间，换向元件 1 中的电流将从 $+i_a$变为 $-i_a$。在理想情况下，若换向元件中无任何电动势的作用，且电刷与换向片间的接触电阻与接触面积成反比，则换向元件中的电流从 $+i_a$变为 $-i_a$的变化规律大致为一直线，如图 1-74 中的直线 1 所示，这种情况称为直线换向。直线换向是良好的换向。实际上在换向过程中，换向元件中会出现电抗电动势和电枢反应电动势，它们将会影响换向电流的变化。

（1）电抗电动势　在换向过程中，换向元件中的电流要发生变化。由于换向元件本身是一个线圈，因而自身电流的变化必然会引起漏磁自感电动势；另外，在电机中同时处于换向的元件不止一个，它们之间还会产生漏磁互感电动势。漏磁自感和互感电动势之和总称为电抗电动势。根据楞次定律，电抗电动势将阻碍换向元件中电流的变化，使换向延迟。

（2）电枢反应电动势 通常电刷放置在几何中性线上，所以换向元件的有效边将处于几何中性线上。虽然在几何中性线处主磁极的气隙磁通密度等于零，但该处交轴电枢磁场的磁通密度却不等于零，因此换向元件的有效边切割此电枢磁场后将产生旋转电动势，称为电枢反应电动势。根据楞次定律可以判定，电枢反应电动势也将阻碍换向元件中电流的变化，使换向延迟，如图 1-74 中曲线 2 所示。严重的延迟换向，会使后电刷边出现火花，损伤换向器表面。

2. 改善换向的方法

改善换向的方法，均从减小和消除上述这两种电动势着手。常用的方法有以下几种。

（1）装换向极 换向极安装在相邻的主磁极之间的几何中性线处，且换向极绕组与电枢绕组串联，如图 1-75 所示。换向极的磁动势除了应抵消交轴电枢磁动势外，还应当剩余一部分磁动势以产生换向极磁场，使换向元件切割该磁场后产生一个换向电动势，以抵消电抗电动势的作用，从而使换向元件中的合成电动势为零，使换向变为良好的直线换向。

换向极磁场的极性与该处的电枢磁场的极性相反，如图 1-75 所示。对发电机而言，换向极的极性应与顺着旋转方向的下一个主磁极的极性相同；电动机则相反。另外，为了使负载变化时，换向极磁动势也能相应变化，以便在带任何负载时换向元件中的合成电动势始终为零，故要求换向极绕组必须与电枢绕组串联，并保证换向极磁路不饱和。

（2）正确选用电刷 电刷的接触电阻主要与电刷材料有关，目前常用的电刷有石墨电刷、电化石墨电刷和金属石墨电刷等。电化石墨电刷的接触电阻最大，石墨电刷的接触电阻次之，金属石墨电刷的接触电阻最小。从改善换向的角度来看，似乎应该采用接触电阻较大的电刷，但接触电阻大，则接触电压降也升高，使电刷损耗和换向器的发热加剧，对换向不利，所以应当合理选用电刷。

3. 补偿绕组

当直流电机的换向不利时，电刷下的火花与换向片间的电位差火花汇合在一起，可使换向器表面的整个圆周上发生环火。环火能把电刷和换向器表面烧坏，因此，在大容量直流电机中，常常在主磁极极靴上专门冲出一些均匀分布的槽，槽中放置一套补偿绕组。补偿绕组与电枢绕组串联，其磁动势方向与交轴电枢磁动势相反，以消除电位差火花和环火。

本章小结

直流电机的工作原理建立在电磁感应定律和电磁力定律的基础之上。由于电刷和换向器的作用，使直流发电机电枢绕组感应的交变电动势，在电刷端输出时变成直流电动势；也使直流电动机电刷两端的直流电流，被引入到电枢绕组中后产生恒定方向的电磁转矩。换言之，无论是在直流电动机还是在直流发电机中，电机外电路中的电压、电流及电动势都是直流性质，而电枢元件中的电流及电动势均是交变性质。在学习过程中应注意以下几点。

1）旋转电机的实物模型主要包括静止的定子和旋转的转子，以及定子和转子之间的气隙。定子主磁极的作用是建立主磁场；转子电枢的作用是产生电磁转矩和感应电动势，实现机电能量转换。

直流电机的铭牌数据包括额定功率、额定电压、额定电流、额定转速及额定励磁电流等，他们是正确选择电机的依据，所以要充分理解其含义。

直流电机按照励磁方式可分为他励和自励两大类，其中自励又分为并励、串励和复励。励磁方式不同，直流电机的特性也就不同。

2）直流电机的物理模型指电机内的电路和磁场分布。

电枢绕组是直流电机的主电路。直流电枢绕组是由许多相同的绕组元件，按一定规律连接起来的闭合绕组。单叠绕组是将同一个主磁极下的元件串联起来构成一条支路，所以单叠绕组的支路数与电机的主磁极数相等。

直流电机内部的磁场是主磁极磁场和电枢磁场的合成。电机空载时，只有主磁极磁场，其气隙磁感应强度是空间位置静止的平顶波；当电机带上负载后，电枢绕组中有电流流过，载流绕组将产生电枢磁场，交轴电枢磁场在气隙中的分布是空间位置静止的马鞍形波；利用叠加原理将两个空间位置相对静止的磁感应强度波叠加，可得到带负载时电机的气隙磁场。电枢磁场对主磁极磁场的影响称为电枢反应。当电刷位于几何中性线时，电枢反应是交轴性质的，其作用是使主磁极磁场发生畸变，饱和时还有一定的去磁作用。

直流电机负载运行时，电枢绕组将产生感应电动势和电磁转矩。感应电动势的计算公式为 $E_a = C_e \Phi n$，电磁转矩的计算公式为 $T_e = C_T \Phi I_a$。

3）直流电机的基本方程是指直流电机的电压方程式、转矩方程和功率方程，利用它们可以分析电机的运行特性。

直流发电机运行时，感应电动势大于电枢端电压（即 $E_a > U$）。它的运行特性有空载特性、外特性和调节特性。他励直流发电机的空载特性与磁路的磁化曲线相似，可以用试验或计算法获得。他励直流发电机的外特性是略微下垂的曲线，同样可以用试验方法测得。并励直流发电机的自励必须满足三个条件：①电机的主磁路必须有剩磁；②电枢绕组与励磁绕组的连接必须正确，使励磁电流产生的磁动势与剩磁方向一致；③励磁回路的总电阻必须小于该转速下的临界电阻。

直流电动机运行时，电枢端电压大于电枢绕组感应电动势（即 $U > E_a$）。其运行特性包括工作特性和机械特性。不同励磁方式的直流电动机的工作特性是不同的。他励直流电动机的工作特性是指电枢电压为额定电压、电枢回路无外串电阻、励磁电流为额定励磁电流时，电机的转速、电磁转矩和效率与电机输出功率（或电枢电流）之间的关系，即 $n, T_e, \eta = f(P_2)$ 或 $n, T_e, \eta = f(I_a)$。他励直流电动机的特点是：在励磁电流不变时，磁通基本不变，所以负载变化时电机转速变化很小，基本是一种恒速电动机，且电磁转矩也基本正比于电枢电流。并励直流电动机的特性与他励直流电动机的特性相似，而串励直流电动机的特性与他励和并励时有很大差别。随着负载的变化，串励直流电动机励磁电流和主磁通同时变化，所以负载增大时，转速下降很快，电磁转矩近似正比于电枢电流的二次方，所以串励直流电动机的起动转矩和过载能力较大，但在空载时会产生“飞车”现象，所以串励直流电动机不允许空载或轻负载运行。

4）直流电动机的机械特性是指电动机的转速与电磁转矩之间的关系曲线，即 $n = f(T_e)$。他励直流电动机的机械特性表达式为

$$n = \frac{U}{C_e \Phi} - \frac{R_a + R_\Omega}{C_e C_T \Phi^2} T_e$$

当 $U = U_N$，$\Phi = \Phi_N$，$R_\Omega = 0$ 时，为固有机械特性；分别改变 U、Φ 和 R_Ω 时，可得到不同的人为机械特性，故人为机械特性有三种。

5）直流电动机的起动需要有足够大的起动转矩和尽可能小的起动电流，所以中、大型

直流电动机一般不允许直接起动。为限制起动电流，可在电枢回路中串联电阻或降低电源电压。

6）直流电动机处于制动状态运行的特征是：电动机的电磁转矩与旋转方向相反。制动的目的是为了迅速停车或把拖动系统的转速限定在一定范围内。他励直流电动机常用的电气制动方法有三种：能耗制动、反接制动和回馈制动。应该理解制动运行状态的三个方面：一是制动产生的条件；二是制动过程中工作点的移动过程；三是制动的稳态运行。

7）调速是电动机应用的重要方面，他励直流电动机有三种调速方法：电枢回路串联电阻调速、降低电源电压调速和弱磁调速。

习　题

1. 请判定在下列情况下，直流发电机电刷两端电压的性质？

1）磁极固定，电刷与电枢同时旋转。

2）电枢固定，电刷与磁极同时旋转。

3）电刷固定，磁极与电枢以不同的速度同时旋转。

2. 一台直流电动机，其额定数据为：$P_N = 17kW$，$U_N = 220V$，$n_N = 1500r/min$，$\eta_N = 0.85$，试求该电动机的额定电流 I_N、额定输入功率 P_1 各为多少？

3. 一台直流发电机，其额定数据为：$P_N = 100kW$，$U_N = 230V$，$n_N = 1450r/min$，试求该发电机的额定电流 I_N 为多少？

4. 什么是电枢反应？它对直流电机的运行有何影响？对并励直流电动机的转矩-转速特性有何影响？

5. 直流电机中有哪些损耗？

6. 直流电动机的速度变化率是如何定义的？

7. 如何调节并励直流电动机的转速？

8. 他励和并励直流电动机运行特性的区别有哪些？

9. 并励直流电动机正常运行时，其励磁电路突然开路，将产生什么后果？

10. 直流电动机起动时，为何要在起动电路中串入起动电阻？

11. 如何改变他励直流电动机的旋转方向？

12. 如何改变并励直流电动机的旋转方向？

13. 并励直流发电机的空载电压是如何建立的？

14. 电枢反应对他励直流发电机的端电压有何影响？

15. 旋转电机中，电角度与机械角度有何关系？

16. 一台4极直流发电机，电枢绕组为单叠整距绕组，每极磁通 $\Phi = 3.5 \times 10^{-2} Wb$，总导体数 N = 157，求：

1）当电机转速为1200r/min时，电机的空载电动势为多少？

2）当每条支路电流为50A时，电机的电磁转矩为多少？

17. 一台并励直流电动机，已知 $U_N = 220V$，$I_N = 80A$，$R_a = 0.1\Omega$，额定励磁电压 $U_f = 220V$，励磁绕组电阻 $R_f = 88.8\Omega$，一对电刷接触压降为2V，附加损耗为额定功率的1%，$\eta_N = 0.85$，试求：

1）电动机的额定输入功率为多少？

2）电动机的额定输出功率为多少？

3）电动机的总损耗为多少？

4）电枢回路的铜损为多少？

5）电动机的附加损耗为多少？

6）电动机的机械损耗与铁损之和为多少？

18. 一台并励直流发电机，已知 $P_N = 6kW$，$U_N = 230V$，$n_N = 1450r/min$，$R_a = 0.921\Omega$，励磁绕组电阻 $R_f = 177\Omega$，$2\Delta U_S = 2V$，铁损和机械损耗之和为 313.9W，附加损耗为 60W，试求：

1）发电机带额定负载时的输入功率为多少？

2）发电机额定运行时的电磁功率为多少？

3）发电机额定运行时的电磁转矩为多少？

4）发电机额定运行时的效率为多少？

19. 一台并励直流发电机，已知 $P_N = 90kW$，$U_N = 230V$，$R_a = 0.04\Omega$，励磁绕组电阻 $R_f = 60\Omega$，铁损和机械损耗之和为 2kW，附加损耗为额定功率的 1%，试求：

1）发电机带额定负载时的输入功率为多少？

2）发电机额定运行时的效率为多少？

20. 一台 8 极，25kW，120V 的直流发电机，其电枢绕组为单叠绕组，电枢绕组有 64 个线圈，每个线圈 16 匝，发电机的额定转速为 2400r/min。

1）要让该发电机在空载时产生额定电压，其每极磁通应该为多少？

2）发电机在额定负载时的支路电流为多少？

3）发电机在额定负载时的电磁转矩为多少？

4）如果每匝导体的电阻为 0.011Ω，则该发电机的电枢电阻为多少？

21. 一台并励直流电动机，已知 $P_N = 17kW$，$U_N = 220V$，$n_N = 3000r/min$，$I_N = 89.9A$，$R_a = 0.114\Omega$，励磁绕组电阻 $R_f = 181.5\Omega$。若忽略电枢反应，试求：

1）电动机的额定输出转矩为多少？

2）电动机带额定负载时的电磁转矩为多少？

3）电动机带额定负载时的效率为多少？

4）当电枢电流 $I_a = 0$ 时，电动机的理想空载转速为多少？

22. 一台并励直流电动机，已知 $U_N = 220V$，$R_a = 0.316\Omega$，理想空载转速 $n_0 = 1600r/min$。试求当电枢电流 $I_a = 50A$ 时，电动机的转速和电磁转矩各为多少？

23. 一台并励直流发电机，铭牌数据为：$P_N = 23kW$，$U_N = 230V$，$n_N = 1500r/min$，$R_a = 0.1\Omega$，励磁绕组电阻 $R_f = 57.5\Omega$。若不计电枢反应和磁路饱和，将这台发电机改为并励直流电动机运行，把电枢两端和励磁绕组两端都接在 220V 的直流电源上，运行时维持电枢电流为额定值，试求：

1）电动机的转速为多少？

2）电动机的电磁功率为多少？

3）电动机的电磁转矩为多少

24. 一台他励直流电动机的额定数据为：$U_N = 220V$，$I_N = 41.1A$，$n_N = 1500r/min$，$R_a = 0.4\Omega$，保持额定负载转矩不变，求：

1）若在电枢回路串入 1.65Ω 电阻，问串入的瞬间，电枢电流的数值为多少？当电动机达到稳定运行后，其电枢电流和转速各为多少？

2）若将电源电压下降为110V，问降低的瞬间，电枢电流的数值为多少？当电动机达到稳定运行后，其电枢电流和转速各为多少？

3）若将磁通减小为额定磁通的90%，问减小的瞬间，电枢电流的数值为多少？当电动机达到稳定运行后，其电枢电流和转速各为多少？

25. 一台5.5kW，115V的并励直流电动机的额定转速为1750r/min，额定线电流为57.6A，$R_a = 0.5\Omega$，$R_f = 144\Omega$。在电动机额定运行时，突然在电枢回路串入0.5Ω的电阻，计算串入电阻瞬间的电枢感应电动势、电枢电流和电动机的电磁转矩，并计算电动机达稳态运行时的转速。

26. 一台5.5kW，115V的并励直流电动机的额定转速为1750r/min，额定线电流为57.6A，$R_a = 0.5\Omega$，$R_f = 144\Omega$。在电动机额定运行时，突然将每极磁通减小15%，计算磁通减小瞬间的电枢感应电动势、电枢电流和电动机的电磁转矩，并计算电动机达稳态运行时的转速。

27. 一台120V的并励直流电动机，其额定转速为900r/min，额定电枢电流为150A，电枢回路电阻为0.025Ω，计算该电动机带恒转矩负载，且磁通保持恒定时，在下列情况下的稳定转速：

1）电枢电压增加15%。

2）电枢电压降低15%。

28. 一台220V的并励直流电动机运行于1000r/min时的电枢电流为25A，电动机的电枢电阻$r_a = 0.2\Omega$。当电动机带恒转矩负载、电枢回路串入0.8Ω的电阻时，电动机的稳定转速是多少？

29. 一台他励直流电动机，已知$P_N = 2.2\text{kW}$，$U_N = 220\text{V}$，$n_N = 1500\text{r/min}$，$I_N = 11.26\text{A}$，$R_a = 0.5\Omega$，试求：

1）电动机的理想空载转速为多少？

2）电动机在额定负载下的电磁转矩为多少？

3）当$I_a = 0.5I_N$时，电动机的转速为多少？

4）当转速$n_N = 1530\text{r/min}$时，电动机的电枢电流为多少？

30. 一台他励直流电动机，已知$P_N = 10\text{kW}$，$U_N = 220\text{V}$，$n_N = 1500\text{r/min}$，$I_N = 53.4\text{A}$，$R_a = 0.4\Omega$，忽略电刷的接触压降，电动机带额定负载运行。试求：

1）额定运行时的电磁转矩、输出转矩和空载转矩。

2）理想空载转速和实际空载转速。

3）当电枢端电压为额定电压的一半时电动机稳定时的转速和电枢电流。

4）当磁通为额定磁通的4/5时，电动机稳定后的转速和电枢电流。

5）当电枢回路串入1.5Ω电阻时，电动机稳定后的转速和电枢电流。

31. 一台他励直流电动机，已知$P_N = 22\text{kW}$，$U_N = 220\text{V}$，$n_N = 1500\text{r/min}$，$I_N = 115\text{A}$，$R_a = 0.1\Omega$，当电动机拖动额定转矩负载$T_L = T_N$运行时，要求把转速降到1000r/min，不计电动机的空载转矩。试求：

1）当采用电枢回路串电阻调速时，应串入多大的电阻？

2）当采用降低电源电压调速时，应将电源电压降为多少？

3）上述两种情况下，拖动系统输入的电功率和输出的机械功率为多少？

32. 一台他励直流电动机，已知$P_N = 30\text{kW}$，$U_N = 220\text{V}$，$n_N = 1000\text{r/min}$，$I_N = 158.5\text{A}$，

$R_a=0.1\Omega$，负载转矩 $T_L=0.8T_N$，试求：

1）电动机的转速为多少？

2）电枢回路串入0.3Ω电阻时，电动机的稳定转速为多少？

3）电压降到188V时，降压瞬间的电枢电流和降压后电动机的稳定转速为多少？

4）将磁通减弱为额定磁通的0.8倍时，电动机的稳定转速和电枢电流？

33. 一台5.5kW，115V的并励直流电动机，额定转速为1750r/min，额定线电流为57.6A，励磁绕组电阻为144Ω，电枢回路电阻为0.15Ω。要使该电动机带15N·m的负载运行于900r/min时，需要在电枢回路串入多大的电阻？

34. 一台19kW，120V的并励直流电动机的额定转速为900r/min，额定电枢电流为175A，电枢回路电阻 $r_a=0.026\Omega$。计算：

1）该电动机的电磁转矩为多少？

2）要使电动机在半载时运行于700r/min，需要在电枢回路串入多大的电阻（假定磁通保持恒定）？

第2章 变 压 器

电力变压器是一种静止的电气设备，用来将某一数值的交流电压变成频率相同的另一种或几种数值不同的电压。电力变压器由铁心和套于其上的两个或多个绕组组成。

变压器的用途很广泛。在电力系统中需要长距离输送电能时，要用升压变压器提高系统的电压，这样在输送功率相同时可以降低线路上的电流，从而降低线路中的损耗，实现电能经济输送。在用电地区再用降压变压器将电能逐步从输电电压下降到配电电压，供给用户安全使用。因此变压器的数量和总容量比发电机的装机容量大得多。自问世以来变压器的基本工作原理没有改变，但技术上取得了长足的进步，性能、容量都显著提高，单台变压器的容量已超过1000MV·A。

本章主要以单相双绕组变压器为例，介绍变压器的工作原理和结构，采用方程、等效电路和相量图等方法对其空载和负载运行进行分析，这些方法也适用于对称运行的三相变压器的任意一相；然后介绍变压器的运行特性、三相变压器的特点和变压器的并联运行；最后，对自耦变压器和仪用互感器做简单介绍。

2.1 变压器的基本结构与额定值

2.1.1 变压器的基本结构

变压器是利用电磁感应原理工作的。最简单的变压器是由一个铁心和套在其上的两个绕组组成，如图2-1所示。通常，变压器的一个绕组接电源，称为一次绕组，该侧称为一次侧，其匝数用N_1表示；另一个绕组接负载，称为二次绕组，该侧称为二次侧，其匝数用N_2表示。

变压器主要部件是铁心和绕组，它们构成变压器的器身。变压器铁心的结构通常有心式和壳式两种。心式结构的铁心柱被绕组包围，如图2-1和图2-2a所示；壳式结构则是铁心包围在绕组的顶面、底面和侧面，如图2-2b和图2-3所示。图2-4为一台三相心式油浸电力变压器的结构示意图。

铁心是构成变压器磁路的主要部件，它由铁心柱和铁轭组成。铁心柱上套装绕组，铁轭为闭合磁路的其他部分。为了提高磁路的导磁性能，减少磁滞和涡流损耗，铁心通常用高磁导率、厚度为0.35mm或0.5mm、表面涂有绝缘漆的硅钢片叠成。

绕组是变压器的电路部分，由用绝缘材料包扎的铜线或铝线绕成，线型有扁线和圆线。对最常见的双绕组变压器，按连接对象的不同，接电源的绕组称为一次绕组，接负载的绕组称为二次绕组。按电压等级的不同，绕组可分为高压绕组和低压绕组。

从高、低压绕组的相对位置来看，变压器的绕组可分为同心式和交叠式两类。同心式绕组的高、低压绕组同心地套装在铁心柱上，如图2-2a所示。交叠式绕组的高、低压绕组沿铁心柱高度方向相互交叠地放置，如图2-2b所示。同心式绕组结构简单、制造方便，电力变压器中常采用这种结构。交叠式绕组多用于特种变压器中。

如图2-4所示，油浸式变压器除器身外，还包含油箱、储油柜、分接开关等部件，以及各种保护装置。

2.1.2 额定值

额定值是保证变压器能长期可靠工作，且具有良好性能的量值，标注在变压器的铭牌上，额定值符号下标用N来表示。变压器的额定值主要有以下几个。

1. 额定容量（S_N，单位为kV·A或MV·A）

额定容量表示变压器在额定工况下运行时输出功率的保证值，它指的是视在功率。对于三相变压器，额定容量指的是三相的总容量。

2. 额定电压（U_N，单位为V或kV）

额定电压表示变压器空载时，在额定分接头下各绕组端电压的保证值（有效值）。对于三相变压器，额定电压指的是线电压。

3. 额定电流（I_N，单位为A）

额定电流表示变压器带额定负载时各绕组长期运行允许通过的电流（有效值）。对于三相变压器，额定电流指的是线电流。

对单相变压器，额定容量、额定电压和额定电流之间的关系为

$$S_N = U_{1N}I_{1N} = U_{2N}I_{2N} \tag{2-1}$$

对三相变压器，则有

$$S_N = \sqrt{3}U_{1N}I_{1N} = \sqrt{3}U_{2N}I_{2N} \tag{2-2}$$

4. 额定频率（f_N，单位为Hz）

我国规定的标准工业用电频率为50Hz。

例2-1 一台三相变压器$S_N=200\text{kV}\cdot\text{A}$，$U_{1N}/U_{2N}=10/6.3\text{kV}$，Yd联结，求其高、低压绕组的额定电流$I_{1N}$和$I_{2N}$。

解：

$$I_{1N} = \frac{S_N}{\sqrt{3}U_{1N}} = \frac{200}{\sqrt{3}\times 10}\text{A} = 11.5\text{A}$$

$$I_{2N} = \frac{S_N}{\sqrt{3}U_{2N}} = \frac{200}{\sqrt{3}\times 6.3}\text{A} = 18.3\text{A}$$

所求均为线电流，如果需要额定的相电流，则按星形联结（用Y或y表示）和三角形联结（用D或d表示）的不同规律进行换算：

$$I_{1N\phi} = I_{1N} = 11.5A, I_{2N\phi} = I_{2N}/\sqrt{3} = 10.6\text{A}$$

2.2 变压器的空载运行

变压器一次绕组接额定功率、额定电压的交流电源，二次绕组开路的运行状态称为空载运行。

2.2.1 空载运行时的物理过程

如图2-5所示，变压器一次绕组接入正弦交流电压$\dot{U}_1$，一次绕组中将流过空载电流$\dot{I}_0$，

该电流将产生交变的磁通 $\dot{\Phi}$。因铁心的磁导率远大于空气的磁导率，绝大部分磁通将沿铁心闭合，并同时交链一次和二次绕组，该磁通称为主磁通。除主磁通外，有少量磁通主要沿非铁磁材料闭合，只交链一次绕组，称为一次绕组的漏磁通 $\dot{\Phi}_{1\sigma}$。根据电磁感应定律，主磁通在一次和二次绕组中分别产生感应电动势 $\dot{E}_1$ 和 $\dot{E}_2$，漏磁通只在一次绕组中感应漏感电动势 $\dot{E}_{1\sigma}$。且有

$$e_1 = -N_1\frac{\mathrm{d}\Phi}{\mathrm{d}t}, e_2 = -N_2\frac{\mathrm{d}\Phi}{\mathrm{d}t} \tag{2-3}$$

由于一次绕组和二次绕组的匝数不同，因此感应电动势也不同，忽略其他次要因素，则一次绕组和二次绕组的电压也不同，这就是变压器能实现变压的原理。

另外，空载电流在一次绕组的电阻 r_1 上将产生电阻压降 I_0r_1。二次绕组的开路电压为 $\dot{U}_{20}$。

图 2-5 中各物理量用相量表示，正方向选取惯例为：一次侧电流 $\dot{I}_0$ 的正方向与电压 $\dot{U}_0$ 的正方向符合电动机惯例；二次侧电流与电压的正方向符合发电机惯例；主磁通和漏磁通的正方向与产生它的电流符合右手螺旋关系；感应电动势的正方向与产生它的磁通符合右手螺旋关系。变压器空载运行时各物理量的关系如图 2-6 所示。

2.2.2 空载运行时的主要物理量

1. 空载电流

空载电流 $\dot{I}_0$ 由两部分组成，一部分是产生交变磁通的无功分量，又称磁化分量，起励磁作用，它与主磁通同相位，用 $\dot{I}_\mu$ 表示；另一部分是有功分量，又称铁损分量，它与磁滞和涡流损耗有关，铁损分量超前主磁通 90°，用 $\dot{I}_{Fe}$表示。从数量上看，有功分量仅为无功分量的 10% 左右，因此空载电流基本上是属于感性无功性质，通常称为励磁电流。空载电流的数值很小，一般仅占额定电流的 1% ~ 10%。

2. 主磁通及其感应电动势

设主磁通 Φ 随时间正弦变化，即 $\Phi = \Phi_m \sin\omega t$，其中 Φ_m表示主磁通的幅值，则有

$$\begin{cases} e_1 = -N_1\dfrac{\mathrm{d}\Phi}{\mathrm{d}t} = -N_1\omega\Phi_m\cos\omega t = \sqrt{2}E_1\sin(\omega t - 90°) \\ e_2 = -N_2\dfrac{\mathrm{d}\Phi}{\mathrm{d}t} = -N_2\omega\Phi_m\cos\omega t = \sqrt{2}E_2\sin(\omega t - 90°) \end{cases} \tag{2-4}$$

由式（2-4）知，e_1、e_2在相位上滞后主磁通 90°，其有效值 E_1、E_2为

$$\begin{cases} E_1 = \dfrac{\omega N_1\Phi_m}{\sqrt{2}} = \dfrac{2\pi f N_1\Phi_m}{\sqrt{2}} = 4.44fN_1\Phi_m \\ E_2 = \dfrac{\omega N_2\Phi_m}{\sqrt{2}} = \dfrac{2\pi f N_2\Phi_m}{\sqrt{2}} = 4.44fN_2\Phi_m \end{cases} \tag{2-5}$$

根据式（2-5），变压器的电压比为

$$k = \frac{E_1}{E_2} = \frac{N_1}{N_2} \tag{2-6}$$

一、二次绕组感应电动势的相量表达式分别为

$$\dot{E}_1 = -\mathrm{j}4.44 f N_1 \dot{\Phi}_m \tag{2-7}$$

$$\dot{E}_2 = -\mathrm{j}4.44 f N_2 \dot{\Phi}_m \tag{2-8}$$

由式（2-7）和式（2-8）可知，正弦变化的磁通产生的感应电动势大小与电源频率、绕组匝数和主磁通的幅值成正比，其相位滞后主磁通 90°。

3. 一次绕组的漏感电动势

一次绕组漏磁通 $\dot{\Phi}_{1\sigma}$ 在一次绕组中感应的漏感电动势 $\dot{E}_{1\sigma}$ 为

$$\dot{E}_{1\sigma} = -\mathrm{j}4.44 f N_1 \dot{\Phi}_{1\sigma} \tag{2-9}$$

由于漏磁通 $\dot{\Phi}_{1\sigma}$ 主要通过非铁磁材料闭合，磁路为线性，所以 $\dot{\Phi}_{1\sigma}$ 的大小与产生它的电流 $\dot{I}_0$ 成正比，且相位相同。于是 $\dot{E}_{1\sigma}$ 与 $\dot{I}_0$ 之间大小成正比，相位上滞后 90°，若比例系数用 x_1 表示，则式（2-9）可写作

$$\dot{E}_{1\sigma} = -\mathrm{j}\,\dot{I}_0 x_1 \tag{2-10}$$

式中，x_1 称为一次绕组的漏电抗，即可将漏磁通所感应的电动势用漏电抗压降的形式来表示。

根据基尔霍夫第二定律，可得一次和二次绕组的电压方程。对一次绕组有

$$\dot{U}_1 = -\dot{E}_1 - \dot{E}_{1\sigma} + \dot{I}_0 r_1 = -\dot{E}_1 + \mathrm{j}\dot{I}_0 x_1 + \dot{I}_0 r = -\dot{E}_1 + \dot{I}_0 z_1 \tag{2-11}$$

式中，z_1 为一次绕组漏阻抗，$z_1 = r_1 + \mathrm{j}x_1$，$|z_1|$ 数值上很小，若忽略 $\dot{I}_0 z_1$，则有

$$\dot{U}_1 \approx -\dot{E}_1 = \mathrm{j}4.44 f N_1 \dot{\Phi}_m \tag{2-12}$$

由式（2-12）可知，如果忽略一次绕组的漏阻抗压降，当电源频率和绕组匝数不变时，主磁通的大小主要由电源电压的大小 U_1 决定，当 U_1 一定时，主磁通将基本不变。这一点在变压器负载运行时也基本成立，对于分析变压器的运行很重要。

对二次绕组有

$$\dot{U}_{20} = \dot{E}_2 \tag{2-13}$$

2.2.3 变压器的空载等效电路和相量图

主磁通产生的感应电动势 $\dot{E}_1$ 也希望仿照漏感电动势，用励磁电流 $\dot{I}_m$ 所产生的电压降来表示。考虑到主磁通会引起铁损，因此不仅要引入一个与主磁通磁路所对应的电抗，还要引入一个表征铁损的等效电阻，即引入一个阻抗 z_m。也就是说，$\dot{E}_1$ 可以用 $\dot{I}_m$（空载时即为 $\dot{I}_0$）在阻抗 z_m 上的电压降来表示，即

$$-\dot{E}_1 = \dot{I}_0 z_m = \dot{I}_0 (r_m + \mathrm{j}x_m) \tag{2-14}$$

式中，z_m 称为励磁阻抗，$z_m = r_m + \mathrm{j}x_m$；$x_m$ 为励磁电抗，与主磁通的磁路相对应；r_m 为励磁电阻，是模拟铁损的等效电阻（$I_0^2 r_m$ 等于铁损）。

励磁阻抗随着铁心主磁通的饱和程度而变化，通常变压器运行时电源电压保持在额定值，所以励磁阻抗取额定电压时的值，可认为是常量。另外，由于主磁路的磁阻远远小于漏磁路的磁阻，因此 $x_m \gg x_1$，可得变压器空载时的等效电路如图 2-7a 所示。

变压器空载时的相量图可直观地反应各物理量之间的相位关系，如图 2-7b 所示，作图步骤如下：

1）以主磁通 $\dot{\Phi}_m$ 作为参考相量，绘于水平位置。

2）$\dot{E}_1$ 和 $\dot{E}_2$ 落后主磁通 90°，$\dot{E}_2 = \dot{U}_{20}$。

3）励磁电流 $\dot{I}_m$ 的无功分量 $\dot{I}_\mu$ 与主磁通同相位，有功分量 $\dot{I}_{Fe}$ 超前主磁通 90°，二者相加为 $\dot{I}_m$，空载时 $\dot{I}_m$ 即为空载电流 $\dot{I}_0$。

4）$-\dot{E}_1$ 依次加上 $\dot{I}_0 r_1$（与 $\dot{I}_0$ 平行）和 $\mathrm{j}\dot{I}_0 x_1$（领先 $\dot{I}_0$ 90°），得到 $\dot{U}_1$。

2.3 变压器的负载运行

变压器一次绕组接额定频率、额定电压的交流电源，二次绕组接入负载、且有电流流过时的运行状态，称为变压器的负载运行。

2.3.1 负载运行的物理过程

1. 各物理量之间的电磁关系

如图 2-8 所示，当变压器二次绕组接入负载 z_L 时，二次绕组中就有电流 $\dot{I}_2$ 流过，该电流产生磁动势 $\dot{F}_2$（$= N_2 \dot{I}_2$），此磁动势也作用在主磁路上，试图改变主磁通 $\dot{\Phi}$。

由空载运行分析可知，当电源电压 $\dot{U}_1$ 不变时，主磁通 $\dot{\Phi}$ 基本保持不变，因此一次绕组的电流必将相应地由 $\dot{I}_0$ 变为 $\dot{I}_1$，一次绕组的磁动势也由 $\dot{F}_0$ 变为 $\dot{F}_1$（$= N_1 \dot{I}_1$），以抵消磁动势 $\dot{F}_2$ 对主磁通的影响，从而保持主磁通基本不变。此外，$\dot{F}_1$ 和 $\dot{F}_2$ 还将分别产生仅与各自绕组交链的漏磁通 $\dot{\Phi}_{1\sigma}$ 和 $\dot{\Phi}_{2\sigma}$，并在一次和二次绕组中感应漏感电动势 $\dot{E}_{1\sigma}$ 和 $\dot{E}_{2\sigma}$。它们分别可以用漏抗压降的形式来表示，即 $\dot{E}_{1\sigma} = -\mathrm{j}\dot{I}_1 x_1$，$\dot{E}_{2\sigma} = -\mathrm{j}\dot{I}_2 x_2$，其中 x_2 为二次绕组的漏磁电抗（简称漏抗）。x_2 与二次绕组的漏磁通相对应，其大小与二次绕组的频率和匝数的二次方成正比，与二次绕组的漏磁通所经磁路的磁阻（$R_{2\sigma}$）成反比，即 $x_2 = 2\pi f \frac{N_2^2}{R_{2\sigma}}$，$x_2$ 也是常量。最后，一次和二次绕组的电流将分别在一次和二次绕组的电阻中产生电阻压降 $\dot{I}_1 r_1$ 和 $\dot{I}_2 r_2$，其中 r_2 为二次绕组的电阻。变压器负载运行时各物理量的关系如图 2-9 所示。

2. 磁动势平衡方程

由前可知，当电源电压 $\dot{U}_1$ 不变时，主磁通 $\dot{\Phi}$ 基本保持不变，因此一次绕组的电流需要由 $\dot{I}_0$ 变为 $\dot{I}_1$，磁动势也由 $\dot{F}_0$ 变为 $\dot{F}_1 = N_1 \dot{I}_1$，以抵消磁动势 $\dot{F}_2$ 对主磁通的影响，从而保持主磁通基本不变。有

$$N_1\dot{I}_{\mathrm{m}} = N_1\dot{I}_1 + N_2\dot{I}_2 \tag{2-15}$$

将上式两边同除以 N_1 并移项可得

$$\dot{I}_1 = \dot{I}_{\mathrm{m}} + \left(-\frac{1}{k}\dot{I}_2\right) = \dot{I}_{\mathrm{m}} + \dot{I}_{1\mathrm{L}} \tag{2-16}$$

式中，$\dot{I}_{1\mathrm{L}} = -\frac{1}{k}\dot{I}_2$。

由式（2-16）可知，变压器负载运行时，一次绕组电流包含两个分量：一个是产生主磁通的励磁分量 $\dot{I}_{\mathrm{m}}$；另一个是负载分量 $\dot{I}_{1\mathrm{L}}$，用以抵消二次侧电流磁动势对主磁通的影响。

负载运行时，励磁电流 $\dot{I}_{\mathrm{m}}$ 所占份额很小，若将其忽略，则式（2-16）可近似写成

$$\dot{I}_1 \approx -\frac{1}{k}\dot{I}_2 \tag{2-17}$$

上式表明，$\dot{I}_1$ 和 $\dot{I}_2$ 在相位上相差近 180°，数值上为 $\frac{1}{k}$ 倍，即一次和二次绕组的电流与其匝数近似成反比关系，即变压器在变压的同时也改变电流的大小，这从能量守恒的原理看也是必然的。

3. 电压方程

一次和二次绕组中漏磁通产生的感应电动势分别用电流流过漏电抗的压降表示，则根据基尔霍夫第二定律，可得一次和二次绕组的电压方程分别为

$$\dot{U}_1 = -\dot{E}_1 - \dot{E}_{1\sigma} + \dot{I}_1 r_1 = -\dot{E}_1 + \mathrm{j}\dot{I}_1 x_1 + \dot{I}_1 r = -\dot{E} + \dot{I}_1 z_1 \tag{2-18}$$

$$\dot{U}_2 = \dot{E}_2 + \dot{E}_{2\sigma} - \dot{I}_2 r_2 = \dot{E}_2 - \mathrm{j}\dot{I}_2 x_2 - \dot{I}_2 r_2 = \dot{E}_2 - \dot{I}_2 z_2 = \dot{I}_2 z_{\mathrm{L}} \tag{2-19}$$

式中，z_2 称为二次绕组漏阻抗，$z_2 = r_2 + \mathrm{j}x_2$；$z_{\mathrm{L}}$ 是负载阻抗，$z_{\mathrm{L}} = R_{\mathrm{L}} + \mathrm{j}X_{\mathrm{L}}$。

2.3.2　变压器的等效电路和相量图

依据式（2-18）和式（2-19），可以得到变压器负载时一、二次侧的电路耦合关系如图 2-10 所示。

1. 绕组折算

由于变压器一次和二次绕组之间只有磁的耦合，为了得到一次和二次绕组间有电联系的等效电路，需要引入绕组折算的概念。

在变压器中，通常把二次侧折算到一次侧，用一个与一次绕组匝数相等的假想二次绕组来代替实际的二次绕组，且不改变变压器原有的电磁关系，即折算应满足“等效”的原则。习惯用原物理量右上角加一硬撇号（′）来表示折算后的物理量。

由于折算后二次绕组的匝数成为 N_1，因此可得

$$\dot{E}'_2 = -\mathrm{j}4.44fN_1\dot{\Phi}_{\mathrm{m}} = \dot{E}_1 = k\dot{E}_2 \tag{2-20}$$

二次绕组对一次绕组的影响是由于磁动势的作用，折算前、后应保证二次绕组的磁动势不变，可得

$$N_1\dot{I}' = N_2\dot{I}_2 \tag{2-21}$$

故

$$\dot{I}'_2 = \frac{N_2}{N_1}\dot{I}_2 = \frac{1}{k}\dot{I}_2 \tag{2-22}$$

再根据折算前、后二次绕组的有功功率损耗、无功功率损耗和输出功率不变，可得

$$I'^2_2 r'_2 = I_2^2 r_2, I'^2_2 x'^2_2 = I_2{}^2 x_2{}^2, I'^2_2 Z'^2_L = I_2^2 Z_L^2 \tag{2-23}$$

即

$$r'_2 = \left(\frac{I_2}{I'_2}\right)^2 r_2 = k^2 r_2, x'_2 = \left(\frac{I_2}{I'_2}\right)^2 x_2 = k^2 x_2, Z'_L = \left(\frac{I_2}{I'_2}\right)^2 Z_L = k^2 Z_L \tag{2-24}$$

或

$$z'_2 = k^2 z_2 \tag{2-25}$$

折算以后，负载的电压应为

$$\dot{U}'_2 = \dot{I}'_2 Z'_L = \frac{1}{k}\dot{I}_2(k^2 Z_L) = k\dot{U}_2 \tag{2-26}$$

综上所述，把二次侧折算到一次侧时，电动势和电压乘以变比 k，电流除以变比 k，阻抗乘以 k^2。

2. 等效电路和相量图

经过绕组折算后，变压器的基本方程变为

$$\begin{cases} \dot{U}_1 = -\dot{E}_1 + \dot{I}_1 z_1 \\ \dot{U}'_2 = \dot{E}'_2 - \dot{I}'_2 z'_2 = \dot{I}'_2 Z'_L \\ \dot{E}_1 = \dot{E}'_2 = -\dot{I}_m z_m \\ \dot{I}_1 = \dot{I}_m + (-\dot{I}'_2) \end{cases} \tag{2-27}$$

由折算后的基本方程（2-27），可画出变压器的 T 形等效电路，如图 2-11a 所示。

T 形等效电路的运算较麻烦。考虑到一次绕组漏阻抗压降很小，可将励磁支路移到一次绕组漏阻抗前，这样就得到图 2-11b 所示的近似等效电路，又称Γ形等效电路。进一步，若忽略励磁电流，可得如图 2-11c 所示的简化等效电路，计算十分方便，而且在许多场合，用简化等效电路就可满足要求。

在图 2-11c 中，$r_k = r_1 + r'_2$，称为短路电阻；$x_k = x_1 + x'_2$，称为短路电抗；$z_k = r_k + jx_k$，称为短路阻抗。短路阻抗是变压器的重要参数之一，从正常运行角度来看，希望它小一些，即变压器的漏阻抗压降小一些，使二次侧电压随负载变化的波动程度小一些；而从限制短路电流的角度来看，又希望它大一些。

若已知负载的情况和变压器的参数，即已知负载的电压 $\dot{U}'_2$、电流 $\dot{I}'_2$ 和功率因数角 φ_2，变压器的参数 $z_1 = r_1 + jx_1$、$z'_2 = r'_2 + jx'_2$、$z_m = r_m + jx_m$，根据基本方程（2-27），即可画出相量图，如图 2-12 所示。具体步骤如下：

1）画出负载的端电压 $\dot{U}'_2$ 和电流 $\dot{I}'_2$，$\dot{I}'_2$ 滞后于 $\dot{U}'_2$ 的角度为 φ_2。

2）在 $\dot{U}'_2$ 上依次加 $\dot{I}'_2 r'_2$（与 $\dot{I}'_2$ 平行）和 $j\dot{I}'_2 x'_2$（超前 $\dot{I}'_2$ 90°），可得二次绕组的电动势 $\dot{E}'_2$，由 $\dot{E}_1 = \dot{E}'_2$，即得 $\dot{E}_1$。

3）根据电动势公式（2-7），可画出主磁通 $\dot{\Phi}$，$\dot{\Phi}$ 超前 $\dot{E}_1$ 90°。

4）根据式（2-14），作励磁电流 $\dot{I}_m = -\dot{E}_1/z_m$，$\dot{I}_m$ 超前 $\dot{\Phi}$ 以铁损角 α，且 $\alpha = \arctan(r_m/x_m)$。

5）作 $-\dot{I}'_2$，并与 $\dot{I}_m$ 相加得 $\dot{I}_1$。

6）作 $-\dot{E}_1$，在 $-\dot{E}_1$ 上依次加上 $\dot{I}_1 r_1$ 和 $j\dot{I}_1 x_1$，可得电源电压 $\dot{U}_1$。$\dot{U}_1$ 与 $\dot{I}_1$ 的夹角为变压器负载运行时一次侧的功率因数角，用 φ_1 表示。

2.4　变压器的参数测定

变压器等效电路不仅具有理论意义，其中的参数可通过试验获得。对已经制造出来的变压器，可用空载试验和短路试验测定等效电路中的各阻抗参数。

2.4.1　空载试验

空载试验可测定变压器的励磁阻抗和电压比，图 2-13 是单相和三相变压器空载试验原理接线图。

试验时，将二次侧开路，一次侧绕组加额定频率的额定电压，测量一次侧的输入功率 p_0（空载损耗）、一次电压 U_{1N} 和电流 I_0，以及二次侧的开路电压 U_{20}。

根据测量数据 U_{1N}（一次侧额定电压）、U_{20}（二次侧开路电压）、I_0（一次侧空载电流）和 p_0（空载损耗），按等效电路图 2-11，可得电压比 k 和励磁阻抗 Z_m 为

$$k = \frac{U_{1N}}{U_{20}} \tag{2-28}$$

$$Z_m = \frac{U_{1N}}{I_0}, r_m = \frac{p_0}{I_0^2}, x_m = \sqrt{Z_m^2 - r_m^2} \tag{2-29}$$

为了试验安全和仪表选择方便，一般在低压侧加试验电压而高压侧开路。由于励磁阻抗的数值与铁心的饱和程度有关，即与外施电压有关，因此试验电压必须是额定电压。

由于空载试验是在低压侧加电压，所以测得的励磁参数是低压侧的数值，如果需要得到高压侧的数值，必须进行折算，即乘以 k^2，$k = N_{高压侧}/N_{低压侧}$。

2.4.2　短路试验

短路试验可测定变压器的短路阻抗，图 2-14 是单相和三相变压器短路试验原理接线图。试验时，将二次侧短路，一次侧加可调低电压，使短路电流达到额定值，测量此时一次侧的电压 U_k 和电流 I_{1N}，以及输入功率 P_k。

根据测量数据 U_k、P_k 和 I_{1N}，利用短路时的简化等效电路，可计算出短路阻抗为

$$Z_k = \frac{U_k}{I_{1N}}, r_k = \frac{P_k}{I_{1N}^2}, x_k = \sqrt{Z_k^2 - r_k^2} \tag{2-30}$$

在 T 形等效电路中，一般可认为 $r_1 = r'_2 = r_k/2$，$x_1 = x'_2 = x_k/2$。

由于电阻与温度有关，按国家标准，应将试验温度下测定的电阻值换算到 75℃时的值。对于铜线变压器按下式换算：

$$r = \frac{234.5 + 75}{234.5 + t} \tag{2-31}$$

式中，t 为试验时的环境温度。

对铝线变压器，式（2-31）中的常数 234.5 应改为 228。

为了便于测量，短路试验一般将变压器高压侧经调压器接入试验电源，而低压侧短路。为了避免试验电流过大，外加试验电压必须降低，以使试验电流为额定电流。

由于短路试验是在高压侧加低电压，所测得数据为高压侧值，故求得的短路阻抗也为高压侧值。如需要低压侧的数值，也要进行折算，即除以 k^2，$k = N_{高压侧}/N_{低压侧}$。

2.4.3 标幺值

在电力工程的计算中，往往不用各物理量的实际值，而是用实际值与同一单位的某一选定的基值之比来表示，称为标幺值，即

$$标幺值 = \frac{实际值}{基值} \tag{2-32}$$

标幺值是个相对值，没有单位。某物理量的标幺值，用原来符号的右上角加“*”表示。

1. 基值的选择

通常取各物理量的额定值作为基值，具体选择方法如下：

1）线电流、线电压的基值，选额定线值；相电流、相电压的基值，选额定相值。

2）电阻、电抗、阻抗共用一个基值，这些都是一相的值，故阻抗基值为

$$z_N = \frac{U_{N\phi}}{I_{N\phi}}$$

3）有功功率、无功功率、视在功率共用一个基值，以额定视在功率即变压器的容量为基值。

2. 标幺值的特点

1）额定电流、额定电压、额定视在功率的标幺值为 1。

2）变压器各物理量在本侧取标幺值和折算到另一侧取标幺值，两者相等，例如

$$U_2'^* = \frac{U_2'}{U_{1N}} = \frac{kU_2}{kU_{2N}} = \frac{U_2}{U_{2N}} = U_2^*$$

3）对于电流、电压，相值的标幺值与线值的标幺值相等；对于功率，单相值的标幺值与三相值的标幺值相等。

4）某些物理量的标幺值具有相同的数值，使计算简化。例如：

$$U_{kN}^* = \frac{U_{kN}}{U_{1N}} = \frac{I_{1N}z_k}{U_{1N}} = \frac{z_k}{U_{1N}/I_{1N}} = \frac{z_k}{z_{1N}} = z_k^*$$

5）便于比较变压器或电机的参数和性能。同类型的变压器，虽然容量和电压可以相差很大，但用标幺值表示的参数和典型的性能数据，通常都在一定的范围以内。例如，对于电力变压器，短路阻抗的标幺值 z_k^* 为 0.04 ~ 0.175；空载电流标幺值 I_0^* 为 0.02 ~ 0.1。

例 2-2 一台三相变压器，$S_N = 320\text{kV·A}$，$U_{1N}/U_{2N} = 6300/400\text{V}$，Y－△联结。空载和短路试验数据如下：

试验名称	线电压/V	线电流/A	总功率/W	备注
空载	400	27.7	1450	电压加在低压侧
短路	284	29.3	5700	电压加在高压侧

试求励磁阻抗、短路阻抗的标幺值和折算到高压侧的实际值。

解：

方法一：实际值计算法

（1）由空载试验数据求励磁阻抗　空载试验是在低压侧加电压，低压侧绕组采用三角形联结，换算为相值的空载电压、空载电流和空载损耗分别为

$$U_{2N\phi} = U_{2N} = 400\text{V}$$

$$I_{0\phi} = \frac{I_0}{\sqrt{3}} = \frac{27.7}{\sqrt{3}}\text{A} = 15.99\text{A}$$

$$p_{0\phi} = \frac{p_0}{3} = \frac{1450}{3}\text{W} = 483.33\text{W}$$

将以上各值代入式（2-29），得低压侧的励磁阻抗为

$$Z_m = \frac{U_{2N\phi}}{I_{0\phi}} = \frac{400}{15.99}\Omega = 25.02\Omega$$

$$r_m = \frac{p_{0\phi}}{I_{0\phi}^2} = \frac{483.33}{15.99^2}\Omega = 1.89\Omega$$

$$x_m = \sqrt{Z_m^2 - r_m^2} = \sqrt{25.02^2 - 1.89^2}\Omega = 24.95\Omega$$

（2）求电压比 k 和折算到高压侧的励磁阻抗值

$$k = \frac{U_{1N\phi}}{U_{2N\phi}} = \frac{U_{1N}}{\sqrt{3}U_{2N}} = \frac{6300}{\sqrt{3}\times 400} = 9.09$$

$$Z'_m = k^2 Z_m = 9.09^2 \times 25.02\Omega = 2067.35\Omega$$

$$r'_m = k^2 r_m = 9.09^2 \times 1.89\Omega = 156.17\Omega$$

$$x'_m = k^2 x_m = 9.09^2 \times 24.95\Omega = 2061.57\Omega$$

（3）由短路试验求短路阻抗　短路试验是在高压侧加电压，高压侧绕组采用星形联结，则换算为相值的短路电压、短路电流和负载损耗分别为

$$U_{k\phi} = \frac{U_k}{\sqrt{3}} = \frac{284}{\sqrt{3}}\text{V} = 163.97\text{V}$$

$$I_{k\phi} = I_k = 29.3\text{A}$$

$$p_{k\phi} = \frac{p_k}{3} = \frac{5700}{3}\text{W} = 1900\text{W}$$

将以上各值代入式（2-30），可得高压侧的短路阻抗为

$$Z_k = \frac{U_{k\phi}}{I_{k\phi}} = \frac{163.97}{29.3}\Omega = 5.60\Omega$$

$$r_k = \frac{p_{k\phi}}{I_{k\phi}^2} = \frac{1900}{29.3^2}\Omega = 2.21\Omega$$

$$x_k = \sqrt{Z_k^2 - r_k^2} = \sqrt{5.6^2 - 2.21^2}\Omega = 5.145\Omega$$

（4）求高压侧阻抗基值 z_{1N} 和阻抗的标幺值

$$I_{1N} = \frac{S_N}{\sqrt{3}U_{1N}} = \frac{320\times 10^3}{\sqrt{3}\times 6300}\text{A} = 29.33\text{A}$$

$$Z_{1N} = \frac{U_{1N\phi}}{I_{1N\phi}} = \frac{U_{1N}}{\sqrt{3}I_{1N}} = \frac{6300}{\sqrt{3} \times 29.33}\Omega = 124.02\Omega$$

$$Z_m^* = \frac{Z'_m}{Z_{1N}} = \frac{2067.35}{124.02} = 16.67$$

$$r_m^* = \frac{r'_m}{Z_{1N}} = \frac{156.17}{124.02} = 1.26$$

$$x_m^* = \frac{x'_m}{Z_{1N}} = \frac{2061.57}{124.02} = 16.62$$

$$Z_k^* = \frac{Z_k}{Z_{1N}} = \frac{5.6}{124.02} = 0.0451$$

$$r_k^* = \frac{r_k}{Z_{1N}} = \frac{2.21}{124.02} = 0.0178$$

$$x_k^* = \frac{x_k}{Z_{1N}} = \frac{5.145}{124.02} = 0.0414$$

方法二：标幺值计算法

(1) 空载试验求励磁阻抗　空载试验是在低压侧加电压，选低压侧额定值为基值。低压侧的额定电流为

$$I_{2N} = \frac{S_N}{\sqrt{3}U_{2N}} = \frac{320 \times 10^3}{\sqrt{3} \times 400}A = 461.89A$$

换算为标幺值时的空载电压、空载电流和空载损耗分别为

$$U_{2N}^* = 1$$

$$I_0^* = \frac{I_0}{I_{2N}} = \frac{27.7}{461.89} = 0.06$$

$$p_0^* = \frac{p_0}{S_N} = \frac{1450}{320 \times 10^3} = 0.00453$$

将上述标幺值代入式（2-29），得励磁阻抗的标幺值为

$$Z_m^* = \frac{U_{2N}^*}{I_0^*} = \frac{1}{0.06} = 16.67$$

$$r_m^* = \frac{p_0^*}{I_0^{*2}} = \frac{0.00453}{0.06^2} = 1.26$$

$$x_m^* = \sqrt{Z_m^{*2} - r_m^{*2}} = \sqrt{16.67^2 - 1.26^2} = 16.62$$

(2) 短路试验求短路阻抗　短路试验是在高压侧加电压，选高压侧额定值为基值。高压侧的额定电流为

$$I_{1N} = \frac{S_N}{\sqrt{3}U_{1N}} = \frac{320 \times 10^3}{\sqrt{3} \times 6300}A = 29.33A$$

换算为标幺值时的短路电压、短路电流和负载损耗分别为

$$U_k^* = \frac{U_k}{U_{1N}} = \frac{284}{6300} = 0.0451$$

$$I_k^* = \frac{I_k}{I_{1N}} = \frac{29.3}{29.33} = 0.999$$

$$p_k^* = \frac{p_k}{S_N} = \frac{5700}{320 \times 10^3} = 0.0178$$

将上述标幺值代入式（2-30），得短路阻抗的标幺值为

$$Z_k^* = \frac{U_k^*}{I_k^*} = 0.0451$$

$$r_k^* = \frac{p_k^*}{I_k^{*2}} = 0.0178$$

$$x_k^* = \sqrt{Z_k^{*2} - r_k^{*2}} = \sqrt{0.0451^2 - 0.0178^2} = 0.414$$

（3）高压侧阻抗的实际值　由高压侧阻抗基值 Z_{1N}，可得高压侧阻抗的实际值为

$$Z_{1N} = \frac{U_{1N\phi}}{I_{1N\phi}} = \frac{U_{1N}}{\sqrt{3}I_{1N}} = \frac{6300}{\sqrt{3} \times 29.33}\Omega = 124.02\Omega$$

$$Z'_m = Z_m^* Z_{1N} = 16.67 \times 124.02\Omega = 2067.41\Omega$$

$$r'_m = r_m^* Z_{1N} = 1.26 \times 124.02\Omega = 156.27\Omega$$

$$x'_m = x_m^* Z_{1N} = 16.62 \times 124.02\Omega = 2061.21\Omega$$

$$Z_k = Z_k^* Z_{1N} = 0.0451 \times 124.02\Omega = 5.59\Omega$$

$$r_k = r_k^* Z_{1N} = 0.0178 \times 124.02\Omega = 2.21\Omega$$

$$x_k = x_k^* Z_{1N} = 0.414 \times 124.02\Omega = 5.13\Omega$$

2.5 变压器的运行特性

变压器的运行特性主要指外特性和效率特性，表征变压器运行性能的主要指标是电压变化率和效率。电压变化率是反映变压器供电电压质量的指标，效率则是反映变压器运行时的经济指标。

2.5.1 外特性和电压变化率

当变压器一次电压为额定电压、负载的功率因数一定时，二次电压随负载电流的变化曲线称为变压器的外特性曲线。

图 2-15 给出了不同性质负载的外特性曲线。由图可知，变压器二次电压的变化不仅与负载电流的大小有关，而且还与负载的功率因数有关。

负载运行时，二次电压从空载到负载时所变化的相对值，称为电压变化率 ΔU，有

$$\Delta U = \frac{U_{20} - U_2}{U_{2N}} = \frac{U_{2N} - U_2}{U_{2N}} = 1 - U_2^* \tag{2-33}$$

根据以标幺值表示的变压器简化等效电路的相量图，如图 2-16 所示，$U_{1N}^* = 1$。变压器带感性负载 $I_2^* = I_1^*$。由于短路阻抗压降很小（图中被放大了），则 $\dot{U}_{1N}^*$ 与 $-\dot{U}_2^*$ 的夹角就很小，因此可得电压变化率 ΔU 为

$$\Delta U = 1 - U_2^* \approx \overline{AC} = \overline{AB} + \overline{BC}$$

易得$\overline{AB}=I_1^* r_k^* \cos\varphi_2$，$\overline{BC}=I_1^* x_k^* \sin\varphi_2$，于是

$$\Delta U \approx I_1^* (r_k^* \cos\varphi_2 + x_k^* \sin\varphi_2) \tag{2-34}$$

由式（2-34）可知，电压变化率ΔU与负载I_1^*的大小和负载的性质（功率因数$\cos\varphi_2$）以及短路阻抗标幺值的大小有关。

当变压器带纯电阻负载和感性负载时，$\varphi_2>0$，ΔU为正值，故外特性曲线是下降的；当带容性负载时，$\varphi_2<0$，当$r_k^* \cos\varphi_2<|x_k^* \sin\varphi_2|$时，$\Delta U$为负值，此时外特性曲线将是上升的，说明二次电压$U_2$将高于空载电压$U_{20}$。

例 2-3 变压器参数同例 2-2，求变压器在额定负载、功率因数$\cos\varphi_2=0.8$（滞后）和$\cos\varphi_2=0.8$（超前）时的电压变化率。

解：

由例 2-2 知，$r_k^*=0.0178$，$x_k^*=0.0414$，

当$\cos\varphi_2=0.8$（滞后）时，$\sin\varphi_2=0.6$。代入式（2-34）可得

$$\Delta U_N \approx r_k^* \cos\varphi_2 + x_k^* \sin\varphi_2 = 0.0178\times0.8+0.0414\times0.6=0.03908\approx3.91\%$$

当$\cos\varphi_2=0.8$（超前）时，$\sin\varphi_2=-0.6$，则

$$\begin{aligned}\Delta U_N &\approx r_k^* \cos\varphi_2 + x_k^* \sin\varphi_2\\ &=0.0178\times0.8+0.0414\times(-0.6)=-0.0106\approx-1.06\%\end{aligned}$$

2.5.2 效率和效率特性

效率η是指变压器输出的有功功率P_2和输入的有功功率P_1的比值，通常用百分值来表示，即

$$\eta=\frac{P_2}{P_1}\times100\% \tag{2-35}$$

效率与变压器的损耗有关，损耗主要包括铁损和铜损两类。铁损与主磁通有关，因电源电压稳定时主磁通基本不变，所以铁损也基本不变，称为不变损耗。通常取空载损耗p_0近似等于铁损p_{Fe}。铜损是电流流过变压器绕组时产生的损耗，它与电流的大小和绕组电阻的大小（即短路电阻r_k）有关，即随负载变化而变化，故称为可变损耗。铜损p_{Cu}可通过负载电流I_2^*和额定电流时的短路损耗p_{kN}求得，即

$$p_{Cu}=I_2^2 r_k=\left(\frac{I_2}{I_{2N}}\right)^2 I_{2N}^2 r_k=I_2^{*2} p_{kN} \tag{2-36}$$

变压器的总损耗为

$$\sum p=p_{Fe}+p_{Cu}=p_0+I_2^{*2}p_{kN} \tag{2-37}$$

变压器输出的有功功率P_2为

$$P_2=U_2 I_2\cos\varphi_2\approx U_{2N}I_2\cos\varphi_2=I_2^* S_N\cos\varphi_2 \tag{2-38}$$

故变压器的效率η可由式（2-35）改写为

$$\eta=\frac{P_2}{P_2+\sum p}\times100\%=\frac{I_2^* S_N\cos\varphi_2}{I_2^* S_N\cos\varphi_2+p_0+I_2^{*2}p_{kN}}\times100\% \tag{2-39}$$

由式（2-39）可知，变压器的效率η与负载I_2^*的大小和负载的性质（功率因数$\cos\varphi_2$）有关。当负载功率因数$\cos\varphi_2$一定时，效率η与负载I_2^*的关系曲线（即效率特性曲线）如图 2-17 所示。

由图 2-17 可知，当变压器空载时，输出功率为零，$I_2^* = 0$，效率 $\eta = 0$；随着负载的增大，在效率公式（2-39）中，分子显著增大，分母中因可变损耗 $I_2^{*2} p_{kN}$ 较小，不变损耗 p_0 起主要作用，使其增加较慢，因此效率 η 将逐步增大；当负载 $I_2^* > I_{2(\max)}^*$（$I_{2(\max)}^*$ 为达到最大效率时的负载）时，随着负载的增大，在效率公式（2-39）中，分母中的可变损耗 $I_2^{*2} p_{kN}$ 较大，起主要作用，它使分母的增大较分子快，故效率 η 反而减小。

将式（2-39）对 I_2^* 求导，令其等于零，得对应于最大效率 $\eta_{\max}$ 时的负载电流 $I_{2(\max)}^* = \sqrt{p_0 / p_{kN}}$，此时可变损耗（铜损）恰好等于不变损耗（铁损）。

2.6　三相变压器

目前电力系统均采用三相制供电，故三相变压器得到广泛的应用。从运行原理来看，三相变压器对称运行时，各相的电压及电流大小相等，相位彼此相差 120°，任意一相都可以沿用前述单相变压器的分析方法。本节主要讨论三相变压器的特殊问题，如三相变压器的磁路系统和绕组联结组。

2.6.1　三相变压器的磁路系统

如图 2-18 所示，三台单相变压器联结为三相系统，称为三相变压器组，三相的磁路彼此独立。

如果将三台单相变压器的铁心合并成图 2-19a 所示的形式，当外加三相对称电压时，三相主磁通是对称的，则中间公共心柱的磁通为 $\dot{\Phi}_A + \dot{\Phi}_B + \dot{\Phi}_C = 0$，故可以取消公共铁心柱，如图 2-19b 所示。为了便于制造，将三相铁心柱放在同一平面内，得到图 2-19c 的形式，这就是三相心式变压器的铁心结构。

三相心式变压器比同容量的三相变压器组节省材料、占地少、维护方便，得到广泛应用，但对于大型变压器，多采用三相变压器组，因其中的每个单相变压器体积小、重量轻、便于运输，并可减少备用容量。

2.6.2　三相变压器的联结组

三相变压器绕组不同的联结方式会影响到高、低压绕组线电压的相位关系，可用联结组来予以区别。

1. 绕组的端头标志与极性

变压器高、中、低压绕组的首、尾端的标志有统一的规定，见表 2-1。这些标志都标在变压器的出线套管上。

表 2-1　变压器首尾端标志

绕组名称	单相变压器		三相变压器		中性点
	首端	尾端	首端	尾端	
高压绕组	A	X	A、B、C	X、Y、Z	N
低压绕组	a	x	a、b、c	x、y、z	n
中压绕组	Am	Xm	Am、Bm、Cm	Xm、Ym、Zm	Nm

由于变压器同一相的高、低压绕组都被同一主磁通所交链，当高压绕组某一端头电位高于另一端头时，低压绕组在该瞬时也有一个端头电位高于另一端头。这两个具有高电位的端头就是同极性端（或称同名端），一般在同极性端的端头标上“·”标记。当然另外两个低电位的端头也是同极性端，但无需再标记。

对于三相变压器的任意一相，相电压相量规定正方向为绕组首端指向尾端，以 A 相为例，若高压和低压绕组的首端 A 和 a 是同极性端，则相电压 $\dot{U}_{AX}$与 $\dot{U}_{ax}$同相位，否则为反相位。

2. 三相变压器的联结组及其标号

三相变压器的联结组标号是反映三相变压器高、低压绕组的联结方式及其对应线电压的相位关系的。通常采用时钟表示法，即把高压绕组的线电压相量看作时钟的长针并固定指向 12 点，低压绕组的同名线电压相量看作时钟的短针，它所指的小时数就是三相变压器的联结组别数。三相变压器的联结组标号不仅与绕组的绕向（同名端）有关，而且还与首尾端的标志和三相绕组的联结方式有关，下面分别介绍。

（1）Y-Y联结　Y-Y联结表示三相变压器高、低压绕组均采用星形联结，如图 2-20a 所示。图中将高、低压绕组的同极性端标为首端，故高、低压绕组对应的相电压与线电压均为同相位（图中为简明起见，将 $\dot{U}_{AX}$简写为 $\dot{U}_{A}$，其余类推）。故当 $\dot{U}_{AB}$指向 12 点时，$\dot{U}_{ab}$也指向 12 点（即 0 点），所以联结组标号为 Yy0。

如果将图 2-20a 中高、低压绕组换标为异极性端为首端，如图 2-20b 所示，则高、低压绕组对应的各相电压为反相位，且高、低压绕组对应的各线电压也反相位，即 $\dot{U}_{ab}$逆时针转过 180°，指向 6 点，所以联结组标号为 Yy6。

如果保持图 2-20a 中高、低压绕组首端为同极性端，保持高压绕组的端头标志不变，把低压绕组端头标志顺相序左移一个端头，即由 abc 改为 bca，尾端也相应改变，如图 2-21a 所示，则 $\dot{U}_{ab}$逆时针转过 120°，指向 8 点，则联结组标号为 Yy8；若再将图 2-21a 中高、低压绕组换标为异极性端为首端，则 $\dot{U}_{ab}$逆时针再转过 180°，指向 2 点，则联结组标号为 Yy2。

如果保持图 2-20a 中高、低压绕组首端为同极性端，保持高压绕组的端头标志不变，把低压绕组端头标志顺相序右移一个端头，即由 abc 改为 cab，尾端也相应改变，如图 2-21b 所示，则 $\dot{U}_{ab}$顺时针转过 120°，指向 4 点，则联结组标号为 Yy4；若再将图 2-21b 中高、低压绕组换标为异极性端为首端，则 $\dot{U}_{ab}$再逆时针转过 180°，指向 10 点，则联结组标号为 Yy10。

综上所述，对Y-Y联结，可得到 0、2、4、6、8、10 共 6 种偶数的联结组标号。

（2）Y-△联结　Y-△联结表示高压绕组采用星形联结，低压绕组采用三角形联结，如图 2-22a 所示。图中将高、低压绕组的同极性端标为首端，故高、低压绕组对应的各相电压为同相位，保持 A-B-C 和 a-b-c 的相序，对各个相电压的相量进行平移，把电气上联结在一起的端点进行组合：高压绕组的 XYZ 为同一点，低压绕组 ay、bz、cx 分别为同一点，可得到高、低压绕组的相量组合图，如图 2-22a 所示，$\dot{U}_{ab} = -\dot{U}_{b}$，即当 $\dot{U}_{AB}$指向 12 点时，$\dot{U}_{ab}$指向 11 点，所以该联结组标号为 Yd11。

如果低压侧接成图2-22b所示的三角形联结，仍按上述方法平移相电压相量，把az、bx、cy分别置于同一点，高压绕组相量组合图不变，如图2-22b所示，此时 $\dot{U}_{ab}=\dot{U}_{a}$，即当 $\dot{U}_{AB}$ 指向12点时，$\dot{U}_{ab}$ 指向1点，所以联结组标号为Yd1。

与丫-丫联结的规律相同，丫-△联结还可以得到3、5、7、9的联结组。

为了便于制造和并联运行，我国规定三相电力变压器的标准联结组标号为：Yyn0、YNy0、Yy0、Yd11、YNd11共5种。

2.7　变压器的并联运行

变压器的并联运行是指两台或多台变压器的一次和二次绕组分别接在一次和二次公共母线上，共同向负载供电的运行方式。变压器并联运行具有提高供电的可靠性和经济性、减少总备用容量等优点。

变压器并联运行的理想情况是：各台并联变压器的一次和二次绕组中没有环流，负载能按各台变压器的容量成正比地分配，且各台变压器二次电流最好是同相位。只有这样，才能避免因并联引起的附加损耗，并且充分利用变压器的容量。要达到理想情况，并联运行的变压器要满足以下3个条件：

1）各台变压器的额定电压和电压比应相等。

2）各台变压器具有相同的联结组。

3）各台变压器的短路阻抗标幺值要相等，短路阻抗角也相等。

如果变压器并联运行时不满足以上条件，就会产生不良的后果。下面逐一讨论当某一条件不满足时会产生的后果。

1. 电压比不等时的并联运行

以两台电压比不等的单相变压器并联运行为例来分析。

如图2-23所示，两台变压器一次绕组都接在电源 $\dot{U}_1$ 上，假设电压比 k_{I} 小于 k_{II}，变压器二次绕组的空载电压分别为 $\dot{U}_{20\mathrm{I}}=\dot{U}_1/k_{\mathrm{I}}$，$\dot{U}_{20\mathrm{II}}=\dot{U}_1/k_{\mathrm{II}}$，则 $\dot{U}_{20\mathrm{I}}>\dot{U}_{20\mathrm{II}}$。因此，在开关 S_1 两端存在电压差 $\Delta\dot{U}_{20}=\dot{U}_{20\mathrm{I}}-\dot{U}_{20\mathrm{II}}$。若合上 S_1，使两台变压器并联运行，二次绕组中会有环流 $\dot{I}_{2h}$ 产生。设两台变压器折算到二次侧的短路阻抗为 $z_{k\mathrm{I}}$ 和 $z_{k\mathrm{II}}$，则有

$$\dot{I}_{2h}=\frac{\Delta\dot{U}_{20}}{z_{k\mathrm{I}}+z_{k\mathrm{II}}} \tag{2-40}$$

由于短路阻抗值很小，即使电压差 $\Delta\dot{U}_{20}$ 不大，也能引起较大的环流，造成变压器空载运行时的额外损耗。

2. 联结组标号不同时的并联运行

如果两台联结组标号不同的变压器并联运行，则二次侧线电压的相位至少相差30°。此时 $\Delta\dot{U}_{20}$ 的大小为 $\Delta U_{20}=2U_{20\mathrm{I}}\sin15°=0.518U_{20\mathrm{I}}$。即 ΔU_{20} 达到额定电压的51.8%，由于变压器短路阻抗值很小，会产生数倍于额定电流的大环流，致使变压器严重发热，甚至烧毁，因此联结组不同的变压器绝对不允许并联运行。

3. 短路阻抗标幺值不等时的并联运行

假定两台变压器的电压比和联结组都相同，则并联运行时一次和二次电压都相等，故短路阻抗压降相等，用标幺值表示为

$$\dot{I}_{\mathrm{I}}^{*} z_{\mathrm{kI}}^{*} = \dot{I}_{\mathrm{II}}^{*} z_{\mathrm{kII}}^{*} \tag{2-41}$$

于是

$$\dot{I}_{\mathrm{I}}^{*} : \dot{I}_{\mathrm{II}}^{*} = \frac{1}{z_{\mathrm{kI}}^{*}} : \frac{1}{z_{\mathrm{kII}}^{*}} \tag{2-42}$$

也就是说，并联运行的各台变压器的负载分配与短路阻抗标幺值成反比。这样，就有可能出现短路阻抗标幺值小的变压器满载时，短路阻抗标幺值大的变压器却欠载，使变压器容量得不到充分利用。

2.8 自耦变压器和仪用互感器

在电力系统中，除了大量采用前面介绍的三相双绕组变压器外，还常采用多种其他用途的变压器。本节主要介绍较常用的自耦变压器和仪用互感器的工作原理及特点。

2.8.1 自耦变压器

自耦变压器的结构特点是一次和二次绕组中有一部分是共用绕组，因此一次和二次绕组之间不仅有磁的耦合关系，而且还有电的联系，即它不仅通过电磁耦合和感应作用传递功率，而且还可以从一次侧直接把功率传导到二次侧，如图 2-24 所示（习惯以降压变压器为例，且只分析一相）。图中 AC 为低压绕组，它又是高压绕组的一部分，又称公共绕组；BC 是与公共绕组相串联的绕组，称为串联绕组；AB 为高压绕组。自耦变压器输出端电压有效值为

$$U_2 = (N_2/N_1) \times U_1 \tag{2-43}$$

当自耦变压器负载运行时，如图 2-25 所示。此时，设 BC 部分绕组中的电流为 I_1，负载电流为 I_2，则 CA 部分绕组中的电流为 $I_2 - I_1$。相应的磁动势关系为 $I_1(N_1 - N_2) = (I_2 - I_1) N_2$。

270 化简后得到磁动势方程为

$$I_1 N_1 = I_2 N_2 \tag{2-44}$$

由式（2-44）可变形得到绕组电流与自耦变压器的电压比 k_a 之间的关系为

$$I_1 = I_2 N_2/N_1 = I_2/k_a \tag{2-45}$$

当忽略自耦变压器的损耗和励磁电流时，一次绕组从电源吸收的视在功率应等于二次绕组输出的视在功率，即

$$I_1 U_1 = I_2 U_2 \tag{2-46}$$

由于自耦变压器一次和二次绕组之间有直接的电联系，因此为防止一次侧过电压时引起二次侧严重过电压，要求自耦变压器的中性点必须可靠接地，并且一次和二次侧都要装避雷器。

2.8.2 仪用互感器

在大电流、高电压的电力系统中常用互感器来进行辅助测量，一方面可使测量回路与被测回路隔离，以保证人员和设备的安全；二是可以减小所用表计的量程。仪用互感器分为电流互感器和电压互感器两大类。

1. 电压互感器

电压互感器是用来测量高电压的仪用互感器，其原理接线如图2-26所示。电压互感器的一次绕组匝数很多，并且并联接在被测线路上；二次绕组匝数较少，与阻抗很大的仪表（电压表或功率表的电压线圈）连接组成闭合回路，因此二次电流很小。电压互感器相当于空载运行的降压变压器。

因为 $U_1/U_2 = N_1/N_2$，电压互感器利用一次和二次绕组匝数不同，可将线路的高电压转换成二次侧的低电压来测量。通常将电压互感器二次绕组的额定电压设计为100V。电压互感器的准确度分为0.2、0.5、1.0和3.0共4个等级。

电压互感器运行时二次绕组不允许短路。如果二次绕组发生短路，就会产生很大的短路电流而烧坏电压互感器。铁心和二次绕组的一端必须可靠接地。

2. 电流互感器

电流互感器是用来测量大电流的仪用互感器，其原理接线如图2-27所示。电流互感器均制成单相的，一次绕组由一匝或几匝粗导线组成，串接在被测回路中；二次绕组由匝数较多的细导线组成，与阻抗很小的仪表（电流表、功率表的电流线圈或继电器的线圈）串联组成闭合回路。因此，电流互感器相当于短路运行的升压变压器。

忽略励磁电流，就有 $I_1/I_2 = N_2/N_1$，电流互感器利用一次和二次绕组匝数不同，可将线路的大电流转换成小电流测量。通常电流互感器二次绕组的额定电流为5A，准确度可分为0.2、0.5、1.0、3.0和10.0共5个等级。

电流互感器运行时二次绕组不允许开路。如果二次绕组开路，被测回路的大电流就成为互感器的励磁电流，使铁心严重饱和且过热，并且在二次绕组感应产生过电压，危及人员和仪表的安全。铁心和二次绕组的一端必须可靠接地。

本 章 小 结

变压器是利用电磁感应原理变压的，它通过一次和二次绕组的电磁感应作用和磁动势平衡关系将一次侧的电功率传递到二次侧。本章的分析思路是，通过对双绕组变压器空载和负载的电磁过程分析，推导出变压器的电压方程、等效电路和相量图，并对变压器的运行性能进行分析。在学习过程中要注意以下几点：

1）变压器的主要结构部件包括铁心和绕组。

2）变压器空载运行时，注意空载电流、主磁通、感应电动势和电源电压这些主要物理量之间的关系。空载电流主要用以产生交变的主磁通，主磁通又在一次和二次绕组中感应电动势，其中一次绕组中的感应电动势加上漏阻抗压降后将与电源电压相平衡。因此，空载电流的大小不仅与一次绕组的匝数和磁路磁阻有关，而且还与电源电压有关。另外，由感应电动势公式 $\dot{E}_1 = -j4.44fN_1\dot{\Phi}_m$ 和一次绕组的电压方程可知：当电源频率和绕组匝数不变时，若电源电压 $\dot{U}_1$ 不变，主磁通 $\dot{\Phi}_m$ 将基本不变。这对分析变压器的负载运行很重要。变压器的电磁过程分析中，既有电量又有磁量，比较复杂。为了简化分析，引入了励磁阻抗 z_m 以及一次和二次绕组的漏电抗参数 x_1 和 x_2，以取代相应的磁通和磁路，使电磁问题简化为等效电路问题。注意：励磁电抗 x_m 与主磁通相对应，一次和二次绕组的漏电抗 x_1 和 x_2 与各自的漏磁通相对应。

3）变压器负载运行时，通过电磁过程的分析可以得到负载运行时的基本方程、等效电路和相量图。在推导等效电路时，引入了绕组折算的概念，且折算必须满足“等效”的原则。基本方程表达电路和磁路各物理量之间的数学关系；等效电路便于分析计算；相量图可以直观地反映各物理量的大小和相位关系，常用于定性分析。

4）变压器等效电路参数可通过空载试验和短路试验来进行测定和计算。注意：计算励磁参数需要采用额定电压时的数据，计算短路参数需要采用额定电流时的数据。

5）掌握计算变压器的电压变化率和效率的方法。电压变化率的计算公式为 $\Delta U \approx I_1^*(r_k^*\cos\varphi_2 + x_k^*\sin\varphi_2)$，影响变压器二次电压变化的因素有负载的大小、性质（负载功率因数）和变压器的短路阻抗。变压器的效率 $\eta = \dfrac{I_2^* S_N\cos\varphi_2}{I_2^* S_N\cos\varphi_2 + p_0 + I_2^{*2}p_{kN}} \times 100\%$，影响效率的因素有负载的大小、性质（负载功率因数）和变压器自身的参数（铁损 p_0 和额定负载时的铜损 p_{kN}）。

6）在变压器的计算中，常采用标幺值进行运算，这有利于比较不同容量、不同电压等级变压器的参数和性能。

7）三相变压器的磁路系统分为三相磁路彼此独立的三相变压器组和三相磁路彼此相关的三相心式变压器。三相变压器的联结组除与绕组的绕向和首尾端标志有关外，还与三相绕组的联结方式有关。

8）变压器并联运行应满足三个条件：

① 各台变压器一次和二次额定电压分别要相等，即电压比相等；

② 各台变压器具有相同的联结组；

③ 各台变压器的短路阻抗标幺值要相等，短路阻抗角也相等。

9）掌握自耦变压器、电压互感器、电流互感器的运行与使用特点。自耦变压器是两绕组变压器的一种特殊联结方式，其一次和二次绕组之间具有电的联系。自耦变压器传递的功率中，除感应功率外，还有传导功率。使用电压互感器时，其二次侧不能短路；使用电流互感器时，其二次侧不能开路。

习　题

1. 变压器空载运行时是否从电源吸收功率？

2. 单相变压器和三相变压器一、二次侧的额定电压是如何定义的？

3 变压器励磁电流有哪几种分量？这些分量在等效电路中是如何表示的？

4. 什么是变压器的漏磁通？它们在等效电路中是用什么参数表示的？

5. 变压器正常运行时，存在哪几种损耗？

6. 一台额定容量 S_N 为 5000kV·A 的三相变压器，额定电压为 U_{1N}/U_{2N} = 10/6.3kV，Yd 联结，试求其高、低压绕组的额定电流、额定相电流（$I_{1N\phi}$ 和 $I_{2N\phi}$）和额定相电压（$U_{1N\phi}$ 和 $U_{2N\phi}$）

7. 一台额定频率为 60Hz 的单相变压器接在 50Hz 的电网上运行，若额定电压不变，问主磁通、空载电流、铁损、漏抗和励磁电抗各如何变化？

8. 一台 220/110V 单相变压器，电压比为 2，能否设计为一次绕组 2 匝，二次绕组 1 匝，为什么？

9. 当变压器的一次侧电压升高时，该变压器的励磁阻抗、励磁电流和铁损将如何变化？

10. 一台单相变压器的额定数据为：20kV·A、10 000/500V、60Hz，该变压器能给 50Hz

的 15kV·A、420V 的负载安全供电吗？为什么？

11. 一台单相变压器的额定数据为：18kV·A、20 000/480V、60Hz，该变压器能给 50Hz、15kV·A、415V 的负载安全供电吗？为什么？

12. 为什么变压器的空载试验时只存在铁损，而没有铜损？而短路试验时只存在铜损，却没有铁损？

13. 为什么变压器二次侧电压随负载的大小和性质（负载功率因数）的变化而改变？变压器额定运行时其电压变化率能为 0 吗？

14. 为什么负载的功率因数（负载的性质）会影响变压器的电压变化率？

15. 变压器的效率由哪些因素决定？在什么情况下，变压器的效率可取得最大值？

16. 变压器并联运行时必须满足哪些条件？其中哪个条件必须绝对满足？

17. Y-Y联结的三相变压器在运行时会存在哪些问题？

18. 一台 2kV·A、220/110V 的单相变压器，其等效电路参数 $r_1=1\Omega$，$x_1=6\Omega$，$r_2=0.25\Omega$，$x_2=1.5\Omega$。当负载阻抗 $z_L=28+j28\Omega$ 时，计算变压器一、二次侧的电流各为多少？此时变压器的电压变化率为多少？

19. 一台单相变压器，$S_N=4.6$kV·A，$U_{1N}/U_{2N}=380/115$V。空载和短路试验数据如下：

试验名称	电压/V	电流/A	功率/W	备注
空载试验	115	3	60	电压加在低压侧
短路试验	15.6	12.1	172	电压加在高压侧

试求：

（1）折算到高压侧的励磁阻抗和短路阻抗；

（2）额定负载且功率因数为 0.8（滞后）和功率因数为 0.8（超前）两种情况的电压变化率和效率；

（3）功率因数为 0.8（滞后）时变压器的最大效率。

20. 一台三相变压器，额定容量 $S_N=125000$kV·A、50Hz，额定电压 $U_{1N}/U_{2N}=110/11$kV，Yd 联结。在低压侧加额定电压做空载试验，测得空载电流 $I_0=131.22$A，空载损耗 $p_0=133$kW；在高压侧加电源做短路试验，当短路电流为额定电流时，测得短路电压 $U_{kN}=11\,550$V，负载损耗 $p_{kN}=600$kW，试求：

（1）励磁阻抗和短路阻抗的标幺值以及折算到高压侧的实际值；

（2）半载（$I_2^*=0.5$）时，功率因数分别为 1、0.8（滞后）和 0.8（超前）这 3 种情况下的电压变化率。

（3）功率因数为 0.8（滞后）时，半载和满载两种情况下的效率和功率因数为 0.8（滞后）时的最大效率。

21. 一台单相变压器的额定数据为 20kV·A、20 000/480V、60Hz，其空载试验与短路试验的数据如下表所示，计算变压器等效电路参数的标幺值（60Hz）。

试验名称	电压/V	电流/A	功率/W	备注
空载试验	480	1.6	305	电源加在低压侧
短路试验	1130	1.0	260	电压加在高压侧

22. 一台单相变压器的额定数据为：1000V·A、230/115V，空载试验与短路试验的数据如下表所示，且所有试验均在一次侧（高压侧）进行。

试验名称	电压/V	电流/A	功率/W	备注
空载试验	230	0.45	30	电源加在高压侧
短路试验	19.1	8.7	42.3	电压加在高压侧

（1）计算折算到低压侧的等效电路参数。

（2）计算该变压器在负载功率因数分别为0.8（滞后）、1.0、0.8（超前）时的电压变化率。

（3）计算变压器在额定负载且功率因数为0.8（滞后）时的效率。

23. 一台5000kV·A、230/13.8kV的单相变压器，其短路电阻的标幺值为0.01，短路电抗的标幺值为0.05。在低压侧进行空载试验的数据：空载电压为13.8kV，空载电流为15.1A，空载功率为44.9kV。

（1）计算折算到低压侧的变压器的等效电路参数；

（2）如果二次侧所接负载的电压为13.8kV，功率为4000kW，功率因数为0.8（滞后），计算变压器的电压变化率和效率各为多少?

24. 一台750kV·A、10/0.4kV、Yy0联结的三相变压器，其空载试验和短路试验的数据如下表所示，计算该变压器等效电路参数的标幺值。

	线电压/V	线电流/A	三相功率/W	实验侧
空载试验	400	60	3800	低压侧
短路试验	440	43.3	10 900	高压侧

25. 一台5600kV·A，10/6.3kV，Yd11联结的三相变压器，其空载试验与短路试验的数据如下表所示，计算该变压器等效电路参数的标幺值。

	线电压/V	线电流/A	三相功率/W	实验侧
空载试验	6300	7.4	6800	低压侧
短路试验	550	323	17 900	高压侧

26. 三相变压器组由三台单相变压器组成，已知三相变压器的容量为600kV·A，高、低压侧的电压分别为34.5/13.8kV。当将此组成三相变压器组的三个单相变压器分别接成丫-丫、丫-△、△-丫、△-△联结时，计算接成后变压器组一、二次侧的额定电压、变压器的额定容量和电压比。

27. 一台100 000kV·A、230/115kV，△-△联结的三相变压器，其短路电阻标幺值为0.02，短路电抗的标幺值为0.055。励磁支路的励磁电阻标幺值为110，励磁电抗标幺值为20。

（1）当变压器的负载功率为80MV·A，功率因数为0.85（滞后）时，绘制该变压器的相量图。

（2）在上述负载条件下，该变压器的电压变化率为多少?

（3）计算折算到低压侧的等效电路参数，并绘制该等效电路。

28. 三台单相变压器连接成△-丫的三相变压器组，单相变压器的额定数据为25kV·A、24 000/277V。在三相变压器组中进行空载试验和短路试验的数据为：

试验名称	线电压/V	线电流/A	三相总功率/W	备注
空载试验	480	4.10	945	电源加在低压侧
短路试验	1600	2.0	1150	电压加在高压侧

（1）计算变压器组等效电路参数的标幺值。

（2）当变压器组接功率因数为0.9（滞后）的额定负载时，计算变压器的电压变化率；

（3）计算变压器组在上述负载条件下的效率。

29. 一台三相变压器，$S_N = 100\text{kV}\cdot\text{A}$，$U_{1N}/U_{2N} = 6000/400\text{V}$，联结组标号为Yy0。空载试验时测得空载电流为额定电流的6.5%，$p_0 = 600\text{W}$；短路时测得 $U_{kN} = 5\%$，$p_{kN} = 1800\text{W}$。试求：

（1）励磁阻抗和短路阻抗的标幺值以及折算到低压侧的实际值；

（2）满载时，功率因数为0.8（滞后）的二次电压和效率；

（3）功率因数为0.8（滞后）时的最大效率和此时的负载电流。

30. 一台三相变压器，其额定数据为：1000kV·A、10/6.3kV、Yd11。变压器额定运行时的铁损为4.9kW，负载损耗为15kW。计算该变压器在额定负载且 $\cos\varphi_2 = 0.8$（滞后）时的效率和最大效率各是多少。

31. 分别画出联结组标号为Yy2、Yd3、Yd5、Yd7的三相变压器的绕组联结图和相量图。

32. 用相量图判别下图2-28中所示的三相变压器的联结组标号。

第3章 交流电机的共同理论

通常将产生和使用交流电能的旋转电机称为交流电机，交流电机中通过交流电流的绕组称为交流绕组。交流电机分同步电机和感应电机两大类，尽管在工作原理、基本结构、电磁关系以及运行特性等方面，这两种电机有很大区别，但在定子绕组结构、感应电动势以及磁动势的性质和特点等方面有许多共同之处，通常将这部分内容称为交流电机的共同问题。本章在介绍同步发电机工作原理的基础上，分析交流绕组的绕制方法及其感应电动势，并且，在介绍感应电动机工作原理的基础上，分析交流绕组磁动势的性质和特点。

3.1 三相同步发电机的工作原理

图3-1为一台两极三相同步发电机模型的剖面图，发电机由定子和转子组成。定子槽中有6个间隔均匀的导体，通过端部连接成为最简单的三相对称绕组AX、BY、CZ，这三相绕组每相只有一匝，三相绕组轴线在空间上互差120°。转子上装有励磁绕组，通过集电环和电刷引入直流励磁电流，在转子上产生N、S极，形成磁极，建立气隙磁场。

假设磁极产生的气隙磁场沿空间按正弦规律分布，如图3-2所示。当原动机拖动转子以恒定转速n逆时针方向旋转时，定子槽内导体“切割”气隙磁场后，将产生随时间按正弦规律变化的感应电动势（见图3-3），并且各导体感应电动势的有效值相等。显然，转子旋转一周，定子槽内的6个导体的感应电动势变化一个周期，但各导体电动势出现最大值的时刻不同，且相邻导体电动势出现最大值的时间间隔为60°，6根导体电动势的相量关系如图3-4所示。

如图3-5所示，将A、X两根导体用端接线连接成整距线圈AX，则该线圈的电动势$\dot{E}_{AX}$与导体A的电动势$\dot{E}_{A}$同相；用同样方法构成另外两个线圈BY、CZ，其电动势$\dot{E}_{BY}$、$\dot{E}_{CZ}$分别与导体B、C的电动势$\dot{E}_{B}$、$\dot{E}_{C}$同相。这样，就构成了最简单的三相对称绕组AX、BY、CZ，并且这三相绕组中将产生三相对称感应电动势（见图3-6）。

将定子绕组接成星形，三个出线端对外接入三相对称负载，发电机就可以向负载输出三相对称交流电流，将原动机输入的机械能转换为电能。

3.2 交流绕组

交流绕组按槽内层数分为单层绕组和双层绕组，按相数分为单相绕组、两相绕组、三相绕组和多相绕组。此外，还有其他分类方式。

对交流绕组的基本要求如下：

1）在一定的材料消耗下，力求得到较大的基波电动势和基波磁动势。

2）要求电动势和磁动势的波形尽量接近正弦波，其中的谐波成分尽量小。

3）三相绕组的感应电动势必须对称，各相阻抗要平衡。

4）绕组的绝缘和机械强度可靠，制造工艺简单，散热条件好，维修方便。

3.2.1　交流绕组的基本概念

1. 电角度

在分析多极电机时，为了方便，引入了空间电角度的概念。对于沿空间按正弦规律分布的磁场，一对磁极对应着一个完整的正弦磁密波，因此，通常把一对磁极范围所对应的空间角度定义为 360°电角度。如果是 p 对极电机，则整个定子圆周相当于 $p \times 360°$ 电角度，因此有

$$电角度 = p \times 机械角度 \tag{3-1}$$

2. 线圈

构成交流绕组的基本单位是线圈，线圈可以是单匝的，也可以是多匝的，线圈的匝数一般用 N_c 表示，线圈数目用 S 表示，当定子绕组为双层绕组时，线圈数等于槽数 Q；当定子绕组为单层绕组时，线圈数为槽数的一半。电机定子线圈的总匝数为

$$N = 2SN_c \tag{3-2}$$

3. 极距

相邻两个磁极轴线之间沿定子内圆周的距离称为极距 τ。τ 既可用槽数，也可用长度来表示，用槽数表示时，有

$$\tau = \frac{Q}{2p} \tag{3-3}$$

式中，Q 为总槽数。

一个极距对应 180°空间电角度。

4. 线圈节距

同一个线圈的两个线圈边之间的距离称为线圈节距 y_1，y_1 通常用槽数来表示。$y_1 = \tau$ 称为整距线圈，$y_1 > \tau$ 称为长距线圈，$y_1 < \tau$ 称为短距线圈。

5. 槽距角

相邻两槽之间的距离用电角度表示，称为槽距角 α。

$$\alpha = \frac{p \times 360°}{Q} \tag{3-4}$$

6. 每极每相槽数

每极每相槽数用 q 表示，有

$$q = \frac{Q}{2mp} \tag{3-5}$$

式中，m 为电机的相数。

3.2.2　三相单层叠绕组

现以一实例说明三相单层叠绕组的有关概念、连接方法及其特点。

已知定子槽数 $Q = 36$，极数 $2p = 4$，试绘出并联支路数 $a = 1$ 的三相单层叠绕组的展开图。

1. 槽电动势星形图

在图 3-7 中，设气隙磁密波沿气隙圆周按正弦规律分布，磁极以转速 n_s 逆时针方向旋

转，此时，定子槽内的导体将感应出正弦交流电动势，各导体感应电动势的有效值相同，相邻两槽内导体的感应电动势相位相差 α；绘出所有导体感应电动势的相量图，得到图 3-8，此图称为槽电动势星形图。因为电机为 4 极，每两个极下的导体电动势形成一个电动势星形图，因此 4 个极下将形成两个重叠在一起的电动势星形图。

2. 分相

对本例来说，$q=3$，在第一对极距内（图 3-7 中 N_1 到 S_1），将 1、2、3 和 10、11、12 号槽内的导体分给 A 相，在第二对极距内将 19、20、21 和 28、29、30 号槽内的导体分给 A 相，同样，将 7、8、9，16、17、18，25、26、27 和 34、35、36 号槽内的导体分给 B 相；将 13、14、15，4、5、6，31、32、33 和 22、23、24 号槽内的导体分给 C 相，如图 3-8 所示。再依据槽电动势星形图，将 1、2、3 号槽内的导体分别与 10、11、12 号槽内的导体连接起来，构成属于 A 相的三个整距线圈。同样，将分属 B、C 相的导体连接起来，可以构成分属三相的共 18 个整距线圈。

可见，单层绕组是将一对极距内属于同一相的 q 个线圈串联起来组成线圈组，整个电机共 p 对极，故每相有 p 个线圈组。这 p 个线圈组的电动势大小相等、相位相同，因而可以单独作为该相绕组的一条并联支路，于是每相绕组的最大并联支路数为 $a_{max}=p$。本例题要求并联支路数 =1，故可按照电动势顺极性叠加的原则，将 A 相的这两个线圈组串联起来，得到 A 相绕组。同样方法可以构成 B 相、C 相绕组，图 3-9 为三相绕组展开图。单层绕组具有结构简单、生产率较高等优点，在 10kW 以下的异步电机中广泛使用。

3.2.3 三相双层叠绕组

为了改善绕组的电动势和磁动势波形，功率较大的交流电机通常采用双层绕组。双层绕组分为叠绕组和波绕组两大类，本书只介绍叠绕组的连接方法。

已知 $2p=4$，$Q=36$，绘制并联支路数 $a=1$ 的三相双层叠绕组展开图。

1. 线圈电动势星形图

双层绕组通常为短距绕组，节距 y_1 按 $5/6\tau$ 取整数（后面解释其原理）。对本例来说，$\tau=9$，$y_1=8$。对于每个线圈，其一个线圈边放在某槽的上层，称为上层边；其另一个线圈边放在另一槽的下层，称为下层边。为了描述方便，按如下方式进行编号：以某线圈上层边所在槽的号码代表该线圈的号码，即 1 号线圈的上层边放在 1 号槽的上层，按照 $y_1=8$，1 号线圈的下层边放在 9 号槽的下层。按照这样的方式，形成总共 36 个线圈。槽电动势星形图仍然如图 3-8 所示。根据槽电动势星形图，通过相量求和，可得所有 36 个线圈各自的电动势相量。不难得出，相邻两线圈电动势的相位差仍为槽距角 α_1，由此可绘出由线圈电动势相量组成的线圈电动势星形图，其图形与槽电动势星形图相同，现在将图 3-8 看成是线圈电动势星形图。

2. 分相

与单层绕组相似，根据线圈电动势星形图，将 36 个线圈分配给 A、B、C 三相。属于 A 相线圈的号码是 1、2、3 和 10、11、12，以及 19、20、21，28、29、30；属于 B 相线圈的号码是 7、8、9，16、17、18，以及 25、26、27，34、35、36；属于 C 相线圈的号码是 13、14、15，4、5、6，以及 31、32、33，22、23、24。

3. 构成相绕组

如图 3-10 所示，在属于 A 相的 12 个线圈中，将 1、2、3 号线圈串联，组成另一个线圈组。同样，将10、11、12 号线圈串联，组成另一个线圈组；将 19、20、21 号线圈串联，组成一个线圈组；将 28、29、30 号线圈串联，组成又一个线圈组。由此，形成 A 相的 4 个线圈组。

由线圈电动势星形图不难得出，这 4 个线圈组的感应电动势大小相同、相位相同或相反，每个线圈组可作为一条并联支路。注意，由图 3-10 可见，第一个线圈组的 1、2、3 号线圈的上层边在 N_1 极下，其他线圈组的 10、11、12 号线圈，19，20、21 号线圈，28、29、30 号线圈的上层边分别在 S_1、N_2、S_2 极下，因此对于双层绕组，每相的线圈组的数量等于极数 $2p$，每相的最大并联支路数为 $a_{max}=2p$。当 $a=1$ 时，应将 A 相的 4 个线圈组串联起来，构成 A 相绕组，如图 3-10 所示。图中实线表示线圈的上层边，虚线表示线圈的下层边。同理可构成 B 相、C 相绕组。

3.3　交流绕组的感应电动势

本节主要讨论在正弦分布的气隙磁场下，交流绕组每相感应电动势有效值的计算方法，即计算展开图中某相绕组从首端到末端（如 A 到 X）之间的电动势有效值。在本节中，根据绕组的构成，按照导体电动势、线圈电动势、线圈组电动势、相电动势的顺序，逐步进行分析，最后讨论气隙磁场非正弦分布时产生的谐波电动势。

首先考虑磁场为正弦分布的情况下，交流绕组的感应电动势。

3.3.1　导体电动势

设槽内导体的有效长度为 l，磁场相对于导体运动的线速度为 v，磁感应强度沿定子内圆按正弦规律分布，磁感应强度幅值为 B_1，用电角度 θ_s 表示定子内圆上的位置，则各位置上的磁感应强度可表示为 $b=B_1\sin\theta_s$，如图 3-11 所示。

导体感应电动势的有效值为

$$E_1=\frac{E_{m1}}{\sqrt{2}}=\frac{1}{\sqrt{2}}B_1lv \tag{3-6}$$

由于 $B_{av}=\dfrac{2}{\pi}B_1$ 为每个磁极下磁感应强度的平均值，τl 为一个磁极下的面积，因此，每极磁通为 $\Phi_1=B_{av}\tau l=\dfrac{2}{\pi}B_1\tau l$，则

$$B_1=\frac{\pi}{2}\frac{\Phi_1}{\tau l} \tag{3-7}$$

而转速为 n 时，磁场相对导体运动的线速度 v 可以表示为

$$v=\frac{2p\tau n}{60}=2\tau f \tag{3-8}$$

将式（3-7）和式（3-8）代入式（3-6），得到导体电动势的有效值 E_1 为

$$E_1=\frac{\pi}{\sqrt{2}}f\Phi_1=2.22f\Phi_1 \tag{3-9}$$

3.3.2　线圈电动势和节距因数

1. 单匝整距线圈

单匝整距线圈只有两个导体（见图 3-12a），如图 3-12b 中实线所示，$y_1=\tau$，两线圈边

的感应电动势大小相等、相位相反（见图 3-12c 中的右图），则有

$$\dot{E}_{c1} = \dot{E}_1 - \dot{E}_1' = 2\dot{E}_1 \tag{3-10}$$

因此单匝整距线圈的电动势有效值为

$$E_{c1} = 2E_1 = 4.44f\Phi_1 \tag{3-11}$$

2. 单匝短距线圈

短距时，$y_1 < \tau$，如图 3-12b 中虚线所示。此时，两线圈边感应电动势 $\dot{E}_1$、$\dot{E}_1'$ 的相位差为 γ，以电角度表示其大小

$$\gamma = \frac{y_1}{\tau} \times 180° \tag{3-12}$$

由于 $\dot{E}_{c1} = \dot{E}_1 - \dot{E}'_1$，根据图 3-12c 中的右图，线圈电动势的有效值为

$$E_{c1} = 2E_1\cos\frac{180° - \gamma}{2} = 2E_1\sin\frac{\gamma}{2} = 2E_1\sin\frac{y_1}{\tau} \times 90° = 4.44fk_{p1}\Phi_1 \tag{3-13}$$

式中，k_{p1} 称为线圈的基波节距因数，其物理意义为

$$k_{p1} = \frac{\text{两个导体的电动势的几何和}}{\text{两个导体的电动势的算术和}}$$

对于短距线圈，$k_{p1} < 1$，而整距线圈的 $k_{p1} = 1$，因此短距线圈的电动势比整距线圈的电动势有所减小。长距线圈的 $k_{p1} < 1$，但由于长距线圈的端部较长，用铜量较多，所以很少采用。

3. 多匝线圈的电动势

当线圈匝数为 N_c 时，各匝的感应电动势大小相等、相位相同，因此多匝线圈的电动势应为单匝电动势的 N_c 倍，其有效值为

$$E_{c1} = 4.44fN_ck_{p1}\Phi_1 \tag{3-14}$$

3.3.3 线圈组电动势和分布因数

已知线圈组由 q 个线圈串联组成，因此线圈组电动势为各线圈电动势的相量和。另外，相邻两线圈电动势的相位差为 α。设 $q = 3$，将 q 个线圈的电动势相加，得线圈组电动势 $\dot{E}_{q1} = \dot{E}_{c1} + \dot{E}_{c2} + \dot{E}_{c3}$，如图 3-13 所示。

根据图 3-13 的特征不难看出，将 q 个线圈电动势相量首尾相接，即构成了一个正多边形的一部分。作正多边形的外接圆，设 R 为外接圆的半径，则各边所对应的圆心角均为 α，由几何关系可得线圈组电动势的有效值 E_{q1} 为 $E_{q1} = 2R\sin\frac{q\alpha}{2}$，而 $R = \dfrac{E_{c1}}{2\sin\frac{\alpha}{2}}$，因此得

$$E_{q1} = \frac{E_{c1}\sin\frac{q\alpha}{2}}{\sin\frac{\alpha}{2}} = qE_{c1}\frac{\sin q\frac{\alpha}{2}}{q\sin\frac{\alpha}{2}} = qE_{c1}k_{d1} \tag{3-15}$$

式中，k_{d1} 称为绕组的基波分布因数，有

$$k_{d1} = \frac{E_{q1}}{qE_{c1}} = \frac{\sin q\frac{\alpha}{2}}{q\sin\frac{\alpha}{2}} \tag{3-16}$$

其物理意义为

$$k_{d1}=\frac{q\text{ 个线圈的电动势的几何和}}{q\text{ 个线圈的电动势的算术和}}$$

将式（3-14）代入式（3-15）中得

$$E_{q1}=4.44fN_cqk_{p1}k_{d1}\Phi_1=4.44fN_cqk_{w1}\Phi_1 \tag{3-17}$$

式中，k_{w1}称为基波绕组因数，有

$$k_{w1}=k_{p1}k_{d1} \tag{3-18}$$

其物理意义为

$$k_{w1}=\frac{\text{线圈组所有导体电动势的几何和}}{\text{线圈组所有导体电动势的算术和}}$$

3.3.4　相绕组的基波电动势

相电动势指的是支路电动势。每相绕组有 $2p$（双层绕组）或 p（单层绕组）个线圈组，并且按照要求形成 a 条并联支路。因此，每条并联支路由 $2p/a$（双层）或 p/a（单层）个线圈组串联组成，所以相电动势等于线圈组电动势的 p/a 倍（单层绕组）或 $2p/a$ 倍（双层绕组）。因此，单层绕组的相电动势为

$$E_{\phi1}=\frac{p}{a}E_{q1}=4.44f\frac{p}{a}qN_ck_{w1}\Phi_1=4.44fN_1k_{w1}\Phi_1 \tag{3-19}$$

双层绕组的相电动势为

$$E_{\phi1}=4.44f\frac{2p}{a}qN_ck_{w1}\Phi_1=4.44fN_1k_{w1}\Phi_1 \tag{3-20}$$

上两式中，N_1 称为每相串联匝数，实际代表每条并联支路的总匝数。对于单层绕组

$$N_1=\frac{p}{a}qN_c \tag{3-21}$$

而对于双层绕组

$$N_1=\frac{2p}{a}qN_c \tag{3-22}$$

式（3-19）和式（3-20）就是相绕组基波电动势有效值的计算公式。

定子三相绕组可以为丫形或△形联结。当为丫联结时，绕组端电压为相电压 $E_{\phi1}$ 的$\sqrt{3}$倍；当为△联结时，绕组端电压等于相电压 $E_{\phi1}$。

定子三相感应电动势的瞬时表达式为

$$\begin{cases}e_{AA'}(t)=\sqrt{2}E_{\phi1}\sin\omega t\\ e_{BB'}(t)=\sqrt{2}E_{\phi1}\sin(\omega t-120°)\\ e_{CC'}(t)=\sqrt{2}E_{\phi1}\sin(\omega t-240°)\end{cases} \tag{3-23}$$

3.3.5　相绕组的谐波电动势

在同步电机中，磁极建立的气隙磁场并非严格按正弦规律分布。磁密波中，除基波分量外，还含有一系列高次谐波分量，其对应的磁场称为谐波磁场。谐波磁场相对于绕组运动

时，会在绕组中产生相应的谐波电动势，这会影响电动势的波形并产生其他不良影响。

由于每个磁极下的磁密波对磁极中心线是对称的，所以谐波磁场中只含有奇次谐波分量。设ν代表谐波次数，则$\nu=1, 3, 5, \cdots\cdots$，其中，$\nu=1$时为基波。$\nu$次谐波磁场的极对数为基波磁场极对数的$\nu$倍，即$p_\nu=\nu_p$，其极距为基波极距的$1/\nu$，即$\tau_\nu=\tau/\nu$。同时，各次谐波的转速都等于磁极的转速，即$n_\nu=n_s$，故$\nu$次谐波磁场产生的$\nu$次谐波电动势的频率为$f_\nu=\dfrac{p_\nu n_\nu}{60}=\dfrac{\nu p n_s}{60}=\nu f_1$。

同样可以得到相绕组ν次谐波电动势有效值的计算公式为

$$E_{\phi\nu}=4.44 f_\nu N_1 k_{w\nu}\Phi_\nu \tag{3-24}$$

式中，Φ_ν为ν次谐波磁场的每极磁通；$k_{w\nu}$为ν次谐波的绕组因数，$k_{w\nu}=k_{p\nu}k_{d\nu}$，$k_{p\nu}$、$k_{d\nu}$分别为$\nu$次谐波的节距因数和分布因数，且$k_{p\nu}=\sin\nu\dfrac{y_1}{\tau}\times 90°$，$k_{d\nu}=\dfrac{\sin q\nu\dfrac{\alpha}{2}}{q\sin\nu\dfrac{\alpha}{2}}$。

例3-1 一台三相两极50Hz的同步发电机，定子槽数$Q=48$，定子绕组为双层叠绕组，线圈节距$y_1=20$，每相串联匝数$N_1=16$，星形联接，基波磁通$\Phi_1=1.11\text{Wb}$，试计算该发电机定子绕组的：

1）基波绕组因数；

2）基波相电动势有效值。

解：

1）发电机的极距为

$$\tau=\frac{Q}{2p}=\frac{48}{2}=24$$

槽距角为

$$\alpha=\frac{p\times 360°}{Q}=\frac{1\times 360°}{48}=7.5°$$

每极每相槽数为

$$q=\frac{Q}{2mp}=\frac{48}{2\times 3}=8$$

基波节距因数为

$$k_{p1}=\sin\frac{y_1}{\tau}\times 90°=\sin\frac{20}{24}\times 90°=0.966$$

基波分布因数为

$$k_{d1}=\frac{\sin\dfrac{q\alpha}{2}}{q\sin\dfrac{\alpha}{2}}=\frac{\sin\dfrac{8\times 7.5°}{2}}{8\sin\dfrac{7.5°}{2}}=0.956$$

基波绕组因数为

$$k_{w1}=k_{p1}k_{d1}=0.966\times 0.956=0.923$$

2）基波相电动势为

$$E_{\phi1} = 4.44fN_1k_{w1}\Phi_1 = 4.44 \times 50 \times 16 \times 0.923 \times 1.11\text{V} = 3639\text{V}$$

3.4　三相感应电动机的工作原理

图 3-14 为一台三相感应电动机模型，其定子槽内嵌放三相对称绕组，转子槽内放置导条，将所有导条的两端分别用两个短路环短接，形成闭合的转子绕组。

当定子三相对称绕组中通入三相对称电流时，三相电流瞬时值的表达式为

$$\begin{cases} i_A = I_m\cos\omega t \\ i_B = I_m\cos(\omega t - 120°) \\ i_C = I_m\cos(\omega t - 240°) \end{cases} \tag{3-25}$$

规定电流瞬时值为正时，电流从每相绕组的尾端（X、Y、Z）流入，首端（A、B、C）流出。

当 $\omega t = 0°$时，$i_A = I_m$，$i_B = i_C = -\dfrac{1}{2}I_m$，将各相电流的实际方向表示在各相绕组的导体中，根据右手螺旋定则，三个绕组中的电流产生的合成磁场将如图 3-15a 所示。同样，考察 $\omega t = 120°$、240°时的三相电流方向，分别得到如图 3-15b、3-15c 所示的磁场分布。由图可见，当三相电流随时间连续变化时，三相绕组产生的合成磁场的轴线位置（即 N、S 极的位置）将沿着电动机圆周移动，在气隙中形成按逆时针方向“旋转”的磁场。

如果开始时转子转速为零，转子导条切割旋转磁场会产生感应电动势 e_2，由于转子导条是短路的，因而转子导体中将会出现短路电流 i_2。根据右手定则：N 极下的转子导体中的感应电动势的方向以及电流有功分量的方向为进入纸面；S 极下为流出纸面。转子电流 i_2 与气隙磁场相互作用，将产生电磁力，其方向由左手定则确定。电磁力产生的转矩，称为电磁转矩。在电磁转矩的作用下，转子就会顺着旋转磁场的方向旋转起来。

由于转子方的电动势和电流是靠电磁感应作用产生的，故这种电动机称为感应电动机。另外，如果转子转速与旋转磁场的同步转速相同，转子导条中就不会出现感应电动势，也就没有电磁力和电磁转矩了，所以感应电动机正常运行时，转子转速与旋转磁场转速之间总是有差异的，因此感应电动机也称为异步电动机。三相感应电动机的气隙旋转磁场是由定子三相交流绕组的合成磁动势产生的，为此需要研究交流绕组的磁动势。

3.5　交流绕组的磁动势

本节首先分析单相绕组中通过交流电流时产生的磁动势，再分析三相对称绕组中通过三相对称电流时产生的磁动势。

为了简化分析，假定：

1）绕组中的电流随时间按正弦规律变化。

2）槽内电流集中在槽中心处。

3）气隙均匀，忽略齿槽的影响。

4）不计铁心中的磁位降，不计磁饱和的影响。

3.5.1 单相交流绕组的磁动势——脉振磁动势

1. 整距线圈的磁动势

以下计算整距线圈中通过交流电流时产生的每极磁动势。

图 3-16 表示一个整距线圈通入电流后形成的两极磁场分布情况，设线圈匝数为 N_c，电流为 $i_c=\sqrt{2}I_c\cos\omega t$。根据安培环路定律，沿着任一闭合磁力线的磁位降等于该磁力线所包围的电流值 N_ci_c。由于忽略了铁心中的磁位降且气隙为均匀，因而线圈产生的磁动势 N_ci_c 将消耗在 N 极和 S 极下的两个气隙上，且每个气隙上的磁位降都等于每极磁动势 $N_ci_c/2$。将图 3-16 展开后，得到磁动势沿气隙分布的波形（见图 3-17）。图 3-17 中的正半波对应于图 3-16 中磁通由定子进入转子的情况，负半波对应于磁通由转子进入定子的情况。当电流随时间变化时，磁动势波的空间位置不变，幅值大小和正负随时间按余弦规律变化，变化的频率与电流的频率相同。这种空间位置不变，幅值大小和正负随时间变化的磁动势称为脉振磁动势。以线圈轴线位置作为空间坐标的原点，磁动势波的表达式为

$$f_c(\theta_s,t)=\begin{cases}\dfrac{1}{2}i_cN_c=\dfrac{\sqrt{2}}{2}I_cN_c\cos\omega t & \left(-\dfrac{\pi}{2}<\theta_s<\dfrac{\pi}{2}\right)\\[2ex] -\dfrac{1}{2}i_cN_c=-\dfrac{\sqrt{2}}{2}I_cN_c\cos\omega t & \left(\dfrac{\pi}{2}<\theta_s<\dfrac{3\pi}{2}\right)\end{cases} \tag{3-26}$$

将矩形波分解为富氏级数，得

$$\begin{aligned}f_c(\theta_s,t)&=\frac{\sqrt{2}}{2}N_cI_c\left[\frac{4}{\pi}\left(\cos\theta_s-\frac{1}{3}\cos3\theta_s+\frac{1}{5}\cos5\theta_s-\cdots\right)\right]\cos\omega t\\&=\left(F_{c1}\cos\theta_s-F_{c3}\cos3\theta_s+F_{c5}\cos5\theta_s-\cdots\right)\cos\omega t\end{aligned} \tag{3-27}$$

$$F_{c1}=\frac{4\sqrt{2}}{2\pi}N_cI_c=0.9N_cI_c \tag{3-28}$$

$$F_{c\nu}=\frac{1}{\nu}F_{c1} \tag{3-29}$$

式中，F_{c1} 为基波磁动势的振幅，即电流瞬时值达到最大值时每极基波磁动势的幅值，$F_{c\nu}$ 为 ν 次谐波磁动势的振幅。

基波磁动势和谐波磁动势都是脉振磁动势，且脉振频率都与电流频率相同，如图 3-18 所示。基波磁动势的瞬时表达式为

$$f_{c1}(\theta_s,t)=F_{c1}\cos\theta_s\cos\omega t \tag{3-30}$$

2. 整距线圈组的磁动势

以下计算整距线圈组中通过交流电流时产生的每极磁动势。

整距线圈组由 q 个整距线圈串联组成，相邻两线圈的轴线在空间相差 α 电角度。图 3-19 表示 $q=3$ 的整距线圈组，将 3 个线圈的矩形磁动势波叠加，得到线圈组的合成磁动势波 f_q。f_q 为阶梯波，同样是脉振磁动势。为了便于对 f_q 的大小进行计算，这里要引入空间矢量的概念。以后会发现，在对交流电机进行分析的过程中，空间矢量将起到非常重要的作用，因此应该扎实地掌握其有关概念。

对于空间上按正弦（余弦）规律分布的物理量（如线圈的基波磁动势），在波长一定的情况下，可以用空间矢量来表示，这样的物理量与表示它的空间矢量之间的对应关系为：矢量的长度表示该物理量的正弦分布波幅值的大小，矢量的位置或方向表示其正波幅所在的位

置，这样，空间矢量和正弦分布波就是一一对应的关系。借助空间矢量，可将磁动势波的叠加过程转换为矢量运算，用矢量相加的方法得到线圈组磁动势的基波和各次谐波的振幅。在图 3-19 中f_{c11}、f_{c12}、f_{c13}为 3 个线圈的基波磁动势，f_{q1}为线圈组的基波磁动势。图 3-20 中$\boldsymbol{F}_{c11}$、$\boldsymbol{F}_{c12}$、$\boldsymbol{F}_{c13}$为与 3 个线圈的基波磁动势对应的空间矢量，相邻两线圈的基波磁动势的空间相位相差 α，$\boldsymbol{F}_{q1}$为线圈组磁动势的空间矢量，参照线圈组电动势的计算方法，得到线圈组每极基波磁动势的振幅 F_{q1}为

$$F_{q1} = qF_{c1}k_{d1} = 0.9N_cI_cqk_{d1} \tag{3-31}$$

其中，k_{d1}为基波磁动势的分布因数，$k_{d1} = \dfrac{\sin q\dfrac{\alpha}{2}}{q\sin\dfrac{\alpha}{2}}$。

由图 3-19 可见，线圈组基波磁动势幅值的位置在线圈组的轴线上。同理可得线圈组每极谐波磁动势的振幅 $F_{q\nu}$为，

$$F_{q\nu} = qF_{c\nu}k_{d\nu} = \frac{1}{\nu}0.9N_cI_cqk_{d\nu} \tag{3-32}$$

式中，$k_{d\nu}$为 ν 次谐波磁动势的分布因数，$k_{d\nu} = \dfrac{\sin q\nu\dfrac{\alpha}{2}}{q\sin\nu\dfrac{\alpha}{2}}$。

线圈组基波磁动势的瞬时表达式为

$$f_{q1}(\theta_s, t) = F_{q1}\cos\theta_s\cos\omega t \tag{3-33}$$

3. 相绕组的磁动势

（1）单层绕组的磁动势　可认为各种形式的单层绕组在电气上都属于整距绕组，节距因数为 1。每相单层绕组包括 p 个线圈组，产生 p 对极磁场，因此单层绕组的相绕组每极磁动势事实上就等于线圈组的每极磁动势。因此，相绕组基波磁动势的振幅 $F_{\phi1}$和谐波磁动势的振幅 $F_{\phi\nu}$分别为

$$\begin{cases} F_{\phi1} = F_{q1} = 0.9qN_cI_ck_{d1} \\ F_{\phi\nu} = F_{q\nu} = \dfrac{1}{\nu}0.9qN_cI_ck_{d\nu} \end{cases} \tag{3-34}$$

（2）双层绕组的磁动势　双层绕组通常为短距绕组。以三相绕组中的 A 相绕组为例，图 3-21 为在一对极距范围内，A 相绕组的两个 $q=3$ 的线圈组在槽内的分布情况（与图 3-10 对应），具体地说，图 3-10 中的 1、2、3 号线圈的上层边放在图 3-21 中的 1、2、3 号槽的上层（图 3-21 中的 1、2、3），图 3-10 中的 1、2、3 号线圈的下层边放在图 3-21 中的 9、10、11 号槽的下层（图 3-21 中的 9′、10′、11′）；图 3-10 中的 10、11、12 号线圈的上层边放在图 3-21 中的 10、11、12 号槽的上层（图 3-21 中的 10、11、12），图 3-10 中的 10、11、12 号线圈的下层边放在图 3-21 中的 18、1、2 号槽的下层（图 3-21 中的 18′、1′、2′）。现在，将这样的两个短距线圈组看成两个等效的整距线圈组，其中：一个整距线圈组由原来的两个短距线圈组的下层边（即处在 18′、1′、2′和 9′、10′、11′的线圈边）构成，其基波磁动势用空间矢量 $\boldsymbol{F}_{q11}$表示；另一个整距线圈组由原来的两个短距线圈组的上层边（即处在 1、2、3 和 10、11、12 的线圈边）构成，其基波磁动势用空间矢量 $\boldsymbol{F}_{q12}$表示。这两个整距线圈组的轴线在空间上相差的电角度，正好等于短距线圈节距 y_1 较极距 τ 缩短的电角度 β，那么两整

距线圈组的基波磁动势 $\boldsymbol{F}_{q11}$、$\boldsymbol{F}_{q12}$ 的空间相位也相差 β 电角度。与求短距线圈电动势的方法相似，相绕组每极磁动势的振幅为

$$F_{\phi1}=2F_{q1}k_{p1}=2\times0.9qN_cI_ck_{p1}k_{d1}=0.9\times2qN_cI_ck_{w1}\tag{3-35}$$

式中，F_{q1} 为整距线圈组每极磁动势的振幅；k_{p1} 为基波磁动势的节距因数；k_{w1} 为基波磁动势的绕组因数。

此时，$F_{\phi1}$ 的位置在相绕组的轴线上。

同样，两等效整距线圈组的 ν 次谐波磁动势在空间相差 $\nu\beta$ 电角度，所以 ν 次谐波磁动势的振幅为

$$F_{\phi\nu}=2F_{q\nu}k_{p\nu}=\frac{1}{\nu}0.9\times2qN_cI_ck_{q\nu}k_{d\nu}=\frac{1}{\nu}0.9\times2qN_cI_ck_{w\nu}\tag{3-36}$$

式中，$F_{q\nu}$ 为整距线圈组 ν 次谐波磁动势的振幅；$k_{p\nu}$ 为 ν 次谐波磁动势的节距因数；$k_{d\nu}$ 为 ν 次谐波磁动势的分布因数；$k_{w\nu}$ 为 ν 次谐波磁动势的绕组因数。

设 N 为每相串联匝数，I_ϕ 为相电流的有效值。

对于单层绕组，有

$$qN_c=\frac{N_1a}{p}\tag{3-37}$$

对于双层绕组，有

$$qN_c=\frac{N_1a}{2p}\tag{3-38}$$

又考虑到支路电流有效值 I_c 与相电流有效值 I_ϕ 的关系为

$$I_c=\frac{I_\phi}{a}\tag{3-39}$$

可将相绕组基波磁动势和谐波磁动势的幅值统一表达为

$$F_{\phi1}=0.9\frac{N_1I_\phi}{p}k_{w1}\tag{3-40}$$

$$F_{\phi\nu}=\frac{1}{\nu}0.9\frac{N_1I_\phi}{p}k_{w\nu}\tag{3-41}$$

再考虑到电流随时间变化，以及磁动势沿空间按余弦规律分布等因素，可得到相绕组基波磁动势和谐波磁动势的瞬时表达式为

$$\begin{cases}f_1(\theta_s,t)=F_{\phi1}\cos\theta_s\cos\omega t\\f_\nu(\theta_s,t)=F_{\phi\nu}\cos\nu\theta_s\cos\omega t\end{cases}\tag{3-42}$$

注意，在某一瞬间（如电流为最大值时），相绕组合成磁动势的分布波是由布置在槽中的众多导体产生的，仍然是阶梯波；相绕组磁动势也属于脉振磁动势。

下面将相绕组磁动势随时间和空间的变化规律归纳如下：

1）单相绕组的磁动势是脉振磁动势，脉振的频率与电流的频率相同，该磁动势波沿气隙圆周按阶梯形分布；在空间上可分解为基波磁动势和一系列奇次谐波磁动势，其振幅分别由式（3-40）和式（3-41）表示。基波磁动势和谐波磁动势均为脉振磁动势，脉振的频率与电流的频率相同，沿气隙圆周按正弦规律分布。

2）基波磁动势波幅的位置在相绕线轴线上，各次谐波磁动势的波幅亦在相绕组的轴线上。

3.5.2 三相交流绕组的合成磁动势——旋转磁动势

1. 三相合成基波磁动势

当三相对称绕组中通入三相对称电流

$$\begin{cases} i_A = I_m\cos\omega t \\ i_B = I_m\cos(\omega t - 120°) \\ i_C = I_m\cos(\omega t - 240°) \end{cases} \tag{3-43}$$

以 A 相绕组轴线的位置作为空间坐标的原点，以 A 相电流达到最大的时刻作为时间的起点，按照 A →B →C 相序，三相绕组的基波磁动势可以表示为

$$\begin{cases} f_{A1}(\theta_s,t) = F_{\phi1}\cos\theta_s\cos\omega t \\ f_{B1}(\theta_s,t) = F_{\phi1}\cos(\theta_s - 120°)\cos(\omega t - 120°) \\ f_{C1}(\theta_s,t) = F_{\phi1}\cos(\theta_s - 240°)\cos(\omega t - 240°) \end{cases} \tag{3-44}$$

利用三角函数的“积化和差”公式可将上式变换为

$$\begin{cases} f_{A1}(\theta_s,t) = \frac{1}{2}F_{\phi1}\cos(\theta_s - \omega t) + \frac{1}{2}F_{\phi1}\cos(\theta_s + \omega t) \\ f_{B1}(\theta_s,t) = \frac{1}{2}F_{\phi1}\cos(\theta_s - \omega t) + \frac{1}{2}F_{\phi1}\cos(\theta_s + \omega t - 240°) \\ f_{C1}(\theta_s,t) = \frac{1}{2}F_{\phi1}\cos(\theta_s - \omega t) + \frac{1}{2}F_{\phi1}\cos(\theta_s + \omega t - 120°) \end{cases} \tag{3-45}$$

将以上三式相加，可得三相绕组合成磁动势的基波为

$$\begin{aligned} f_1(\theta_s,t) &= f_{A1}(\theta_s,t) + f_{B1}(\theta_s,t) + f_{C1}(\theta_s,t) \\ &= \frac{3}{2}F_{\phi1}\cos(\omega t - \theta_s) = F_1\cos(\omega t - \theta_s) \end{aligned} \tag{3-46}$$

式中，F_1 为三相合成磁动势基波的幅值

$$F_1 = \frac{3}{2}F_{\phi1} = \frac{3}{2}\times 0.9\frac{N_1k_{w1}}{p}I_\phi = 1.35\frac{N_1k_{w1}}{p}I_\phi \tag{3-47}$$

这里称 F_1 为幅值而不是振幅，是由于式（3-46）表示的 $f_1(\theta_s,t)$ 是一个幅值为恒定不变的旋转波，下面对 f_1（θ_s，t）随时间和空间的变化规律进行分析。

1）当 $\omega t = 0°$ 时，A 相电流为最大值。此时 $f_1(\theta_s,t) = F_1\cos\theta_s$，是一个幅值为 F_1，沿气隙圆周按余弦规律分布的磁动势波，其正波幅位于 $\theta_s = 0°$ 处，与 A 相绕组的轴线重合，如图 3-22a 所示。

2）当 $\omega t = 120°$ 时，B 相电流为最大值。此时 $f_1(\theta_s,t) = F_1\cos(\theta_s - 120°)$，是一个幅值为 F_1，沿气隙圆周按余弦规律分布的磁动势波，其正波幅位于 $\theta_s = 120°$ 处，与 B 相绕组轴线重合，如图 3-22b 所示。与 A 相电流最大时相比，时间上经过了 120°，而合成磁动势正波幅的位置在空间上也转过了 120°电角度。

3）当 $\omega t = 240°$ 时，C 相电流为最大值。此时 f_1（θ_s，t）$= F_1\cos$（$\theta_s - 240°$），是一个幅值为 F_1，沿气隙圆周按余弦规律分布的磁动势波，其正波幅位于 $\theta_s = 240°$ 处，与 C 相绕组轴线重合，如图 3-22c 所示。与 B 相电流最大时相比，时间上经过了 120°，而合成磁动势正波幅的位置在空间上同样转过了 120°电角度。

4）当时间再经过 120°，重新出现 A 相电流为最大值时，合成磁动势正波幅的位置重新

与 A 相绕组轴线重合。

下面进一步对 $f_1(\theta_s,t)$ 的移动速度进行讨论。由于合成磁动势正波幅转过的电角度与电流变化的时间角度相同，因此电流每变化一个周期，即时间上经过 360°，磁动势波在空间也移动 360°电角度，即一对极距。这样，电流每秒变化 f 次，磁动势波在空间移动 f 对极距，而对于 p 对极电机，旋转磁动势波的转速（单位为 r/min）为

$$n = \frac{60f}{p} \tag{3-48}$$

即等于同步转速 n_s。

下面对三相绕组合成磁动势的基波随时间和空间的变化规律归纳如下：

1）三相绕组合成磁动势的基波是一个旋转磁动势波，转速为同步转速。

2）当某相电流达到最大值时，合成磁动势基波的正波幅与该相绕组的轴线重合，因此磁动势的转向是从电流超前的相绕组轴线转向电流滞后的相绕组轴线。改变电流的相序即可改变磁动势的旋转方向。

3）合成磁动势的基波幅值不随时间变化，由式（3-47）表示。

2. 三相合成磁动势的高次谐波

将 A、B、C 三相绕组所产生的高次谐波磁动势相加，得到三相合成磁动势的高次谐波：

$$\begin{aligned} f_\nu(\theta_s,t) &= f_{A\nu}(\theta_s,t) + f_{B\nu}(\theta_s,t) + f_{C\nu}(\theta_s,t) \\ &= F_{\phi\nu}\cos\nu\theta_s\cos\omega t + F_{\phi\nu}\cos\nu(\theta_s - 120°)\cos(\omega t - 120°) + \\ &\quad F_{\phi\nu}\cos\nu(\theta_s - 240°)\cos(\omega t - 240°) \end{aligned} \tag{3-49}$$

可知：

1）当 $\nu = 3k\ (k = 1,3,5,\cdots)$ 时，$f_\nu(\theta_s,t) = 0$，即三相合成磁动势中不存在 3 次及 3 的倍数次谐波磁动势。

2）当 $\nu = 6k + 1\ (k = 1,3,5,\cdots)$ 时，$f_\nu(\theta_s,t) = \frac{3}{2}F_{\phi\nu}\cos(\omega t - \nu\theta_s)$，即 $\nu = 6k+1$ 次谐波的合成磁动势为转向与基波磁动势的转向相同，转速为 n_s/ν，幅值为 $3/2F_{\phi\nu}$ 的旋转磁动势。

3）当 $\nu = 6k - 1\ (k = 1,3,5,\cdots)$ 时，$f_\nu(\theta_s,t) = \frac{3}{2}F_{\phi\nu}\cos(\omega t + \nu\theta_s)$，即 $\nu = 6k-1$ 次谐波的合成磁动势转向与基波磁动势的转向相反，转速为 n_s/ν，幅值为 $3/2F_{\phi\nu}$ 的旋转磁动势。

谐波磁动势的存在，会对交流电机的运行带来不利影响，因此需要采取各种措施来削弱谐波磁动势。当采取了短距绕组、分布绕组等措施之后，由于谐波磁动势的绕组因数很小，因而幅值被大大削弱，因此，在今后的分析中，主要考虑电机中的基波磁动势和基波磁场。

本章小结

本章主要介绍了三相对称绕组的结构、感应电动势的计算和磁动势的计算这 3 部分内容，这些内容涉及同步电机和感应电机，是后面分析这两种电机的理论基础。在学习过程中要注意以下几点：

1）在构成交流绕组时，首先绘出槽电动势星形图。再按照对交流绕组的基本要求进行分相。根据分相的结果，构成线圈、线圈组和相绕组，最终构成三相对称绕组。单层绕组结构简单，但不能利用短距措施改善波形，因而与双层绕组相比，电气性能稍差。

2）在计算感应电动势的过程中，由于短距和分布的影响，引入了节距因数、分布因数以及绕组因数，其中，整距绕组（包括单层绕组）的节距因数为1，集中绕组（$q=1$）的分布因数为1，整距集中绕组的绕组因数为1。与整距集中绕组相比，在合理选择短距和分布的情况下，虽然短距分布绕组的基波电动势略有减小，但明显削弱了谐波电动势，因而改善了电动势波形。

3）分析磁动势时特别要注意磁动势的性质以及随时间、沿空间的变化规律。单相绕组通过交流电流时，产生脉振磁动势，其运动规律类似于“驻波”。三相对称绕组通过三相对称电流时，产生旋转磁动势，其运动规律类似于“行波”。对旋转磁动势，要还注意其转向、转速以及幅值的位置与三相电流瞬时值之间的关系。在计算磁动势时也出现了节距因数和分布因数，这在一定程度上反映了电与磁之间的联系，利用短距和分布措施同样能够改善磁动势的波形。

习　题

1. 异步电机与同步电机的基本差别是什么？

2. 说明单相绕组磁动势基波的性质和特点？

3. 说明三相绕组合成磁动势基波的性质和特点？

4. 交流电机中磁场的旋转速度与交流电流的频率之间有何关系？

5. 三相单层绕组，$Q=24$，$2p=4$，$a=2$，画出三相绕组的展开图并计算绕组因数。

6. 三相双层短距叠绕组，$Q=24$，$2p=2$，$y_1=\frac{5}{6}\tau$，$a=1$，画出 A 相绕组的展开图并计算绕组因数。

7. 如何改变感应电动机的旋转方向？

8. 说明节距因数和分布因数的物理意义，交流电机中为何采用短距、分布绕组？

9. 在对称的两相绕组（绕组轴线空间差 90°电角度）内通入对称的两相电流（时间上差 90°），试分析产生的合成磁动势的基波。

10. 交流绕组中的谐波电动势是如何产生的？如何消除或削弱谐波电动势？

11. 说明交流电机中产生谐波磁动势的原因。

12. 一台三相同步发电机，$2p=6$，定子为双层叠绕组，$Q_1=54$，$y_1=7$，$a=2$，星形联接，每个线圈的匝数为 10 匝，试求：

（1）若在绕组中通入频率为 50Hz、有效值为 10 A 的三相对称电流，求其基波旋转磁动势的幅值和转速。

（2）若每极磁通 $\Phi_m=0.11$Wb，其相电动势的有效值。

13. 一台三相感应电动机，$2p=4$，定子为双层叠绕组，$Q_1=24$，$y_1=5$，$a=1$，星形联接，每个线圈的匝数为 10 匝。当通入对称三相电流，每相电流的有效值为 20A 时，试求三相合成磁动势的基波的幅值和转速。

14. 一台三相、星形联结、50Hz 的 2 极同步电机，定子每相串联总匝数为 2000 匝，绕组因数 $k_{w1}=0.966$，要使电机的电枢端电压为 6kV，每极磁通应该为多少？

15. 一台三相、星形联结、60Hz 的 2 极同步发电机，定子每相串联总匝数为 18 匝，绕组因数 $k_{w1}=0.933$。每极的基波磁通为 3.31Wb，电机转速为 3600 r/min。计算发电机空载端电压的有效值。

16. 一台60Hz、星形联结的三相2极同步发电机，定子每相串联总匝数为45匝，绕组因数$k_{w1}=0.93$，每极磁通为2.1Wb，电枢电流为20A。试求

(1) 电枢绕组感应电动势的幅值和有效值。

(2) 电枢一相绕组基波磁动势的幅值。

(3) 电枢三相绕组基波磁动势的幅值。

17. 如果在三相对称绕组中通入大小及相位均相同的电流，试分析三相合成基波磁动势的情况。

18. 三相异步电动机，定子绕组为星形联结，若定子绕组有一相断线，将产生什么性质的磁动势？

第4章 感 应 电 机

感应电机是一种交流旋转电机，主要作为电动机使用。感应电动机的主要优点在于：结构简单、运行可靠、效率较高、价格便宜、坚固耐用。感应电动机的不足之处在于：起动和调速性能较差，运行时需要从电网吸收感性无功功率。感应电动机是各行各业及日常生活中应用最广泛的一种电动机，随着电力电子技术的发展，各种交流调速技术日益完善，传统的直流调速系统正逐步被交流调速系统所取代，因而感应电机的应用领域还将进一步扩大。

本章首先介绍感应电机的基本结构，分析电机运行时各物理量之间的电磁关系，然后建立电机的基本方程和等效电路，并在此基础上研究电机的运行特性。

4.1 感应电机的基本结构和额定值

4.1.1 感应电机的基本结构

感应电机根据转子结构的不同可分为笼型感应电机和绕线型感应电机两类，分别如图4-1和图4-2所示。在结构上，感应电机主要包括定子和转子两大部件，定子、转子之间留有0.2～2mm的气隙。

1. 定子

感应电机的定子主要包括定子铁心、定子绕组、机座和端盖等部件。

（1）定子铁心 定子铁心一般由厚度为0.5mm的硅钢片叠成，以使铁心的导磁性能优良并抑制铁损。定子铁心位于机座内，构成主磁路的一部分。定子铁心内圆均匀布置定子槽，其中嵌入定子绕组。

（2）定子绕组 定子绕组为三相对称绕组，嵌放在定子槽内。定子绕组分为散嵌绕组和成型绕组。散嵌绕组也称为软绕组，由圆截面、高强度漆包线制成，用于100kW以下的小型感应电机。100kW以上的大、中型感应电机的定子绕组一般为成型绕组，用矩形截面的扁铜线制成，线圈预制成形，也称为硬绕组。

（3）机座和端盖 机座包在定子铁心外，起支撑、防护作用。端盖通过止口和螺栓固定在机座的两端，端盖上安装有轴承，用以支承转子。机座和端盖一般由铸铁或钢板焊接而成，应有足够的强度和刚度。此外，机座上还装有铭牌、出线盒、风扇罩、底脚和吊攀等。

2. 转子

转子由转子铁心、转子绕组和转轴等部件组成。

（1）转子铁心 转子铁心固定在转轴上，是电机主磁路的一部分，一般用厚度为0.5mm的硅钢片叠成。在转子铁心外圆均匀布置转子槽，用于安放转子绕组。

（2）转子绕组 感应电机的转子绕组分为笼型转子和绕线转子两类。

1）笼型转子。笼型转子绕组由转子槽内的导条和连接导条两端的端环构成，因这样的绕组外形很像一个笼子而被称为笼型绕组，如图4-3所示。小型笼型感应电机多以铸铝方式形成笼型绕组，优点是用铜量小、生产率高。对于容量在100kW以上的感应电机，由于铸铝质量难以保证，多采用将铜条插入转子槽，并在两端焊上端环的方式，构成笼型绕组。

2）绕线转子。绕线转子绕组采用三相对称绕组，极数应与定子绕组极数相同，3个尾端连在一起接成星形联结，3个首端接到转轴的3个集电环上，通过电刷可与外电路连接，如图4-4所示。与笼型转子相比，绕线转子的优点在于可通过电刷在转子回路串入外加电阻，改善电机的起动性能和调速性能；缺点在于结构复杂、价格较高、可靠性稍差。

3）转轴。转轴由钢制成，一端装有离心式风扇，通过风扇罩的导向作用，将风向由径向改为轴向，冷却机座表面。转轴的另一端从端盖伸出，称为轴伸，用以连接机械负载，电机的机械功率从转轴输入或输出。

4.1.2 感应电机的运行状态

在不同的转速范围内，感应电机有3种运行状态。感应电机运行时，转子转速 n 与旋转磁场的同步转速 n_s 总是不相等的，以旋转磁场的转向作为转速 n 的正方向，二者之差称为转差，用 Δn 表示

$$\Delta n = n_s - n \tag{4-1}$$

转差与同步转速之比称为转差率，用 s 表示

$$s = \frac{n_s - n}{n_s} \tag{4-2}$$

转差率是反映感应电机运行状态的一个重要变量，s 的大小和正负既反映了转子不同的转速范围，也反映了感应电机的3种运行状态。

1. 电动机状态

当转子正向旋转，转速低于同步转速时，$0 < n < n_s$，$0 < s < 1$，此时转子导体的感应电动势及转子电流有功分量的方向如图4-5b所示，转子电流有功分量与气隙磁场相互作用产生电磁力，使转子受到与 n_s 同方向的电磁转矩 T_e，拖动转子及转轴所带的机械负载以转速 n 旋转，将电源输入的电能转换为机械能，感应电机运行于电动机状态。

2. 发电机状态

如果由原动机将转子转速提高到 $n > n_s$，则 $s < 0$，如图4-5c所示，与电动机状态时相比，由于旋转磁场相对转子的旋转方向改变，转子导体中感应电动势和转子电流有功分量的方向改变，因此，电磁转矩的方向与 n 的方向相反。此时感应电机运行于发电机状态，吸收原动机输入的机械功率，将机械能转换为电能。

3. 电磁制动状态

当转子转向与旋转磁场的转向相反时，$n < 0$，$s > 1$。如图4-5a所示，此时电磁转矩的方向与旋转磁场的方向相同，说明电机吸收电功率，而电磁转矩的方向与转向相反，说明吸收机械功率，这两项功率都转化为热能，消耗在电机内部，感应电机运行于电磁制动状态。

感应电机主要作为电动机使用，其他两种运行状态只在少数情况中出现。由于感应发电机的性能稍差，因此只在一些小水电站和风力发电站中使用。

4.1.3 感应电动机的额定值

感应电动机的铭牌上一般标有如下额定数据：

(1) 额定功率 P_N　指电动机在额定状态下运行时，电动机轴端输出的机械功率，单位为kW。

（2）额定电压 U_N　指电动机在额定状态下运行时，加在定子绕组上的线电压，单位为 V。

（3）额定电流 I_N　指电动机在额定状态下运行时，定于绕组的线电流，单位为 A。

（4）额定转速 n_N　指电动机在额定状态下运行时，转子的转速，单位为 r/min。

（5）额定频率 f_N　指电动机在额定情况下运行时，定子供电电源的频率，单位为 Hz。我国电网的标准频率为 50Hz。

感应电动机除以上额定数据外，还有额定效率 η_N、额定功率因数 $\cos\varphi_N$ 和温升等。

根据额定数据，三相感应电动机的输入功率可表示为

$$P_1 = \sqrt{3}U_N I_N \cos\varphi_N \tag{4-3}$$

三相感应电动机的输出功率可表示为

$$P_N = P_1\eta_N = \sqrt{3}U_N I_N \eta_N \cos\varphi_N \tag{4-4}$$

4.2　感应电动机内部的电磁关系

4.2.1　感应电动机的主磁通和漏磁通

感应电动机是以磁场为媒介进行能量转换的。电动机工作时，其内部的磁场由定、转子双方的电流共同建立，磁场的分布比较复杂，一般按各部分磁通的作用和性质不同，分为主磁通和漏磁通。

1. 主磁通

定、转子电流单独或共同建立的、与定子绕组和转子绕组同时交链的基波旋转磁场的每极磁通称为主磁通，用 Φ 表示。主磁通的路径称为主磁路，如图 4-6 所示，主磁路由气隙、定子齿、定子轭、转子齿、转子轭 5 部分组成。由于主磁通与定、转子绕组同时交链，因此，感应电动机主要是通过主磁通的作用传递功率的，主磁路磁阻的大小受铁心饱和程度的影响，饱和程度越高，磁阻越大。

2. 漏磁通

感应电动机中除主磁通以外的磁通统称为漏磁通，漏磁通又分为定子漏磁通 $\Phi_{1\sigma}$和转子漏磁通 $\Phi_{2\sigma}$，分别由定、转子电流产生。漏磁通的分布非常复杂，包括：槽漏磁通，指横穿槽部闭合的漏磁通，如图 4-7a 所示；端部漏磁通，指环绕绕组端部闭合的漏磁通，如图 4-7b 所示；谐波漏磁通，指高次谐波磁动势产生的磁通，这部分漏磁通在绕组中产生的感应电动势的频率与主磁通产生的感应电动势的频率相同，即

$$f_\nu = \frac{p_\nu n_\nu}{60} = \frac{\nu p \dfrac{n_s}{\nu}}{60} = \frac{p n_s}{60} = f_1 \tag{4-5}$$

谐波漏磁通虽然也通过气隙并进入转子，但因极数和转速与基波不同，并在定子绕组内产生电抗压降，因而将其归入漏磁通。一般认为漏磁路不饱和，漏磁路的磁阻基本不变。

若定、转子漏磁通大小都以频率 f_1 变化，则在定、转子绕组中产生频率为 f_1 的漏磁电动势 $\dot{E}_{1\sigma}$和 $\dot{E}_{2\sigma}$，将 $\dot{E}_{1\sigma}$和 $\dot{E}_{2\sigma}$用负的漏电抗压降表示

$$\begin{aligned}\dot{E}_{1\sigma} &= -\mathrm{j}\dot{I}_1 2\pi f L_{1\sigma} = -\mathrm{j}\dot{I}_1 X_{1\sigma}\\ \dot{E}_{2\sigma} &= -\mathrm{j}\dot{I}_2 2\pi f L_{2\sigma} = -\mathrm{j}\dot{I}_2 X_{2\sigma}\end{aligned} \tag{4-6}$$

式中，$L_{1\sigma}$和$L_{2\sigma}$为定、转子的漏电感；$X_{1\sigma}$和$X_{2\sigma}$为定、转子的漏电抗。

由于漏磁路的磁阻基本不变，因此$L_{1\sigma}$和$L_{2\sigma}$以及$X_{1\sigma}$和$X_{2\sigma}$为常数。

4.2.2 感应电动机的电磁关系

1. 转子绕组开路时

下面以一台三相绕线转子感应电动机为例，分析感应电动机静止时的运行情况。在分析中，定、转子双方的各种变量分别用下标“1”和“2”表示。设三相对称电源的相电压（有效值）为U_1，频率为f_1，定、转子绕组的极对数为p。

由于转子绕组开路，转子电流$\dot{I}_2=0$，转速$n=0$，转差率$s=1$，此时类似于变压器空载运行。如图4-8所示，在$\dot{U}_1$的作用下，定子绕组中通过频率为f_1的三相对称电流$\dot{I}_1$（定子三相电流为$\dot{I}_{1A}$、$\dot{I}_{1B}$、$\dot{I}_{1C}$），按电动机惯例标出定子电流$\dot{I}_1$的正方向。$\dot{I}_1$产生的基波旋转磁动势$\boldsymbol{F}_1$的转速为$n_s=60f_1/p$，转向为逆时针方向。由于$I_2=0$，由$\boldsymbol{F}_1$单独建立气隙中的旋转磁场$\boldsymbol{B}_m$，因此，此时的$\dot{I}_1$为单纯的励磁电流$\dot{I}_m$，$\boldsymbol{F}_1$为单纯的励磁磁动势$\boldsymbol{F}_m$，如图4-9所示。由于铁心磁滞和涡流的影响，磁密波在空间上滞后磁动势波一个铁损角α_{Fe}，即当气隙上某点磁动势达到最大值时，该点的磁感应强度尚未达到最大值，等磁密波转过α_{Fe}电角度后，该点的磁感应强度才达到最大值。因此，在空间上，$\boldsymbol{B}_m$落后$\boldsymbol{F}_m$的电角度为α_{Fe}。$\boldsymbol{B}_m$相对于定、转子绕组旋转，使定、转子各相绕组交链的磁通$\boldsymbol{\Phi}_m$的大小随时间按正弦规律变化。定、转子绕组中分别感应出三相对称电动势E_1和E_2，频率为$f=\dfrac{pn_s}{60}=f_1$，即此时E_1和E_2的频率等于电源的频率，相序分别为A→B→C和a→b→c，E_1和E_2（有效值）的大小分别为

$$\begin{aligned}E_1&=4.44f_1N_1k_{w1}\boldsymbol{\Phi}_m\\E_2&=4.44f_1N_2k_{w2}\boldsymbol{\Phi}_m\end{aligned}\tag{4-7}$$

式中，$\boldsymbol{\Phi}_m$为每极主磁通的幅值，也就是每相绕组交链磁通的最大值。

$$\frac{E_1}{E_2}=\frac{N_1k_{w1}}{N_2k_{w2}}=k_e\tag{4-8}$$

式中，k_e称为电压比。

将感应电动势$\dot{E}_1$用负的阻抗压降来表示

$$\dot{E}_1=-\dot{I}_mZ_m=-\dot{I}_m(R_m+jX_m)\tag{4-9}$$

式中，Z_m称为励磁阻抗，是表示主磁路的磁化特性和铁损的综合参数；R_m称为励磁电阻，是表征铁损的等效参数；X_m称为励磁电抗，是表征主磁路的磁化特性的等效参数。一般将这3个参数统称为励磁参数，它们的大小与铁心的饱和程度有关，铁心越饱和，励磁电抗越小。

考虑到定子电阻R_1和定子漏电抗$X_{1\sigma}$，可画出图4-10所示的定子一相的电路，由此列出定子电压方程为

$$\dot{U}_1=-\dot{E}_1+\dot{I}_1(R_1+jX_{1\sigma})\tag{4-10}$$

式中，R_1为定子每相的电阻。

2. 转子绕组闭合时

将转子绕组闭合后，转子绕组中出现转子电流，产生与旋转磁场方向相同的电磁转矩，

使转子沿着旋转磁场的转向，以转速 $0 < n < n_s$ 旋转起来，感应电机运行于电动机状态。与转子静止时相比，旋转磁场相对于转子的转速减小为

$$\Delta n = n_s - n = sn_s \tag{4-11}$$

因而转子电动势和电流的频率应减小为

$$f_2 = \frac{psn_s}{60} = sf_1 \tag{4-12}$$

f_2 称为转差频率。

转子磁动势相对于转子的转速为

$$n_2 = \frac{60f_2}{p} = \frac{60sf_1}{p} = sn_s = \Delta n \tag{4-13}$$

考虑到转子相对于定子的转速为 n，因此，转子磁动势相对于定子的转速为

$$n_2 + n = \Delta n + n = (n_s - n) + n = n_s \tag{4-14}$$

即转子磁动势与定子磁动势转速相同，定、转子磁动势保持相对静止，如图 4-11 所示。

定、转子磁动势保持相对静止是产生恒定电磁转矩的必要条件。

此时，电机内部的磁场应该由 $\boldsymbol{F}_1$ 和 $\boldsymbol{F}_2$ 的合成磁动势 $\boldsymbol{F}_1 + \boldsymbol{F}_2$ 建立，即励磁磁动势为

$$\boldsymbol{F}_m = \boldsymbol{F}_1 + \boldsymbol{F}_2 \tag{4-15}$$

式（4-15）为感应电机的磁动势方程。

上式亦可写为

$$\boldsymbol{F}_1 = \boldsymbol{F}_m + (-\boldsymbol{F}_2) = \boldsymbol{F}_m + \boldsymbol{F}_{1L} \tag{4-16}$$

即认为 $\boldsymbol{F}_1$ 由两个分量构成：一个是励磁分量 $\boldsymbol{F}_m$，用于产生主磁通；另一个是负载分量 $\boldsymbol{F}_{1L}$，用以补偿转子磁动势。

此外，认为 $\dot{I}_1$ 由两个分量构成，即

$$\dot{I}_1 = \dot{I}_m + \dot{I}_{1L} \tag{4-17}$$

其中，$\dot{I}_m$ 为 $\dot{I}_1$ 的励磁分量，用于产生 $\boldsymbol{F}_1$ 的励磁分量 $\boldsymbol{F}_m$，即

$$F_m = 0.9\frac{m_1}{2}\frac{N_1k_{w1}I_m}{p} \tag{4-18}$$

$\dot{I}_{1L}$为 $\dot{I}_1$ 的负载分量，用于产生 $\boldsymbol{F}_1$ 的负载分量 $\boldsymbol{F}_{1L}$，即

$$F_{1L} = 0.9\frac{m_1}{2}\frac{N_1k_{w1}I_{1L}}{p} \tag{4-19}$$

此时，由主磁通感应的转子一相的电动势的有效值为

$$E_{2s} = 4.44f_2N_2k_{w2}\Phi_m = 4.44sf_1N_2k_{w2}\Phi_m = sE_2 \tag{4-20}$$

式中，E_{2s}中的下标“s”表示转子电动势的频率为转差频率。

转子漏电抗的有效值为

$$X_{2\sigma s} = 2\pi f_2L_{2\sigma} = sX_{2\sigma} \tag{4-21}$$

图 4-12 表示转子旋转时的定、转子电路，其中转子电路的频率为转差频率。由图 4-12 可列出定子、转子电压方程

$$\dot{U}_1 = -\dot{E}_1 + \dot{I}_1(R_1 + jX_{1\sigma}) \tag{4-22}$$

$$\dot{E}_{2s} = \dot{I}_{2s}(R_2 + jX_{2\sigma s}) \tag{4-23}$$

式中，R_2 为转子一相的电阻。

定、转子各量之间的关系可以归纳为图 4-13。

与变压器相似，感应电机的定、转子绕组之间通过气隙磁场耦合，在电路上没有直接联系。从以上分析可知，转子通过转子磁动势 $\boldsymbol{F}_2$ 对定子起作用，或者说影响定子。因此，转子对定子的影响取决于 $\boldsymbol{F}_2$ 的大小和空间相位，已知 $\boldsymbol{F}_2$ 的大小取决于转子电流，因而下面对 $\boldsymbol{F}_2$ 的空间相位进行分析。

图 4-14 示出了转子三相绕组在气隙磁场中的情况。若转子漏抗 $X_{2\sigma s}=0$，则情况如图 4-14a 所示，图中示出了转子各线圈边中的电动势方向，图示瞬间，转子 a 相感应电动势为正的最大值。由于转子的阻抗角 ψ_2 为 0°，转子电流没有无功分量，故此时转子电流与转子电动势相位相同，即此时转子 a 相感应电流为正的最大值，转子电流所产生的转子基波磁动势 F_2 波的幅值位置与气隙磁场 B_m 波的幅值位置相差 90°电角度，即在空间相位上，$\boldsymbol{F}_2$ 落后 $\boldsymbol{B}_m$90°。

若转子漏抗 $X_{2\sigma s}\neq 0$，则情况如图 4-14b 所示。图示瞬间，仍然是转子 a 相感应电动势为最大值，各线圈边中的电动势方向与图 4-14a 的情况相同，但电流的情况不同。若假设转子的阻抗角 ψ_2 为 60°，则此时转子 b 相电流为负的最大值，由图可见，在空间相位上，$\boldsymbol{F}_2$ 落后 $\boldsymbol{B}_m$（90°+60°）。一般来说，如果转子阻抗角为 ψ_2，$\boldsymbol{F}_2$ 落后 $\boldsymbol{B}_m$（$90°+\psi_2$）。注意，$\boldsymbol{F}_2$ 落后 $\boldsymbol{B}_m$ 的电角度与转子转速没有直接关系，而只取决于 ψ_2 的大小，即使转子静止，$\boldsymbol{F}_2$ 仍然落后 $\boldsymbol{B}_m$（$90°+\psi_2$）。

4.3 感应电动机的等效电路

为了便于对感应电动机进行分析计算，需要推导感应电动机的等效电路。

4.3.1 频率归算

频率归算的目的是要解决定、转子频率不同的问题。

将式（4-23）写成

$$\dot{E}_{2s}e^{j\omega_2 t} = \dot{I}_{2s}e^{j\omega_2 t}(R_2 + jX_{2\sigma s}) \tag{4-24}$$

将上式中的两边同时乘以 $\frac{1}{s}e^{j(\omega_1-\omega_2)t}$，得到

$$\dot{E}_2 e^{j\omega_1 t} = \dot{I}_2 e^{j\omega_1 t}(\frac{R_2}{s} + jX_{2\sigma}) \tag{4-25}$$

或

$$\dot{E}_2 = \dot{I}_2(\frac{R_2}{s} + jX_{2\sigma}) \tag{4-26}$$

与变换之前相比：转子频率从 $f_2=sf_1$ 变为 f_1；转子电动势和漏电抗已经变为转子静止时的形式；转子电阻值从 R_2 变为 R_2/s；转子电流写成定子频率的形式 $\dot{I}_2$，有效值不变，即 $I_2=I_{2s}$；转子回路的阻抗角不变，即 $\psi_2=\arctan\frac{X_{2\sigma s}}{R_2}=\arctan\frac{X_{2\sigma}}{R_2/s}$。以上变换称为频率归算，图 4-15 为经过频率归算后的定、转子电路。频率归算的物理含义是：用静止的、转子电阻

为 R_2/s 的等效转子代替旋转的、电阻为 R_2 的实际转子。由于变换前后转子电流的有效值和转子阻抗角不变，因此转子磁动势的大小、转速和空间位置不变，转子方对定子方的影响不变，因而定子方的所有物理量以及定、转子之间的功率传递情况不变。

将进行频率归算之后的转子电阻写成如下形式

$$\frac{R_2}{s} = R_2 + \frac{1-s}{s}R_2 \tag{4-27}$$

式（4-27）中等号左边的电阻上消耗的功率为通过磁场从定子传递到转子的功率，即电磁功率频率归算前后，电磁功率大小不变；等号右边的第一项为转子本身的电阻，其上消耗的功率为转子铜损，频率归算前后，转子铜损大小不变；等号右边的第二项为与转子旋转时所产生的机械功率对应的电阻，其上消耗的功率等于转子旋转时产生的总机械功率。

综上所述，进行频率归算有两方面的意义，一是使定、转子频率相等；二是推导出了代表机械功率的等效电阻。

根据 $F_1 = 0.9\frac{m_1}{2}\frac{N_1k_{w1}I_1}{p}, F_2 = 0.9\frac{m_2}{2}\frac{N_2k_{w2}I_2}{p}, F_m = 0.9\frac{m_1}{2}\frac{N_1k_{w1}I_m}{p}$ 以及磁动势方程 $\boldsymbol{F}_1 = \boldsymbol{F}_m + (-\boldsymbol{F}_2)$，又根据时空矢量图的基本概念，即当某相时轴与该相相轴重合时，该相电流的相量与其所属的三相电流系统所产生的基波合成磁动势的空间矢量重合，因此 $\boldsymbol{F}_1$、$\boldsymbol{F}_2$ 和 $\boldsymbol{F}_m$ 在空间上的相位关系与 $\dot{I}_1$、$\dot{I}_2$ 和 $\dot{I}_m$ 在时间上的相位关系相同，由此写出磁动势方程的电流形式

$$\dot{I}_1 = \dot{I}_m + \left(-\frac{\dot{I}_2}{k_i}\right) \tag{4-28}$$

式中，k_i 称为电流变比。$k_i = \frac{m_1N_1k_{w1}}{m_2N_2k_{w2}}$。

4.3.2 绕组归算

绕组归算是要解决定、转子绕组匝数、相数和绕组因数不同的问题。进行绕组归算就是：在转子对定子影响不变的原则下，用匝数、相数和绕组因数与定子绕组的匝数、相数和绕组因数相同的等效静止转子绕组代替经频率归算得到的转子绕组。为了保持转子对定子影响不变，绕组归算前后，转子磁动势的大小和空间相位不变。绕组归算后，转子方的各物理量均加“′”，以示区别，即

$$m_2' = m_1, N_2' = N_1, k_{w2}' = k_{w1} \tag{4-29}$$

1. 转子电流的归算

为使绕组归算前后转子磁动势 $\boldsymbol{F}_2$ 的大小和空间相位不变，应使

$$0.9\frac{m_2'}{2}\frac{N_2'k_{w2}'\dot{I}_2'}{p} = 0.9\frac{m_1}{2}\frac{N_1k_{w1}\dot{I}_2'}{p} = 0.9\frac{m_2}{2}\frac{N_2k_{w2}\dot{I}_2}{p} \tag{4-30}$$

因此有

$$\dot{I}_2' = \frac{m_2N_2k_{w2}}{m_1N_1k_{w1}}\dot{I}_2 = \frac{\dot{I}_2}{k_i} \tag{4-31}$$

2. 转子电动势的归算

由于 $\boldsymbol{F}_2$ 的大小和相位不变，因而主磁通 $\dot{\Phi}_m$ 不变，归算后转子电动势为

$$\dot{E}_2' = 4.44 f_1 N_2' k_{w2}' \dot{\Phi}_m = 4.44 f_1 N_1 k_{w1} \dot{\Phi}_m = k_e \dot{E}_2 = \dot{E}_1 \tag{4-32}$$

3. 转子阻抗的归算

绕组归算前后，转子上消耗的有功功率和无功功率应保持不变，因此有

$$\begin{aligned} m_2' I_2'^2 \frac{R_2'}{s} &= m_1 I_2'^2 \frac{R_2'}{s} = m_2 I_2^2 \frac{R_2}{s} \\ m_2' I_2'^2 X_{2\sigma}' &= m_1 I_2'^2 X_{2\sigma}' = m_2 I_2^2 X_{2\sigma} \end{aligned} \tag{4-33}$$

因此，转子漏阻抗的归算值分别为

$$R_2' = \frac{m_2}{m_1}\left(\frac{I_2}{I_2'}\right)^2 R_2 = k_e k_i R_2, \quad X_{2\sigma}' = \frac{m_2}{m_1}\left(\frac{I_2}{I_2'}\right)^2 X_{2\sigma} = k_e k_i X_{2\sigma} \tag{4-34}$$

经归纳可得到感应电动机的基本方程为

$$\begin{cases} \dot{U}_1 = -\dot{E}_1 + \dot{I}_1 (R_1 + jX_{1\sigma}) \\ \dot{E}_2' = \dot{I}_2' (R_2'/s + jX_{2\sigma}') \\ \dot{I}_1 = \dot{I}_m + (-\dot{I}_2') \\ \dot{E}_1 = -\dot{I}_m Z_m \\ \dot{E}_1 = \dot{E}_2' \end{cases} \tag{4-35}$$

感应电动机经过频率和绕组归算后的定、转子电路如图 4-16 所示。

4.3.3 等效电路和相量图

将归算后的感应电动机基本方程与归算后的变压器基本方程相比，不难发现，两者的数学结构完全一样，因此，只需将变压器的 T 形等效电路的 Z_L' 换成$\frac{1-s}{s}R_2'$，便可得到感应电动机的 T 形等效电路，如图 4-17 所示，并且不难得到感应电动机的相量图，如图 4-18 所示。

应该注意，对感应电动机的分析是按照一相进行的，因此利用等效电路计算的电压和电流均为相电压和相电流。

基本方程、等效电路和相量图是反映感应电动机内部各物理量电磁关系的 3 种不同方式，它们在本质上是一致的。实际运用中，可根据具体情况灵活选择，例如，等效电路便于进行计算，而相量图则清楚地反映出各量之间的相位关系。

4.3.4 感应电动机负载变化时的物理过程

感应电动机空载运行时，只需要克服电动机自身很小的空载制动转矩，因此转子电流很小，转子转速接近同步转速，即 $n \approx n_s$，$s \approx 0$，$R_2'/s \to \infty$，转子近似于开路，转子功率因数 $\cos\varphi_2 \to 1$。此时，定子电流的负载分量很小，主要是励磁电流，即 $\dot{I}_1 \approx \dot{I}_m$，定子功率因数角 φ_1 很大，定子功率因数很低。此外，由于定子漏阻抗压降 $\dot{I}_1$（$R_1 + jX_{1\sigma}$）很小，因此有 $U_1 \approx E_1$。

若增加感应电动机的负载，则随着负载转矩的增加，转子电流增大，转速降低，转差率增大，R_2'/s 减小，转子功率因数下降。此时，与空载时相比，定子电流的负载分量增加，因

而定子电流增大，定子漏阻抗压降 $\dot{I}_1$（$R_1+jX_{1\sigma}$）增大，因此 E_1 减小，主磁通减小。同时，定子电流的有功分量增加，因此定子功率因数增大。但是，当负载较大时，负载再增加，转子电流继续增大，由于转子转速下降明显，转子频率明显增加，ψ_2 显著增大，使 φ_1 增大，定子功率因数减小。

当负载转矩持续增大，转速下降到 $n=0$，$s=1$ 时，定、转子电流都很大，转子的功率因数较低，导致定子的功率因数也较低，定子漏阻抗压降很大，主磁通值约为空载时的 50% ~ 60% 。

4.3.5　感应电动机的近似等效电路

T 形等效电路能够准确、全面地反映感应电动机的基本方程，属于精确等效电路，但电路结构稍显复杂。为了便于使用，对 T 形等效电路进行如下简化。

由图 4-17，定、转子电流可分别表示为

$$\left.\begin{aligned}&\dot{I}_1=\frac{\dot{U}_1}{Z_{1\sigma}+\dfrac{Z_mZ'_2}{Z_m+Z'_2}},\dot{I}'_2=-\dot{I}_1\frac{Z_m}{Z_m+Z'_2}=-\frac{\dot{U}_1}{Z_{1\sigma}+\dot{c}Z'_2}\\&\dot{I}_m=\dot{I}_1\frac{Z'_2}{Z_m+Z'_2}=\frac{\dot{U}_1}{Z_m}\frac{1}{\dot{c}+\dfrac{Z_{1\sigma}}{Z'_2}}\end{aligned}\right\}\tag{4-36}$$

式中，$Z_{1\sigma}=R_1+jX_{1\sigma}$，为定子漏阻抗；$Z'_2=\dfrac{R'_2}{s}+jX'_{2\sigma}$，为转子等效阻抗；$\dot{c}$ 为一个复数系数，$\dot{c}=1+\dfrac{Z_{1\sigma}}{Z_m}\approx1+\dfrac{X_{1\sigma}}{X_m}$。

感应电动机正常运行时，$|Z_{1\sigma}|<<|Z'_2|$，即 $Z_{1\sigma}/Z'_2\approx0$，则式（4-36）中：

$$\dot{I}_m\approx\frac{\dot{U}_1}{\dot{c}Z_m}=\frac{\dot{U}_1}{Z_{1\sigma}+Z_m}\tag{4-37}$$

$$-\dot{I}'_2=\frac{\dot{U}'_1}{Z_{1\sigma}+\dot{c}Z'_2},\dot{I}_1=\dot{I}_m+(-\dot{I}'_2)\tag{4-38}$$

由此可画出近似等效电路，如图 4-19 所示。

例 4-1　一台三相感应电动机：$U_N=380\text{V}$，$f_N=50\text{Hz}$，$n_N=1455\text{r/min}$，定子绕组为三角形联结，$R_1=1.42\Omega$，$X_{1\sigma}=5.41\Omega$，$R'_2=1.25\Omega$，$X'_{2\sigma}=7.75\Omega$，$X_m=90.4\Omega$，$R_m=10.6\Omega$，试计算转子频率，并用 T 形等效电路计算额定转速时的定子电流、定子功率因数及输入功率。

解：

1）转子频率：由于感应电动机的额定转速接近同步转速，因此同步转速应为 1500r/min，额定转差率为

$$s_N=\frac{n_s-n_N}{n_s}=\frac{1500-1455}{1500}=0.03$$

转子频率为

$$f_2 = sf_1 = 0.03 \times 50\text{Hz} = 1.5\text{Hz}$$

2）额定转速时，转子的等效阻抗为

$$Z_2' = \frac{R_2'}{s_N} + jX_{2\sigma}' = \left(\frac{1.26}{0.03} + j7.75\right)\Omega = (42 + j7.75)\Omega = 42.71\angle 10.45°\Omega$$

励磁阻抗为

$$Z_m = R_m + jX_m = (10.6 + j90.4)\Omega = 91.02\angle 83.31°\Omega$$

设定子电压相量为 $\dot{U}_1 = 380\angle 0°\text{V}$，定子电流为

$$\dot{I}_1 = \frac{\dot{U}_1}{Z_1 + \dfrac{Z_m Z_2'}{Z_m + Z_2'}} = \frac{380\angle 0°}{(1.42 + j4.51) + \dfrac{91.02\angle 83.31° \times 42.71\angle 10.45°}{(42 + j7.75) + (10.6 + j90.4)}}\text{A}$$

$$= 9.84\angle -36.51°\text{A}$$

定子功率因数为

$$\cos\varphi_1 = \cos 36.51° = 0.804$$

输入功率为

$$P_1 = 3U_1 I_1 \cos\varphi_1 = 3 \times 380 \times 9.84 \times 0.804\text{W} = 9019\text{W}$$

4.4 感应电动机的功率方程和转矩方程

感应电动机运行时，通过磁场将电源输入的电功率传递到转子，转换成机械功率输出。下面利用等效电路，按照功率的传递过程，对各种功率和损耗进行分析，并建立功率方程和转矩方程。

4.4.1 功率方程

如图 4-20 所示，感应电动机从电源输入的电功率为

$$P_1 = m_1 U_1 I_1 \cos\varphi_1 \tag{4-39}$$

式中，U_1、I_1 分别为定子相电压和相电流的有效值，$\cos\varphi_1$ 为定子的功率因数。

定子绕组的电阻产生定子铜损，铁心中产生铁损，分别为

$$p_{Cu1} = m_1 I_1^{\,2} R_1, p_{Fe} = m_1 I_m^2 R_m \tag{4-40}$$

由于感应电动机正常运行时，旋转磁场相对于转子的转速很低，因此转子铁损很小，感应电动机的铁损主要是定子铁损。

从输入功率中减去定子铜损和铁损后，其余部分通过旋转磁场传递到转子上，这部分利用电磁感应作用传递到转子方的功率称为电磁功率，用 P_e 表示，因此

$$P_e = P_1 - p_{Cu1} - p_{Fe} \tag{4-41}$$

根据等效电路，电磁功率可表示为

$$P_e = m_1 E_2' I_2' \cos\psi_2 = m_1 I_2^{\,2} \frac{R_2'}{s} \tag{4-42}$$

从传递到转子的电磁功率中减去转子铜损 p_{Cu2}，可以得到转换为机械能的总机械功率，用 P_Ω 表示，这部分功率称为转换功率：

$$p_{\mathrm{Cu2}} = m_1 I_2'^2 R_2' \tag{4-43}$$

$$P_\Omega = P_{\mathrm{e}} - p_{\mathrm{Cu2}} = m_1 I_2'^2 \frac{1-s}{s} R_2' \tag{4-44}$$

考虑到式（4-42）~式（4-44），有如下关系：

$$p_{\mathrm{Cu2}} = sP_{\mathrm{e}}, P_\Omega = (1-s)P_{\mathrm{e}} \tag{4-45}$$

可见，传递到转子的电磁功率中，s 部分转变成转子铜损，$(1-s)$ 部分转换为机械功率。感应电动机的额定转差率 s_{N} 都很小，一般为 2% ~5%。再从总机械功率中扣除机械损耗 p_Ω 和杂散损耗 p_Δ，得到感应电动机输出的机械功率 P_2

$$P_2 = P_\Omega - p_\Omega - p_\Delta \tag{4-46}$$

小型笼型感应电动机满载时的杂散损耗可达输出功率的 1% ~ 3%；大型感应电动机满载时的杂散损耗可达输出功率的 0.5%。

图 4-21 为感应电动机的功率流程。当感应电动机的负载变化时，定、转子电流会跟着变化，但感应电动机的主磁通及转速基本不变。因此，认为机械损耗和铁损为不变损耗，定、转子铜损和杂散损耗为可变损耗。

4.4.2　转矩方程

感应电动机运行时，转子上受 3 个转矩的作用：拖动性质的电磁转矩 T_{e}；制动性质的负载转矩 T_2；制动性质的空载转矩 T_0。当感应电动机稳态运行时，拖动性质的电磁转矩应与制动性质负载转矩和空载转矩相平衡，即

$$T_{\mathrm{e}} = T_2 + T_0 \tag{4-47}$$

各转矩与功率、损耗之间的关系为

$$T_{\mathrm{e}} = \frac{P_\Omega}{\Omega}, T_0 = \frac{p_\Omega + p_\Delta}{\Omega}, T_2 = \frac{P_2}{\Omega} \tag{4-48}$$

式中，Ω 为转子的机械角速度。

4.4.3　电磁转矩

由于 $\Omega = (1-s)\Omega_{\mathrm{s}}$，因此电磁转矩可表示为

$$T_{\mathrm{e}} = \frac{P_\Omega}{\Omega} = \frac{(1-s)P_{\mathrm{e}}}{(1-s)\Omega_{\mathrm{s}}} = \frac{P_{\mathrm{e}}}{\Omega_{\mathrm{s}}} \tag{4-49}$$

即既可以通过将总机械功率除以机械角速度，也可以通过将电磁功率除以同步角速度来计算电磁转矩。

将 $P_{\mathrm{e}} = m_1 E_2' I_2' \cos\psi_2$，$E_2' = \sqrt{2}\pi f_1 N_1 k_{\mathrm{w1}} \Phi_{\mathrm{m}}$，$\Omega_{\mathrm{s}} = \dfrac{2\pi f_1}{P}$代入式（4-49），得到

$$T_{\mathrm{e}} = \left(\frac{pm_1 N_1 k_{\mathrm{w1}}}{\sqrt{2}}\right)\Phi_{\mathrm{m}} I_2' \cos\psi_2 = \left(\frac{pm_2 N_2 k_{\mathrm{w2}}}{\sqrt{2}}\right)\Phi_{\mathrm{m}} I_2 \cos\psi_2 = C_{\mathrm{T}}' \Phi_{\mathrm{m}} I_2 \cos\psi_2 \tag{4-50}$$

式中，C_{T}' 为转矩常数，$C_{\mathrm{T}}' = \dfrac{pm_2 N_2 k_{\mathrm{w2}}}{\sqrt{2}}$，对于已制成的电动机，$C_{\mathrm{T}}'$ 是一个常数。

式（4-50）说明，电磁转矩与主磁通 Φ_m 和转子电流的有功分量 $I_2\cos\psi_2$ 成正比。

4.5 笼型转子绕组的相数和极数

任何旋转电机在正常运行时，其定、转子的极对数都应相等，否则就不会产生平均电磁转矩，电机就不能正常工作。笼型转子绕组由于结构特殊，因而没有明显的极数和相数，本节将讨论如何确定其极数和相数。

4.5.1 笼型转子的相数

笼型转子的结构是对称的，转子导条在铁心表面均匀分布。设转子槽数（即导条数）为 Q_2。当极对数为 p 的旋转磁场相对转子旋转时，各导条感应电动势的有效值相等，相邻两导条感应电动势的相位差为

$$\alpha_2 = \frac{p \times 360°}{Q_2} \tag{4-51}$$

由于在交流绕组中，属于同一相的绕组，其中的电流是同相位的，因此，笼型转子的相数，取决于一对极下有多少根不同电流相位的导条。

当每对极下的导条数 $Q_2/p=$ 整数，则此 Q_2/p 根导条的电动势相量将构成一个均匀分布的电动势星形图，如图 4-22 所示。此时每对极下的每一根导条将构成一相，故转子的相数 $m_2=\dfrac{Q_2}{p}$；每相有 p 根并联导条。当每对极下的导条数 $Q_2/p\neq$ 整数，则转子的相数 $m_2=Q_2$，此时，转子极数等于 2，每相有 2 根并联导条。

由于一根导条相当于半匝，所以笼型绕组的每相串联匝数为

$$N_2 = \frac{1}{2} \tag{4-52}$$

这样也就没有短距和分布的问题了，节距因数和分布因数都等于 1，故绕组因数为

$$k_{w2} = 1 \tag{4-53}$$

4.5.2 笼型转子的极数

图 4-23 表示极对数为 1 的旋转磁场相对于转子运动时的情况。各导条在某一瞬间感应电动势的大小与该瞬间其切割磁感应强度大小成正比，因而气隙磁感应强度的正弦分布波与导条感应电动势瞬时值的包络线形成的正弦曲线在空间上同相位。由于转子漏抗的影响，导条中的电流在时间相位上将滞后其感应电动势 φ_2，因此，导条电流瞬时值的包络线形成的正弦曲线在空间上落后感应电动势瞬时值的包络线 φ_2 电角度，即当某一导条感应电动势达到最大值时，其电流尚未达到最大值，等 ωt 经过了 φ_2，感应电动势瞬时值的包络线形成的正弦曲线在空间移过了 φ_2 电角度以后，该导条中的电流才达到最大值，如图 4-23 所示。根据导条电流的分布，在 2 极磁场的作用下，导条电流形成的转子基波磁动势的极数也为 2。同理可知，如果气隙磁场的极数为 4，则导条电流将形成 4 极的基波磁动势。因此，笼型转子的极数不是固定的，而是自动与定子绕组的极数相同。笼型转子的极数与导条的数量无关，并且定、转子磁动势波的转速相等，始终为同步转速。

4.6 感应电动机的参数测定

如要运用等效电路对感应电动机进行分析计算，则可通过空载试验和堵转（短路）试验来测定等效电路中的有关参数。

在进行空载和堵转试验前，先根据图4-24的连接，或用电桥测量三相感应电动机定子相电阻 R_1：

$$R_1=\frac{V_{DC}}{2I_1} \tag{4-54}$$

4.6.1 空载试验

空载试验的目的是测定感应电动机的励磁电阻 R_m、励磁电抗 X_m，以及铁损 p_{Fe} 和机械损耗 p_Ω。

进行空载试验的接线如图4-25a所示，通过调压器将感应电动机接到额定频率的电源上，逐渐增加定子电压 U_1，转子转速将提高至接近同步转速。使定子端电压 U_1 从（1.1 ~ 1.2）U_N 开始单调下降，测量 U_1、空载电流 I_0 和空载输入功率 P_{10}，一般测8~10个点。

根据试验数据，画出感应电动机的空载特性曲线 I_{10}，$P_{10}=f(U_1)$，如图4-25b所示。根据感应电动机的功率平衡方程，如忽略空载时的杂散损耗和转子铜损，空载功率全部用于产生定子铜损、铁损和机械损耗，即

$$P_{10}=p_{cu1}+p_{Fe}+p_\Omega \tag{4-55}$$

因此，$P'_{10}=P_{10}-p_{cu1}=P_{10}-m_1I_{10}^2R_1=p_{Fe}+p_\Omega$

由于 U_1 变化时，电动机的转速变化很小，机械损耗基本不变，可认为 p_Ω 与 U_1 无关。又由于铁损与 ${U_1}^2$ 成正比，P'_{10} 与 ${U_1}^2$ 的关系曲线 $P'_{10}=f({U_1}^2)$ 近似为一条直线，如图4-25c所示。将该线延长到 ${U_1}^2=0$，即可把铁损和机械损耗分离，进而得到不同电压下的铁损。

空载时，$s\to0$，可认为转子开路，T形等效电路变为如图4-26所示。根据分离出的铁损可以计算励磁电阻：

$$R_m=\frac{p_{Fe}}{m_1{I_{10}}^2} \tag{4-56}$$

根据空载时的等效电路，定子的空载阻抗 Z_0 为

$$|Z_0|=\frac{U_1}{I_{10}} \tag{4-57}$$

而定子的空载电阻 $R_0=R_1+R_m$，因此空载电抗 X_0 为

$$X_0=\sqrt{|Z_0|^2-{R_0}^2} \tag{4-58}$$

励磁电抗为

$$X_m=X_0-X_{1\sigma} \tag{4-59}$$

其中的定子漏阻抗可由短路试验确定。

4.6.2 堵转试验

堵转试验又称短路试验，其目的是测定感应电动机的短路阻抗 Z_k、短路电抗 X_k 和短路

电阻 R_k，并且可以进一步求出转子电阻 R'_2。

进行堵转试验前，先要将转子堵住（卡住）。堵转试验的接线如图 4-27a 所示，通过调压器将感应电动机接到额定频率的电源上，将定子电压 U_1 从 0 逐渐增加到 $0.4U_N$ [对小型电动机，若条件允许，最好加到 $(0.9\sim1.0)U_N$]，此时转子不转，$s=1$，然后逐步降低 U_1，并记录 U_1、定子电流 I_{1k} 和功率 P_{1k}。

将记录的试验数据整理成曲线，得到感应电动机的短路特性 I_{1k}，$P_{1k}=f(U_1)$，如图 4-27b 所示。

由于 $s=1$，因此 $Z_m \gg Z'_2$，此时可以认为励磁支路开路，T 形等效电路变为图 4-28 所示。另外，由于电压 U_1 较低并且转速为 0，因此可以不计铁损和机械损耗，认为全部输入功率都变成定、转子铜损。因此，可以按照以下公式求出有关参数：

$$|Z_k| = \frac{U_1}{I_k}, R_k = \frac{p_k}{m_1 I_k{}^2}, X_k = \sqrt{|Z_k|^2 - R_k{}^2} \tag{4-60}$$

由于已经测量出了定子电阻 R_1，因此

$$R'_2 = R_k - R_1 \tag{4-61}$$

定、转子漏电抗难以分离，一般认为

$$X_{1\sigma} \approx X_{2\sigma} \approx \frac{1}{2}X_k \tag{4-62}$$

4.7 感应电动机的工作特性

感应电动机的工作特性是指：在额定电压和额定频率下，感应电动机的转速 n、电磁转矩 T_e、定子电流 I_1、功率因数 $\cos\varphi_1$、效率 η 与输出功率 P_2 的关系。即，当 $U_1=U_N$、$f_1=f_N$ 时，n、T_e、I_1、$\cos\varphi_1$、$\eta=f(P_2)$，如图 4-29 所示。

1. 转速特性 $n=f(P_2)$

当输出功率 $P_2=0$ 时，$n\approx n_s$；P_2 增加时，负载转矩增加，转速下降，转子电流增加，电磁转矩增加，以克服负载转矩；当 $P_2=P_N$ 时，$s_N=2\%\sim5\%$，因此，$n_N=(1-s_N)\ n_s=(95\%\sim98\%)\ n_s$，如图 4-29 所示。

2. 定子电流特性 $I_1=f(P_2)$

根据 $\dot{I}_1=\dot{I}_m+(-\dot{I}'_2)$，当输出功率 $P_2=0$ 时，$I_1\approx I_m$。P_2 增加时，转子电流增大，定子电流随之增大，如图 4-29 所示。

3. 功率因数特性 $\cos\varphi_1=f(P_2)$

对电网来说，感应电动机属于感性负载，运行时必须从电网中吸收感性无功功率，功率因数恒小于 1，且滞后。

当输出功率 $P_2=0$ 时，$I_1\approx I_m$，此时定子功率因数很低，$\cos\varphi_1$ 约为 0.2。当 P_2 刚开始增加时，s 较小，转子功率因数 $\cos\varphi_2$ 较大，定子电流中的有功分量增加，$\cos\varphi_1$ 增大。P_2 增加到一定程度时，随着 s 增大，转子功率因数减小，导致 $\cos\varphi_1$ 又开始下降，$\cos\varphi_1$ 通常在 P_2 达到 P_N 之前出现最大值，如图 4-29 所示。

4. 电磁转矩特性 $T_e=f(P_2)$

由转矩方程 $T_e=T_0+T_2=T_0+\dfrac{P_2}{\Omega}$ 可见，$P_2=0$ 时，$T_e=T_0$。当 P_2 增加时，转速 n 和 Ω

略有减小，$T_e=f(P_2)$ 为略微上翘的曲线，如图4-29所示。

5. 效率特性 $\eta=f(P_2)$

与其他电机相似，可以依据不变损耗和可变损耗，对感应电动机的效率特性进行分析。感应电动机的效率特性同样是：当 P_2 从空载开始增加时，效率迅速增加；在（0.8～1.1）P_N 范围内，效率出现最大值；此后 P_2 再增加，效率开始下降，如图4-29所示。不同容量的感应电动机的额定效率在76%～94%之间，电动机容量越大，额定效率越高。

感应电动机的工作特性反映了在额定电压和额定频率下，感应电动机的主要性能指标随输出功率的变化。另外，由工作特性可知，在额定负载附近，感应电动机具有较高的效率和功率因数，运行的经济性较好，因此，在选用电动机时，应使电动机的容量与负载的功率相匹配。如所选的电动机功率过大，除设备投资较大外，电动机长期轻载运行，效率和功率因数都较低。反之，如所选的电动机的容量小于负载的功率，则电动机长期过载运行，寿命会大为缩短。

4.8 感应电动机的机械特性

三相感应电动机拖动生产机械运行时，必须满足机械负载对电动机转矩和转速的要求，为此，需要分析其机械特性。三相感应电动机的机械特性是：在电源电压、电源频率和电动机参数不变的情况下，电动机的转速与电磁转矩之间的关系，即 $n=f(T_e)$。可利用感应电动机的T形等效电路来推导机械特性的参数表达式。

根据式（4-49）和式（4-42），电磁转矩与转子电流的关系为

$$T_e=\frac{P_e}{\Omega_s}=\frac{m_1}{\Omega_s}I_2'^2\frac{R_2'}{s}=\frac{m_1I_2'^2\dfrac{R_2'}{s}}{\dfrac{2\pi n_s}{60}}=\frac{m_1I_2'^2\dfrac{R_2'}{s}}{\dfrac{2\pi f_1}{p}} \tag{4-63}$$

在感应电机动的T形等效电路中，由于励磁阻抗 Z_m 比定、转子漏阻抗 z_1 和 z_2' 大很多，如把励磁阻抗支路认为开路，就有

$$I_2'\approx\frac{U_1}{\sqrt{\left(R_1+\dfrac{R_2'}{s}\right)^2+(X_1+X_2')^2}} \tag{4-64}$$

将式（4-64）代入到式（4-63）中，可得

$$T_e=\frac{m_1U_1^2\dfrac{R_2'}{s}}{\dfrac{2\pi n_s}{60}\left[\left(R_1+\dfrac{R_2'}{s}\right)^2+(X_1+X_2')^2\right]}=\frac{m_1pU_1^2\dfrac{R_2'}{s}}{2\pi f_1\left[\left(R_1+\dfrac{R_2'}{s}\right)^2+(X_1+X_2')^2\right]} \tag{4-65}$$

式（4-65）就是机械特性的参数表达式。当电源电压 U_1、频率 f_1 和阻抗等为固定时，把不同的转差率 s 代入式（4-65），算出对应的电磁转矩，可得到电磁转矩 T_e 与转差率 s 的曲线，即 T_e-s 曲线。图4-30为典型的感应电动机的机械特性曲线，图4-31为感应电机的转矩-转差率曲线。在图4-31中 $0<s<1$ 时是电动机状态，$s<0$ 时是发电机状态，$s>1$ 时是电磁制动状态。

1. 最大转矩

由图4-30可知，机械特性曲线上转矩有一个最大值，对于式（4-65），令 $\mathrm{d}T_e/\mathrm{d}s=0$，

即可求得产生最大电磁转矩的转差率 s_m 为

$$s_m = \pm \frac{R_2'}{\sqrt{R_1^2 + (X_{1\sigma} + X_{2\sigma}')^2}} \tag{4-66}$$

s_m 称为临界转差率。

将式（4-66）代入式（4-65），可得最大电磁转矩 T_{max} 为

$$T_{max} = \pm \frac{m_1 p}{2\pi f_1} \times \frac{U_1^2}{2[\pm R_1 + \sqrt{R_1^2 + (X_{1\sigma} + X_{2\sigma}')^2}]} \tag{4-67}$$

式（4-66）和式（4-67）中，"+"号对应电动机状态；"-"号对应发电机状态。考虑到 $R_1 \ll X_{1\sigma} + X_{2\sigma}'$，可忽略 R_1 而得到近似表达式：

$$s_m \approx \pm \frac{R_2'}{X_{1\sigma} + X_{2\sigma}'} \tag{4-68}$$

$$T_{max} \approx \pm \frac{m_1 p U_1^2}{4\pi f_1 (X_{1\sigma} + X_{2\sigma}')} \tag{4-69}$$

分析式（4-68）和式（4-69）可知：

1）当电源频率 f_1 和电动机参数不变时，最大电磁转矩 T_{max} 与定子相电压 U_1 的二次方成正比，即 $T_{max} \propto U_1^2$，与定、转子漏抗之和近似成反比。临界转差率 s_m 与电源电压无关，因此，当改变电源电压时，最大电磁转矩 T_m 随电源电压的减小而降低；而临界转差率 s_m 不变，可画出降低电源电压时的机械特性曲线如图 4-32 所示，它们是一组通过同步转速点，且临界转差率保持不变，而最大电磁转矩减小的曲线族。

2）最大电磁转矩 T_{max} 与转子电阻值 R_2' 无关，临界转差率 s_m 与 R_2' 成正比，即 $s_m \propto R_2'$。当 R_2' 增大时，s_m 增大，但 T_{max} 保持不变，此时机械特性的最大值将向下偏移，如图 4-33 所示。

最大电磁转矩 T_{max} 与额定转矩 T_N 之比称为电动机的过载能力，用 λ_m 表示，即

$$\lambda_m = T_{max}/T_N \tag{4-70}$$

λ_m 是感应电动机的一个重要性能指标，它反映了电动机承受短时过载的能力。如果负载转矩大于电动机的最大电磁转矩，电动机就会停转。为了保证电动机不因短时过载而停机，要求电动机具有一定的过载能力，一般三相感应电动机的 λ_m 为 1.6 ~ 2.5。

2. 起动转矩

电动机处于静止状态，即 $n=0$ 时，电动机的电磁转矩称为起动转矩 T_{st}，又称为堵转转矩。将 $s=1$ 代入式（4-65），可求得起动转矩 T_{st} 为

$$T_{st} = \frac{m_1}{2\pi f_1} \frac{p U_1^2 R_2'}{(R_1 + R_2')^2 + (X_{1\sigma} + X_{2\sigma}')^2} \tag{4-71}$$

由式（4-71）可知：

1）当电源频率 f_1 和电动机参数不变时，起动转矩 T_{st} 与定子相电压 U_1（电源电压）的二次方成正比，即 $T_{st} \propto U_1^2$（见图 4-32）。

2）当定子相电压 U_1（电源电压）和频率 f_1 一定时，如果 $(R_1 + R_2') \ll (X_{1\sigma} + X_{2\sigma}')$，则起动转矩近似与 $(X_{1\sigma} + X_{2\sigma}')^2$ 成反比，与转子回路电阻 R_2' 成正比，即 $T_{st} \propto \frac{R_2'}{(X_{1\sigma} + X_{2\sigma}')^2}$。因此对于绕线转子感应电动机，在转子回路串联适当的电阻 R_{st}，可提高起动转矩。若使起

动时达到最大转矩，应使 $s_m = 1$，将其代入式（4-66）和式（4-68），可得

$$R_2' + R_{st}' = \sqrt{R_1^2 + (X_{1\sigma} + X_{2\sigma}')^2} \approx X_{1\sigma} + X_{2\sigma}' \tag{4-72}$$

$$R_{st}' = X_{1\sigma} + X_{2\sigma}' - R_2' \quad R_{st}' = R_{st} k_e k_i \tag{4-73}$$

式中，R_{st}'为转子串联电阻折算到定子侧时的数值。

感应电动机的起动转矩与额定转矩之比称为起动转矩倍数，即

$$K_{st} = T_{st}/T_N \tag{4-74}$$

K_{st}也是感应电动机的一个重要性能指标，它反映了电动机的起动能力。普通笼型感应电动机的起动转矩倍数 K_{st}为0.8～1.2。

例4-2 一台6极三相笼型感应电动机，定子绕组为星形联结，额定电压 $U_N = 380V$，额定转速 $n_N = 965r/min$，电源频率$f_1 = 50Hz$，定子电阻 $R_1 = 2.1\Omega$，定子漏抗 $X_{1\sigma} = 3.08\Omega$，转子电阻的折算值 $R_2' = 1.48\Omega$，转子漏抗的折算值 $X_{2\sigma}' = 4.2\Omega$。试计算该电动机的：

1）额定电磁转矩；

2）最大电磁转矩及过载能力；

3）临界转差率；

4）起动转矩和起动转矩倍数。

解：

该电动机的同步转速 n_s为

$$n_s = \frac{60f_1}{p} = \frac{60 \times 50}{3} \text{r/min} = 1000\text{r/min}$$

额定转差率 s_N为

$$s_N = \frac{n_s - n_N}{n_s} = \frac{1000 - 965}{1000} = 0.035$$

定子绕组额定相电压 U_1为

$$U_1 = \frac{380}{\sqrt{3}}\text{V} = 220\text{V}$$

1）额定转矩 T_N为

$$T_N = \frac{m_1 p U_1^2 \frac{r_2'}{s_N}}{2\pi f_1 \left[\left(r_1 + \frac{r_2'}{s_N} \right)^2 + (x_1 + x_2')^2 \right]}$$

$$= \frac{3 \times 3 \times 220^2 \times \frac{1.48}{0.035}}{2\pi \times 50 \times \left[\left(2.1 + \frac{1.48}{0.035} \right)^2 + (3.08 + 4.2)^2 \right]} \text{N} \cdot \text{m}$$

$$= 35\text{N} \cdot \text{m}$$

2）最大电磁转矩 T_{max}为

$$T_{max} = \frac{m_1 p U_1^2}{4\pi f_1 (x_1 + x_2')} = \frac{3 \times 3 \times 220^2}{4\pi \times 50 \times (3.08 + 4.2)} \text{N} \cdot \text{m} = 95.23\text{N} \cdot \text{m}$$

过载能力 λ_m为

$$\lambda_m = \frac{T_{max}}{T_N} = \frac{95.23}{35} = 2.72$$

3）临界转差率s_m为

$$s_m = \frac{R_2'}{X_{1\sigma} + X_{2\sigma}'} = \frac{1.48}{3.08 + 4.2} = 0.203$$

4）起动转矩T_{st}为

$$T_{st} = \frac{m_1}{2\pi f_1} \frac{pU_1^2 r_2'}{(r_1 + r_2')^2 + (x_1 + x_2')^2}$$

$$= \frac{3 \times 3 \times 220^2 \times 1.48}{2\pi \times 50 \times [(2.1 + 1.48)^2 + (3.08 + 4.2)^2]} \text{N} \cdot \text{m} = 31.18 \text{ N} \cdot \text{m}$$

起动转矩倍数k_{st}为

$$k_{st} = \frac{T_{st}}{T_N} = \frac{31.18}{35} = 0.89$$

4.9 三相感应电动机的起动

电动机从静止状态到稳态运行的过渡过程称为起动过程，简称起动。衡量三相感应电动机起动性能的主要指标有起动转矩倍数$K_{st} = T_{st}/T_N$、起动电流倍数$K_I = I_{st}/I_N$、起动时间和起动设备。通常希望起动转矩足够大，而起动电流不要太大，以免电动机绕组过热以及线路上产生过大的电压降而影响到同一电网上的其他用电设备，另外，还希望起动设备简单、经济、可靠。

感应电动机定子绕组接入额定电压直接起动时，起动瞬间，$n=0$，$s=1$，从等效电路可见，此时定子电流（也称为起动电流）I_{st}仅受定、转子漏阻抗之和$Z_1 + Z_2'$的限制，由于漏阻抗较小，故起动电流很大，一般可达到额定电流I_N的4～7倍。

起动转矩T_{st}受电磁转矩公式$T_e = C_T \Phi_m I_2' \cos\varphi_2$约束，式中转子功率因数角$\varphi_2 = \arctan(sX_2'/R_2')$。起动时，$s=1$，$X_2' \gg R_2'$，故$\varphi_2$接近于90°，转子的功率因数$\cos\varphi_2$很低。因此，虽然起动电流$I_{st}$很大，但转子电流的有功分量$I_2'\cos\varphi_2$并不大，且由于起动电流$I_{st}$大，使定子绕组的漏阻抗压降增大，感应电动势$E_1$和主磁通$\Phi_m$减小，结果使起动转矩并不大。一般$T_{st}$为额定转矩$T_N$的0.8～1.2倍。通常当起动转矩倍数$K_{st} = T_{st}/T_N \geq 1.1$时，电动机就可以起动起来。图4-34为三相感应电动机直接起动时的固有机械特性与定子电流特性，其中I_1为定子相电流。

4.9.1 三相感应电动机的起动方法

三相笼型感应电动机的起动方法主要有直接起动、降压起动两种；三相绕线转子感应电动机的起动方法主要有在转子回路串电阻或频敏变阻器。

1. 直接起动

直接起动就是将电动机定子绕组直接接在额定电压的电源上起动，也称为全压起动。直接起动的优点是操作简单，不需要复杂的起动设备；缺点是起动电流大，对电动机本身和电网电压均会带来不利影响。因此，通常只有小容量的电动机采用直接起动。较大容量的电动机能否直接起动，与电网容量的大小和电动机的起动电流倍数有关，通常采用下面的经验公式来判断：

$$K_{st} = \frac{I_{st}}{I_N} \leqslant \frac{1}{4}\left[3 + \frac{\text{电源总容量}}{\text{起动电动机容量}}\right] \tag{4-75}$$

当起动电流倍数满足式（4-75）时，电动机可以直接起动，否则必须采取降低电压的办法来限制起动电流。式（4-75）中的“电源总容量”是指为该电动机供电的变压器的容量（即供电变压器的视在功率），单位为kV·A；“起动电动机容量”是指该电动机的视在功率，单位为kV·A，与电动机铭牌数据的关系为

$$\text{起动电动机容量} = \frac{P_N}{\eta_N \cos\varphi_{1N}} = \sqrt{3}U_{1N}I_{1N} \tag{4-76}$$

2. 降压起动

在感应电动机的简化等效电路中，若忽略励磁电流 I_m，则

$$I_1 \approx I_2' = \frac{U_1}{\sqrt{(R_1 + R_2'/s)^2 + (X_{1\sigma} + X_{2\sigma}')^2}} \tag{4-77}$$

起动时，$s=1$，代入上式，可得起动电流为

$$I_{st} \approx \frac{U_1}{\sqrt{(R_1 + R_2')^2 + (X_{1\sigma} + X_{2\sigma}')^2}} \tag{4-78}$$

由式（4-78）可知，起动电流大小与电源电压成正比，因此降低加在电动机定子绕组上的电压，可以减小起动电流，但起动转矩将随电压的二次方成正比地下降，因此降压起动只适用于空载或轻载起动。

（1）定子回路串接对称电阻或电抗降压起动　感应电动机起动时，在定子电路中串接对称电阻或电抗，可降低电动机定子绕组的端电压，从而使起动电流减小。

假设串接电阻或电抗后，加在定子绕组上的电压减小到原来的 $1/k$，由于起动电流与电压成正比，因此起动电流也减小到原来的 $1/k$，而起动转矩与电压的二次方成正比，故起动转矩将减小到原来的 $1/k^2$。可见，这种起动方法的缺点是起动转矩小，只适用于空载或轻负载起动，且起动时电能损耗较大；其优点是起动平稳、设备简单，若采用串接电阻起动时，还能使起动阶段的功率因数得到提高。串电抗降压起动通常用于高压电动机，电阻降压起动一般用于低压电动机。

（2）自耦变压器降压起动　自耦变压器降压起动也称为自耦补偿起动。假设自耦变压器的电压比为 k，则加在电动机上的电压将减小到直接起动时的 $1/k$，使得起动转矩减小到直接起动时的 $1/k^2$，电动机的起动电流也减小到直接起动时的 $1/k$。而电网供给的起动电流是自耦变压器高压侧的电流，它是电动机起动电流即自耦变压器低压侧电流的 $1/k$，所以，实际电网供给的起动电流减小到直接起动时的 $1/k^2$。

可见，如果一台三相感应电动机允许的起动电流一定时，用自耦变压器降压起动时的起动转矩，大于定子回路串接对称电阻或电抗器起动时的起动转矩，这是它的一个优点。自耦变压器起动的另一个优点是，可以灵活地选择抽头，得到允许的起动电流值和所需的起动转矩值。这种起动方法常用于10kW以上的电动机中。自耦变压器起动的缺点是体积大、价格高，且不能带重载起动。

（3）星形/三角形换接降压起动　星形/三角形换接起动也是一种降压起动法。这种方法只适合于正常运行时定子三相绕组是三角形连接的感应电动机，且定子绕组的6个出线端都要引出。

假设电源电压为 U_1，电动机每相阻抗为 z。电动机作三角形联结直接起动时，每相电压为 U_1，每相起动电流为 U_1/z，如图 4-35a 所示；由于三角形联结时线电流是相电流的$\sqrt{3}$倍，因此电网供给的起动电流为 $I_{stD}=\sqrt{3}U_1/z$。电动机作星形联结降压起动时，如图 4-35b 所示，相电压为 $U_1/\sqrt{3}$，每相起动电流为 $U_1/\sqrt{3}z$，由于星形联结时线电流等于相电流，因此电网供给的起动电流 $I_{stY}=U_1/\sqrt{3}z$。所以星形联结降压起动与三角形联结直接起动时，起动电流的比值为

$$\frac{I_{stY}}{I_{stD}}=\frac{U_1/\sqrt{3}z}{\sqrt{3}U_1/z}=\frac{1}{3} \tag{4-79}$$

由于起动转矩与相电压的二次方成正比，故星形联结降压起动与三角形联结直接起动时，起动转矩的比值为

$$\frac{T_{stY}}{T_{stD}}=\left(\frac{U_1/\sqrt{3}}{U_1}\right)^2=\frac{1}{3} \tag{4-80}$$

可见，这种起动方法的效果相当于电压比 $k=\sqrt{3}$的自耦变压器降压起动。其优点是起动设备简单，只需要一个星形/三角形转换开关，成本低、运行可靠。缺点是定子绕组的 6 个出线端都要引出，对于高压电动机有一定的困难，且只适合于正常运行时定子绕组是三角形联结的感应电动机。低压电动机空载或轻载起动时，应优先选用星形/三角形换接降压起动。

（4）转子回路串接对称电阻起动　三相绕线转子感应电动机的转子回路中可串接三相对称电阻，以增大电动机的起动转矩，并减小起动电流。如果串接大小合适的电阻 R_{st}，可以使起动转矩等于电动机的最大转矩，同时使起动电流明显下降。起动结束后，可以切除串接的电阻，电动机的运行效率不受影响。因此在需要重载起动或频繁起动的场合，一般常采用绕线转子感应电动机。

（5）转子串接频敏变阻器起动　绕线转子感应电动机转子回路串接电阻起动，可以增大起动转矩，同时能减小起动电流，但每切除一级电阻，电磁转矩会突然增大，造成机械冲击，虽然这可通过增加起动级数来使起动平滑，但起动设备相应会变得很复杂。为了克服这个缺点，设计出转子回路串接频敏变阻器的起动方法。

频敏变阻器相当于只有一次绕组的三相心式变压器，绕组为星形联结，通过电刷和集电环与转子绕组相接。电动机起动时，$n=0$，$s=1$，转子电流频率 $f_2=sf_1=f_1$ 较高，因此频敏变阻器中铁损和反映铁损的励磁电阻 R_m 较大，相当于在转子回路串接了一个较大的电阻，从而限制了起动电流，增大了起动转矩。随着转速 n 的上升，转差率 s 逐渐减小，转子频率 f_2 也相应减小，频敏变阻器中反映铁损的励磁电阻 R_m 随之不断减小，相当于在起动过程中逐渐切除了转子回路串入的起动电阻，实现了电动机的无级平滑起动。选择适当的频敏变阻器参数，可以得到恒转矩的起动特性。

频敏变阻器的结构简单、维护方便，能够实现无级平滑起动，但频敏变阻器具有一定的电抗，与转子回路串接对称电阻起动的方法相比，在同样起动电流下，起动转矩要小一些。

4.9.2 高起动转矩的三相笼型感应电动机

笼型感应电动机具有结构简单、造价低、坚固耐用、便于维护等优点，因而得到了广泛应用，但由于起动转矩小，只适合于空载或轻载起动，从而限制了它的使用。如果在起动时能降低起动电流，又能保证足够大的起动转矩，就比较理想了。

由式（4-71）可知，在一定范围内增大转子电阻，可以加大起动转矩，并减小起动电流。但转子电阻增大，转子铜损将增大，电动机的运行效率将降低。为了解决这些问题，工程技术人员从改进转子槽形等入手，研制出了高起动转矩的笼型感应电动机，下面将分别予以介绍。

1. 转子电阻值较大的笼型感应电动机

当笼型感应电动机的笼型转子采用合金铝浇注，或者同时采用转子小槽（见图4-36a），减小导条截面积，这样可以增大转子电阻 R_2，在电动机起动时，就可以增大起动转矩，能改善电动机的起动性能。一般高转差率感应电动机和起重冶金用感应电动机都是这种类型的。

转子电阻大，直接起动时起动转矩大，最大转矩也大，但同时额定转差率较大，运行段的机械特性较软。电动机转子电阻大，正常运行时效率较低，而且电动机价格较贵。

2. 深槽式笼型感应电动机

深槽式笼型感应电动机的转子冲片形状如图4-36b所示，转子槽形深而窄，其槽深与槽宽之比为10~20，而普通笼型感应电动机的这个值不超过5。

图4-37a、b分别表示电流流经深槽转子上部和下部的情况，可见，槽电流流经上部时产生的磁通与定子耦合紧密，因而深槽转子上部区域的漏电感较小，而深槽转子下部区域的漏电感较大。

深槽式笼型感应电动机运行时，转子导条中有电流通过，其槽漏磁通分布如图4-38a所示。把沿槽高方向的转子导条看成是由许多根小的股线并联组成，则越靠近槽底的小导体交链的漏磁通越多，其漏电抗也越大；接近于槽口部分的小导体交链的漏磁通少，其漏电抗也越小。

电动机起动时，$n=0$，$s=1$，转子电流频率 $f_2=sf_1=f_1$（较高），转子漏电抗 $sx_2=x_2\gg r_2$，因此各小导体中的电流分布主要取决于漏电抗的大小。因为槽底小导体的漏电抗大，槽口部分小导体的漏电抗小，所以在相同气隙磁通所感应的电动势 E_2 的作用下，导条中靠近槽底处的电流密度将很小，而越靠近槽口处，电流密度越大，这种现象称为电流的趋肤效应（也称为挤流效应），如图4-38b所示；其结果相当于减小了导体的有效截面积，如图4-38c所示，因此转子有效电阻增大，起动电流减小，起动转矩增大。

当起动完毕，电动机正常运行时，转子电流频率 $f_2=sf_1$ 为1~3Hz（很低），转子漏电抗 $sx_2\ll r_2$，因此各小导体中的电流分布主要取决于电阻的大小，此时趋肤效应基本消失，导条中的电流按电阻均匀分布，导条的有效截面积又恢复到原来的截面积，转子电阻自动减小，使电动机的铜损减小，并提高了电动机的运行效率。

深槽式笼型感应电动机转子槽漏抗较大，功率因数稍低，最大转矩倍数稍小。

3. 双笼型转子感应电动机

双笼型转子感应电动机的结构特点是：转子上有两套笼型导条，分为上笼和下笼，如图4-36c所示。上笼导条截面积小，用电阻系数较大的黄铜或铝青铜制成；下笼导条截面积大，用电阻系数较小的纯铜制成，因此上笼电阻大，下笼电阻小，且下笼的漏电抗比上笼的大很多。

电动机起动时，$n=0$，$s=1$，转子电流频率 $f_2=sf_1=f_1$（较高），转子漏电抗较大，转子电流的分布主要取决于漏电抗的大小。因为下笼的漏电抗大，电流主要从漏电抗小的上笼流过，因此上笼又称为起动笼。由于上笼电阻大，能有效地限制起动电流，并产生较大的起动转矩。

当起动完毕，电动机正常运行时，转子电流频率f_2为1~3Hz（很低），转子漏电抗很小，转子电流的分布主要取决于电阻的大小，此时电流主要从电阻较小的下笼流过，产生正常运行时的电磁转矩，因此下笼又称为运行笼。

双笼转子感应电动机的优点是，改变上、下笼的几何尺寸和所用的材料，可以灵活地获得所需的机械特性，以满足不同负载的要求。双笼型转子感应电动机的起动性能比深槽式笼型感应电动机好，但后者结构简单、造价低，使用更为广泛。这两种感应电动机的共同优点是起动性能好，共同的缺点是由于转子漏电抗比普通笼型感应电动机大，导致功率因数和最大转矩（过载能力）比普通笼型感应电动机低。

4.10 三相感应电动机的调速

近年来，随着电力电子技术、微电子技术、计算机技术以及自动控制技术的飞速发展，三相感应电动机的调速技术也迅速发展，并日趋完善，大有取代用直流电动机调速的趋势。在工业应用中，凡是能用直流电动机调速的场合，都能改用交流调速。由于感应电动机一般不带换向器，能制成直流电动机达不到的大容量、高转速、高电压的调速系统，另外，对大部分不需调速的风机、泵类负载，采用交流调速后，可以大幅度节能，因而对感应电动机进行调速、节能的研究具有重要意义。

由感应电动机的转速公式

$$n = n_s(1-s) = \frac{60f_1}{p}(1-s) \tag{4-81}$$

可知，要改变感应电动机的转速n，可以采用以下3种方法：

1）变极调速。改变极对数p，以改变定子旋转磁场的同步转速n_s来实现调速。

2）变频调速。改变供电电源频率f_1，以改变定子旋转磁场的同步转速n_s来实现调速。

3）变转差率调速。改变电动机的转差率s来实现调速，分为降低定子电压调速、转子回路串接电阻调速和串级调速等方法。

4.10.1 变极调速

改变三相感应电动机定子绕组的极对数p时，定子旋转磁场的同步转速n_s将随之改变，于是转子转速n也将相应改变。在电动机中，仅当定、转子极对数相等时才能产生平均电磁转矩，实现机电能量的转换，因此改变定子绕组极对数时，转子绕组的极对数必须作相应改变。对于绕线转子感应电动机，改变定子绕组极对数时，就必须同时改变转子绕组的极对数，从而使接线变得很复杂；而对于笼型感应电动机，其转子极对数总能自动和定子极对数保持相等，所以变极调速多应用于笼型感应电动机。下面介绍变极原理。

三相感应电动机通常是通过改变定子绕组的接线方式来改变极对数的。图4-40是感应电动机的变极原理。假设电动机定子每相绕组都由两个完全对称的半相绕组组成，以A相绕组为例，图4-40a表示两个半相绕组A_1X_1和A_2X_2正向串联，形成4极磁场；图4-40b表示两个半相绕组A_1X_1和A_2X_2反向串联，A_2X_2中的电流反向，形成2极磁场；图4-40c表示两个半相绕组A_1X_1和A_2X_2反向并联，A_2X_2中的电流反向，形成2极磁场。可见，将定子绕组从正向串联改接成反向串联或反向并联，使得两个半相绕组中的任意一个半相绕组的电流反向，即可将极对数减少一半，因而使同步转速成倍地变化。这种调速为有级调速，改接方式是单绕组倍极比的方式。所谓倍极比是指：多速电动机变极前后的极数比为整数倍，如 2/4 极、4/8 极和 6/12 极等；除此之外，还有非倍极比的变极方法，如 4/6 极、6/8 极等；另外还可在定子上嵌放两套绕组，每一套都能改变极数，以得到三速或四速电动机。

4.10.2 变频调速

改变三相感应电动机的供电频率 f_1，可以改变定子旋转磁场的同步转速 n_s，于是转子转速 n 也相应改变，可实现转速的平滑调节。通常将额定频率称为基频，电源频率既可以从基频向上调，也可以从基频向下调。

1. 从基频向下的变频调速

当忽略三相感应电动机的定子漏抗和定子绕组电阻时，定子的感应电动势为

$$U_1 \approx E_1 = 4.44 f_1 N_1 k_{w1} \Phi_m$$

可见，若保持电源电压 U_1 为额定值，则随着频率 f_1 的下降，气隙每极磁通 Φ_m 将增加。电动机磁路本来就比较饱和，Φ_m 增加时，磁路将过于饱和，使励磁电流急剧增加，这是不允许的。因此降低电源频率 f_1 时，必须同时降低电源电压 U_1，即

$$U_1/f_1 = 常数 \tag{4-82}$$

如果降低电源频率 f_1 的同时保持 $U_1/f_1 = 常数$，则 $\Phi_m \approx 常数$，这时电动机的电磁转矩为

$$T_e = \frac{m_1 p}{2\pi}\left(\frac{U_1}{f_1}\right)^2 \frac{f_1 \dfrac{R_2'}{s}}{\left(R_1 + \dfrac{R_2'}{s}\right)^2 + (X_{1\sigma} + X_{2\sigma}')^2} \tag{4-83}$$

最大电磁转矩为

$$\begin{aligned} T_{max} &= \frac{1}{2} \frac{m_1 p U_1^2}{2\pi f_1 \left[R_1 + \sqrt{R_1^2 + (X_{1\sigma} + X_{2\sigma}')^2}\right]} \\ &= \frac{1}{2} \frac{m_1 p}{2\pi}\left(\frac{U_1}{f_1}\right)^2 \frac{f_1}{R_1 + \sqrt{R_1^2 + (X_{1\sigma} + X_{2\sigma}')^2}} \end{aligned} \tag{4-84}$$

从式（4-84）可知，当改变电源频率 f_1 时，若保持 $U_1/f_1 = 常数$，则最大转矩 T_{max} 稍有变化；在 f_1 接近额定频率时，$R_1 \ll (X_{1\sigma} + X_{2\sigma}')$，随着 f_1 的减小，最大转矩 T_{max} 减小得不多，但是当 f_1 较低时，$(X_{1\sigma} + X_{2\sigma}')$ 比较小，R_1 相对较大，随着 f_1 的降低，最大转矩 T_{max} 减小得较多。

根据式（4-83）可画出保持 $U_1/f_1 = 常数$时，变频调速的机械特性，如图 4-41 中实线所示。此时的变频调速近似属于恒转矩调速。

2. 从基频向上的变频调速

从基频向上调节时，升高电源电压是不允许的，所以应采用恒压控制方式，即保持 $U_1 = U_{1N}$ 不变。根据式（4-83），当频率 f_1 增大时，磁通 Φ_m 将减小，因此从基频向上的变频调速是一种恒压弱磁的调速方式，这种情况与直流电动机（恒压）弱磁调速很相似。图 4-42 为保持电源电压为 U_{1N} 不变时变频调速的机械特性曲线。此时的变频调速近似属于恒功率调速。

综上所述，三相感应电动机变频调速的特点为：

1）从基频向下调速，为恒转矩调速方式；从基频向上调速，近似为恒功率调速方式。

2）调速范围大。

3）转速稳定性好。

4）运行时 s 小，效率高。

5）频率 f_1 可以连续调节，变频调速为无级调速。

4.10.3 降低定子电压调速

降低定子电压时的人为机械特性是一组通过同步转速点，且临界转差率不变的曲线族，如图 4-43 所示。若电动机拖动恒转矩负载 T_{L1}，工作点分别为 A 点、B 点和 C 点，对应的转速分别是 n_A、n_B和 n_C。可见，降低定子电压时，转速将下降，但降压调速的调速范围很窄，对于一般的笼型感应电动机，没有多大的实用价值。若电动机拖动通风机类负载 T_{L2}，降压调速的范围就比较大，但应注意感应电动机低速运行时，会出现过电流和功率因数较低的现象。

降低定子电压调速既不是恒转矩调速，也不是恒功率调速，它最适合于转矩随转速降低而减小的负载（如通风机类负载），也可用于恒转矩负载，最不适合于恒功率负载。

4.10.4 转子回路串接电阻调速

在绕线转子感应电动机的转子回路串接适当电阻，也可以调节电动机的转速。转子回路串接电阻的人为机械特性是一组通过同步转速点，且最大转矩恒定不变的曲线族，如图 4-44 所示。

假设电动机拖动恒转矩负载 T_L运行于固有机械特性的 A 点，转速为 n_A。当转子回路串接附加电阻 R_{ad1}时，转子电流 I_2瞬间减小，电磁转矩 T_e也相应减小，工作点从 A 点水平跃变至 A'点，此时 $T_e < T_L$，电动机减速，转差率 s 增大，于是转子电动势、转子电流和电磁转矩 T_e开始增大，直到 $T_e = T_L$为止，电动机稳定运行于 B 点，转速为 n_B。显然，转子回路串接电阻后，转速将下降，所串电阻越大，转速越低、机械特性越软。这种调速属于恒转矩调速，适合于拖动恒转矩负载。

4.10.5 串级调速

为了克服在转子回路串电阻调速的缺点，可以在转子回路中串联一个与转子电动势 $s\dot{E}_2$ 同频率的附加电动势 $\dot{E}_{ad}$，通过改变 $\dot{E}_{ad}$的大小和相位，改变电动机的转差率 s，从而达到调速的目的。这种调速方法称为串级调速。若串入的附加电动势 $\dot{E}_{ad}$与转子电动势 $s\dot{E}_2$ 同相时，电动机的转速上升；反之，电动机转速下降。

串级调速的基本原理如下，当转子回路未串入附加电动势 $\dot{E}_{ad}$时，转子电流为

$$I_2 = \frac{sE_2}{\sqrt{R_2^2 + (sX_{2\sigma})^2}} \tag{4-85}$$

当转子回路串入反相附加电动势 $\dot{E}_{ad}$时，转子电流为

$$I_2 = \frac{sE_2 - E_{ad}}{\sqrt{R_2^2 + (sX_{2\sigma})^2}} \tag{4-86}$$

可见，此时转子电流I_2将减小，电动机的电磁转矩T_e也随之减小，出现$T_e < T_L$，电动机开始减速，转差率s增大。由式（4-86）可知，随着s的增大，转子电流I_2开始逐渐回升，电磁转矩T_e也随之升高，直到$T_e = T_L$为止，减速过程结束，电动机在此低速下稳态运行。串入反相附加电动势$\dot{E}_{ad}$调速时的机械特性如图4-45所示。图中曲线1是电动机的固有机械特性，曲线2为加有反相附加电动势$\dot{E}_{ad}$时的机械特性。由图可见，串入反相附加电动势$\dot{E}_{ad}$后，电动机的理想空载转速n'_s将低于原来的同步转速n_s，所以常称为低同步串级调速。串入的反相附加电动势$\dot{E}_{ad}$的幅值越大，则电动机的稳定转速就越低，电动机的理想空载转速也就越低，如图4-45所示。

当在转子回路串入同相附加电动势$\dot{E}_{ad}$时，转子电流为

$$I_2 = \frac{sE_2 + E_{ad}}{\sqrt{R_2^2 + (sX_{2\sigma})^2}} \tag{4-87}$$

可见，此时转子电流I_2将增大，电动机的电磁转矩T_e也随之增大，出现$T_e > T_L$，电动机开始加速，转差率s减小。由式（4-87）可知，随着s的减小，转子电流I_2开始逐渐下降，电磁转矩T_e也随之减小，直到$T_e = T_L$为止，加速过程结束，电动机在此高速下稳定运行。这种串入同相附加电动势$\dot{E}_{ad}$调速的机械特性如图4-46所示。图中曲线1是电动机的固有机械特性，曲线2为串入同相附加电动势$\dot{E}_{ad}$后的机械特性。这种情形常称为超同步串级调速。

在低同步串级调速过程中，提供附加电动势$\dot{E}_{ad}$的装置从转子电路吸收电能并回馈到电网；而在超同步串级调速过程中，提供附加电动势$\dot{E}_{ad}$的装置向转子电路输入电能，同时电源也向定子电路输入电能，因此又称其为电动机的双馈运行。

串级调速克服了转子串电阻调速的缺点。它的优点是，机械特性较硬，调速的平滑性好，能实现无级调速，且能将转差损耗回馈给电网，效率较高；其缺点是，获得附加电动势的装置较复杂，成本较高，且过载能力随转速降低而降低。因此，串级调速比较适合于调速范围不太大（一般为2～4）的场合，如通风机、轧钢机、矿井提升机等生产机械。

本章小结

本章介绍了感应电机的基本结构、工作原理和3种运行状态。转差率是感应电机的基本变量，反映了感应电机的不同运行状态。通过对感应电动机空载和负载时内部电磁关系的分析，得到了感应电动机的基本方程。通过频率归算和绕组归算，推导出感应电动机的等效电路和相量图。无论是频率归算还是绕组归算，都是在保持转子磁动势的大小和空间相位不变的原则下，即在转子对定子的影响不变的情况下，用等效转子代替实际转子。

基本方程、等效电路和相量图是分析感应电动机的3种不同表示方式，计算时用得最多的是等效电路。

根据等效电路，推出了感应电动机的功率方程和转矩方程，并对感应电动机的运行特性进行了分析。

从分析方法和推导过程看，感应电机与变压器有许多相似之处，但是，必须注意到它们

之间的差别，主要表现在：

1）感应电机的主磁场是旋转磁场，而变压器中是脉振磁场。

2）感应电机使用短距分布绕组，而变压器使用整距集中绕组。

3）由于感应电机中有气隙，因此励磁电流较大。

4）作为旋转电机，感应电机产生电磁转矩，涉及机电能量转换。

5）感应电机的定、转子频率不同，需要进行频率归算。

三相感应电动机的固有机械特性是一条非线性曲线，其最大转矩与电压的二次方成正比，与转子电阻无关，而临界转差率与转子电阻成正比。当转子电阻增大时，临界转差率增大，而最大转矩不变。

三相笼型感应电动机的起动方法分为直接起动、降压起动。限制起动电流的常用降压起动方法有定子回路串对称电阻或电抗降压起动、星形/三角形换接降压起动、自耦变压器降压起动。三相绕线转子感应电动机可采用转子回路串电阻或频敏变阻器起动，其起动转矩大，起动电流小。

三相感应电动机的调速方法可分为变极调速、变频调速、降低定子电压调速、转子回路串联电阻和串级调速。

习 题

1. 感应电机的转差率和转差速度是如何定义的？

2. 一台绕线转子感应电动机，定子绕组短路，转子绕组通入频率为 f_1 的三相对称交流电流，产生的旋转磁场相对于转子以转速 $n_s=\dfrac{60f_1}{p}$（单位为 r/min）逆时针方向旋转，问此时转子转向如何？转差率如何计算？

3. 在感应电动机中，转子静止与转动时相比，转子侧的电量和参数有何变化？

4. 感应电动机工作时，转子转速下降，转子频率是增大还是减小？

5. 感应电动机工作时，如果电源频率增加而电源电压不变，主磁通如何变化？

6. 为什么说感应电动机从空载到满载，主磁通的大小基本不变？

7. 感应电动机转速变化时，为什么定、转子磁动势之间没有相对运动？

8. 三相感应电动机的电压比 k_e 为什么和电流比 k_i 不同？

9. 说明感应电动机工作时的能量传递过程。

10. 为什么感应电动机正常运行时，转差率很小？

11. 感应电动机在额定电压下起动时（$n=0$），定子电流很大，为什么？可以采取什么措施限制定子电流？

12. 为什么笼型感应电动机的转子极数与定子极数自动相等？

13. 进行空载试验时，电源电压从额定电压开始下降，空载电流开始下降，电压下降到一定程度，空载电流反而上升，为什么？

14. 三相感应电动机在进行短路试验时，若定子电流达到额定值，电磁转矩是否也达到额定值？

15. 感应电动机的电磁转矩是如何产生的？

16. 感应电动机的转速为何不能等于同步转速？

17. 感应电动机空载试验的目的是什么？

18. 感应电动机短路试验的目的是什么？

19. 绘制感应电动机典型的转矩-转速特性曲线，并说明其特点？

20. 什么是深槽式笼型转子？在电机运行中有何优点？

21. 为什么感应电动机（笼型转子或绕线转子）在转差率较大时效率很低？

22. 感应电动机有哪些调速方法？

23. 感应电动机变频调速时，为什么在降低频率的同时需要降低电源电压？

24. 一台感应电动机在额定状态运行，当负载转矩增大时，以下物理量将如何变化？①电机转速；②转差率；③转子感应电压；④转子电流；⑤转子电流频率；⑥同步转速；⑦转子铜损。

25. 一台220V、三相2极、50Hz的感应电动机，当其转差率为5%时，试求：

1）旋转磁场每分钟的转速为多少？

2）转子每分钟的转速为多少？

3）转子的转差速度为多少？

4）转子电流的频率为多少？

26. 一台480V、三相4极、60Hz的感应电动机，当其转差率为3.5%时，试求：

1）旋转磁场每分钟的转速为多少？

2）转子每分钟的转速为多少？

3）转子的转差速度为多少？

4）转子电流的频率为多少？

27. 一台60Hz三相感应电动机，空载运行转速为890r/min，满载运行转速为840r/min。试求：

1）该电动机的极数是多少？

2）该电动机满载时的转差率为多少？

3）该电动机带1/4额定负载时的转速是多少？

4）该电动机带1/4额定负载时的转子电流的频率是多少？

28. 一台60Hz的三相感应电动机，空载运行转速为1198 r/min，满载运行转速为1112 r/min 。试求：

1）该电动机的极数是多少？

2）该电动机满载时的转差率为多少？

3）转子电流的频率？

4）转子磁场相对于转子的转速和相对于定子的转速各为多少？

29. 一台50kW、440V、50Hz、6极感应电动机，额定运行时的转差率为6%，摩擦损耗为300W，铁损为600W。试求该电动机在额定负载下：

1）转子的转速。

2）输出功率。

3）输出转矩。

4）电磁转矩。

5）转子电流的频率。

30. 一台三相感应电动机：$P_N=7.5\text{kW}$，$U_N=380\text{V}$，$I_N=15.6\text{A}$，$n_N=1426\text{r/min}$，$f_N=50\text{Hz}$，$\cos\varphi_N=0.87$（滞后），试求：

1）电动机的极数是多少？

2）额定负载下的转差率是多少？

3）额定负载下的效率是多少？

31. 一台460V、37kW、60Hz、4极感应电动机，额定转速为1755r/min，当该电动机带额定负载运行时，试求：

1）转子的转差率。

2）转子电流的频率。

3）定子旋转磁场相对于定子的转速及定子旋转磁场相对于转子的转速。

4）转子旋转磁场相对于定子的转速及转子旋转磁场相对于转子的转速。

32. 一台三相感应电动机：$P_N=150kW$，$f_N=50Hz$，$2p=4$，满载运行时的转子铜损为4.8kW，机械损耗为700W，附加损耗为1500W，求电动机的电磁转矩和输出转矩。

33. 一台三相感应电动机：$U_N=380V$，$f_N=50Hz$，$n_N=1426r/min$，定子绕组为三角形联结，$R_1=2.865\Omega$，$X_{1\sigma}=7.71\Omega$，$R_2'=2.82\Omega$，$X_{2\sigma}'=11.75\Omega$，$X_m=202\Omega$，R_m忽略不计，试利用T形等效电路计算额定负载时的I_1、P_1、$\cos\varphi_1$和I_2'。

34. 一台三相感应电动机：$P_N=7.5kW$，$U_N=380V$，$f_N=50Hz$，$n_N=962\ r/min$，定子绕组为三角形联结，带额定负载时$\cos\varphi_N=0.827$，定子铜损为470W，铁损为234W，机械损耗为45W，附加损耗为80W。试计算在带额定负载时的转差率、转子电流的频率、转子铜损、效率、定子电流。

35. 一台三相感应电动机：$P_N=17kW$，$U_N=380V$，$f_N=50Hz$，定子绕组为三角形联结，$R_1=0.715\Omega$，$X_{1\sigma}=1.74\Omega$，$R_2'=0.416\Omega$，$X_{2\sigma}'=3.03\Omega$，$X_m=75\Omega$，$R_m=6.2\Omega$，机械损耗$p_\Omega=139W$，额定负载时的附加损耗$p_\Delta=320W$。试计算额定负载时的转差率、定子电流、定子功率因数、电磁转矩、输出转矩和效率。

36. 一台三相感应电动机：$P_N=10kW$，$U_{1N}=380V$，$f_N=50Hz$，定子绕组为星形联结，$I_N=19.8A$。已知$R_1=0.5\Omega$；空载试验数据为$U_N=380V$，$I_{10}=5.4A$，$P_{10}=425W$，$p_\Omega=170W$；短路试验数据如下：

U_{1k}（线）/V	200	160	120	80	40
I_{1k}（线）/A	36	27	18.1	10.5	4
P_{1k}（三相）/W	3680	2080	920	290	40

试求：

（1）X_m、R_m、$X_{1\sigma}$、$X_{2\sigma}'$和R_2'（设$X_{1\sigma}=X_{2\sigma}'$）。

（2）用T形等效电路确定额定电流I_{1N}和额定功率因数$\cos\varphi_N$（设$p_\Delta=1\%P_N$）。

37. 一台11kW、208V、2极、60Hz、星形联结的绕线转子感应电动机，其等效电路参数为：$R_1=0.2\Omega$，$R_2'=0.120\Omega$，$X_m=15.0\Omega$，$X_1=0.410\Omega$，$X_2'=0.410\Omega$；机械损耗为250W，铁损为180W，不计附加损耗，当电动机的转差率为0.05时，试求：定子线电流、定子铜损、电磁功率、总机械功率、电动机的电磁转矩、电动机的输出转矩、电动机的效率、电动机最大电磁转矩时的临界转差率、电动机的最大电磁转矩。

38. 一台定子绕组星形联结的三相6极笼型感应电动机，额定电压$U_N=380V$，额定转速$n_N=957r/min$，额定频率$f_N=50Hz$，定子电阻$R_1=2.08\Omega$，转子电阻折算值$R_2'=1.53\Omega$，定

子漏电抗 $X_{1\sigma}=3.12\Omega$，转子漏电抗折算值 $X'_{2\sigma}=4.25\Omega$。试求：额定电磁转矩、最大转矩及过载能力、临界转差率。

39. 一台三相8极笼型感应电动机，额定功率 $P_N=260kW$，额定电压 $U_N=380V$，额定转速 $n_N=722r/min$，额定频率 $f_N=50Hz$，过载能力 $\lambda_m=2.13$。试求：额定转差率、额定电磁转矩、最大转矩、临界转差率、当 $s=0.02$ 时的电磁转矩。

40. 一台三相笼型感应电动机，额定功率 $P_N=320kW$，额定电压 $U_N=600V$，额定转速 $n_N=740r/min$，额定电流 $I_N=40A$，额定功率因数 $\cos\varphi_N=0.83$，起动转矩倍数 $k_T=1.93$，起动电流倍数 $k_I=5.04$，最大转矩倍数 $\lambda_m=2.2$，定子绕组为星形联结。试求：

1）直接起动时的起动电流与起动转矩。

2）把起动电流限定在160A时，应在定子回路每相串入多大的电阻？此时的起动转矩是多少？

41. 一台三相2极笼型感应电动机，额定功率 $P_N=11kW$，额定电压 $U_N=380V$，额定电流 $I_N=21.8A$，额定转速 $n_N=2930r/min$，额定频率 $f_N=50Hz$，最大转矩倍数 $\lambda_m=2.2$。采用变频调速拖动 $T_L=0.8T_N$ 的恒转矩负载，若采用 $U_1/f_1=$ 常数的变频调速方式，忽略空载转矩 T_0。试求当频率降低为 $0.8f_N$ 时电动机的转速。

第5章 同步电机

同步电机和感应电机都是交流电机。同步电机稳态运行时，其转速 n 与电机内部旋转磁场的转速 n_s 相同，即 $n=n_s=60f/p$，因此称为同步电机，而感应电机的转速 $n \neq n_s$，所以感应电机又称为异步电机。

同步电机一般作为发电机运行，世界上电力的大部分由同步发电机发出。同步电机也可以用作电动机，它的优点是功率因数可以调节。同步电机还有一种特殊的运行方式，即专门调节电网无功功率来改善电网功率因数的补偿机（又称调相机）。本章主要讲述同步发电机的工作原理和电磁关系，导出其基本方程、等效电路和相量图，再进一步分析同步发电机的运行特性以及同步发电机与电网并联运行时的功率调节问题，最后简单介绍同步电动机。

5.1 同步电机的基本结构、运行状态和额定值

5.1.1 同步电机的基本结构

按照结构型式，同步电机可分为旋转电枢式（简称转枢式）和旋转磁极式（简称转极式）两类。转枢式电机的主磁极装在定子上，电枢绕组装在转子上，电枢绕组通过滑动接触输出或输入电能，对于大容量电机来说，转枢式极不可靠，故转枢式常用作几个千瓦以下的小容量同步电机。而转极式电机的电枢绕组装在定子上，励磁绕组装在转子上，由于励磁部分的容量和电压常较电枢小得多，电刷和集电环的负载大为减轻，故转极式适用于高压、大容量同步电机。

在转极式同步电机中，按照磁极形状，可分为隐极式和凸极式两种基本型式，如图5-1所示。隐极式同步电机的转子做成圆柱形，气隙均匀，励磁绕组分布于转子表面槽内，故转子机械强度高，适合高速（3000r/min）旋转，如汽轮发电机由汽轮机拖动高速旋转；凸极式同步电机的转子有明显凸出的磁极，气隙不均匀，励磁绕组集中放置，制造较为简单，适合中、低速（1000r/min及以下）旋转，如水轮发电机由水轮机拖动低速旋转。

无论转子结构为隐极式还是凸极式，同步发电机的基本结构都包括静止的定子和旋转的转子，且定子和转子之间存在气隙。

1. 定子

同步发电机的定子由定子铁心、定子绕组、机座和端盖等部件组成。图5-2为一台200MW汽轮发电机的定子，图5-3为一台大型水轮发电机的分瓣定子。

定子铁心一般用厚度为0.5mm、两面涂有绝缘漆的硅钢片叠成，沿轴向分成好几叠，每叠厚度为3～6cm，叠与叠之间留有宽0.8～1cm的通风沟。整个铁心用拉紧螺杆和非磁性端压板压紧成整体后，固定在定子机座上。

同步电机中，定子绕组常被称为电枢绕组，一般为三相、双层短距叠绕组，三相绕组采用星形连接，中性点接地。大容量汽轮发电机的电枢线圈由于尺寸较大，大都制成半匝式

（线棒），每个线棒由若干股铜线并在一起，分成一排或两排，把两个线棒的一端焊在一起，即成为一个线圈。

机座起固定和支撑定子铁心的作用，并形成风道。

2. 转子

隐极式同步发电机的转子由转子铁心、励磁绕组、护环和风扇等部件组成。图5-4为一台大型汽轮发电机的转子铁心。由于汽轮发电机的转速很高，所以转子铁心一般用整块导磁性好的高强度合金钢锻成。转子表面约2/3部分铣有轴向凹槽，用于嵌放励磁绕组，不铣槽的约1/3部分形成大齿，即磁极。励磁绕组是用扁铜线绕成的同心式线圈，嵌放在转子槽中并用非磁性硬铝槽楔压紧，励磁绕组端部还要套上用高强度非磁性钢锻成的护环。

水轮发电机的转子由磁极、励磁绕组、磁轭、转轴和阻尼绕组等部件组成。图5-5为一台320MW水轮发电机的转子。

图5-6为一台水轮发电机转子上的磁极和绕组，转子磁极一般由厚度为1～1.5mm的钢板冲片叠压而成，极身上套装同心式励磁绕组，极靴上开有槽，插入极靴槽中的铜条和两端的端环焊成一个闭合绕组称为阻尼绕组，如图5-6所示。当发电机发生振荡时，阻尼绕组可使振荡衰减；当发电机不对称运行时，阻尼绕组起削弱负序旋转磁场的作用。

5.1.2　同步电机的运行状态

当同步电机带负载对称稳态运行时，定子（电枢）三相绕组中将流过三相对称电流，在电机气隙内产生一个以同步转速旋转的定子磁场；同时转子励磁绕组中通入的直流励磁电流建立了一个主极磁场，也以同步转速同向旋转。因此定子旋转磁场与转子主极磁场在空间上保持相对静止，二者合成了一个幅值恒定且以同步转速旋转的气隙合成磁场。

主极磁场与气隙合成磁场之间形成了一个夹角δ，如图5-7所示，磁力线从主极发出后扭斜进入定子，使转子受到磁拉力的作用，形成电磁转矩作用在转子上，从而实现了机电能量转换。δ越大，磁场产生的切向力及电磁转矩和电磁功率就越大，故把δ称为功率角。

如同转差率s是反映感应电机运行状态的一个重要变量一样，功率角δ是反映同步电机运行状态的一个重要变量。功率角δ的数值代表同步电机电磁功率的大小，其正负则决定同步电机的运行状态。同步电机有3种运行状态：发电机、电动机和补偿机，如图5-7所示。

当主极磁场超前气隙合成磁场时，$\delta>0$，磁力线从主极发出后向后扭斜进入定子，转子上受到一个制动电磁转矩，原动机必须输入驱动转矩来克服制动的电磁转矩，此时转子输入机械功率，定子向电网输出电功率，同步电机处于发电机状态；当主极磁场轴线与气隙合成磁场的轴线重合时，$\delta=0$，磁力线从主极发出后垂直进入定子，电磁转矩为零，同步电机不发生机电能量转换，但可以发出或吸收无功功率专门调节电网功率因数，同步电机处于补偿机状态或空载状态；当主极磁场滞后气隙合成磁场时，$\delta<0$，磁力线从主极发出后向前扭斜进入定子，转子上受到一个驱动电磁转矩，拖动转轴上的负载旋转而输出机械功率，此时定子必须从电网吸收电功率，同步电机处于电动机状态。

5.1.3　同步电机的额定值

同步电机的额定值一般有如下几个：

（1）额定容量S_N（或额定功率P_N）　指额定运行时同步电机的输出功率。同步发电机可以用输出的视在功率 S_N（kV·A）表示，也可以用输出的有功功率 P_N（kW）表示；同步电动机指轴上输出的机械功率，用 P_N（kW）表示；同步补偿机指输出的最大无功功率，用 S_N（kV·A）表示。

（2）额定电压 U_N　指额定运行时定子绕组的线电压，单位为伏（V）。

（3）额定电流 I_N　指额定运行时定子绕组中的线电流，单位为安（A）。

（4）额定功率因数 $\cos\varphi_N$　指额定运行时同步电机的功率因数。

（5）额定频率 f_N　指额定运行时定子电压的频率，我国标准工频规定为50Hz。

（6）额定转速 n_N　指额定运行时同步电机的转速，即同步转速 n_s，单位为 r/min。

除以上额定值外，同步电机的铭牌上还标有绝缘等级、允许温升、额定励磁电压 U_{fN} 和额定励磁电流 I_{fN} 等。

同步电机的励磁方式指同步电机获得直流励磁电流的方式，而供给励磁电流的装置称为励磁系统。励磁系统主要分为直流励磁机励磁系统和整流器励磁系统两大类，前者用同轴直流发电机供给励磁电流；后者又分为静止整流器励磁系统和旋转整流器励磁系统两种。静止整流器励磁系统用同轴交流主励磁机发出交流电，经静止半导体整流器整流后供给励磁电流。旋转整流器励磁系统是用同轴转枢式交流主励磁机发出交流电，经同步发电机转子上的整流器整流后供给励磁电流，由于取消了集电环和电刷，故又称为无刷励磁系统。图5-8为带有无刷励磁机的凸极式同步电机的转子，图5-9为一台8极转子带同轴励磁机的凸极式同步电机的剖面图。

5.2　同步发电机的空载和负载运行

5.2.1　同步发电机的空载运行

同步发电机用原动机拖动以同步转速旋转，励磁绕组通入直流励磁电流，电枢绕组开路或电枢电流为零的情况，称为同步发电机的空载运行。

图5-10为一台4极同步发电机的空载磁路，图中 Φ_0 表示通过气隙同时与励磁绕组和电枢绕组交链的磁通，称为主磁通；$\Phi_{f\sigma}$ 表示不通过气隙仅与励磁绕组本身相交链的磁通，称为主极漏磁通。主磁路包括气隙、电枢齿、电枢轭、主磁极和磁轭五部分。

1. 空载运行时的磁动势和磁场

同步发电机空载运行时，转子励磁电流 I_f 产生主极磁动势 F_f，由于原动机拖动转子以同步转速旋转，因此主极磁动势是一个旋转磁动势，其基波用 $\boldsymbol{F}_{f1}$ 表示。主极磁动势基波 $\boldsymbol{F}_{f1}$ 将在气隙中建立一个旋转磁场 $\boldsymbol{B}_0$，定子导体切割气隙旋转磁场，在定子三相绕组中产生三相对称感应电动势，称为励磁电动势。

忽略高次谐波时，励磁电动势（相电动势）的有效值 E_0 为

$$E_0 = 4.44fN_1k_{w1}\Phi_0 \tag{5-1}$$

通过以上分析可知，同步发电机空载时的磁动势和磁场以及感应电动势之间的电磁关系为

$$I_f \rightarrow \boldsymbol{F}_{f1} \rightarrow \boldsymbol{B} \rightarrow \dot{\Phi}_0 \rightarrow \dot{E}_0$$

其中 $\dot{\Phi}_0$ 为一相电枢绕组交链的每极主磁通，它是一个随时间变化的时间相量。

2. 空载特性

同步电机以同步转速稳态运行时，励磁电动势 E_0 与励磁电流 I_f 之间的关系曲线 $E_0=f(I_f)$，称为同步电机的空载特性，如图5-11所示。由于 E_0 正比于 Φ_0，I_f 正比于 F_f，所以空载特性 $E_0=f(I_f)$ 与磁化曲线 $\Phi_0=f(F_f)$ 的形状相似。空载特性实际上反映的是电机磁路的饱和情况，它是同步电机的一条重要特性。

当主磁通 Φ_0 较小时，整个磁路处于不饱和状态，所以空载特性的起始段是一条直线，其延长线称为气隙线。随着主磁通 Φ_0 的增大，铁心逐渐饱和，空载特性就逐渐弯曲。空载电压等于额定电压时，主磁路的饱和系数 $k_\mu=I_{f0}/I'_{f0}=E_0/U_{N\varphi}\approx1.1\sim1.25$。在研究同步电机的许多问题时，为了避免求解非线性问题带来的复杂性，常常不计铁心的磁饱和，此时空载特性就成为一条理想化的直线——气隙线。

空载特性可以通过计算或试验得到，表5-1给出了一条典型的同步发电机空载特性的数据。

表5-1　典型同步发电机空载特性

E_0^*	0.58	1.0	1.21	1.33	1.40	1.46	1.51
I_f^*	0.5	1.0	1.5	2.0	2.5	3.0	3.5

5.2.2　同步发电机的负载运行

1. 带负载运行时的磁动势和磁场

当同步发电机带对称负载运行时，电枢绕组中流过三相对称电流，产生以同步转速旋转的电枢磁动势，其基波用 $\boldsymbol{F}_a$ 表示。主极磁动势基波 $\boldsymbol{F}_{f1}$ 与电枢磁动势基波 $\boldsymbol{F}_a$ 同向同速旋转，二者在空间保持相对静止，合成为一个幅值恒定且以同步转速旋转的合成磁动势 $\boldsymbol{F}$。负载时气隙内的旋转磁场 $\boldsymbol{B}$ 由合成磁动势 $\boldsymbol{F}$ 建立，定子导体切割气隙旋转磁场，在定子三相绕组中产生三相对称感应电动势，称为气隙电动势（也叫合成电动势）。

同步发电机带负载运行时磁动势和磁场以及感应电动势之间的电磁关系为

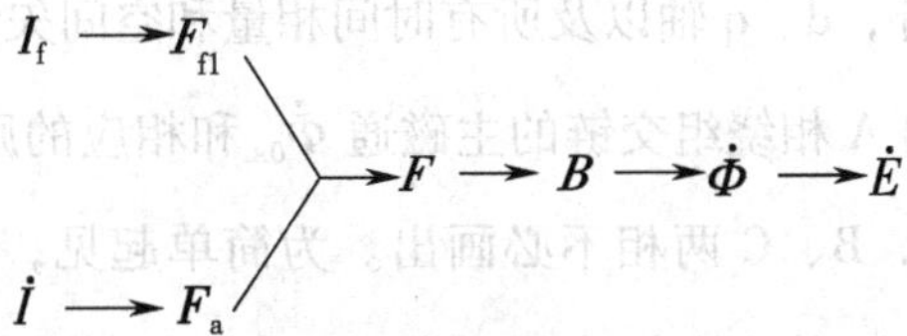

其中 $\dot{I}$ 为一相电枢绕组电流，$\dot{\Phi}$ 为合成磁动势产生的气隙磁场与一相电枢绕组交链的每极磁通，$\dot{E}$ 为一相电枢绕组内感应的气隙电动势。

2. 带对称负载运行时的电枢反应

同步电机带对称负载运行时，电枢磁动势的基波对主极磁场的影响称为电枢反应。电枢反应对同步电机的运行性能有很大影响，为研究电枢反应，让我们先复习一下时空矢量图。

时空矢量图就是把时间坐标系下的时间相量图和空间坐标系下的空间矢量图画在一起，人为地使时间相量和空间矢量有了相位关系（此相位并无物理意义，取决于时间参考轴与绕组相轴之间人为设定的相位差）。

在时空矢量图中，对于对称多相系统，经常把时间参考轴（时轴）与某一相绕组的轴线（相轴）取为同一方向（同相），此时存在下列关系：

1）多相对称电流产生的磁动势矢量 $\boldsymbol{F}$ 与该相电流相量 $\dot{I}$ 同相（当某相电流达到最大值（转到时轴上）时，合成磁动势基波的正波幅转到该相绕组的轴线上）。

2）忽略磁滞和涡流损耗时，旋转磁场的磁感应强度矢量 $\boldsymbol{B}$ 与产生它的磁动势矢量 $\boldsymbol{F}$ 同相（实际上磁密波在空间上滞后磁动势波一个铁损角）。

3）该相绕组交链的磁通相量 $\dot{\Phi}$ 与产生它的旋转磁场磁感应强度矢量 $\boldsymbol{B}$ 同相。

根据上述关系画出的时空矢量图如图 5-12 所示。

因为时间相量和空间矢量的旋转角速度 ω_s 相等，因此任何时刻它们之间的相位关系不变。主极磁动势 $\boldsymbol{F}_{f1}$ 和主磁场 $\boldsymbol{B}_0$ 以及与某相绕组交链的主磁通 $\dot{\Phi}_0$ 之间也存在上述相位关系。

下面就利用时空矢量图来分析三相同步发电机的电枢反应。电枢反应的性质（分为增磁、去磁和交磁 3 种）取决于电枢磁动势基波 $\boldsymbol{F}_a$ 与主磁场磁感应强度 $\boldsymbol{B}_0$（或主极磁动势基波 $\boldsymbol{F}_{f1}$）在空间的相对位置。分析表明，这一相对位置取决于内功率因数角 ψ_0。励磁电动势 $\dot{E}_0$ 与电枢电流 $\dot{I}$ 的时间相位差称为内功率因数角，通常用 ψ_0 表示，它不仅与负载的大小和性质（负载功率因数）有关，还与发电机的内阻抗有关。

以两极凸极同步发电机为例，取主极轴线超前 A 相绕组轴线 90°电角度的时刻来分析电枢反应，如图 5-13a 所示。通常把转子主极的轴线称为直轴，用 d 轴表示，与直轴正交的轴线称为交轴，用 q 轴表示。d、q 轴是旋转直角坐标系，与转子主极同步旋转。在图 5-13a 所示瞬间，虽然 A 相绕组交链的主磁通 $\dot{\Phi}_{0A}$ 为零，但 A 相绕组切割的主极磁力线最多，因此 A 相绕组中感应的励磁电动势 $\dot{E}_{0A}$ 达到最大值，即 $\dot{E}_{0A}$ 转到了时轴上，$\dot{\Phi}_{0A}$ 超前 $\dot{E}_{0A}$ 90°电角度。一般情况下，同步发电机带感性负载运行，电枢电流 $\dot{I}_A$ 通常滞后 $\dot{E}_{0A}$ 一个锐角，即 $0° < \psi_0 < 90°$。若把时轴与 A 相相轴取为同一方向，可得到如图 5-13b 所示的时空矢量图。图中除了相轴和时轴静止不动外，d、q 轴以及所有时间相量和空间矢量都以同步转速旋转。

在图 5-13b 中，仅画出 A 相绕组交链的主磁通 $\dot{\Phi}_{0A}$ 和相应的励磁电动势 $\dot{E}_{0A}$ 以及电枢电流 $\dot{I}_A$ 就能够分析电枢反应，B、C 两相不必画出。为简单起见，可把下标 A 省略，分别用 $\dot{\Phi}_0$、$\dot{E}_0$ 和 $\dot{I}$ 来表示 A 相绕组交链的主磁通、励磁电动势和电枢电流。此时三相电流产生的电枢磁动势 $\boldsymbol{F}_a$ 与 A 相电枢电流 $\dot{I}$ 同相，主极磁动势 $\boldsymbol{F}_{f1}$ 与 A 相绕组交链的主磁通 $\dot{\Phi}_0$ 同相，因此电枢磁动势 $\boldsymbol{F}_a$ 与主极磁动势 $\boldsymbol{F}_{f1}$ 的空间相位差等于主磁通 $\dot{\Phi}_0$ 与电枢电流 $\dot{I}$ 的时间相位差即 $90° + \psi_0$，可见同步发电机的电枢反应取决于内功率因数角 ψ_0。

下面根据 ψ_0 分析几种典型的电枢反应。

(1) $\psi_0 = 0°$（电枢电流 $\dot{I}$ 与励磁电动势 $\dot{E}_0$ 同相）　对应的时空矢量图如图 5-14 所示。此时电枢磁动势 $\boldsymbol{F}_a$ 作用在 q 轴上，而且主极磁动势 $\boldsymbol{F}_{f1}$ 与电枢磁动势 $\boldsymbol{F}_a$ 及 d、q 轴同步旋转，电枢磁动势 $\boldsymbol{F}_a$ 始终与主极磁动势 $\boldsymbol{F}_{f1}$ 正交，称为交轴电枢反应。

从图 5-14 可见，交轴电枢反应使气隙合成磁场（或合成磁动势）与主极磁场（或主极磁动势）在空间形成了一个夹角。对于同步发电机来说，转子主极磁场超前于气隙合成磁场，磁力线从主极发出后将向后扭斜进入定子，转子上受到一个制动的电磁转矩，原动机必须输入驱动转矩来克服制动的电磁转矩，从而将机械能转变成电能。所以，同步发电机电磁转矩的产生和机电能量转换与交轴电枢反应有关。

（2）$\psi_0=90°$（电枢电流 $\dot{I}$ 滞后励磁电动势 $\dot{E}_0$ 90°）　对应的时空矢量图如图5-15所示。此时电枢磁动势 $\boldsymbol{F}_a$ 滞后q轴90°，作用在d轴的反方向上，而且主极磁动势 $\boldsymbol{F}_{f1}$ 与电枢磁动势 $\boldsymbol{F}_a$ 及d、q轴同步旋转，电枢磁动势 $\boldsymbol{F}_a$ 始终与主极磁动势 $\boldsymbol{F}_{f1}$ 方向相反，气隙磁场被削弱，称为直轴去磁电枢反应。

从图5-15可见，此时主极磁场轴线与气隙合成磁场轴线重合，磁力线从主极发出后垂直进入定子，电磁转矩为零，同步发电机不发生机电能量转换，但可以发出无功功率。

（3）$\psi_0=-90°$（电枢电流 $\dot{I}$ 超前励磁电动势 $\dot{E}_0$ 90°）　此时电枢磁动势 $\boldsymbol{F}_a$ 超前q轴90°，作用在d轴上（见图5-16），而且主极磁动势 $\boldsymbol{F}_{f1}$ 与电枢磁动势 $\boldsymbol{F}_a$ 及d、q轴同步旋转，电枢磁动势 $\boldsymbol{F}_a$ 始终与主极磁动势 $\boldsymbol{F}_{f1}$ 方向相同，气隙磁场被增强，称为直轴增磁电枢反应。同步发电机不发生机电能量转换，但可以发出无功功率。

（4）$0°<\psi_0<90°$（电枢电流 $\dot{I}$ 滞后励磁电动势 $\dot{E}_0$ 一个锐角）　对应的时空矢量图如图5-13b所示。此时电枢磁动势 $\boldsymbol{F}_a$ 既不作用在d轴上也不作用在q轴上，可以将 $\boldsymbol{F}_a$ 分解成直轴电枢磁动势 $\boldsymbol{F}_{ad}$ 和交轴电枢磁动势 $\boldsymbol{F}_{aq}$ 两个分量，即

$$\boldsymbol{F}_a=\boldsymbol{F}_{ad}+\boldsymbol{F}_{aq} \tag{5-2}$$

其中

$$\begin{cases}F_{ad}=F_a\sin\psi_0\\F_{aq}=F_a\cos\psi_0\end{cases} \tag{5-3}$$

相应地，电枢电流也可分解成直轴电枢电流 $\dot{I}_d$ 和交轴电枢电流 $\dot{I}_q$ 两个分量，即

$$\dot{I}=\dot{I}_d+\dot{I}_q \tag{5-4}$$

其中

$$\begin{cases}I_d=I\sin\psi_0\\I_q=I\cos\psi_0\end{cases} \tag{5-5}$$

由图5-13b可见，电枢电流 $\dot{I}$ 滞后励磁电动势 $\dot{E}_0$ 一个锐角时，同时发生交轴和直轴去磁电枢反应。

同理，当电枢电流 $\dot{I}$ 超前励磁电动势 $\dot{E}_0$ 一个锐角时，同时发生交轴和直轴增磁电枢反应。

5.3　同步发电机的电压方程、相量图和等效电路

由于隐极式电机和凸极式电机的转子结构明显不同，因此它们的分析方法也有所不同，下面分别对隐极式同步发电机和凸极式同步发电机进行分析。另外，分析每一类发电机时又按照不考虑铁心的磁饱和与考虑铁心磁饱和两种情况分开来进行分析。

5.3.1 隐极式同步发电机的电压方程、相量图和等效电路

1. 不考虑磁饱和

(1) 负载时各物理量之间的电磁关系 不考虑磁饱和时，隐极式同步发电机的磁路为线性。可应用叠加原理，将同步发电机负载运行时的主极磁动势 $\boldsymbol{F}_{f1}$ 与电枢磁动势 $\boldsymbol{F}_a$ 的作用分别单独考虑。主极磁动势 $\boldsymbol{F}_{f1}$ 产生主磁通 $\dot{\Phi}_0$，并在定子绕组中感应励磁电动势 $\dot{E}_0$，而电枢磁动势 $\boldsymbol{F}_a$ 产生电枢反应磁通 $\dot{\Phi}_a$，并在定子绕组中感应电枢反应电动势 $\dot{E}_a$，将 $\dot{E}_0$ 和 $\dot{E}_a$ 相量相加，可得定子每相合成电动势 $\dot{E}$ 。发电机电枢电流 $\dot{I}$ 还产生电枢漏磁通 $\dot{\Phi}_\sigma$，并在定子绕组中感应电枢漏感电动势 $\dot{E}_\sigma$。通过以上分析可知，同步发电机负载运行时各物理量之间的电磁关系如下：

$$
\begin{aligned}
&I_f \rightarrow F_{f1} \rightarrow \dot{\Phi}_0 \rightarrow \dot{E}_0 \searrow \\
&\qquad\qquad\qquad\qquad\qquad\quad \dot{E} \\
&\dot{I} \rightarrow F_a \rightarrow \dot{\Phi}_a \rightarrow \dot{E}_a \nearrow \\
&\quad \searrow \dot{\Phi}_\sigma \rightarrow \dot{E}_\sigma
\end{aligned}
$$

(2) 电压方程 根据上述分析，并规定同步发电机各物理量的正方向如图 5-17 所示，可得出定子每相电压 $\dot{U}$ 为定子每相绕组内的所有感应电动势 $\dot{E}_0+\dot{E}_a+\dot{E}_\sigma$ 与电枢电阻压降 $\dot{I}R_a$ 的相量差，即

$$\dot{E}_0+\dot{E}_a+\dot{E}_\sigma-\dot{I}R_a=\dot{U} \tag{5-6}$$

因为电枢反应电动势 E_a 与电枢反应磁通 Φ_a 成正比，电枢磁动势 F_a 与电枢电流 I 成正比，当不计磁饱和时，可认为 Φ_a 与 F_a 成正比，于是得到以下关系

$$E_a \propto \Phi_a \propto F_a \propto I$$

在时间相位上，$\dot{E}_a$ 滞后 $\dot{\Phi}_a 90°$，不计定子铁损时，电枢反应磁场 $\boldsymbol{B}_a$ 与电枢磁动势 $\boldsymbol{F}_a$ 同相，由时空矢量图得到 $\dot{\Phi}_a$ 与 $\dot{I}$ 同相，因此得出 $\dot{E}_a$ 滞后 $\dot{I}$ 90°，于是 $\dot{E}_a$ 可表示为负电抗压降的形式，即

$$\dot{E}_a=-\mathrm{j}\dot{I}X_a \tag{5-7}$$

式中，X_a 称为电枢反应电抗，是一个反映电枢反应磁通 Φ_a 的参数，$X_a=E_a/I$

式 (5-7) 将电枢反应电动势 $\dot{E}_a$ 和电枢电流 $\dot{I}$ 直接联系起来，简化了分析计算。

同理，可引入一个反映电枢漏磁通 Φ_σ 的参数，将电枢漏感电动势 $\dot{E}_\sigma$ 和电枢电流 $\dot{I}$ 直接联系起来，即

$$\dot{E}_\sigma=-\mathrm{j}\dot{I}X_\sigma \tag{5-8}$$

式中，X_σ 称为电枢绕组的漏电抗。$X_\sigma=E_\sigma/I$。

将式 (5-7) 和式 (5-8) 代入式 (5-6)，可得

$$\dot{E}_0=\dot{U}+\dot{I}R_a+\mathrm{j}\dot{I}X_a+\mathrm{j}\dot{I}X_\sigma=\dot{U}+\dot{I}R_a+\mathrm{j}\dot{I}X_s \tag{5-9}$$

式中，X_s 称为同步电抗，是一个反映电枢总磁通（包括电枢反应磁通和电枢漏磁通）的参数，$X_s = X_a + X_\sigma$。它与电枢绕组匝数的二次方和电枢总磁通所经磁路的磁导成正比。

$$X_s \propto N_1^2 \Lambda_s \propto N_1^2(\Lambda_a + \Lambda_\sigma)$$

式中，N_1为电枢绕组每相串联匝数；Λ_a为电枢反应磁通所经磁路的等效磁导；Λ_σ 为电枢漏磁通所经磁路的等效磁导；Λ_s为电枢总磁通的等效磁导。

（3）等效电路　根据式（5-9）可画出对应的等效电路，如图 5-18 所示。定子绕组接成 Y 形的三相等效电路如图 5-18a 所示，单相等效电路如图 5-18b 所示。

（4）相量图　根据式（5-9）可画出隐极式同步发电机的相量图，如图 5-19 所示。其中图 5-19a 为隐极式同步发电机带纯阻性负载时的相量图，图 5-19b 为隐极式同步发电机带感性负载（滞后）时的相量图，图 5-19c 为隐极式同步发电机带容性负载（超前）时的相量图。图中，励磁电动势 $\dot{E}_0$ 与端电压 $\dot{U}$ 的夹角称为功率角 δ，显然 $\delta = \psi_0 - \varphi$。

2. 考虑磁饱和

考虑磁饱和时，隐极式同步发电机的磁路是非线性的，叠加原理不再适用。此时应把主极磁动势 $\boldsymbol{F}_{f1}$ 与电枢磁动势 $\boldsymbol{F}_a$ 合成为一个合成磁动势 $\boldsymbol{F}$，由它产生气隙合成磁场的磁通 $\dot{\Phi}$，并在定子绕组中感应气隙电动势 $\dot{E}$。发电机电枢电流 $\dot{I}$ 还产生电枢漏磁通 $\dot{\Phi}_\sigma$，并在定子绕组中感应电枢漏电动势 $\dot{E}_\sigma$。上述各物理量之间的电磁关系可表示为

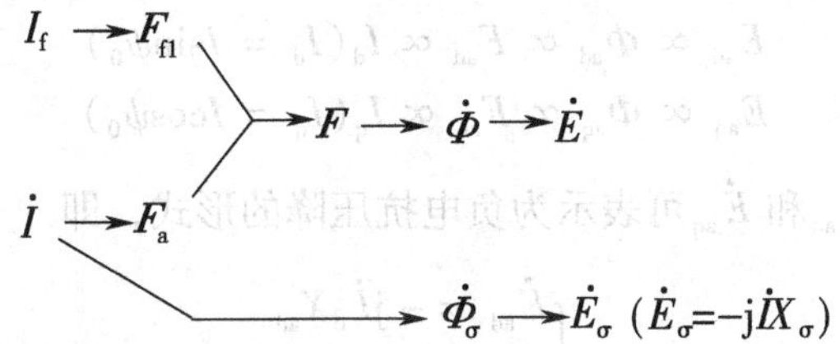

根据上述分析，得到的定子电压方程为

$$\dot{E} + \dot{E}_\sigma - \dot{I}R_a = \dot{U} \tag{5-10}$$

整理得

$$\dot{E} = \dot{U} + \dot{I}(R_a + jX_\sigma) \tag{5-11}$$

5.3.2　凸极式同步发电机的电压方程和相量图

凸极式同步电机的气隙是不均匀的，需要运用双反应理论来分析。

1. 双反应理论

在凸极同步电机中，当电枢电流为零，电机中只有主极励磁绕组产生的主极磁场，图 5-20a 为转子主极励磁磁动势 F_f、主极磁场 B_f 和主极基波磁场 B_{f1} 的分布示意图。由于凸极同步电机的气隙不均匀，在磁极下的气隙 δ 较小，而在两极之间的气隙 δ 较大，气隙的比磁导（即单位面积的气隙磁导 $\lambda = \mu_0/\delta$）是变化的，因此同样大小的电枢磁动势作用在气隙的不同位置时电枢反应就不相同。当电枢磁动势恰好作用在直轴或交轴位置时，电枢磁场的波形是对称的，其基波磁场的幅值也容易确定，如图 5-20b、c 所示。

一般情况下，电枢反应既不作用在直轴也不作用在交轴，电枢磁场的分布是不对称的，

电枢反应难以直接确定。此时可把电枢磁动势分解成直轴和交轴两个分量，然后分别求出直轴和交轴电枢磁动势所产生的电枢反应，再把它们的效果叠加起来。这种由于电机气隙不均匀把电枢反应分成直轴和交轴电枢反应来分别处理的方法，称为双反应理论。

2. 不考虑磁饱和

（1）负载时各物理量之间的电磁关系　不考虑磁饱和时，凸极式同步发电机的磁路为线性，此时可应用叠加原理和双反应理论，将主极磁动势 $\boldsymbol{F}_{f1}$、电枢磁动势 $\boldsymbol{F}_a$ 的直轴分量 $\boldsymbol{F}_{ad}$ 和交轴分量 $\boldsymbol{F}_{aq}$ 的作用分别单独考虑，由此得到负载时各物理量之间的电磁关系如下：

$$I_f \longrightarrow F_{f1} \longrightarrow \dot{\Phi}_0 \longrightarrow \dot{E}_0$$

$$\dot{I} \longrightarrow \dot{I}_d \longrightarrow F_{aq} \longrightarrow \dot{\Phi}_{ad} \longrightarrow \dot{E}_{ad}$$

$$\dot{I} \longrightarrow \dot{I}_q \longrightarrow F_{aq} \longrightarrow \dot{\Phi}_{aq} \longrightarrow \dot{E}_{aq}$$

$$\dot{E}_0,\ \dot{E}_{ad},\ \dot{E}_{aq} \longrightarrow \dot{E}$$

$$\dot{I} \longrightarrow \dot{\Phi}_\sigma \longrightarrow \dot{E}_\sigma\ (\dot{E}_\sigma = -j\dot{I}X_\sigma)$$

（2）电压方程　根据上述分析，可得

$$\dot{E}_0 + \dot{E}_{ad} + \dot{E}_{aq} + \dot{E}_\sigma - \dot{I}R_a = \dot{U} \tag{5-12}$$

与隐极式同步发电机类似，不计磁饱和时，有以下关系：

$$E_{ad} \propto \Phi_{ad} \propto F_{ad} \propto I_d (I_d = I\sin\psi_0)$$

$$E_{aq} \propto \Phi_{aq} \propto F_{aq} \propto I_q (I_q = I\cos\psi_0)$$

若不计定子铁损，$\dot{E}_{ad}$ 和 $\dot{E}_{aq}$ 可表示为负电抗压降的形式，即

$$\begin{cases} \dot{E}_{ad} = -j\dot{I}_d X_{ad} \\ \dot{E}_{aq} = -j\dot{I}_q X_{aq} \end{cases} \tag{5-13}$$

式中，X_{ad} 称为直轴电枢反应电抗，$X_{ad} = \dfrac{E_{ad}}{I_d}$；$X_{aq}$ 称为交轴电枢反应电抗，$X_{aq} = \dfrac{E_{aq}}{I_q}$。

将式（5-13）和式（5-8）代入式（5-12），可得

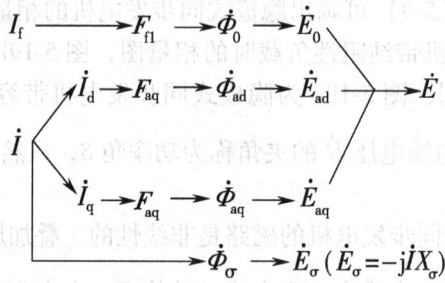

$$\begin{aligned} \dot{E}_0 &= \dot{U} + \dot{I}R_a + j\dot{I}_d X_{ad} + j\dot{I}_q X_{aq} + j\dot{I}X_\sigma \\ &= \dot{U} + \dot{I}R_a + j\dot{I}_d (X_{ad} + X_\sigma) + j\dot{I}_q (X_{aq} + X_\sigma) \\ &= \dot{U} + \dot{I}R_a + j\dot{I}_d X_d + j\dot{I}_q X_q \end{aligned} \tag{5-14}$$

式中，X_d 称为直轴同步电抗，是一个反映直轴电枢电流产生直轴电枢总磁通（包括直轴电枢反应磁通和电枢漏磁通）的参数，$X_d = X_{ad} + X_\sigma$；X_q 称为交轴同步电抗，是一个反映交轴电枢电流产生交轴电枢总磁通（包括交轴电枢反应磁通和电枢漏磁通）的参数，$X_q = X_{aq} + X_\sigma$。

（3）相量图　与式（5-14）对应的相量图如图 5-21 所示。需要说明的是在画此相量图时，为将电枢电流 $\dot{I}$ 分解成直轴分量 $\dot{I}_d$ 和交轴分量 $\dot{I}_q$，要先确定内功率因数角 ψ_0。实际上励磁电动势 $\dot{E}_0$（空载时存在）与电枢电流 $\dot{I}$（负载时存在）之间的夹角 ψ_0 是无法测出

的，只能通过几何作图方法来确定。

假设已画出凸极式同步发电机的相量图，如图5-22所示。如果从图中R点画垂直于相量$\dot{I}$的直线并与相量$\dot{E}_0$交于Q点，则得到一个虚拟电动势相量$\dot{E}_Q$，$\dot{E}_Q$必然与$\dot{E}_0$同相。不难看出线段RQ与相量$j\dot{I}_qX_q$之间的夹角就是ψ_0，线段RQ的长度为$I_qX_q/\cos\psi_0=IX_q$，由此可得虚拟电动势为

$$\dot{E}_Q = \dot{U} + \dot{I}R_a + j\dot{I}X_q \tag{5-15}$$

根据以上分析，凸极式同步发电机相量图的实际画法如下：

1）根据已知条件画出端电压相量$\dot{U}$和电枢电流相量$\dot{I}$。

2）在相量$\dot{U}$的末端叠加与相量$\dot{I}$平行的电枢电阻压降$\dot{I}R_a$和超前相量$\dot{I}$ 90°的相量$j\dot{I}X_q$，得到虚拟电动势相量$\dot{E}_Q=\dot{U}+\dot{I}R_a+j\dot{I}X_q$，由于$\dot{E}_Q$与未知的$\dot{E}_0$同相，因此$\dot{E}_Q$与$\dot{I}$之间的夹角即为$\psi_0$。

3）根据ψ_0，把相量$\dot{I}$分解成直轴分量$\dot{I}_d$和交轴分量$\dot{I}_q$。

4）在电枢电阻压降$\dot{I}R_a$的末端叠加超前交轴电流$\dot{I}_q$ 90°的交轴同步电抗压降$j\dot{I}_qX_q$和超前直轴电流$\dot{I}_d$ 90°的直轴同步电抗压降$j\dot{I}_dX_d$，得到终点T，把起点O和终点T连接起来，得到线段$\overline{OT}$即为励磁电动势相量$\dot{E}_0$。

从图5-22还可得出

$$\psi_0 = \arctan\frac{U\sin\varphi + IX_q}{U\cos\varphi + IR_a} \tag{5-16}$$

$$E_0 = E_Q + I_d(X_d - X_q) \tag{5-17}$$

例5-1 已知一台凸极式同步发电机，其直轴和交轴同步电抗的标幺值分别为$X_d^*=1.0$，$X_q^*=0.7$，忽略电枢电阻，不计磁饱和。试计算该发电机在额定电压、额定电流且$\cos\varphi=0.8$（滞后）时的励磁电动势E_0^*和功率角δ。

解：

取端电压作为参考相量，即设$\dot{U}^*=1\angle0°$，电枢电流为

$$\dot{I}^* = 1\angle -36.87°$$

虚拟电动势为

$$\dot{E}_Q^* = \dot{U}^* + j\dot{I}^*X_q^* = 1 + j0.7\angle -36.87° = 1.526\angle 21.52°$$

故功率角$\delta=21.52°$，于是内功率因数角为

$$\psi_0 = \delta + \varphi = 21.52° + 36.87° = 58.39°$$

电枢电流的直轴分量和交轴分量分别为

$$I_d^* = I^*\sin\psi_0 = 0.8516$$

$$I_q^* = I^*\cos\psi_0 = 0.5241$$

励磁电动势的标幺值为

$$E_0^* = E_Q^* + I_d^*(X_d^* - X_q^*) = 1.526 + 0.8516\times(1-0.7) = 1.781$$

5.4 同步发电机的功率方程和转矩方程

同步发电机利用气隙磁场将原动机输入的机械功率传递到定子，转换成电功率输出。下面按照功率的传递过程，对同步发电机内的各种有功功率和损耗进行分析，并导出功率方程和转矩方程。

1. 功率方程

原动机拖动同步发电机旋转，向发电机输入机械功率 P_1，扣除发电机的机械损耗 p_Ω、定子铁损 p_{Fe}和杂散损耗 p_Δ 后，余下部分将通过气隙磁场和电磁感应作用，转换成定子的电功率，这部分转换功率就是电磁功率 P_e，即

$$P_e = P_1 - p_\Omega - p_{Fe} - p_\Delta \tag{5-18}$$

若同步发电机带同轴励磁机，则上式中的 P_e 还应扣除转子励磁损耗（转子铜损）p_{Cuf}。

从电磁功率 P_e中扣除定子铜损 p_{Cua}后，可得发电机端口输出的电功率 P_2，即

$$P_2 = P_e - p_{Cua} \tag{5-19}$$

式中，$P_2 = mUI\cos\varphi$；$p_{Cua} = mI^2R_a$，m 为定子相数。

式（5-18）和式（5-19）即为同步发电机的功率方程。图 5-23 为同步发电机的功率流程。

2. 电磁功率

由式（5-19）可得

$$P_e = mUI\cos\varphi + mI^2R_a = mI(U\cos\varphi + IR_a)$$

由图 5-24 可知，$U\cos\varphi + IR_a = E\cos\psi = E_Q\cos\psi_0$，因此电磁功率可表示为

$$P_e = mEI\cos\psi = mE_QI\cos\psi_0 = mE_QI_q \tag{5-20}$$

式中，ψ 为气隙电动势 $\dot{E}$ 与电枢电流 $\dot{I}$ 的夹角。

对于隐极式同步发电机，由于 $X_d = X_q$，$E_Q = E_0$，故有

$$P_e = mE_0I\cos\psi_0 = mE_0I_q \tag{5-21}$$

由式（5-20）和式（5-21）可见，电枢电流的交轴分量 I_q越大，电磁功率 P_e就越大，因此同步电机进行机电能量转换时，其电枢电流必须含有交轴分量。交轴电枢反应越强，机电能量转换的电磁功率就越多。

3. 转矩方程

将功率方程式（5-18）两边除以同步机械角速度 Ω_s，即

$$\frac{P_1}{\Omega_s} = \frac{P_e}{\Omega_s} + \frac{p_\Omega + p_{Fe} + p_\Delta}{\Omega_s}$$

可导出同步发电机的转矩方程为

$$T_1 = T_e + T_0 \tag{5-22}$$

式中，T_1 为原动机驱动转矩，$T_1 = P_1/\Omega_s$；T_e 为电磁转矩，$T_e = P_e/\Omega_s$；T_0 为与机械损耗 p_Ω、定子铁损 p_{Fe}和杂散损耗 p_Δ 所对应的阻力转矩，$T_0 = (p_\Omega + p_{Fe} + p_\Delta)/\Omega_s$，若忽略杂散损耗，它就是空载转矩，此时 $T_0 = (p_\Omega + p_{Fe})/\Omega_s$。

5.5　同步发电机的运行特性

同步发电机的运行特性是确定发电机主要参数、评价发电机性能的基本依据。用试验方法测定的同步发电机运行特性包括空载特性、短路特性、外特性和调整特性等。利用空载特性和短路特性可以求得同步发电机的主要参数——同步电抗 X_s 的不饱和值；而从外特性和调整特性中可以确定表征同步发电机运行性能的两个重要数据——电压调整率 Δu 和额定励磁电流 I_{fN}。

1. 空载特性

空载特性可以通过空载试验测得。试验时，将电枢绕组开路，在同步转速下，测得励磁电动势 E_0 随励磁电流 I_f 变化的关系曲线 $E_0=f(I_f)$，即为同步发电机的空载特性，如图 5-25 中的空载特性曲线所示。

2. 短路特性

短路特性可以通过三相稳态短路试验测得。试验时，将电枢绕组三相端头短接，在同步转速下，测得电枢稳态短路电流 I 随励磁电流 I_f 变化的关系曲线 $I=f(I_f)$，即为同步电机的短路特性，如图 5-25 中的短路特性曲线所示。短路时同步发电机的时空矢量图如图 5-26 所示。

由图 5-25 可见，短路特性是一条直线。原因是：短路时 $U=0$，若忽略电枢电阻 R_a，则短路电流仅受同步发电机本身电抗的限制，所以短路电流是纯感性的，$\psi_0=90°$，$\dot{I}_q=0$，$\dot{I}=\dot{I}_d$，此时电压方程为

$$\dot{E}_0 = \dot{U} + \dot{I}R_a + j\dot{I}_dX_d + j\dot{I}_qX_q \approx j\dot{I}X_d \tag{5-23}$$

纯感性短路电流将产生直轴去磁电枢磁动势 $\boldsymbol{F}_{ad}$，使同步发电机合成磁动势 $\boldsymbol{F}$ 很小，故气隙电动势 $\dot{E}$ 很小，短路时

$$\dot{E} = \dot{U} + \dot{I}R_a + j\dot{I}X_\sigma \approx j\dot{I}X_\sigma \tag{5-24}$$

从上式可见，气隙电动势 $\dot{E}$ 仅与电枢漏抗压降相平衡，数值很小，因此短路时同步发电机铁心处于不饱和状态，磁路为线性，于是有 $E_0\propto I_f$，由式（5-23）可知短路电流 $I\propto E_0$，故有 $I\propto I_f$，短路特性是一条直线。

由式（5-23）可得

$$X_s = \frac{E_0}{I} \tag{5-25}$$

式中，E_0 为某一励磁电流下的励磁电动势，可从空载特性上查出；I 为同一励磁电流下对应的短路电流，可从短路特性上查出。

由于短路时同步发电机铁心处于不饱和状态，因此上式中的励磁电动势 E_0 应该从气隙线（不饱和时的空载特性）上查得，如图 5-25 所示，求出的 X_d 为不饱和值。

3. 短路比

短路比是汽轮发电机设计中的一个重要数据。短路比是指产生空载额定电压所需的励磁电流与产生短路额定电流所需的励磁电流之比，用 k_c 表示，即

$$k_c = \frac{I_{f0(U=U_{N\phi})}}{I_{fk(I=I_N)}} \tag{5-26}$$

参照图 5-25 可得

$$k_c = \frac{I_{f0(U=U_{N\phi})}}{I_{fk(I=I_N)}} = \frac{I}{I_N} = \frac{I}{E_0}\frac{E_0}{I_N} = \frac{1}{X_{d(不饱和)}}\frac{E_0}{U_{N\phi}}\frac{U_{N\phi}}{I_N} = \frac{1}{X_d}k_\mu Z_b = k_\mu \frac{1}{X^*_{d(不饱和)}}$$

式中，k_μ 为主磁路的饱和系数，$k_\mu = E_0/U_{N\phi}$；Z_b 为阻抗基值，$Z_b = U_{N\phi}/I_N$。

可见，短路比是一个计及饱和的参数。短路比大，即 X^*_d 小，则负载变化时同步发电机的电压变化较小，并联运行时发电机的稳定度较好，但此时气隙较大，转子的额定励磁安匝和用铜量增多，造价较高。反之，短路比小，则电压调整率大，稳定度较差，但造价较低。对于汽轮发电机，一般取 k_c为 0.4～1.0。

4. 外特性

外特性是指在同步转速下，保持励磁电流和负载功率因数为常值时，同步发电机端电压随电枢电流变化的关系，即 $n=n_s$，$I_f=$常值，$\cos\varphi=$常值时的 $U=f(I)$。

图 5-27 为带不同功率因数负载时同步发电机的外特性。从图 5-27 可见，带阻感性负载和纯电阻负载时，由于电枢反应的去磁作用和定子漏阻抗压降的影响使端电压减小，故外特性是下降的。带阻容性负载且满足内功率因数角 $\psi_0<0$ 时，由于电枢反应的增磁作用及容性电流的漏抗电压使端电压增大，故外特性是上升的。

从外特性可以求出发电机的电压调整率，如图 5-28 所示。调节发电机的励磁电流，使额定负载（$I=I_N$，$\cos\varphi=\cos\varphi_N$）时，发电机的端电压为额定电压 U_N，此励磁电流称为发电机的额定励磁电流 I_{fN}。保持励磁电流为 I_{fN}，转速为同步转速 n_s 不变，卸去负载（即 $I=0$），此时端电压升高的百分值称为同步发电机的电压调整率 Δu，即

$$\Delta u = \frac{E_0 - U_{N\varphi}}{U_{N\varphi}} \times 100\% \tag{5-27}$$

电压调整率是表征同步发电机运行性能的重要数据之一。通常凸极式同步发电机的 Δu 为 18%～30%，隐极式同步发电机的 Δu 为 30%～48%［均为 $\cos\varphi=0.8$（滞后）时的数值］。

5. 调整特性

当发电机的负载变化时，为保持端电压不变，必须调节发电机的励磁电流。调整特性是指在同步转速下，保持端电压为额定电压，功率因数为常值时，励磁电流与电枢电流的关系，即 $n=n_s$，$U=U_N$，$\cos\varphi$ 为常值时的 $I_f=f(I)$。

图 5-29 为带不同功率因数负载时同步发电机的调整特性。从图 5-29 可见，对于阻感性负载和纯电阻负载，为克服电枢反应的去磁作用和定子漏阻抗压降的影响，随着电枢电流的增加，必须相应增大励磁电流，故调整特性是上升的。对于阻容性负载且满足内功率因数角 $\psi_0<0$ 时，为了抵消电枢反应的增磁作用及容性电流的漏抗电压，随着电枢电流的增加，必须相应减小励磁电流，故外特性是下降的。

从调整特性可以确定同步发电机的额定励磁电流 I_{fN}，它是对应于额定电压、额定电流和额定功率因数时的励磁电流。

例 5-2 已知一台星形连接的汽轮发电机，$S_N=15000\text{kV}\cdot\text{A}$，$U_N=6.3\text{kV}$，$\cos\varphi_N=0.8$（滞后），不计磁饱和与电枢电阻，由空载和短路试验得到下列数据：

励磁电流 I_f/A	102
电枢电流 I（从短路特性上查得）/A	887
线电压 U_L（从空载特性上查得）/V	6300
线电压 U_L（从气隙线上查得）/V	8000

试求：

1）同步电抗的实际值和标幺值。

2）带额定负载时发电机的励磁电动势。

解：

1）求同步电抗的实际值和标幺值。从气隙线上查出，$I_f = 102A$ 时，励磁电动势（相电动势）$E_0 = 8000/\sqrt{3} = 4618.8V$；在同一励磁电流下，从短路特性上查出，短路电流 $I = 887A$；

同步电抗为

$$X_d(X_s) = \frac{E_0}{I} = \frac{4618.8}{887}\Omega = 5.207\Omega$$

额定相电流为

$$I_{N\phi} = \frac{S_N}{\sqrt{3}U_N} = \frac{15\ 000 \times 10^3}{\sqrt{3} \times 6300}A = 1374.6A$$

额定相电压为

$$U_{N\phi} = U_N/\sqrt{3} = 6300/\sqrt{3}V = 3637.3V$$

阻抗基值为

$$Z_b = \frac{U_{N\phi}}{I_{N\phi}} = \frac{3637.3}{1374.6}\Omega = 2.646\Omega$$

同步电抗的标幺值为

$$X_d^* = \frac{X_d}{Z_b} = \frac{5.207}{2.646} = 1.968$$

若采用标幺值计算

$$E_0^* = \frac{E_0}{U_{N\phi}} = \frac{8000/\sqrt{3}}{6300/\sqrt{3}} = 1.2698$$

$$I^* = \frac{I}{I_{N\phi}} = \frac{887}{1374.6} = 0.6453$$

$$X_d^* = \frac{E_0^*}{I^*} = \frac{1.2698}{0.6453} = 1.968$$

2）求额定负载时发电机的励磁电动势。取端电压作为参考相量

$$\dot{U}^* = 1\angle 0°, \dot{I}^* = 1\angle -36.87°$$

励磁电动势为

$$\dot{E}_0^* = \dot{U}^* + j\dot{I}^* X_s^* = 1 + j1.968\angle -36.87° = 2.689\angle 35.8°$$

励磁电动势（相电动势）的实际值为

$$E_0 = E_0^* \times U_{N\phi} = 2.689 \times 3637.3V = 9781V$$

5.6 同步发电机与电网的并联运行

现代电力系统一般总是由许多发电厂（包括利用不同能源发电的火电厂、水电站和核电站）并联组成，而每个发电厂或电站又由多台发电机并联在一起运行，因此电网的容量足够大，负载变化对电网电压和频率的影响很小，从而提高了供电的质量。

另外，采用并联运行还可以根据负载的变化统一调度，调整投入运行的机组数目，从而提高了机组的运行效率，特别是对水电站和火电厂并联的系统，可以充分利用水能，合理调度电能。在丰水期，主要由水电站输出大量廉价的电力，火电厂可以少发电；而在枯水季节，主要由火电厂提供电力，水电机组作调峰或作调相机运行，这样总的电能成本减少，从而保证了整个电力系统在最经济的条件下运行。

发电机并联运行，也便于合理安排定期的轮流检修，减少了电机检修和事故的备用容量，而且当一台发电机损坏时不至于引起停电事故，从而提高了供电的可靠性。

5.6.1 并联运行的条件

1. 投入并联的条件

首先分析两台同步发电机的并联运行情况，如图 5-30 所示。假设第一台发电机 G_1 已经通过负荷开关与三相负载连接，运行于额定频率的额定电压条件下，现要把第二台发电机 G_2 与 G_1 并联，以分担其负载。在并联过程中，为了减小对发电机的冲击，必须首先调节 G_2 的转速与转向，使其与 G_1 的转速和转向相同，再调节 G_2 的励磁电流，使 G_2 输出的 a′, b′和 c′三相之间的电压相等，在图 5-30 中，三个电灯 L 分别连接在 a′与 a、b′与 b、c′与 c 之间，当 G_1 与 G_2 之间的电压不等或频率不等时，L 会亮。当频率相等，而线电压的幅值不等时，L 会长亮不闪，且亮度相同，此时，若调节 G_2 的励磁电流，使三个 L 同时熄灭时，表明两台发电机达到了同步，此时，可以闭合并联开关。当两台发电机端电压的幅值相等，但频率不等时，L 会同时忽亮忽暗，发生闪烁，此时，应调节 G_2 原动机的转速，以减小 G_1 与 G_2 的频率差，当三个 L 同时熄灭的瞬间，即可将 G_2 投入与 G_1 并联运行。

将图 5-30 中的 G_1 用无穷大电网代替，即得到如图 5-31 所示的发电机与电网并联运行的示意图。

把同步发电机并联至电网的过程称为投入并联，或称为并车、整步。在投入并联时必须避免巨大的冲击电流，以防止同步发电机受到损坏和电网遭受干扰。为此，投入并联前必须检查发电机和电网是否满足以下 3 个条件：

1）发电机相序与电网相序相同。

2）发电机频率与电网频率相等。

3）发电机励磁电动势 $\dot{E}_0$ 与电网电压 $\dot{U}$ 相等（大小相等且相位相同）。

如果不满足上述条件就把发电机投入并联，将发生严重后果，下面对不满足并联条件的后果进行具体分析。

（1）励磁电动势 $\dot{E}_0$ 与电网电压 $\dot{U}$ 不等　图 5-32a 为发电机投入并联时的单相示意图。若 $\dot{E}_0$ 与 $\dot{U}$ 大小不等（见图 5-32b）或相位不同（见图 5-32c）时把发电机投入并联，则在开关 S 的两端将产生电压差 $\Delta\dot{U}$，如果闭合 S，在发电机和电网组成的回路中必然会出现瞬态冲击电流。冲击电流过大时，电流产生的电磁力将使定子绕组端部受到损伤，转轴上受到冲击力矩。

（2）频率不等　如果发电机与电网的频率不等，则相量 $\dot{E}_0$ 与 $\dot{U}$ 之间的相位差将在 0°~360°之间变化，开关 S 两端的电压差 $\Delta\dot{U}$ 忽大忽小，会产生拍振电流，在发电机内引起功率振荡。

（3）相序不同　如果发电机与电网的相序不同，绝不允许并联。因为这是一种定子端加上一组负序电压的严重故障情况，如果投入并联，发电机内将产生强大环流和机械冲击，从而使发电机遭到毁坏。

5.6.2　凸极式同步发电机的功角特性

同步发电机接在电网上稳态运行时，在恒定励磁和恒定电网电压（即 E_0 = 常值，U = 常值）的条件下，发电机发出的电磁功率 P_e 与功率角 δ 之间的关系 $P_e = f(\delta)$，称为同步发电机的功角特性。利用功角特性可以研究同步发电机与电网并联运行时，有功功率的调节和静态稳定问题。

由于大、中型同步发电机的电枢电阻远小于其同步电抗，因此常常忽略不计。当忽略电枢电阻且不计磁饱和时，同步发电机的相量图如图 5-33 所示，此时电磁功率为

$$\begin{aligned} P_e \approx P_2 &= mUI\cos\varphi = mUI\cos(\psi_0 - \delta) \\ &= mUI(\cos\psi_0\cos\delta + \sin\psi_0\sin\delta) \\ &= mU(I_q\cos\delta + I_d\sin\delta) \end{aligned} \tag{5-28}$$

由图 5-33 可知

$$\begin{cases} I_q X_q = U\sin\delta \\ I_d X_d = E_0 - U\cos\delta \end{cases} \tag{5-29}$$

于是

$$\begin{cases} I_q = \dfrac{U\sin\delta}{X_q} \\ I_d = \dfrac{E_0 - U\cos\delta}{X_d} \end{cases} \tag{5-30}$$

将式（5-30）代入式（5-28），可得

$$P_e = m\frac{E_0 U}{X_d}\sin\delta + m\frac{U^2}{2}\left(\frac{1}{X_q} - \frac{1}{X_d}\right)\sin 2\delta = P_{e1} + P_{e2} \tag{5-31}$$

式中，P_{e1} 为基本电磁功率，$P_{e1} = m\dfrac{E_0 U}{X_d}\sin\delta$；$P_{e2}$ 为附加电磁功率（也称为磁阻功率，直、交轴磁阻不等造成的），$P_{e2} = m\dfrac{U^2}{2}\left(\dfrac{1}{X_q} - \dfrac{1}{X_d}\right)\sin 2\delta$。

根据式（5-31）可绘出同步电机的功角特性，如图 5-34 所示。从图 5-34 可见，$0° \leqslant \delta \leqslant 180°$ 时，电磁功率为正值，对应于发电机状态；$-180° \leqslant \delta \leqslant 0°$ 时，电磁功率为负值，对应于电动机状态。

由式（5-31）和图 5-34 可见，$P_{e1max}=m\frac{E_0U}{X_d}$，发生在 $\delta=90°$时；$P_{e2max}=m\frac{U^2}{2}\left(\frac{1}{X_q}-\frac{1}{X_d}\right)$；发生在 $\delta=45°$时；而最大电磁功率 P_{emax} 发生在 $45°<\delta<90°$之间，且 $P_{emax}\neq P_{e1max}+P_{e2max}$。

对于隐极式同步发电机，由于 $X_d=X_q=X_s$，附加电磁功率 P_{e2}等于零，故电磁功率 P_e等于基本电磁功率 P_{e1}，即

$$P_e = m\frac{E_0U}{X_s}\sin\delta \tag{5-32}$$

$\delta=90°$时，有最大电磁功率

$$P_{emax} = m\frac{E_0U}{X_s} \tag{5-33}$$

如图 5-33 所示，设主磁场 $\boldsymbol{B}_0$ 与气隙合成磁场 $\boldsymbol{B}$ 之间的空间夹角为 δ'，在时空矢量图中，由于主磁场 $\boldsymbol{B}_0$ 超前励磁电动势 $\dot{E}_0$90°电角度，气隙合成磁场 $\boldsymbol{B}$ 也超前气隙电动势 $\dot{E}$90°电角度，因此励磁电动势 $\dot{E}_0$ 与气隙电动势 $\dot{E}$ 之间的时间夹角也为 δ'。由于漏阻抗压降很小，气隙电动势 $\dot{E}$ 与端电压 $\dot{U}$ 之间的夹角常忽略不计，可近似认为 $\delta'\approx\delta$，由于时空矢量图中常用相量 $\dot{U}$ 而不是气隙电动势 $\dot{E}$，因此 δ 比 δ'更实用。通常认为励磁电动势 $\dot{E}_0$ 与端电压 $\dot{U}$ 之间的时间夹角 δ 也是主磁场 $\boldsymbol{B}_0$ 与气隙合成磁场 $\boldsymbol{B}$ 之间的空间夹角，这表明功率角 δ 具有双重物理意义。它既可以表示励磁电动势 $\dot{E}_0$ 与端电压 $\dot{U}$ 之间的时间相位关系，具有时间含义；又可以表示主磁场 $\boldsymbol{B}_0$ 与气隙合成磁场 $\boldsymbol{B}$ 之间的空间相位关系，具有空间含义。

5.6.3 有功功率的调节和静态稳定

同步发电机与电网并联运行时，将向电网输送有功功率和无功功率，下面分别分析如何调节发电机输出的有功功率和无功功率。为简便计，以隐极式同步发电机为例，不计磁饱和与电枢电阻，电网看做“无穷大电网”，即电网电压 U 为常数，频率 f 为常数。

1. 有功功率的调节

当与电网并联运行的发电机的输出有功功率 P_2为零时，通常称发电机处于空载状态。若不计电枢电阻，电磁功率 $P_e=P_2=0$，由式（5-32）可知，功率角 $\delta=0$，如图 5-35a 所示，励磁电动势 $\dot{E}_0$ 与电网电压 $\dot{U}$ 同相且超前电枢电流 $\dot{I}$ 90°，电枢电流全部是无功电流。此时，原动机输入的机械功率 P_1仅用来克服发电机的空载损耗，即 $P_1=p_\Omega+p_{Fe}=p_0$，驱动转矩 T_1等于空载转矩 T_0。

若增加原动机的输入功率 P_1，则驱动转矩 T_1随之增大，产生剩余转矩 T_1-T_0作用在转轴上，使转子加速，于是发电机的主磁场 $\boldsymbol{B}_0$（与转子 d 轴重合）开始超前气隙合成磁场 $\boldsymbol{B}$（该磁场受发电机端口频率不变的约束，转速保持不变）。相应地，励磁电动势 $\dot{E}_0$ 开始超前电机端电压（电网电压）$\dot{U}$，如图 5-35b 所示。此时功率角 $\delta>0$，故电磁功率 $P_e>0$，发电机开始向电网输出有功功率，同时产生制动电磁转矩 T_e作用在转子上，T_e随 δ 增大而增大。

最终功率角稳定在某一数值 δ_a 上，电磁转矩 T_e 与剩余转矩 $T_1 - T_0$ 达到一个新的平衡状态，转子转速仍为同步转速，如图 5-35c 功角特性上的 a 点所示。此时，原动机输入的机械功率 P_1 扣除空载损耗 p_0 后，余下的功率就是电磁功率 P_e，即

$$P_1 - p_0 = P_e = m\frac{E_0 U}{X_s}\sin\delta_a \tag{5-34}$$

忽略电枢电阻时，电磁功率 P_e 即为发电机输出的有功功率 P_2，此时发电机处于负载运行状态。

上述分析表明，增加原动机的输入功率时，功率角 δ 将随之增大，发电机的电磁功率和输出功率便会相应增加。可见，要想增加发电机输出的有功功率，必须增加原动机的输入功率，这是符合能量守恒定律的。但要注意当 $\delta = 90°$ 时，电磁功率达到最大值 $P_{emax} = m\frac{E_0 U}{X_s}$，$P_{emax}$ 称为发电机的功率极限。如果继续增加原动机的输入功率，由于 $\delta > 90°$，发电机输出的有功功率反而减少，此时发电机达不到一个新的平衡状态，最终将失去同步。

如果保持励磁电流 I_f 为常数时调节发电机输出的有功功率，当不计磁饱和时，保持励磁电流 I_f 不变即励磁电动势 E_0 不变，调节发电机有功功率时，功率角 δ 随之改变，发电机输出的无功功率也会发生改变，甚至会变为负值，即无功功率的性质由感性变为容性。如果保持发电机输出无功功率 Q 为常数时调节有功功率，此时励磁电动势 E_0 和功率角 δ 均会发生变化。

2. 静态稳定

与电网并联运行的同步发电机，原先在某一工作点上稳定运行，当外界（电网或原动机）发生微小扰动时，工作点发生了变化，若扰动消失后，发电机能自行回复到原先的状态下运行，就是静态稳定的，反之，则是不稳定的。

下面利用图 5-36 来分析隐极式同步发电机的静态稳定问题。设最初原动机输入的有效机械功率为 $P_T = P_1 - p_0$，它与功角特性有 A、C 两个交点，都满足功率平衡关系 $P_T = P_e$，但 A、C 两点是否是静态稳定的，还需进一步分析。

假定发电机开始时在 A 点运行，功率角为 δ。当某种微小扰动使原动机的有效功率增加 ΔP_T 时，功率角将随之增加 $\Delta\delta$，相应地，电磁功率也增加 ΔP_e，最终在 B 点达到一个新的平衡状态并满足功率平衡关系 $P_T + \Delta P_T = P_e + \Delta P_e$，如图 5-36 中的 B 点所示。当扰动消失后，由于 $P_T < P_e + \Delta P_e$，使输入的有效驱动转矩小于制动的电磁转矩，于是转子减速、功率角又减小，最终使发电机回复到 A 点稳定运行。

若发电机原先在 C 点运行，功率角为 δ'，当某种微小扰动使原动机的有效功率增加 $\Delta P_T'$ 时，功率角将随之增大 $\Delta\delta'$，即图 5-36 中的 D 点所示。电磁功率没有增加反而减少了 $\Delta P'_e$，$P_T + \Delta P_T > P_e - \Delta P'_e$，使输入的有效驱动转矩大于制动的电磁转矩，于是转子继续加速而无法达到新的平衡。即使扰动消失，原动机输入的有效功率 P_T 也大于电磁功率而使转子继续加速，最终导致发电机失去同步，所以 C 点是不稳定的。

上述分析表明，在功角特性的上升部分即 $0° \leqslant \delta < 90°$ 时，发电机是静态稳定的，此时 $\frac{dP_e}{d\delta} > 0$；而在功角特性的下降部分即 $90° < \delta \leqslant 180°$ 时，发电机是不稳定的，此时 $\frac{dP_e}{d\delta} < 0$；在 $\delta = 90°$ 时，发电机处于稳定和不稳定的交界点，此时 $\frac{dP_e}{d\delta} = 0$，称为静态稳定极限。由此可

见，根据$\frac{dP_e}{d\delta}$的正负可以判断同步发电机是否稳定。在$0° \leqslant \delta < 90°$稳定运行区内，工作点离静态稳定极限越远，稳定程度越高，即δ越小，$\frac{dP_e}{d\delta}$越大，发电机稳定性越好。因此根据$\frac{dP_e}{d\delta}$的大小可以衡量同步发电机的稳定程度。通常把$\frac{dP_e}{d\delta}$称为同步电机的比整步功率，用P_{syn}表示，如图 5-36 中的虚线所示。对于隐极式同步发电机：

$$P_{syn} = \frac{dP_e}{d\delta} = m\frac{E_0 U}{X_s}\cos\delta \tag{5-35}$$

为使发电机能够稳定运行，发电机的额定运行点应当距其稳定极限有一定的距离，即发电机的最大电磁功率应比其额定功率大一定的倍数。发电机的最大电磁功率与额定功率之比称为过载能力，用k_p表示。对于隐极式同步发电机：

$$k_p = \frac{P_{emax}}{P_N} \approx \frac{m\frac{E_0 U}{X_s}}{m\frac{E_0 U}{X_s}\sin\delta_N} = \frac{1}{\sin\delta_N} \tag{5-36}$$

式中，δ_N为同步发电机额定运行时的功率角。对于汽轮发电机，一般设计δ_N为30°～40°，k_p为1.6～2.0。

从式（5-33）和式（5-35）可见，增加励磁（即增大E_0）和减小同步电抗X_s对提高同步发电机的功率极限和静态稳定度是有利的，可作为发电机设计和运行的准则之一。

例 5-3 一台8750kV·A、11kV、50Hz、星形连接的三相水轮发电机并联于无穷大电网，额定运行时功率因数为0.8（滞后），每相同步电抗$X_d = 17\Omega$，$X_q = 9\Omega$，忽略电枢电阻，不计磁饱和，试求：

1）同步电抗的标幺值。

2）该发电机在额定运行时的功率角δ和励磁电动势E_0。

3）该发电机的最大电磁功率P_{emax}、过载能力k_p及产生最大电磁功率时的功率角δ。

解：

1）求同步电抗的标幺值。额定相电流为

$$I_{N\phi} = \frac{8750}{\sqrt{3} \times 11}\text{A} = 459.3\text{A}$$

额定相电压为

$$U_{N\phi} = \frac{11\,000}{\sqrt{3}}\text{V} = 6350.9\text{V}$$

同步电抗的标幺值为

$$X_d^* = X_d\frac{I_{N\phi}}{U_{N\phi}} = 17 \times \frac{459.3}{6350.9} = 1.229$$

$$X_q^* = X_q\frac{I_{N\phi}}{U_{N\phi}} = 9 \times \frac{459.3}{6350.9} = 0.6509$$

2）求额定运行时的功率角δ和励磁电动势E_0。

采用标幺值计算，以端电压作为参考相量

$$\dot{U}^* = 1\angle 0°, \dot{I}^* = 1\angle -36.87°$$

虚拟电动势为

$$\dot{E}_Q^* = \dot{U}^* + j\dot{I}^* X_q^* = 1 + j0.6509\angle -36.87° = 1.485\angle 20.53°$$

故功率角 $\delta = 20.53°$，于是内功率因数角为

$$\psi_0 = \delta + \varphi = 20.53° + 36.87° = 57.4°$$

电枢电流的直轴和交轴分量分别为

$$I_d^* = I^* \sin\psi_0 = \sin 57.4° = 0.8425$$

$$I_q^* = I^* \cos\psi_0 = \cos 57.4° = 0.5388$$

励磁电动势的标幺值为

$$E_0^* = U^* \cos\delta + I_d^* X_d^* = \cos 20.53° + 0.8425 \times 1.229 = 1.972$$

励磁电动势（相电动势）的实际值为

$$E_0 = E_0^* U_{N\phi} = 1.972 \times 6350.9\text{V} = 12\ 524\ \text{V}$$

3）求最大电磁功率 P_{emax}、过载能力 k_p 及产生最大电磁功率时的功率角 δ。

电磁功率的标幺值为

$$P_e^* = \frac{E_0^* U^*}{X_d^*}\sin\delta + \frac{U^{*2}}{2}\left(\frac{1}{X_q^*} - \frac{1}{X_d^*}\right)\sin 2\delta$$

$$= \frac{1.972}{1.229}\sin\delta + \frac{1^2}{2}\left(\frac{1}{0.6509} - \frac{1}{1.229}\right)\sin 2\delta$$

$$= 1.605\sin\delta + 0.3613\sin 2\delta$$

注意：用标幺值表示时，式中无相数，因为功率基值是三相额定视在功率 $S_N = 3U_{N\phi}I_{N\phi}$。

令 $\frac{dP_e^*}{d\delta} = 0$，即

$$\frac{dP_e^*}{d\delta} = 1.605\cos\delta + 0.7226\cos 2\delta = 0$$

整理得

$$1.605\cos\delta + 0.7226\ (2\cos^2\delta - 1) = 0$$

$$1.445\cos^2\delta + 1.605\cos\delta - 0.7226 = 0$$

解得

$$\cos\delta = \frac{-1.605 + \sqrt{1.605^2 + 4 \times 1.445 \times 0.7226}}{2 \times 1.445} = \frac{-1.605 + 2.599}{2.89} = 0.3439$$

产生最大电磁功率时的功率角为

$$\delta_m = \arccos 0.3439 = 69.89°$$

对应的最大电磁功率标幺值为

$$P_{emax}^* = 1.605\sin 69.89° + 0.3613\sin 139.78° = 1.74$$

最大电磁功率为

$$P_{emax} = P_{emax}^* S_N = 1.74 \times 8750\text{kW} = 15\ 225\text{kW}$$

过载能力为

$$k_p = \frac{P_{emax}}{P_N} = \frac{1.74}{0.8} = 2.175$$

5.6.4 无功功率的调节和V形曲线

1. 无功功率的调节

调节发电机输出的无功功率时，如果保持原动机的输入功率P_1不变，根据功率平衡关系可知，发电机的电磁功率P_e和输出的有功功率P_2将近似保持不变，即

$$\begin{cases} P_e = m\dfrac{E_0 U}{X_s}\sin\delta = 常数 \\ P_2 = mUI\cos\varphi = 常数 \end{cases} \tag{5-37}$$

由于m、U、X_s均为定值，所以上式可改写为

$$\begin{cases} E_0\sin\delta = 常数 \\ I\cos\varphi = 常数 \end{cases} \tag{5-38}$$

假定保持发电机输出有功功率P_2为常数时调节无功功率，此时若调节发电机的励磁电流I_f，励磁电动势E_0将随之变化，由于$E_0\sin\delta$为常数，因此功率角δ也将发生变化。已知电网电压U及频率f均为常数，即发电机端电压$\dot{U}$不变，根据电压方程式（5-9）可知，E_0变化必将引起电枢电流$\dot{I}$变化，由于$I\cos\varphi$为常数，因此功率因数角φ也会改变，此时发电机输出的无功功率$Q=mUI\sin\varphi$必然会改变。上述分析表明，调节发电机的励磁电流就可以调节其输出的无功功率。

图5-37给出了励磁电流不同时发电机的相量图。当调节励磁电流I_f使励磁电动势E_0变化时，由于$E_0\sin\delta$为常数，故相量$\dot{E}_0$的末端轨迹是一条与$\dot{U}$平行的直线$\overline{AB}$；又由于$I\cos\varphi$为常数，故电枢电流$\dot{I}$的末端轨迹是一条与$\dot{U}$垂直的直线$\overline{CD}$。当励磁电流为I_{f2}时，励磁电动势为$\dot{E}_{02}$，电枢电流为$\dot{I}_2$，此时发电机的功率因数$\cos\varphi=1$，称发电机处于“正常励磁”状态；由于$\dot{I}_2$与$\dot{U}$同相，发电机只输出有功功率。若增加励磁电流到$I_{f1}>I_{f2}$，则励磁电动势增加到$\dot{E}_{01}$，电枢电流变为$\dot{I}_1$，称发电机处于“过励”状态；此时$\dot{I}_1$滞后于$\dot{U}$，发电机功率因数$\cos\varphi<1$（滞后），发电机不仅输出有功功率，还将输出感性无功功率。若减小励磁电流到$I_{f3}<I_{f2}$，则励磁电动势减小到$\dot{E}_{03}$，电枢电流变为$\dot{I}_3$，称发电机处于“欠励”状态；此时$\dot{I}_3$超前于$\dot{U}$，发电机功率因数$\cos\varphi<1$（超前），发电机输出有功功率和容性无功功率。若继续减小励磁电流到I_{f4}使励磁电动势减小到$\dot{E}_{04}$时，$\delta=90°$，发电机将达到静态稳定极限；若进一步减小励磁电流，发电机将失去同步。

2. V形曲线

保持发电机有功功率P_2（或电磁功率P_e）不变，改变励磁电流I_f，可得到电枢电流I随励磁电流I_f变化的关系曲线$I=f(I_f)$，如图5-38所示。因该曲线形似“V”字，故称为V形曲线。每一个恒定的有功功率值P_2，都有一条对应的V形曲线，功率值P_2越大，曲线位置越往上移。在每条曲线的最低点，励磁电流I_f为“正常励磁”值，此时$\cos\varphi=1$，电枢电流I最小，全部为有功分量，这时无论增大还是减小励磁电流，电枢电流都会增加。将各条曲线的最低点连接起来，可得到$\cos\varphi=1$的一条曲线，如图5-38中间的虚线所示，在这条曲线的右侧，发电机处于“过励”状态，功率因数是滞后的，发电机向电网输出感性无功

功率；而在这条曲线的左侧，发电机处于“欠励”状态，功率因数是超前的，发电机向电网输出容性无功功率。V 形曲线左侧还存在着一个不稳定区（对应于 $\delta>90°$），由于欠励区域更靠近不稳定区，因此，同步发电机不宜在过于欠励的状态下运行。

5.7　同步电动机

与其他电机一样，同步电机既可作发电机运行，亦可作电动机运行。同步电动机的优点是，接于频率一定的电网上运行时，其转速恒定，不随负载变化而改变；另外，它的功率因数可以调节。在需要改变功率因数和不需要调速的场合，如大型空气压缩机、粉碎机、离心泵等，常常优先采用同步电动机。同步电动机还有一种特殊的运行方式，即作空载运行、专门调节电网无功功率以此来改善电网功率因数的同步补偿机（又称调相机）。

5.7.1　电压方程、相量图和等效电路

若采用发电机惯例，当功率角 δ 为负值时，主极磁场滞后于气隙合成磁场，励磁电动势 $\dot{E}_0$ 滞后于端电压 $\dot{U}$，同步电机处于电动机状态，此时电磁功率 P_e 为负值，同步电机向电网输出负的电功率 $P_2=mUI\cos\varphi$，故功率因数角 $\varphi>90°$。显然用负的功率角和负的电磁功率分析同步电动机是极不方便的。分析同步电动机时通常采用电动机惯例，规定以输入电流作为电枢电流的正方向，电网向电动机输入正的电功率 $P_1=mUI\cos\varphi$，此时功率因数角 $\varphi>0°$，电磁功率 P_e 和功率角 δ 亦为正值。此时隐极式同步电动机的相量图和等效电路如图 5-39 所示。

根据图 5-39b 所示同步电动机各物理量的正方向，可得隐极式同步电动机的电压方程为

$$\dot{U}=\dot{E}_0+\dot{I}R_a+j\dot{I}X_s \tag{5-39}$$

同理，凸极式同步电动机的电压方程为

$$\dot{U}=\dot{E}_0+\dot{I}R_a+j\dot{I}_dX_d+j\dot{I}_qX_q \tag{5-40}$$

对应的凸极式同步电动机的相量图如图 5-40 所示。

5.7.2　功角特性、功率方程和转矩方程

采用电动机惯例分析同步电动机时，电磁功率 P_e 和功率角 δ 均为正值，即规定励磁电动势 $\dot{E}_0$ 滞后于端电压 $\dot{U}$ 的功率角为正。此时同步电动机的功角特性表达式与同步发电机的功角特性表达式（5-31）完全一样，即

$$P_e=m\frac{E_0U}{X_d}\sin\delta+m\frac{U^2}{2}\left(\frac{1}{X_q}-\frac{1}{X_d}\right)\sin2\delta \tag{5-41}$$

按照电动机惯例，$0°\leqslant\delta\leqslant180°$ 时，电磁功率为正值，对应于电动机状态。同步电动机运行时，电网向电动机输入电功率 P_1，扣除电枢绕组铜损 p_{Cua} 后，余下部分将通过气隙磁场和电磁感应作用，转换为机械功率，这部分功率称为电磁功率 P_e，即

$$P_e=P_1-p_{Cua} \tag{5-42}$$

从电磁功率 P_e 中扣除电动机的机械损耗 p_Ω、定子铁损 p_{Fe} 和杂散损耗 p_Δ 后，余下部分就是转轴上输出的机械功率 P_2，即

$$P_2 = P_e - p_\Omega - p_{Fe} - p_\Delta \tag{5-43}$$

将功率方程（5-43）两边除以同步机械角速度 Ω_s，即

$$\frac{P_e}{\Omega_s} = \frac{P_2}{\Omega_s} + \frac{p_\Omega + p_{Fe} + p_\Delta}{\Omega_s}$$

可导出同步电动机的转矩方程

$$T_e = T_2 + T_0 \tag{5-44}$$

式中，T_e 为电磁转矩，$T_e = \frac{P_e}{\Omega_s}$；$T_2$ 为负载转矩，$T_2 = \frac{P_2}{\Omega_s}$；$T_0$ 为与机械损耗 p_Ω、定子铁损 p_{Fe} 和杂散损耗 p_Δ 对应的阻力转矩，$T_0 = \frac{p_\Omega + p_{Fe} + p_\Delta}{\Omega_s}$，若忽略杂散损耗，它就是空载转矩，此时 $T_0 = \frac{p_\Omega + p_{Fe}}{\Omega_s}$。

5.7.3 无功功率和功率因数的调节

同步电动机与电网并联运行时，将从电网吸收有功功率和无功功率，调节同步电动机吸收有功功率和无功功率的过程与同步发电机类似，下面只分析如何调节同步电动机的无功功率以改善电网的功率因数。为简便计，仍以隐极式同步电动机为例，不计磁饱和与电枢电阻，电网看做“无穷大电网”，即电网电压 U 及频率 f 为常数。

调节同步电动机从电网吸取的无功功率时，如果保持负载转矩 T_2 不变，根据转矩平衡关系可知，同步电动机的电磁转矩 T_e 和相应的电磁功率 P_e 基本不变，忽略电枢绕组铜损，同步电动机从电网输入的有功功率 P_1 也基本不变，即

$$\begin{cases} P_e = m\dfrac{E_0 U}{X_s}\sin\delta = \text{常数} \\ P_1 = mUI\cos\varphi = \text{常数} \end{cases} \tag{5-45}$$

由于 m、U、X_s 均为定值，所以上式可改写为

$$\begin{cases} E_0\sin\delta = \text{常数} \\ I\cos\varphi = \text{常数} \end{cases} \tag{5-46}$$

图 5-41 给出了励磁电流不同时同步电动机的相量图。当调节励磁电流 I_f 使励磁电动势 E_0 变化时，相量 $\dot{E}_0$ 的末端轨迹是一条与 $\dot{U}$ 平行的直线 $\overline{AB}$，而电枢电流 $\dot{I}$ 的末端轨迹是一条与 $\dot{U}$ 垂直的直线 $\overline{CD}$。由图可见，当励磁电流为“正常励磁”值时，励磁电动势为 $\dot{E}_0$，同步电动机的功率因数 $\cos\varphi = 1$，电枢电流 $\dot{I}$ 全部为有功分量，同步电动机只从电网吸收有功功率。当励磁电流大于“正常励磁”值（过励）时，励磁电动势增加到 $\dot{E}_0'$，电枢电流变为超前于 $\dot{U}$ 的 $\dot{I}'$，此时同步电动机功率因数 $\cos\varphi < 1$（超前），同步电动机从电网吸收有功功率和容性无功功率。当励磁电流小于“正常励磁”值（欠励）时，励磁电动势减小到 $\dot{E}''_0$，电枢电流变为滞后于 $\dot{U}$ 的 $\dot{I}''$，此时同步电动机功率因数 $\cos\varphi < 1$（滞后），同步电动机从电网吸收有功功率和感性滞后无功功率。

与同步发电机一样，保持同步电动机电磁功率 P_e 不变，也可以得到一族 V 形曲线 $I=f(I_f)$，如图 5-42 所示。在 $\cos\varphi=1$ 这条曲线的右侧，同步电动机同样处于“过励”状态，但其功率因数为超前（与同步发电机相反）；在这条曲线的左侧，同步电动机同样处于“欠励”状态，但功率因数为滞后（与同步发电机相反）。对应于某一负载，当励磁电流减小到一定数值时，$\delta>90°$，同步电动机将不能稳定运行而失去同步。

上述分析表明，当不调节电动机拖动的负载转矩（T_2 为常数）时，同步电动机从电网吸取的有功功率 P_1 基本不变，而同步电动机从电网吸收的无功功率和同步电动机的功率因数可以通过改变励磁电流来调节，这是同步电动机可贵的特点，常用来改善电网的功率因数。一般情况下，电网的大部分负载是感应电动机和变压器，它们都要从电网吸收感性无功功率，如果使运行在电网上的同步电动机工作在“过励”状态，它从电网吸收容性无功功率，从而提高了电网的功率因数。现代同步电动机的额定功率因数一般设计为 1 ~ 0.8（超前）。

同步补偿机（又称调相机）就是一种不带机械负载，专门用于改善电网功率因数的同步电动机。

5.7.4 同步电动机的起动

同步电动机仅在同步转速时，定子旋转磁场与转子主极磁场才能在空间上保持相对静止，产生恒定的同步电磁转矩。起动时若把定子直接投入电网，转子加上直流励磁，则定子旋转磁场将以同步转速旋转，而转子磁场静止不动，定子、转子磁场之间具有相对运动，因此作用在转子上的电磁转矩正、负交变，平均转矩为零，同步电动机不能自行起动。所以，必须借助于其他方法使同步电动机起动起来。

同步电动机常用的起动方法有下列 3 种。

（1）用辅助电动机起动　通常选用与同步电动机极数相同的感应电动机（容量约为主机的 10% ~ 15%）作为辅助电动机。先用辅助电动机将主机拖动到接近同步转速，再用自整步法将主机投入电网，并切断辅助电动机电源。这种方法只适合于空载起动，而且所需设备多，操作复杂。

（2）变频起动　起动时，同步电动机转子加上励磁，把变频电源的频率调得很低，使同步电动机投入电源后，定子旋转磁场转得极慢。这样，依靠定、转子磁场之间相互作用所产生的同步电磁转矩，使同步电动机开始起动，并在很低的同步转速下运转。之后逐步提高电源的频率，使定子旋转磁场和转子的转速逐步加快，一直到额定转速为止。变频起动过程平稳、性能良好，但这种方法必须有变频电源，而且励磁机必须是非同轴的，否则在最初低速运转时无法产生所需的励磁电压。

（3）异步起动　为实现异步起动，需在同步电动机的主极极靴上装设起动绕组，起动绕组类似于感应电动机转子上的笼型绕组。起动时，先把励磁绕组通过限流电阻短接，再把定子绕组接到三相交流电网上，此时定子旋转磁场与转子起动绕组中的感应电流相互作用，产生异步电磁转矩，使同步电动机起动起来。待转速上升到接近于同步转速时，再把励磁绕组接入励磁电源，使转子建立主极磁场，此时依靠定、转子磁场相互作用所产生的同步电磁转矩，再加上由于凸极效应所引起的磁阻转矩，通常可将转子牵入同步。目前，多数同步电动机都采用异步起动法来起动。

注意，异步起动时励磁绕组不能开路，否则起动时定子旋转磁场会在匝数较多的励磁绕组中感应出高电压，易使励磁绕组击穿或引起人身事故；但也不能直接短路，否则励磁绕组（相当于一个单相绕组）中的感应电流与气隙磁场相互作用，将会产生显著的“单轴转矩”，使合成电磁转矩在$\frac{1}{2}n_s$附近产生明显的下凹，使电动机的转速停滞在$\frac{1}{2}n_s$附近而不能继续上升。为减小单轴转矩，通常将励磁绕组串接一限流电阻，其阻值为励磁绕组本身电阻的5～10倍。

本章小结

同步电机通常作发电机运行。汽轮发电机和水轮发电机在结构上有较大差别：汽轮发电机采用2极隐极式转子结构，机身细长，转动惯量较小，适合于用汽轮机高速拖动；而水轮发电机采用多极凸极式转子结构，机身扁平，适合于用水轮机低速拖动。

同步发电机空载运行时，主极磁动势的基波$\boldsymbol{F}_{f1}$建立主极磁场$\boldsymbol{B}_0$（即空载时的气隙磁场），该磁场以同步转速旋转切割定子导体，在定子三相绕组中产生三相对称感应电动势，称为励磁电动势。励磁电动势（相电动势）的有效值E_0由励磁电流I_f决定，E_0与I_f之间的关系曲线$E_0=f(I_f)$称为同步发电机的空载特性。

同步发电机带对称负载运行时，电枢绕组中流过三相对称电流，产生以同步转速旋转的电枢磁动势，电枢磁动势的基波$\boldsymbol{F}_a$对主极磁场$\boldsymbol{B}_0$的影响称为电枢反应。电枢反应的性质（增磁、去磁或交磁）取决于电枢磁动势基波$\boldsymbol{F}_a$与主磁场$\boldsymbol{B}_0$（或主极磁动势基波$\boldsymbol{F}_{f1}$）的空间相位差$90°+\psi_0$，也就是取决于内功率因数角ψ_0（即励磁电动势$\dot{E}_0$与电枢电流$\dot{I}$的时间相位差）。

不计磁饱和时，可应用叠加原理，认为主极磁动势和电枢磁动势分别产生空载主磁通$\dot{\Phi}_0$和电枢反应磁通$\dot{\Phi}_a$，并各自在电枢绕组中感应相应的励磁电动势$\dot{E}_0$和电枢反应电动势$\dot{E}_a$。对于隐极式同步发电机，$\dot{E}_a=-j\dot{I}X_a$，X_a称为电枢反应电抗，是一个反映电枢反应磁通Φ_a的参数；$\dot{E}_\sigma=-j\dot{I}X_\sigma$，$X_\sigma$称为电枢绕组的漏电抗，是一个反映电枢漏磁通$\Phi_\sigma$的参数。隐极式同步发电机的电压方程为$\dot{E}_0=\dot{U}+\dot{I}R_a+j\dot{I}X_s$，$X_s=X_a+X_\sigma$称为同步电抗。对于凸极式同步电机，运用双反应理论将电枢磁动势$\boldsymbol{F}_a$分解成直轴分量$\boldsymbol{F}_{ad}$和交轴分量$\boldsymbol{F}_{aq}$，对应的电枢反应电动势为$\dot{E}_{ad}=-j\dot{I}_dX_{ad}$和$\dot{E}_{aq}=-j\dot{I}_qX_{aq}$，$X_{ad}$和$X_{aq}$分别称为直轴电枢反应电抗和交轴电枢反应电抗。凸极式同步发电机的电压方程为$\dot{E}_0=\dot{U}+\dot{I}R_a+j\dot{I}_dX_d+j\dot{I}_qX_q$，$X_d=X_{ad}+X_\sigma$和$X_q=X_{aq}+X_\sigma$分别称为直轴同步电抗和交轴同步电抗。根据电压方程可以画出相应的相量图和等效电路。利用功率方程$P_e=P_1-p_\Omega-p_{Fe}-p_\Delta$和$P_2=P_e-p_{Cua}$以及转矩方程$T_1=T_e+T_0$可以分析计算同步发电机的各种有功功率和转矩。利用空载特性和短路特性可以求得同步电机的主要参数——同步电抗X_d的不饱和值；而从外特性和调整特性中可以确定表征同步发电机运行性能的两个重要数据——电压调整率Δu和额定励磁电流I_{fN}。

同步发电机一般与电网并联运行。采用准确整步法把发电机投入并联时必须满足相序相同、频率相等和端电压相等3个条件，并掌握合适的合闸瞬间。同步发电机与无穷大电网并联运行时，其电压和频率将被电网约束为常值。隐极式同步发电机功角特性的表达式

为 $P_e = m\dfrac{E_0U}{X_s}\sin\delta$，凸极式同步发电机为 $P_e = m\dfrac{E_0U}{X_d}\sin\delta + m\dfrac{U^2}{2}\left(\dfrac{1}{X_q}-\dfrac{1}{X_d}\right)\sin2\delta$。功率角 δ 既是励磁电动势 $\dot{E}_0$ 与电枢端电压 $\dot{U}$ 之间的时间夹角，也是主磁场 $\boldsymbol{B}_0$ 与气隙合成磁场 $\boldsymbol{B}$ 之间的空间夹角。当主极磁场超前气隙合成磁场时，δ 为正值，电磁功率也为正值，同步电机处于发电机状态。

同步发电机与无穷大电网并联运行时，通过调节原动机的输入功率，可以实现调节发电机有功功率的目的。对于隐极式同步发电机，最大电磁功率 $P_{emax} = m\dfrac{E_0U}{X_s}$；比整步功率 $P_{syn} = m\dfrac{E_0U}{X_s}\cos\delta$。$P_{syn} > 0$ 时，同步发电机是静态稳定的，且 P_{syn} 越大，发电机的静态稳定度越好。通过调节励磁电流，可以实现调节同步发电机无功功率的目的。“过励”时，同步发电机功率因数 $\cos\varphi < 1$（滞后），同步发电机输出有功功率和感性无功功率；“欠励”时，同步发电机功率因数 $\cos\varphi < 1$（超前），同步发电机输出有功功率和容性无功功率。

同步电机也可以作为电动机运行。与无穷大电网并联运行的同步电动机受电网频率的约束，其转速为常值，不会随负载的变动而变动，而且同步电动机的功率因数可以通过励磁电流来调节，因此在需要改善功率因数和不需要调速的场合，常常优先采用同步电动机。分析同步电动机时，通常采用电动机惯例，此时电磁功率 P_e 和功率角 δ 为正值，气隙合成磁场超前主极磁场。隐极式同步电动机的电压方程为 $\dot{U} = \dot{E}_0 + \dot{I}R_a + j\dot{I}X_s$；凸极式同步电动机的电压方程为 $\dot{U} = \dot{E}_0 + \dot{I}R_a + j\dot{I}_dX_d + j\dot{I}_qX_q$。

同步电动机不能自行起动，一般在同步电动机的主极极靴上装设起动绕组，采用异步起动法来起动，另外还可以采用辅助电动机起动法和变频起动法来起动。

习　题

1. 同步电机的频率、极对数和同步转速之间有什么关系？一台 $f = 50\text{Hz}$、$n = 3000\text{r/min}$ 的汽轮发电机的极数是多少？一台 $f = 50\text{Hz}$、$2p = 100$ 的水轮发电机的转速是多少？

2. 一台三相同步发电机，额定容量 $S_N = 20\text{kV}\cdot\text{A}$，额定电压 $U_N = 400\text{V}$，额定功率因数 $\cos\varphi_N = 0.8$（滞后），试求该发电机的额定电流 I_N 及额定运行时发出的有功功率 P_N 和无功功率 Q_N。

3. 分别画出隐极式同步发电机带三相对称电感负载和三相对称电容负载两种情况下的时空矢量图，忽略电枢绕组电阻，在图上表示出时间相量 $\dot{U}$、$\dot{I}$、$j\dot{I}X_s$、$\dot{E}$ 和空间矢量 $\boldsymbol{F}_{f1}$、$\boldsymbol{F}_a$、$\boldsymbol{F}$。对两种情况分别比较主极磁动势 $\boldsymbol{F}_{f1}$ 与合成磁动势 $\boldsymbol{F}$ 的大小，说明两种情况下电枢反应磁动势 $\boldsymbol{F}_a$ 各起什么作用。

4. 如何测取和计算同步发电机的同步电抗和电枢电阻？

5. 测定同步发电机短路特性时，如果把发电机的转速从额定转速 n_N 降低到 $\dfrac{1}{2}n_N$，对测量结果是否有影响？

6. 请分别画出同步发电机带纯阻性负载、容性负载和感性负载运行时的相量图，即 $\cos\varphi = 1$、$\cos\varphi = 0.8$（滞后）、$\cos\varphi = 0.8$（超前）。

7. 同步发电机带感性负载运行时，为什么发电机的端电压随负载的增大而下降？

8. 同步发电机带容性负载运行时，为什么发电机的端电压随负载的增大而上升？

9. 利用相量图说明，当改变同步发电机的励磁电流，保持发电机输出的有功功率不变时，发电机的无功功率会如何变化？并根据上述相量图绘制出同步发电机的 V 形曲线。

10. 额定频率为 60Hz 的同步发电机用于 50Hz 时，为什么需要降级使用？应该降多少？

11. 额定频率为 400Hz 的同步发电机的体积比额定频率为 60Hz 的同步发电机的体积大还是小？为什么？

12. 某工厂需要同步发电机提供 300kW、60Hz 的电源，但目前只有 50Hz 的电源，技术人员拟进行如下改造：用一台 50Hz 的同步电动机拖动一台同步发电机，使同步发电机输出 60Hz 的电能，请问：两台电机的极数应该各是多少，才能满足上述要求将 50Hz 的电源转换为 60Hz？

13. 同步发电机与电网并联的条件是什么？当某一个并联条件不符合时，会产生什么后果？应采取什么措施使同步发电机满足并联条件？

14. 与无穷大电网并联运行的隐极式同步发电机，当调节发电机输出的有功功率且保持输出的无功功率不变时，功率角 δ 和励磁电流 I_f 是否变化？$\dot{I}$ 与 $\dot{E}_0$ 的变化轨迹是什么（忽略电枢电阻且不计磁饱和）？

15. 在直流电机中，$E_0 > U$ 或 $E_0 < U$ 是判断电机作为发电机还是电动机运行状态的根据之一，在同步电机中这个结论还正确吗？为什么？决定同步电机运行于发电机状态还是电动机状态的条件是什么？

16. 一台同步电机的同步电抗为 2.0Ω，电枢电阻为 0.4Ω，当 $E_A = 460\angle -8°\text{V}$，$U_\phi = 480\angle 0°\text{V}$ 时，请判断该电机是运行于发电机状态还是电动机状态？电机输出或吸收的有功功率和无功功率各为多少？

17. 同步发电机“过励”运行时，向电网输送什么性质的电流和功率？“欠励”时又向电网输送什么性质的电流和功率？同步电动机呢？

18. 同步电动机为什么不能自起动？同步电动机有哪些起动方法？

19. 一台 25MV·A、三相、13.8kV、2 极、60Hz、星形联结同步发电机，其电枢电阻为 0.24Ω。对发电机进行空载试验和短路试验，相关的试验数据如下表所示：

空载试验					
励磁电流/ A	320	365	380	475	570
线电压/kV	13.0	13.8	14.1	15.2	16.0
推算的气隙电压/kV	15.4	17.5	18.3	22.8	27.4
短路试验					
励磁电流/ A	320	365	380	475	570
电枢电流/A	1040	1190	1240	1550	1885

试求：

1）发电机不饱和同步电抗的实际值和标幺值。

2）当励磁电流为 380A 时，发电机饱和同步电抗的实际值和标幺值。

3）当励磁电流为 475A 时，发电机饱和同步电抗的实际值和标幺值。

4）发电机的短路比。

20. 一台 1500kW 的三相水轮发电机，额定电压 $U_N=6300V$（星形联接），额定功率因数 $\cos\varphi_N=0.8$（滞后），发电机参数 $X_d=21.3\Omega$，$X_q=13.7\Omega$，忽略电枢电阻，不计磁饱和，试求：

1）X_d和 X_q的标幺值。

2）额定运行时发电机的功率角 δ 和励磁电动势 E_0。

21. 一台 70 000kV·A、60 000kW、13.8kV（星形联结）的三相水轮发电机，直、交轴同步电抗的标幺值分别为 $X_d^*=1$，$X_q^*=0.7$，试求额定负载时发电机的励磁电动势 E_0^*、功率角 δ 和电压调整率 Δu（不计磁饱和与电枢电阻）。

22. 一台与无穷大电网并联运行的三相隐极式同步发电机，$S_N=31\ 250kV\cdot A$，$U_N=10.5kV$（星形联结），$\cos\varphi_N=0.8$（滞后），定子每相同步电抗 $X_s=7.0\Omega$，忽略电枢电阻且不计磁饱和，发电机额定运行，试求：

1）功率角 δ、电磁功率 P_e、比整步功率 P_{syn}和过载能力 k_p。

2）若维持额定励磁电流不变而输出有功功率减半时，求 δ、P_e和功率因数角 φ，此时输出无功功率怎样变化？

3）若仅将其励磁电流增大 10%，求 δ、P_e和 φ，此时输出无功功率怎样变化？

23. 一台 13.8kV、10MV·A、60Hz、2 极、星形联结、功率因数为 0.8（滞后）的汽轮发电机与无穷大电网并联运行，发电机的同步电抗为 12Ω，每相电枢电阻为 1.5Ω。该发电机额定运行时，试求：

1）励磁电动势 E_A。

2）功率角。

3）如果励磁电流不变，该发电机的最大输出功率是多少？发电机满载时的功率和转矩是多少？

4）在最大功率点，该发电机发出或消耗的无功功率是多少？请画出励磁电流不变的该发电机的相量图。

24. 一台 100MV·A、12.5kV、0.85（滞后）、50Hz、2 极、星形联结的同步发电机，同步电抗的标幺值为 1.1，电枢电阻标幺值为 0.012。试求：

1）发电机同步电抗和电枢电阻的实际值。

2）发电机额定运行时的励磁电动势和功率角。

3）若忽略发电机的损耗，原动机输出给发电机的转矩是多少？

25. 一台三相星形联结同步发电机，额定数据为 120 MV·A、13.2kV、60Hz，功率因数为 0.8（滞后），同步电抗为 0.9Ω，忽略电枢电阻。试求：

1）该发电机的电压调整率。

2）若该发电机工作于 50Hz，且保持与 60Hz 时具有相同的电枢电流和励磁电流，则该发电机的额定端电压和视在功率各为多少？

3）该发电机工作于 50Hz 时的电压调整率为多少？

26. 一台隐极式同步电动机在额定状态下运行时，功率角 δ 为 30°，保持励磁电流不变，运行情况发生下述变化时，功率角有何变化（忽略电枢电阻且不计磁饱和）？ 347

1）电网频率下降 5%，负载转矩不变。

2）电网频率下降 5%，负载功率不变。

3）电网电压和频率各下降 5%，负载转矩不变。

4）电网电压和频率各下降5%，负载功率不变。

27. 一台三相隐极式同步电动机，额定电压 $U_N=380V$（星形联结），同步电抗 $X_s=5\Omega$，忽略电枢电阻，当功率角 $\delta=30°$时，电磁功率 $P_e=16kW$，试求：

1）每相励磁电动势 E_0。

2）保持励磁电流不变，求最大电磁功率。

28. 一台三相同步电动机，额定功率 $P_N=2000kW$，额定电压 $U_N=3000V$（星形联结），额定功率因数 $\cos\varphi_N=0.85$（超前），额定效率 $\eta_N=95\%$，极对数 $p=3$，定子每相电阻 $R_a=0.1\Omega$，试求：

1）额定运行时定子输入的电功率 P_1。

2）额定电流 I_N。

3）额定电磁功率 P_e。

4）额定电磁转矩 T_e。

29. 一台同步电动机在额定电压下运行，从电网吸收功率因数为0.8（超前）的额定电流，该机直、交轴同步电抗的标幺值分别为 $X_d^*=0.8$，$X_q^*=0.5$，试求励磁电动势 E_0^* 和功率角 δ，并说明这台电动机运行于"过励"状态还是"欠励"状态（不计磁饱和与电枢电阻）。

参 考 文 献

[1] Stephen J Chapman. Electric Machinery Fundamentals [M]. 4th ed. New York: McGraw-Hill, 2005.

[2] Theodore Wildi. Electrical Machines, Drives, and Power Systems [M]. 6th ed. New York: Pearson Prentice Hall, 2006.

[3] A E Fitzgerald, Charles Kingsley, Jr., and Stephen D Umans. Electric Machinery [M]. 4th ed. New York: McGraw-Hill, 1983.

[4] A E Fitzgerald, Charles Kingsley, Jr., and Stephen D Umans. Electric Machinery [M]. 6th ed. New York: McGraw-Hill, 2003.

[5] Michael Liwschitz-Garik, Robert T Weil, Jr.. D-C and A-C Machines Based on Fundamental Laws [M]. New York: D Van Nostrand Company, 1952.

[6] Michael Liwschitz-Garik, Clyde C Whipple. Electric Machinery [M]. New York: D Van Nostrand Company, 1946.

[7] 汤蕴璆，史乃. 电机学 [M]. 4 版. 北京：机械工业出版社，2011.

[8] 汤蕴璆，徐德淦. 电机学 [M]. 北京：机械工业出版社，2012.

[9] 顾绳谷. 电机及拖动基础 [M]. 4 版. 北京：机械工业出版社，2004.

[10] 许实章. 电机学 [M]. 3 版. 北京：机械工业出版社，1995.

[11] 李发海，等. 电机学 [M]. 2 版. 北京：科学出版社，1992.

[12] 王毓东. 电机学 [M]. 杭州：浙江大学出版社，1990.

[13] 辜承林，陈乔夫，熊永前. 电机学 [M]. 武汉：华中科技大学出版社，2001.

[14] 王正茂，阎治安，等. 电机学 [M]. 西安：西安交通大学出版社，2000.

[15] 胡虔生，胡敏强，杜炎森. 电机学 [M]. 北京：中国电力出版社，2001.

[16] 周顺荣. 电机学 [M]. 北京：科学出版社，2002.

[17] 刘慧娟，张威. 电机学 [M]. 北京：国防工业出版社，2011.

[18] 孙旭东，冯大钧. 电机学习题与题解 [M]. 北京：科学出版社，2001.

[19] 龚世缨，熊永前. 电机学实例解析 [M]. 武汉：华中科技大学出版社，2001.

[20] 周定颐. 电机及电力拖动 [M]. 2 版. 北京：机械工业出版社，1995.

[21] 汪国梁. 电机学 [M]. 北京：机械工业出版社，1987.

[22] 李发海，等. 电机学 [M]. 北京：科学出版社，1982.